Climate Change and Variability

edited by
Suzanne W. Simard and Mary E. Austin

Climate Change and Variability
Edited by Suzanne W. Simard and Mary E. Austin

CBS, Edition 2016

Published by InTech

Janeza Trdine 9, 51000 Rijeka, Croatia

Publishing Process Manager Jelena Marusic

Technical Editor Sonja Mujacic

Cover Designer Martina Sirotic

Image Copyright kkaplin, 2010. Used under license from Shutterstock.com

Additional hard copies can be obtained from orders@intechopen.com

Climate Change and Variability, Edited by Suzanne W. Simard and Mary E. Austin
p. cm.
ISBN 978-953-307-144-2

INTECH

open science | open minds

Contents

Preface IX

Chapter 1 **A regional approach to the Medieval Warm Period and the Little Ice Age 1**
Fredrik Charpentier Ljungqvist

Chapter 2 **Multi-months cycles observed in climatic data 27**
Samuel Nicolay, Georges Mabille, Xavier Fettweis and M. Erpicum

Chapter 3 **Summer-Time Rainfall Variability in the Tropical Atlantic 45**
Guojun Gu

Chapter 4 **Tropical cyclones, oceanic circulation and climate 65**
Lingling Liu

Chapter 5 **Possible impacts of global warming on typhoon activity in the vicinity of Taiwan 79**
Chia Chou, Jien-Yi Tu and Pao-Shin Chu

Chapter 6 **Influence of climate variability on reactive nitrogen deposition in temperate and Arctic climate 97**
Lars R. Hole

Chapter 7 **Climate change: impacts on fisheries and aquaculture 119**
Bimal P Mohanty, Sasmita Mohanty, Jyanendra K Sahoo and Anil P Sharma

Chapter 8 **Community ecological effects of climate change 139**
Csaba Sipkay, Ágota Drégelyi-Kiss, Levente Horváth, Ágnes Garamvölgyi, Keve Tihamér Kiss and Levente Hufnagel

Chapter 9 **Pelagic ecosystem response to climate variability in the Pacific Ocean off Baja California 163**
Gilberto Gaxiola-Castro, Bertha E. Lavaniegos, Antonio Martínez, Rubén Castro, T. Leticia Espinosa-Carreón

Chapter 10 **Climate change and resilience value of mussel farming for the baltic sea 183**
Ing-Marie Gren

Chapter 11 **Temperate forests and climate change in mexico: from modelling to adaptation strategies 195**
Gómez-Mendoza, Leticia and Galicia, Leopoldo

Chapter 12 **The influence of climate change on tree species distribution in west part of south-east europe 211**
J. Vukelić, S. Vojniković, D. Ugarković, D. Bakšić and S. Mikac

Chapter 13 **Climate change impact on vegetation: lessons from an exceptionally hot and dry decade in south-eastern France 225**
Vennetier Michel and Ripert Christian

Chapter 14 **Climate change, forest fires and air quality in Portugal in the 21st century 243**
Anabela Carvalho

Chapter 15 **The role of mycorrhizas in forest soil stability with climate change 275**
Simard, Suzanne W. and Austin, Mary E.

Chapter 16 **Impact of temperature increase and precipitation alteration at climate change on forest productivity and soil carbon in boreal forest ecosystems in Canada and Russia: simulation approach with the EFIMOD model 303**
Oleg Chertov, Jagtar S. Bhatti and Alexander Komarov

Chapter 17 **Simulating peatland methane dynamics coupled to a mechanistic model of biogeochemistry, hydrology, and energy: Implications to climate change 327**
Takeshi Ise, Allison L. Dunn, Steven C. Wofsy and Paul R. Moorcroft

Chapter 18 **Towards a New Agriculture for the Climate Change Era in West Asia, Iran 337**
Farzin Shahbazi and Diego de la Rosa

Chapter 19 **Simulated potato crop yield as an indicator of climate variability and changes in Estonia 365**
Triin Saue and Jüri Kadaja

Chapter 20 **Determining the relationship between climate variations and wine quality: the WEBSOM approach 389**
Subana Shanmuganathan and Philip Sallis

Chapter 21 **Ecological modernisation and the politics of (un)sustainability in the Finnish climate policy debate 409**
Tuula Teräväinen

Chapter 22 **Transportation and climate change 427**
Yuri Yevdokimov

Chapter 23 **Cost-optimal technology and fuel choices In the transport sector under a stringent climate stabilization target 439**
Takayuki Takeshita

Chapter 24 **Impact of climate change on health and disease in Latin America 463**
Alfonso J. Rodríguez-Morales, Alejandro Risquez and Luis Echezuria

Preface

Global change, including climate change, ecosystem shifts and biodiversity loss as a result of explosive human population growth and consumption, is emerging as one of the most important issues of our time (Vitousek, 1994). Climate change in particular appears to be altering the function, structure and stability of the Earth's ecosystems (Lovelock, 2009). It has been marked by an 80% increase in atmospheric CO2 level and a 0.74 °C increase in average global near-surface temperature over the period 1906–2005, with average temperature projected to increase by an additional 1 to 6oC by 2100 (Intergovernmental Panel on Climate Change [IPCC], 2007). The spring months in 2010, in particular, have been the warmest on record (National Atmospheric and Oceanic Administration, 2010). Warming is expected to continue for centuries, even if greenhouse gas emissions are stabilized, owing to time lags associated with climate processes and feedbacks. There is also evidence that precipitation patterns have changed along with temperature, with average annual increases up to 20% in high-latitude regions but decreases up to 20% in mid- and low-latitudinal regions. The changes in temperature and precipitation have resulted in higher sea levels, reduced extent of snow and ice, earlier timing of species spring events, upward and pole-ward shifts in species ranges, increased and earlier spring run-offs, and increased forest disturbances by fires, insects and diseases (Parmesan, 2006). They have already contributed to increased human morbidity and mortality in some regions of the world (Patz et al., 2005; McMicheal et al., 2006).

Global change is clearly having profound cascading effects on ecosystems and society, but they are poorly understood (Vitousek, 1994; Parmesan, 2006; Fischhoff, 2007). As a result, scientists have mobilized world-wide to more effectively research the impacts of climate change as well as the options for mitigating and adapting to these impacts (IPCC, 2007). New ideas, methods and theories have emerged for investigating and understanding the effects of global change, and for managing our socio-ecological systems sustainably as complex adaptive systems (Root et al., 1995; Folke et al., 2004; Levin, 2005; Puettmann et al., 2009). In spite of these efforts, large gaps in understanding and prediction remain. These gaps have hindered development of strategies for mitigation of global change effects and for improvement of adaptive capacity of ecosystems and society (Lemmen et al, 2008). Moreover, there has been little meaningful action from governments or the public aimed at slowing the pace of climate change (Friedman, 2010), which is apparent in our inability to meet Kyoto Protocol goals, despite the compelling evidence summarized by the IPCC.

In this collection of 24 chapters, we present a cross-section of some of the most challenging issues related to oceans, lakes, forests, and agricultural systems under a changing climate. The authors passionately present evidence for changes and variability in climatic and atmospheric conditions, investigate some the impacts that climate change is having on Earth's ecological and social systems, and provide novel ideas, advances and applications for mitigation and adaptation of our socio-economic systems to climate change. Difficult questions are asked. What are the impacts of climate change on some of our most productive and precious ecosystems? How do we manage for resilient socio-ecological systems? How do we predict the future? What are relevant climatic and management scenarios? How can we shape management regimes to increase our adaptive capacity? These themes are visited across broad spatial and temporal scales, touch on important and relevant ecological socio-political patterns and processes, and represent broad geographic regions, from the tropics, to temperate and boreal regions, to the Arctic. It is the wish of the authors that this information will be used to improve predictions and inform policy for reducing threats and facilitating adaptation and management in the face of our changing climate.

The book is divided into six sections. The first section examines evidence for changes and variability in the Earth's climate and the atmosphere. The first chapter, "A regional approach to the medieval warm period and the little ice age", presents an interesting analysis of temperature variability in the late Holocene. The second chapter, "Multi-months cycles observed in climatic data", uses the wavelet-based methodology to detect and describe cycles in air-surface temperatures, and the authors discuss some mechanisms that may underlie these cycles. Chapter 3, "Summer-time rainfall variability in the Tropical Atlantic", explores summer-time rainfall variations within the tropical Atlantic basin, with a focus on local effects of sea surface temperature and the El Niño-Southern Oscillation. The fourth chapter, "Tropical cyclones, oceanic circulation and climate", examines the role of tropical cyclones in regulating the general oceanic circulation and climate, as well as the effects of the ocean on tropical cyclones. Chapter 5, entitled "Some of the possible impacts of global warming on typhoon activity in the vicinity of Taiwan", presents data indicating an abrupt shift in the typhoon track near Taiwan, discusses its association with global warming, and presents potentially large-scale environmental changes in response to the shift. The final chapter in this section (Chapter 6) examines the influence of climate variability on reactive nitrogen deposition in Norway, and presents evidence that future reductions in nitrogen deposition due to emission reductions in Europe could be partly offset due to increasing precipitation in some regions.

The second section of the book, which includes four chapters, examines some impacts of climate change on aquatic systems and aquaculture. Chapter 7, "Climate change: impacts on fisheries and aquaculture", provides a general review of the impacts of climate change on fisheries and aquaculture. It reaches further to suggest possible mitigation options, and provides guidance on developing suitable monitoring tools. This is followed by Chapter 8, "Community ecological effects of climate change", which is a case study of the effects of warming on phytoplankton activity in the Danube River of Hungary. The authors contrast strategic and tactical models of productivity with other approaches, such as the "geographical analogy" method. Chapter 9, "Pelagic ecosystem response to climate variability in the Pacific Ocean off Baja California", is a second case study examining the associations between large-scale temporal climate physical forcing and the plankton variability off of Baja California. The final chapter in this section, Chapter 10, "Climate change and resilience value of mussel farming for the Baltic Sea", estimates the impacts of global change (specifically nutrient loading) on the resilience values of mussel farming in the Baltic Sea. The authors conclude that mussel farming shows promise for buffering against variable nutrient loads in waters that are becoming increasingly eutrophic with climate change.

The third section of the book examines climate change effects on forests and plant communities. One of the studies was conducted in Mexico and the remaining three chapters report on studies in Europe. In Mexico, the authors of Chapter 11 conducted a modeling study to predict the effects of climate change on temperate forests. They go on to suggest adaptation strategies for dealing with climate change in these forest types. Chapter 12, entitled "The influence of climate change on tree species distribution in west part of south-east Europe", used ecological niche modelling to project climate change scenarios on the distribution of the dominant tree species in south-eastern Europe. Chapter 13 describes a study in France, entitled "Climate change vs vegetation track race: what did we learn from a ten years long anticipated occurrence of 2040 climate in South-eastern France". The authors measured 10-year plant composition changes in a permanent plot network, modeled the potential floristic turnover, and discussed the relationships between observed changes and local site conditions. They found a 14% turnover in species, mainly biased against water-demanding species, and caution that current reserve networks in France may be inadequate to ensure long-term species persistence. The last chapter in this section, Chapter 14, "Climate change, forest fires and air quality in Portugal in the 21st century", focuses on climate change effects on fire disturbance. It evaluates the impacts of the IPCC SRES A2 climatic scenario on forest fire activity and air quality over Portugal. The analysis predicts an increase the fire weather index components and fire severity in all Portuguese districts.

The fourth section examines the some of the effects of climate change on carbon cycling and soil systems, and includes three chapters. Chapter 15, entitled "The role of mycorrhizas in forest soil stability with climate change", reviews the role of mycorrhizas and mycorrhizal networks in the stability of forest ecosystems and forest soils as climate changes. It begins with a review of the role mycorrhizal fungi play in soil carbon flux dynamics, examines the effects of climate change factors on plants and mycorrhizal fungi, and ends with a review of the role of mycorrhizas and mycorrhizal networks in helping mitigate the effects of climate change in on forest ecosystems. Chapter 16 uses a modeling approach to examine potential impacts of climate (temperature and precipitation) changes on forest productivity and soil carbon in boreal forest ecosystems in Canada and Russia. The authors found that all climate change scenarios predicted similar increasing trends of net primary production (NPP) and stand productivity; that disturbances led to a strong decrease in NPP, stand productivity, soil organic matter and nitrogen pools; but that they also led to an increase in CO2 emission to the atmosphere. Chapter 17, "Simulating peatland methane dynamics coupled to a mechanistic model of biogeochemistry, hydrology, and energy: Implications to climate change", also employed simulation modeling. The authors predicted that a 4°C in warming would result in a short term increase in CH4 emission due to a transient increase in microbial activity, followed by a significant long-term decline in CH4 emission caused by a loss in substrate and an increasing prevalence of aerobic conditions.

The fifth set of chapters discusses some of the implications of climate change for agricultural systems. There are three chapters in this section. Chapter 18 is entitled "Towards a New Agriculture for the Climate Change Era in West Asia, Iran". It presents a land evaluation decision support system that can be used for sustainable agriculture planning and management in west Asia under climate change conditions. Chapter 19, entitled "Simulated potato crop yield as an indicator of climate variability and changes in Estonia", applies the meteorologically possible yield (MPY) – the maximum yields under given meteorological conditions (in this case, for potatoes) - to derive qualitatively new information about climate variability. The probable range of temperature and precipitation in years 2050 and 2100 is applied to construct possible distribution of MPY in those years, and in the future. The authors predict that variability of potato yields will decrease slightly due to climate change, but strongly caution that further study, including estimates of meteorological variability, is sorely needed. In Chapter 20, "Determining the relationship between climate variations and wine quality: the WEBSOM approach", the authors first review the literature on potential influence of climate and environmental variation on viticulture. They then use a modeling approach to predict wine quality from grapevine phenology in New Zealand. They conclude with concrete suggestions for future modeling research to predict climate change effects on the world's major wine regions in the southern hemisphere.

We close our book with four studies of some of the socio-economic implications of climate change. Chapter 21 is entitled "Ecological modernisation and the politics of (un)sustainability in the Finnish climate policy debate". This chapter examines the political side of technology of the climate change policy debate in Finland. It provides interesting perspectives at the national and international levels. In Chapter 22, "Transportation and climate change", the authors analyse the contribution of transportation to climate change as well as the impacts of climate change on transportation in Atlantic Canada. Chapter 23 is entitled "Cost-optimal technology and fuel choices in the transport sector under a stringent climate stabilization target". This chapter examines the cost-optimal choice of propulsion systems and fuels for different transport modes over the 21st century under a global mean temperature rise of 2-2.4oC. It also presents the results of a sensitivity analysis with respect to: (1) climate stabilization; (2) technology advances, and (3) demand for supersonic air travel. Our final chapter, Chapter 24, is on the "Impact of climate change on health and disease in Latin America". It explores the implications of climate change on specific communicable and non-communicable diseases in Latin America.

In conclusion, this book presents a broad cross-section of reviews, surveys, experiments, model predictions and discussions focused on climate change and climate variability, the realized and potential impacts of climate change on the Earth's ecological and social systems, and potential mitigation and adaptation strategies for coping with climate change. It is our hope that these studies will provide the basis for future research, policy changes and action for improving the future ecological and human conditions of our home, the planet Earth.

Editor:
Suzanne W. Simard and Mary E. Austin
The University of British Columbia, Vancouver Canada

References:

Fischhoff, B. 2007. Nonpersuasive communication about matters of greatest urgency: climate change. Environmental Science & Technology 41: 7205-7208.

Folke, F.C., Carpenter, B., Walker, M., Scheffer, M., Elmqvist, T., Gunderson, L., Holling, C.S. 2004. Regime shifts, resilience, and biodiversity in ecosystem management. Annual Review of Ecology and Evolution 35: 557-581.

Friedman, T. 2010. We're gonna be sorry. The New York Times. http://www.nytimes.com/2010/07/25/opinion/25friedman.html?_r=1&ref=thomaslfriedman

IPCC, Climate Change 2007: The Physical Science Basis. Fourth Assessment Report. Eds., S. Solomon, D. Qin, M. Manning, Z. Chen, M. Marquis, K.B. Averyt, M. Tignor and H.L. Miller. Cambridge University Press, New York.

Lemmen, D.S., Warren, F.J., Lacroix, J., Bush, E. (eds). 2008. From Impacts to Adaptation: Canada in a Changing Climate 2007; Government of Canada, Ottawa, ON, 448p.

Levin, S.A. 2005. Self-organization and the emergence of complexity in ecological systems. Bioscience 55: 1075-1079.

Lovelock, J. 2009. The vanishing face of Gaia. Basic Books, New York.

McMichael, A.J., Woodruff, R.E., Hales, S. 2006. Climate change and human health: present and future risks. The Lancet 367: 859-869.

National Oceanic and Atmospheric Administration. 2010. State of the Climate, June 2010 Global Analysis Report. National Climatic Data Centre, US Department of Commerce. http://www.ncdc.noaa.gov/sotc/index.php

Parmesan, C. 2006. Ecological and evolutionary responses to recent climate change. Annual Review of Ecology, Evolution and Systematics 37: 637–69.

Patz, P.A., Campbell-Lendrum, D., Holloway, T., Foley, J.A. 2005. Impact of regional climate change on human health. Nature 438: 310-317.

Puettmann, K.J., Coates, K.D., Messier, C. 2009. A Critique of Silviculture: Managing for Complexity. Island Press, Washington.

Root, T.L., Schneider, S.H. 1995. Ecology and climate: research strategies and implications. Science 269: 334-341.

Vitousek, P.M. 1994. Beyond global warming: ecology and global change. Ecology 75: 1861-1876.

A regional approach to the medieval warm period and the little ice age

Fredrik Charpentier Ljungqvist
Stockholm University
Sweden

1. Introduction

In order to gain knowledge of the temperature variability prior to the establishment of a widespread network of instrumental measurements c. AD 1850, we have to draw information from proxy data sensitive to temperature variations. Such data can be extracted from various natural recorders of climate variability, such as corals, fossil pollen, ice-cores, lake and marine sediments, speleothems, and tree-ring width and density, as well as from historical records (for a review, see IPCC 2007; Jones et al. 2009; NRC 2006). Considerable effort has been made during the last decade to reconstruct global or northern hemispheric temperatures for the past 1000 to 2000 years in order to place the observed 20th century warming in a long-term perspective (e.g., Briffa, 2000; Cook et al., 2004; Crowley and Lowery, 2000; D'Arrigo, 2006; Esper et al., 2002; Hegerl et al., 2007; Jones et al., 1998; Jones and Mann, 2004; Juckes et al., 2007; Ljungqvist, 2010; Loehle, 2007; Mann et al., 1999; Mann et al., 2008; Mann et al., 2009; Mann and Jones, 2003; Moberg et al., 2005; Osborn and Briffa, 2006).

Less effort has been put into investigating the key question of to what extent earlier warm periods have been as homogeneous in timing and amplitude in different geographical regions as the present warming. It has been suggested that late-Holocene long-term temperature variations, such as the Medieval Warm Period (MWP) and the Little Ice Age (LIA), have been restricted to the circum-North Atlantic region (including Europe) and have not occurred synchronic in time with warm and cold periods respectively in other regions (Hughes and Diaz, 1994; Mann et al., 1999; Mann and Jones, 2003). This view has, however, been increasingly challenged through the ever growing amount of evidence of a global (or at least northern hemispheric) extent of the MWP and the LIA that have become available (see, for example, Esper and Frank, 2009; Ljungqvist, 2009, 2010; Moberg et al., 2005; Wanner et al., 2008).

A main obstacle in large-scale temperature reconstructions continues to be the limited and unevenly distributed number of quantitative palaeotemperature records extending back a millennium or more. The limited number of records have rendered it impossible to be very selective in the choice of data. Palaeotemperature records used in a large-scale temperature reconstruction should preferably be accurately dated, have a high sample resolution and have a high correlation with the local instrumental temperature record in the calibration period (see the discussion in Jones et al., 2009). The number of long quantitative palaeotemperature records from across the globe, of which a majority are well suited for being used in large-scale temperature reconstructions, have been rapidly increasing in recent years (Ljungqvist, 2009). Thus, it has now become possible to make regional temperature reconstructions for many regions that can help us to assess the spatio-temporal pattern and the MWP and LIA. Only by a regional approach can we truly gain an understanding of the temperature variability in the past 1–2 millennia and assess the possible occurrence of globally coherent warm and cold periods. Presently, only four regional multi-proxy temperature reconstructions exist: two for eastern Asia (Yang et al., 2002; Ge et al., 2010), one for the Arctic (Kaufman et al., 2009), and one for South America (Neukom et al., 2010). Six new quantitative regional multi-proxy temperature reconstructions will here be presented in order to improve our understanding of the regional patterns of past temperature variability.

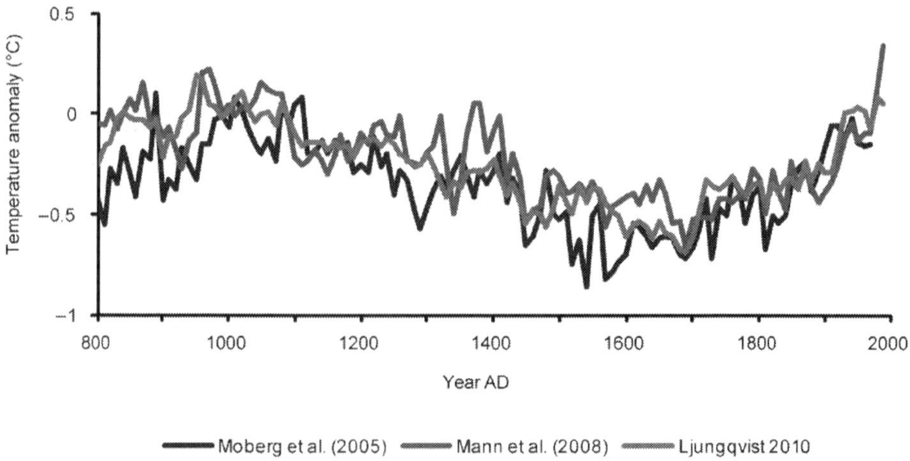

Fig. 1. Comparison of three recent millenium-long multi-proxy Northern Hemisphere temperature reconstructions: decadal means of Moberg et al. (2005), the 'error-in-variables' (EIV) regression method variant of Mann et al. (2008), and the extra-tropical Northern Hemisphere reconstruction by Ljungqvist (2010).

2. New regional temperature reconstructions

Only for limited parts of the Northern Hemisphere is the data coverage sufficient for making quantitative regional temperature reconstructions in order to assess the regional pattern of temperature changes during the last 12 centuries. This has been done for: 1) warm season temperatures of Scandinavia north of 60°N, 2) warm season temperatures for northern Siberia, 3) annual mean temperatures for Greenland, 4) warm season temperatures for the Alp region of Central Europe, 5) annual mean temperatures for China, and 6) annual mean temperatures for the whole of the North American continent. The reconstructions follow the simple but robust method outlined in the new multi-proxy temperature reconstruction for the extra-tropical (90°–30°N) Northern Hemisphere by Ljungqvist (2010). Only proxy records with reasonably high resolution (multi-decadal or better) were utilised and records with lower resolution were instead used for the purpose of verifying the reconstructed low-frequency trends of the reconstructions.

We use the common "composite-plus-scale" method (Lee et al. 2008; von Storch et al. 2004) for creating the different regional temperature reconstructions. First, all records with less than annual resolution were linearly interpolated to have annual resolution and then calculated 10-year-mean values of each record. All the 10-year-mean values were normalized to zero mean and unit standard deviation fitting the decadal mean and variance AD 1000–1899. The arithmetic mean of all the normalized records included in each regional reconstruction was then calculated. Each regional reconstruction was scaled to fit the decadal mean and variance over the longest possible time period in the variance adjusted CRUTEM3 instrumental temperature record (Brohan et al., 2006) and adjusted to have a zero equaling the 1961–1990 mean of this instrumental record. A 2 standard deviation error bar for each regional reconstruction was calculated from the decadal correlation between proxy and instrumental temperature in the calibration period.

Proxy location	Latitude	Longitude	Proxy type*	Sample resolution	Season bias	Reference
1. Finnish Lapland	69.00	25.00	TRW	Annual	Summer	Helama et al. 2009
2. Lake Tsuolbmajavri	68.45	22.05	Lf	Multi-decadal	Summer	Korhola et al. 2000
3. Torneträsk	68.31	19.80	MXD	Annual	Summer	Grudd 2008
4. Jämtland	63.10	13.30	TRW	Annual	Summer	Linderholm and Gunnarson 2005
5. Severnaja	81.00	106.00	Lf	Decadal	Summer	Solomina and Alverson 2004
6. Taimyr peninsula	73.00	105.00	TRW	Annual	Summer	Naurzbaev et al. 2002
7. Indigirka	70.00	149.00	TRW	Annual	Summer	Moberg et al. 2005
8. Yamal			TRW	Annual	Summer	Briffa 2000
9. Polar Urals	66.83	65.75	MXD	Annual	Summer	Esper et al. 2002a
10. Lower Murray Lake	81.21	-69.32	V	Annual	Summer	Cook et al. 2009
11. GISP	72.60	-38.50	Is	Annual	Annual	Johnsen et al. 2001
12. GRIP	72.35	-37.38	Is	Annual	Annual	Grootes and Stuiver 1997
13. Crête	71.12	-37.32	Is	Annual	Annual	Clausen, et al. 1988
14. Nansen Fjord	68.25	-29.60	Sd	Decadal	Summer	Jennings and Weiner 1996
15. Igaliku Fjord	60.40	-46.00	Sd	Decadal	Summer	Jensen et al. 2004
16. Lake Silvaplana	46.45	9.48	Lf	Annual to decadal	Summer	Larocque et al. 2009
17. The Alps	46.30	8.00	MXD	Annual	Summer	Büntgen et al. 2006
18. Mongolia	48.30	98.93	TRW	Annual	Summer	D'Arrigo et al. 2001
19. Shihua Cave	39.47	115.56	Sp	Annual	Summer	Tan et al. 2003
20. Dulan	36.00	98.00	TRW	Annual	Annual	Zhang et al. 2003

21. E. China	35.00	114.00	D	Decadal	Annual	Yang et al. 2002
22. E. China	35.00	114.00	D	Decadal	Winter	Ge et al. 2003
23. Wanxiang	33.19	105.00	Sp	Decadal	Annual	Zhang et al. 2008
24. Zhang Delta	32.00	121.00	D	Decadal	Annual	Zhang et al. 2008
25. Hesheng	19.41	109.36	Sp	Decadal	Annual	Hu et al. 2008
26. Devon Island	75.33	-82.50	Ice-core	5 years	Annual	Fisher et al. 1983
27. Iceberg Lake, Alaska	60.78	-142.95	Lf	Annual	Summer	Loso 2009
28. Gulf of Alaska	60.00	-145.00	MXD	Annual	Summer	D'Arrigo et al. 2006
29. Canadian Rockies	52.15	-117.15	MXD	Annual	Summer	Luckman and Wilson 2005
30. Chesapeake Bay	39.00	-76.40	Md	Multi-decadal	Spring	Cronin et al. 2003
31. Bermuda Rise	33.72	-57.63	Md	Multi-decadal	Annual	Keigwin 1996
32. Nicoa Cave	10.20	-85.30	Sp	Decadal	Annual	Mann et al. 2008
33. Punta Laguna	20.63	-87.50	Lf	Decadal	Annual	Curtis et al. 1996

* D, documentary; Lf, lake/river fossils and sediments; MXD, tree-ring maximum latewood density; Md, marine sediments; Sp, speleothem isotopic analysis; TRW, tree-ring width; V, varved thickness sediments.

Table 1. All temperature proxy records used in the regional temperature reconstructions with information regarding geographical location, latitude and longitude, type of proxy, sample resolution, season bias, and reference to the original publication. The geographical locations of the records are shown on the map in Fig. 2.

Proxy location	Latitude	Longitude	Proxy type*	Sample resolution	Season bias	Reference
A. Lake Gammelheimenvatnet	68.47	17.75	Po	Centennial	Summer	Bjune et al. 2009
B. Lake Sjuodjijaure	67.37	18.07	Lf	Centennial	Summer	Rosén et al. 2003
C. Korallgrottan	64.89	14.16	Sp	Multi-decadal	Annual	Sundqvist et al. 2010
D. Taimyr pollen	70.77	99.13	Po	Centennial	Summer	Andreev et al. 2002
E. Indigirka pollen	70.00	149.00	Po	Centennial	Summer	Velichko et al. 1997
F. Yamal tree-line	67.00	69.00	O	Centennial	Summer	Solomina and Alverson 2004
G. GRIP	72.35	-37.38	B	Centennial	Annual	Dahl-Jensen et al. 1998
H. Dye-3	65.11	-43.49	B	Centennial	Annual	Dahl-Jensen et al. 1998
I. Qipisarqo Lake	61.01	-47.75	Lf	Centennial	Summer	Kaplan et al. 2002
J. Lake Neuchatel	46.80	6.70	Po	Centennial	Summer	Filippi et al. 1999
K. Aletsch Glacier	46.38	7.75	O	Centennial	Summer	Holzhauser et

						al. 2005
L. Groner Glacier	46.05	7.62	O	Centennial	Summer	Holzhauser et al. 2005
M. Lake Qinghai	37.00	100.00	Lf	Multi-decadal	Annual	Liu et al. 2006
N. Hongyuan	32.46	102.3	Lf	Centennial	Annual	Yang et al. 2002
O. Jiaming Lake	25.01	121.3	Lf	Centennial	Annual	Yang et al. 2002
P. Farewell Lake	62.55	-153.63	Lf	Centennial	Summer	Hu et al. 2001
Q. Tebenkof Glacier	60.75	-148.45	O	Centennial	Summer	Barclay et al. 2009
R. North America pollen stack	70–30	-50–170	Po	Centennial	Summer	Viau et al. 2006

* B, borehole; Lf, lake/river fossils and sediments; Md, marine sediments; Sp, speleothem isotopic analysis; O, other types of proxies; Po, pollen.

Table 2. All temperature proxy records used for verifying the low-frequency trends in the regional temperature reconstruction with information regarding geographical location, latitude and longitude, type of proxy, sample resolution, season bias, and reference to the original publication. The geographical locations of the records are shown on the map in Fig. 2.

2.1 Scandinavia

Scandinavia is probably the most data rich region in the world when it comes to palaeotemperature proxy data. Climate and environmental research has a long history in Scandinavia and numerous studies of past climate and vegetation have been conducted, especially in the far north of Scandinavia, for several decades. Many different kinds of archives have been utilised including, but not limited to, tree-ring width and density data, pollen profiles, chironomid records, diatom records, annually-laminated sediments, radiocarbon-dated megafossil tree-remains, and speleothem $\delta^{18}O$ records. Most of this data are, unfortunately, not available from databases and the majority of the records do not, moreover, possess such a high resolution that they are suited for being used in calibrated temperature reconstructions of the last 1200 years. Almost all palaeotemperature proxy data from Scandinavia, especially northern Scandinavia, primarily capture growing season temperatures and are hence biased towards the warm part of the year. In southern Scandinavia the growing season lasts approximately five months (May to September), whereas the growing season in northern Scandinavia lasts three months (June to August) or less on high elevations. We still have a limited possibility to reconstruct cold season or annual mean temperatures for Scandinavia despite the fact that the region is so comparatively rich in data.

Presently, only four records possess such quality that we can use them here for reconstructing the warm season temperature variability in Scandinavia for the last 1200 years: 1) the Torneträsk tree-ring maximum latewood density record (Grudd, 2008), 2) the Finnish Lapland tree-ring width record (Helama et al., 2009), and 3) the Jämtland tree-ring width record (Linderholm and Gunnarson, 2005). The sediment records from Lake Tsuolbmajavri in northernmost Finland (Korhola et al., 2000) also possess such a relatively high resolution and dating control that they can be useful to include in a Scandinavian warm season temperature reconstruction. Other records usually have too low a temporal resolution.

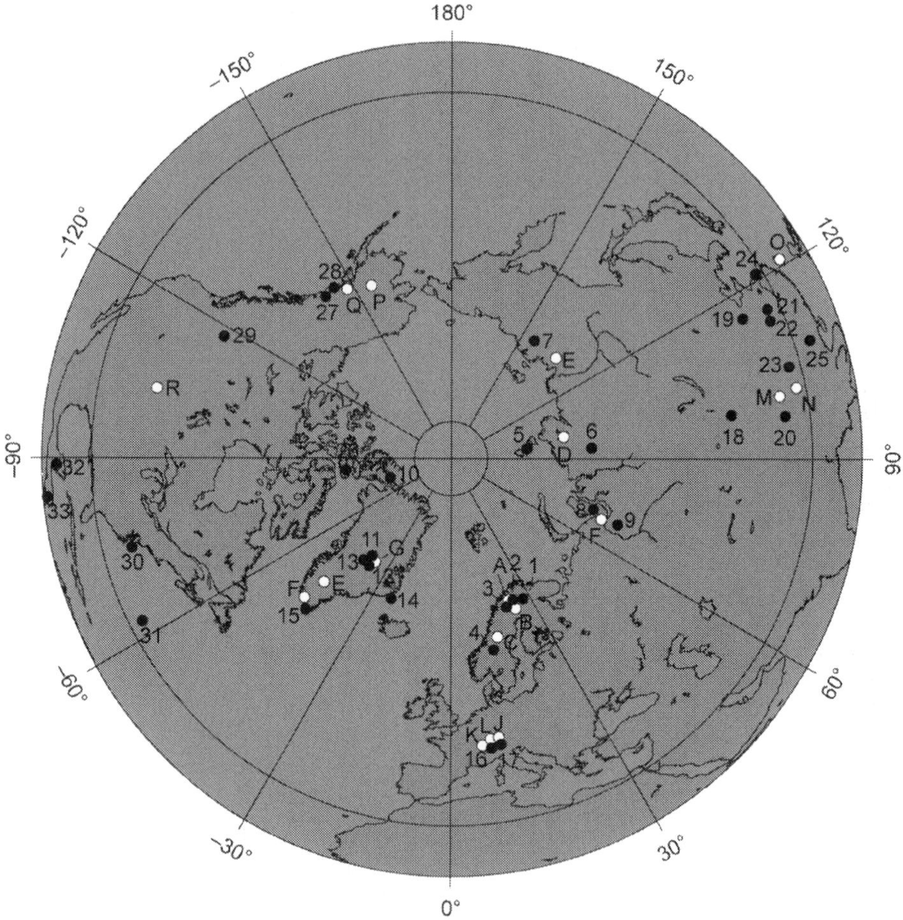

Fig. 2. Map with the geographical location of the proxy records listed in Table 1 and 2.

The Scandinavian warm season temperature reconstruction, consisting of the above-mentioned four records, was calibrated against the May to September mean temperature from the 60–70°N × 5–30°E CRUTEM3 grid cells (Fig. 3). This area represents the central and northern parts of Scandinavia. We have a long and good network of instrumental temperature measurements for Scandinavia and we can thus calibrate the reconstruction over the whole period AD 1850–1999. The correlation coefficient over the calibration period amounts to 0.88 (r^2 = 0.77). Late 20th century warm season temperatures in Scandinavia do not seem to be remarkably warm in a long-term perspective. During the MWP, occurring here c. AD 900–1100, Scandinavian warm season temperatures seem to have exceeded those of recent decades. Temperatures also equalled or exceeded the 1961–1990 reference level in the early 15th century. A LIA cooling is clearly seen approximately AD 1560–1720 and low temperatures are also prevailing c. AD 1350 and c. AD 1900. The total reconstructed decadal temperature variability of the last 12 decades is about 2.5°C, with a centennial variability of as much as 1.5°C.

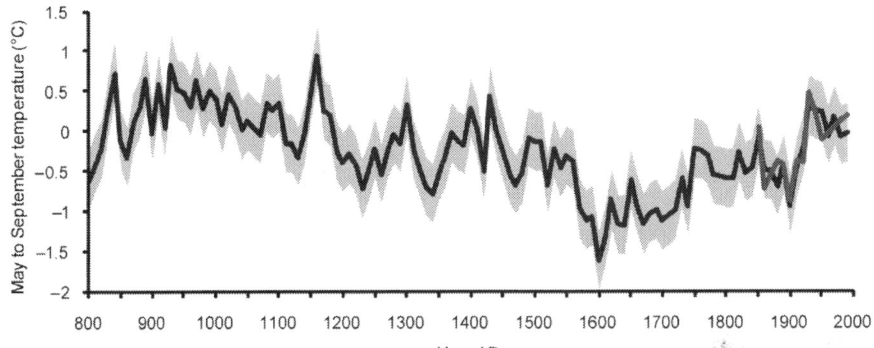

Fig. 3. May to September temperature reconstruction for Scandinavia north of 60°N (blue line) calibrated to instrumental temperatures for the same region (red line) with 2 standard deviation error bars (grey shading).

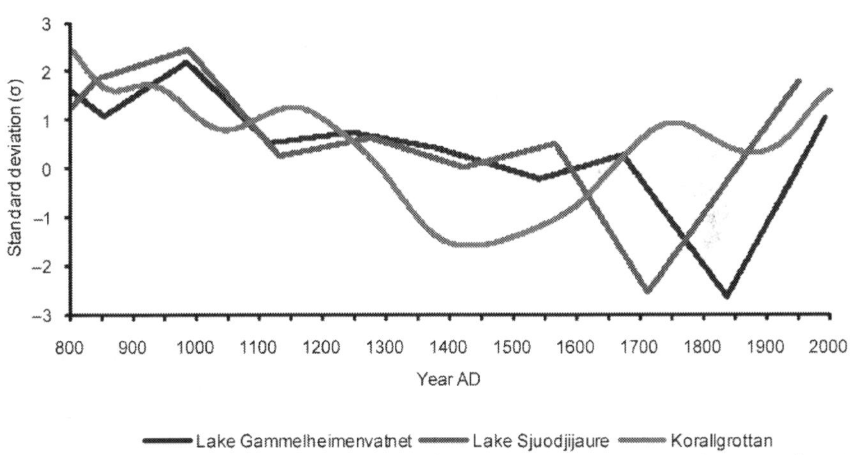

Fig. 4. Three normalised temperature records from Scandinavia shown relative to their AD 1000–1899 mean. The Korallgrottan record has been smoothed with a 100-year cubic spline filter.

As discussed above, there exist large numbers of quantitative temperature reconstructions with lower temporal resolution that can be used to compare and verify the reconstructed low-frequency trend. We have used the Lake Gammelheimenvatnet pollen-based reconstruction (Bjune et al., 2009) and the Lake Sjuodjijaure chironomid-based reconstruction (Rosén et al., 2003) from the northern tree-line area for this purpose together with the speleothem $\delta^{18}O$ record from Korallgrottan in Jämtland in Central Scandinavia (Sundqvist et al., 2010) (Fig. 4). Lake Gammelheimenvatnet and Lake Sjuodjijaure show high summer temperature variability whereas Korallgrottan probably reflects annual mean temperature variability. The general multi-centennial trends of the reconstruction are also seen in the three low-resolution records, and although they do not agree when the maximum LIA cooling occurred, they are consistent with regard to the occurrence of a clear MWP in Scandinavia.

2.2 Northern Siberia

Russia has a long tradition of palaeoclimatology and although much of it is still only available in the Russian language, considerable efforts were made already in the first half of the 1980s in order to incorporate Russian (then Soviet) palaeoclimatology with the Western research community (Velichko, 1984). However, the Russian palaeoclimatology has primarily been focused on long Pleistocene and Holocene perspectives and less focused on the climate variability of the last one or two millennia (Velichko et al., 1997). Siberian, as well as European Russian, pollen-based temperature reconstructions clearly show the occurrence of a MWP and a LIA but they have such a crude resolution that only multi-centennial variations can be detected (Andreev et al., 2000, 2001, 2003, 2004, 2005).

Five records, all primarily reflecting warm season temperatures, with temporal resolution high enough to be used here were found: 1) the Severnaja lake sediment record (Solomina and Alverson, 2004), 2) the Taimyr tree-ring width record (Naurzbaev et al., 2002), 3) the Indigirka tree-ring width record (Moberg et al., 2006), 4) the Yamal tree-ring width record (Briffa, 2000), and 5) the Polar Urals tree-ring maximum latewood density record (Esper et al., 2002). Regular instrumental temperature measurements were started relatively late in Siberia. Our calibration period is therefore limited to AD 1890–1989 and thus we have a limited degree of freedom. The northern Siberia warm season temperature reconstruction has been calibrated against the May to September mean temperature from the 60–80°N × 60–180°E CRUTEM3 grid cells (Fig. 5). During this 10 decade long calibration period, the correlation coefficient amounts to 0.70 (r^2 = 0.48). The relationship between the proxy composite and instrumental temperatures is thus relatively weak although the general trends are in quite good agreement.

The reconstructed Siberian warm season temperatures show that temperatures exceeded the 1961–1990 reference level c. AD 950–1150 and mostly equalled that level from c. AD 800–950 and c. AD 1150–1540. A quite clear LIA is seen c. AD 1540–1920 with temperatures about 0.5°C below the 1961–1990 reference level. Three especially distinct cold spells are seen during the LIA: c. AD 1280, a long cold period c. AD 1600–1750, and c. 1820. The overall amplitude of the reconstructed decadal variability the last 12 centuries well exceeds 1°C.

The reconstructed low-frequency temperature trend agrees well with the normalised values of three warm season temperature reconstructions with lower temporal resolution: the Yamal tree-line record (Solomina and Alverson, 2004), the Indigirka pollen-based temperature reconstruction (Velichko et al., 1997), and the Taimyr pollen-based temperature reconstruction (Andreev et al., 2002) (Fig. 6). The MWP, the LIA, and the modern warming are clearly visible in the low-resolution records, although they are less clear in the Yamal tree-line record. However, maximum LIA cooling seems to occur somewhat earlier in the three low-resolution records than in the quantitative, calibrated temperature reconstruction (Fig. 5).

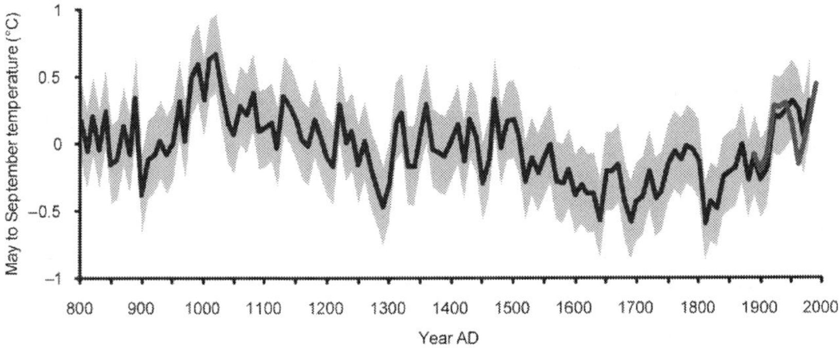

Fig. 5. May to September temperature reconstruction for northern Siberia (blue line) calibrated to instrumental temperatures for the same region (red line) with 2 standard deviation error bars (grey shading).

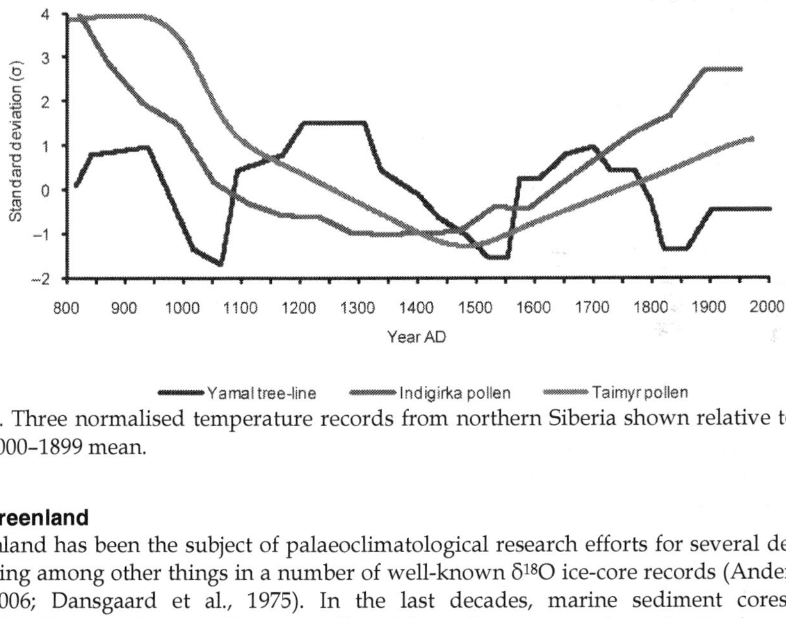

Fig. 6. Three normalised temperature records from northern Siberia shown relative to their AD 1000–1899 mean.

2.3 Greenland

Greenland has been the subject of palaeoclimatological research efforts for several decades, resulting among other things in a number of well-known $\delta^{18}O$ ice-core records (Andersen et al., 2006; Dansgaard et al., 1975). In the last decades, marine sediment cores from Greenland's extensive coastlines as well as lake sediment cores from the ice-free coastal areas have also become available (Andresen et al., 2004; Cremer et al., 2001; Moros et al. 2006; Møller et al., 2006; Roncaglia and Kuijpers, 2004; Seidenkrantz et al., 2007; Wagner and Melles, 2001). Most of the sediment records unfortunately have too low a temporal resolution and/or too large uncertainties in the dating to be useful in a quantitative multi-proxy temperature reconstruction aimed at being calibrated to temperature values. Other potentially useful records end too early to be calibrated to instrumental temperatures (e.g. Alley, 2000). Further records are unsuitable to use in order to reconstruct Greenland's temperature since they either do not have a significant correlation to temperature (as the NorthGRIP $\delta^{18}O$ ice-core record) or stronger correlation to Icelandic rather than Greenlandic temperatures (as the Renland $\delta^{18}O$ ice-core record) (Vinther et al., 2010). Thus, all in all, we find six records useful for our purpose: 1) the Crête $\delta^{18}O$ ice-core record, 2) the GISP2 $\delta^{18}O$ ice-core record, 3) the GRIP $\delta^{18}O$ ice-core record, 4) the Nansen Fjord benthic foraminifera sea floor sediment record (Jennings and Weiner, 1996), 5) the Igaliku Fjord biostratigraphic diatom sea floor sediment record (Jensen et al., 2004), and 6) the annually varved Lower Murray Lake sediment record (Cook et al., 2009). The last record, from Lower Murray Lake, actually comes from Ellesmere Island in the northernmost Canadian Arctic Archipelago but can be used to represent climate conditions in nearby northern Greenland.

The Greenland temperature reconstruction has been calibrated against the annual mean temperature from the 85–60°N × 15–70°W CRUTEM3 grid cells (Fig. 7). The correlation coefficient over the calibration period 1850–1969 amounts to 0.74 (r^2 = 0.55). The reconstruction shows a peak value in the 10th century c. 3°C above the 1961–1990 reference level and c. 1.5°C above the mid-20th century maximum. Temperatures then gradually declined but essentially remained above the 1961–1990 reference level until about AD 1300. We can thus conclude with reasonable safety that the MWP in Greenland well exceeded the observed 20th century warming, although this did not necessarily apply to the Arctic region as a whole (Kaufman et al., 2009). Such a strong regional warming of Greenland is actually well in agreement with global temperature field reconstructions indicating a post-1990 warming exceeding the medieval one on a global scale but not on Greenland (Mann et al., 2009). A pronounced cold period occurred in the mid to late 14th century, probably marking the onset of the LIA, although temperatures then rose again and hovered around the 1961–1990 reference level until the main phase of the LIA commenced in the late 17th century and lasted until c. AD 1920. It may be noted that no late 20th century warming is visible in either the reconstructed or the instrumental temperatures since the recent warming in Greenland only started in the late 1990s and still has not exceeded the level of the mid-20th century Greenland warming (Chylek et al., 2006).

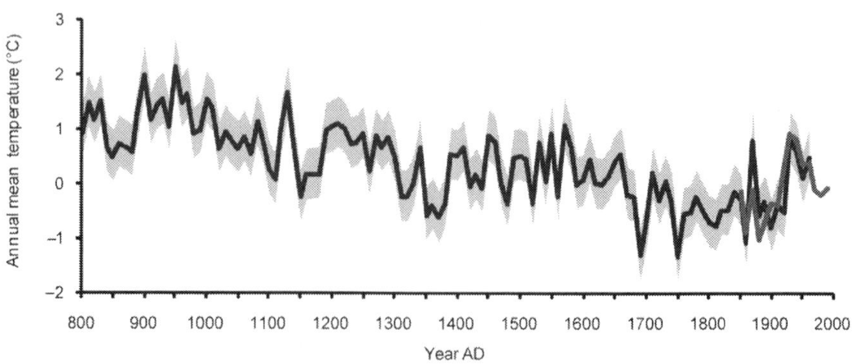

Fig. 7. Annual mean temperature reconstruction for Greenland (blue line) calibrated to instrumental temperatures for the west coast of Greenland (red line) with 2 standard deviation error bars (grey shading).

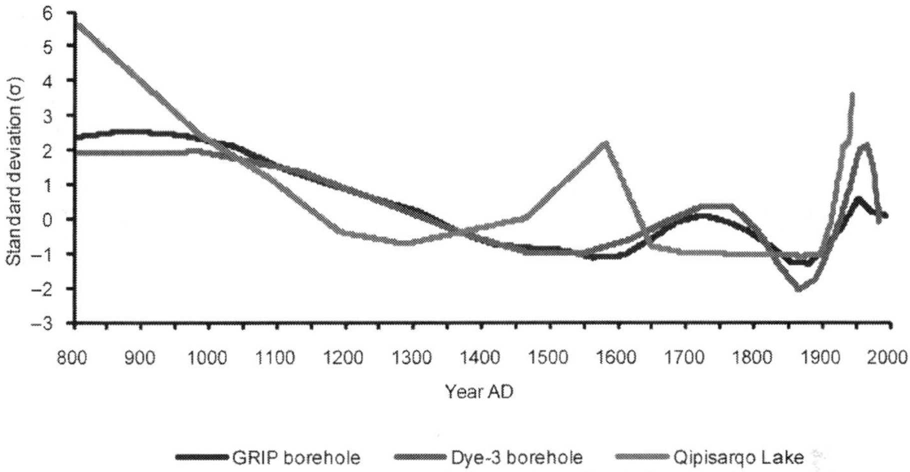

Fig. 8. Three normalised temperature records from Greenland shown relative to their AD 1000–1899 mean.

The reconstructed low-frequency temperature trend agrees well with the normalised values of the Dye-3 and GRIP deep borehole temperature profiles from the Greenland inland ice sheet (Dahl-Jensen et al., 1998) as well as with the normalised values of the biogenic silica sediment record from Qipisarqo Lake (Kaplan et al., 2002) (Fig. 8). Moreover, based on the analysis of the oxygen isotope composition of human tooth enamel ($\delta^{18}O_p$) in teeth from Norse and Inuit skeletons, Fricke et al. (1995) found a similar amplitude in the annual mean temperature variability as we have reconstructed here.

2.4 Central Europe

Considerable palaeoclimatological research has been conducted in the Alp region of Central Europe, aimed at different time perspectives but few useful quantitative temperature reconstructions have so far been produced for the last one or two millennia, although several are under development in the Millennium project founded by the European Union (Gagen et al., 2006). Several of the records from this region that are now available are less suited for being used in a quantitative calibrated temperature reconstruction. The siliceous algae-based Oberer Landschitzsee temperature reconstructions (Schmidt et al., 2007) have a crude resolution and likely contain much non-temperature related information. The Lake Anterne sediment record (Millet et al., 2009) is influenced by anthropogenic activities during the recent centuries, which makes it difficult to compare the level of the medieval warmth to the modern one. The temperature reconstruction from Central Europe by Glaser and Riemann (2009) based on historical documentary sources first starts in the early 1000s and probably underestimates the low-frequency variability in medieval times when the data coverage is sparse. The Spannagel Cave speleothem $\delta^{18}O$ record (Mangini et al., 2005) would be a potentially good record but it unfortunately ends already in the 1930s and thus has too short a calibration period for being really useful. This leaves us with only two records, both reflecting summer temperatures, 1) the varved Lake Silvaplana chironomid-based record

(Larocque et al., 2009), and the tree-ring maximum latewood density record from the Alps by Büntgen et al. (2006).

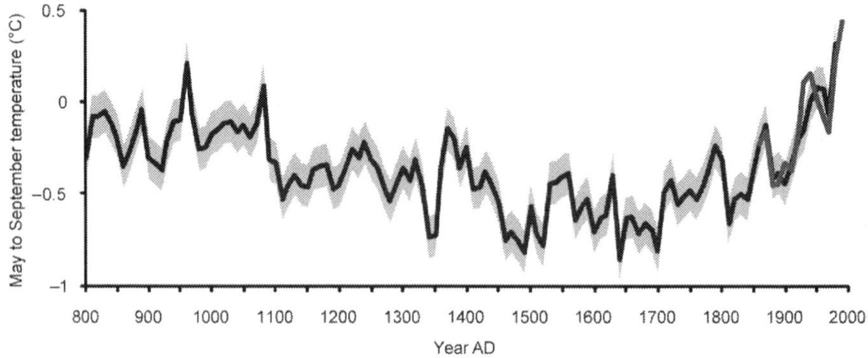

Fig. 9. May to September temperature reconstruction for Central Europe (blue line) calibrated to instrumental temperatures for the Alp region (red line) with 2 standard deviation error bars (grey shading).

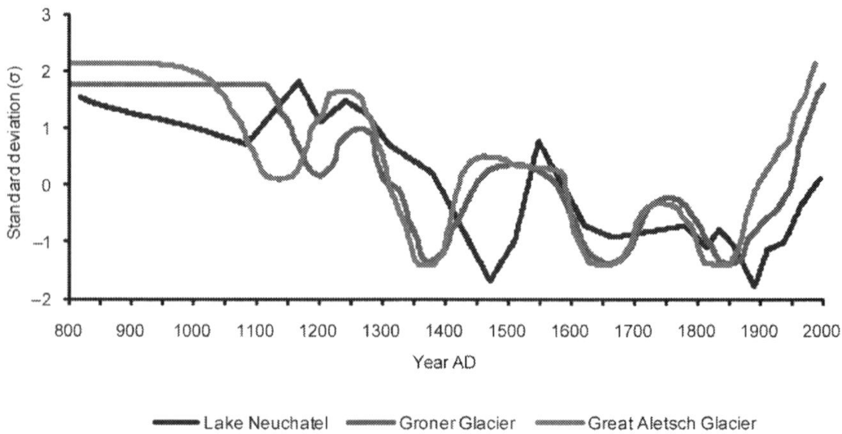

Fig. 10. Three normalised temperature records from Central Europe shown relative to their AD 1000–1899 mean.

Long, homogenised instrumental records are available from Central Europe, going back to ~1760 (Auer et al., 2007; Böhm et al., 2010). Even if we thus have longer instrumental temperature records from Central Europe, we will, for a matter of consistency with the other regional reconstructions, use the May to September 50–45°N × 5–15°E CRUTEM3 grid cells. The correlation coefficient over the calibration period 1850–1999 amounts to 0.84 (r^2 = 0.70). During most of the past 12 centuries, the May to September temperatures of Central Europe have been below the 1961–1990 reference level with the exception for parts of the MWP. In the 10th century, and partly in the 12th century, temperatures seem generally to have been at

or above this reference level. A very clear, albeit somewhat variable, LIA is seen from c. AD 1250 until the mid-19th century with a maximum cooling in the 17th century.
The reconstructed decadal temperature variability of the last 12 decades is about 2°C, with probably the warmest temperatures observed at the end of the 20th century, although this is within the uncertainty level of the medieval warmth (Fig. 9). A medieval warming c. AD 800–1280 is seen, peaking approximately AD 1000, but is interrupted by at least three major cool events: c. AD 1050 and c. AD 1170, and c. 1250. A reconstruction based on only two records cannot be regarded as particularly robust but the low-frequency trend of the reconstruction agrees very well with the with three low-resoled temperature reconstructions from the Alp region (Fig. 10): the glacier length records from the Great Aletsch Glacier and Groner Glacier (Holzhauser et al., 2005) and the pollen-based summer temperature reconstruction from Lake Neuchatel (Filippi et al., 1999) (Fig. 10).

2.5 China

Whereas the coverage of proxy data is very sparse for most of Asia, the coverage for China is rather good. In fact, China is, together with parts of North America, probably the region outside of Europe, with the best palaeoclimatological and palaeoenvironmental records. Previously, much of the research has been published in Chinese, and thus not been easily accessible to Western scholars, but in recent years most Chinese research has been published in English and consequently well incorporated in the international field of palaeoclimatology. The problem remains, however, that relatively few of the reconstructions are available in digital form from public databases. China has the world's longest continuous written historical records that can be used to reconstruct past climate variability. Therefore, much valuable research in the field of historical climatology has been done in China. Many syntheses on past climate in China, using different natural and/or historical archives, have been written and two of them are of relevance to us. Yang et al. (2002) published the first highly resolved annual mean quantitative temperature reconstruction for China covering the last two millennia and Ge et al. (2010) updated this reconstruction and also made reconstructions for different parts of China.

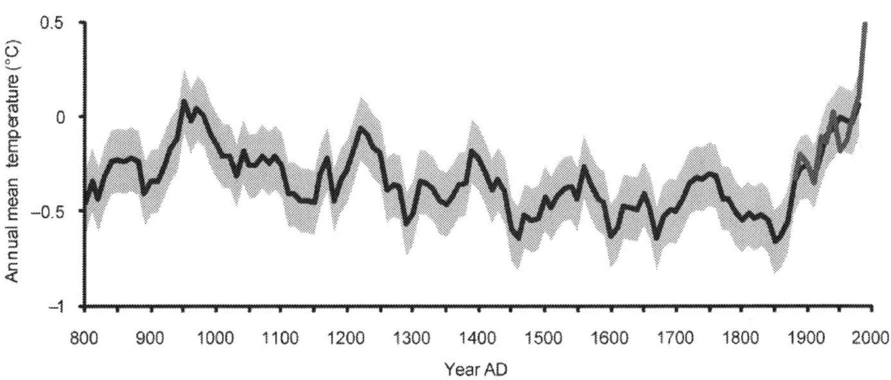

Fig. 11. Annual mean temperature reconstruction for China (blue line) calibrated to instrumental temperatures for China (red line) with 2 standard deviation error bars (grey shading).

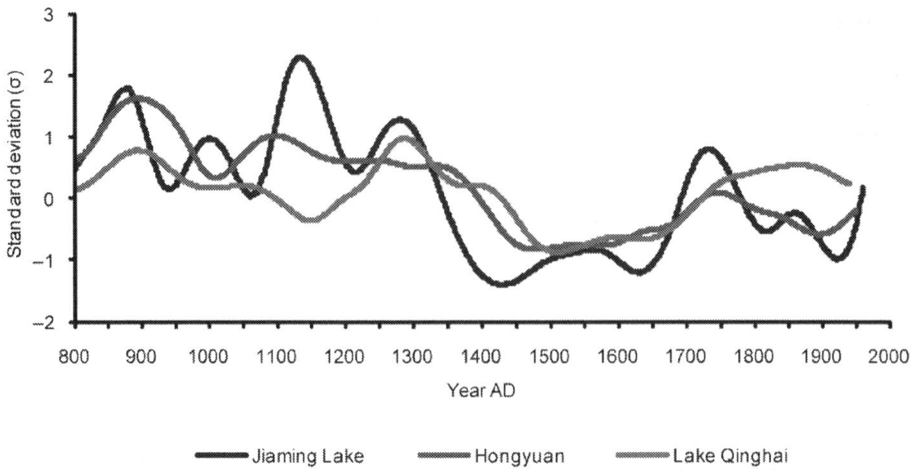

Fig. 12. Three normalised temperature records from China shown relative to their AD 1000–1899 mean. The records have been smoothed with a 100-year cubic spline filter.

Eight records were deemed suitable for reconstructing decadal China temperature variability of the last 12 centuries: 1) the Mongolia tree-ring width record (D'Arrigo et al., 2001), 2) the ShiHua Cave speleothem $\delta^{18}O$ record (Tan et al., 2003), 3) a composite of East China documentary records of annual mean temperature (Yang et al., 2002), 4) a composite of documentary East China records of cold season temperature (Ge et al., 2003), 5) the Dulan, northeast Qinghai-Tibetan Plateau, tree-ring width record (Zhang et al., 2003), 6) the Wanxiang Cave speleothem $\delta^{18}O$ record (Zhang et al., 2008b), 7) a composite of Zangtze Delta region documentary records of annual mean temperature (Zhang et al., 2008a), and 8) the Heshang Cave speleothem $\delta^{18}O$ record (Hu et al., 2008).

The China annual temperature reconstruction has been calibrated against the annual mean temperature from the 55–20°N × 70–135°E CRUTEM3 grid cells. Unfortunately, the meteorological network of instrumental temperature measurements is much shorter and sparser back in time in China than in Europe or most of North America. We can therefore not start the calibration period of the Chinese temperature reconstruction until AD 1880 and the lack of proxy data in the 1990s forces us to end the calibration period AD 1989. Thus, our calibration period for China is just 11 decades. The correlation coefficient over the calibration period amounts to 0.83 (r^2 = 0.68) and thus shows, although the degrees of freedom are limited, a strong relationship between the proxy composite and the instrumental temperatures.

The multi-decadal annual mean temperature variability in China seems to have been slightly less than 1°C and thus smaller then in most other regions (Fig. 11). During most of the past 12 centuries, the temperatures have been well below the 1961–1990 reference level. However, during the second half of the 10th century temperatures equalled or exceeded the 1961–1990 mean. This warm event probably represents the peak of the MWP in China. Warm conditions, similar to those of the 20th century, are also seen in the first half of the 13th century. Five major cooling events during the LIA are visible, centred c. AD 1300, AD 1450, AD 1600, AD 1675 and AD 1850. According to the instrumental record, late 20th century

temperatures may be the highest in the last 12 centuries but this is not seen in the proxy reconstruction itself.
Overall, both the shape and amplitude of the low-frequency temperature variability are well in line with the pollen-inferred temperature reconstructions by He et al. (2004). There also exist some other quantitative temperature reconstructions with lower temporal resolution that can be used to compare and verify the reconstructed low-frequency trend. We have chosen to use the Jiaming Lake (Yang et al., 2002), the Hongyuan Lake (Yang et al., 2002), and the Lake Qinghai (Liu et al., 2006) sediment records for this purpose. They show a quite similar picture of China low-frequency temperature variability, especially the existence of a MWP c. AD 850–1300 and a clear LIA cooling c. AD 1400–1700 (Fig. 12).

2.6 North America

Relatively much data, supposedly reflecting past temperature variability, are available from North America but most of this data consist of tree-ring records. A significant problem is that many of the tree-rings records, especially those from lower latitudes, apparently are more sensitive to precipitation than temperature. All tree-ring width records from high-elevation semi-arid sites in the North American Southwest can be said to possess an ambiguous temperature signal at best, although they have indeed been used in many previous multi-proxy reconstructions (e.g., Cook et al., 2004; Crowley and Lowery, 2000; Esper et al., 2002; Jones and Mann, 2004; Mann et al., 1999; Mann et al., 2008; Mann et al., 2009; Mann and Jones, 2003; Osborn and Briffa, 2006). The potential problems with tree-ring records from semi-arid, high-elevation regions in the North American Southwest have been highlighted in D'Arrigo et al. (2006) and Loehle (2009). Moreover, most of those chronologies have a low replication value back in medieval times. Excluding the tree-ring width records from lower latitudes leaves us with a relatively limited number of palaeotemperature proxy records with adequate resolution and time control to use in a calibrated temperature reconstruction.

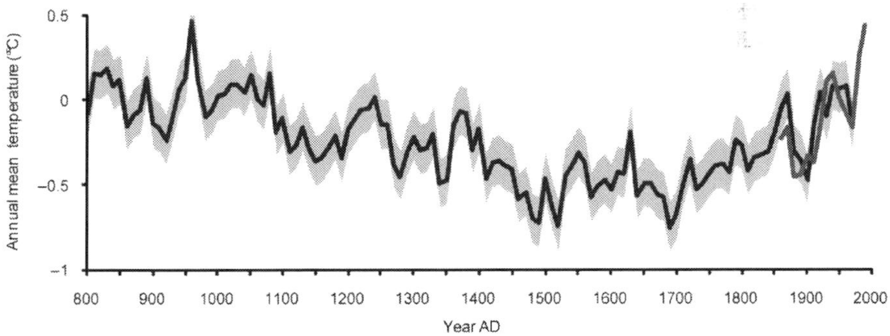

Fig. 13. Annual mean temperature reconstruction for North America (blue line) calibrated to instrumental temperatures for this continent (red line) with 2 standard deviation error bars (grey shading).

Eight records were deemed suitable for reconstructing decadal North American temperature variability of the last 12 centuries: 1) the Devon Island, Canadian Arctic

Archipelago, ice-core δ18O record (Fisher et al., 1983), 2) the Gulf of Alaska tree-ring width record (D'Arrigo et al., 2006), 3) the Iceberg Lake varved sediment record (Loso, 2009), 4) the Canadian Rockies tree-ring width record (Luckman and Wilson, 2005), 5) the Chesapeake Bay sea sediment record (Cronin et al., 2003), 6) the Bermuda Rise sea sediment record (Keigwin, 1996), 7) the Punta Laguna lake sediment record (Curtis et al., 1996), and 8) the Nicoa Cave speleothem δ18O record (Mann et al., 2008). The North America annual temperature reconstruction has been calibrated against the annual mean temperature from the 85–5°N × 50–170°W CRUTEM3 grid cells. The correlation coefficient over the calibration period 1860–1989 amounts to 0.78 ($r^2 = 0.61$).

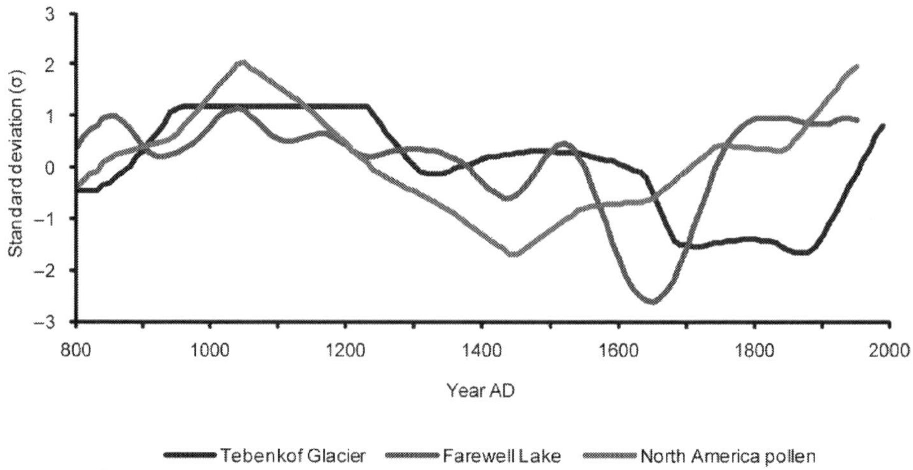

Fig. 14. Three normalised temperature records from North America shown relative to their AD 1000-1899 mean. The Lake Farwell record has been smoothed with a 100-year cubic spline filter.

The reconstructions show that annual mean temperatures have been below the 1961-1990 reference level prior to the 20th century except during the MWP (Fig. 13). A very sharp warm peak during the MWP occur c. AD 960 with temperature levels equalling those of the last two decades. A general cooling, consistent with the LIA, is seen in the North American reconstruction c. AD 1400–1850 with a markedly cold spell also c. AD 1900. The maximum cooling occurs around AD 1500 and AD 1700. The reconstructed decadal variability of the last 12 centuries is slightly more than 1°C.The reconstructed low-frequency trend could be compared to a number of different proxy records with lower temporal resolution. We have chosen to use three that, admittedly, primarily reflect warm season temperatures: The Tebenkof Glacier length record from southern Alaska (Barclay et al., 2009), the Farewell Lake sediment record from central Alaska (Hu et al., 2001), and the entire North American pollen-based temperature reconstruction by Viau et al., 2006. The reconstructed long-term changes are well reflected in the low-resolution records (Fig. 14). The existence of a distinct MWP in North America is verified and also the existence of a LIA, although the three low-resolution records are less consistent when it was coldest.

3. Discussion

The North Atlantic region - from Greenland to Europe - has the best data coverage together with China and the coastal areas of North America. Data are lacking to make meaningful regional reconstructions for Africa, the Middle East or the interior of Asia as well as for all of the Southern Hemisphere (except for, perhaps, South America, see Neukom et al., 2010). Even in relative data rich regions (e.g., Central Europe and Scandinavia) the seasonal bias in the data limits our possibility to make annual mean temperature reconstructions. Most high latitude data, except for the Greenland borehole and $\delta^{18}O$ ice-core records, have a clear bias towards warm season temperatures. We have been able to present regional reconstructions for 1) warm season temperatures of Scandinavia north of 60°N, 2) warm season temperatures for northern Siberia, 3) annual mean temperatures for Greenland, 4) warm season temperatures for the Alp region of Central Europe, 5) annual mean temperatures for China, and 6) annual mean temperatures for the whole of the North American continent.

In all the six reconstructions we can see evidence for the MWP, the LIA and the modern warming. Moreover, in all the six regional reconstructions the medieval warming is peaking in the 10th century. Maximum LIA cooling, however, seems to occur at somewhat different times in the different reconstructions - from the 16th to the 19th centuries - but the 17th century is very cold in all reconstructions. All reconstructions, except the one from central Europe, also agree that the mid to late 19th century was unusually cold, which is of special interest since it is during this period that widespread instrumental temperature measurements start. Late 20th century temperatures are lower than those of the MWP in the Scandinavia, Siberia and especially in the Greenland temperature reconstruction and equal to the medieval warming in the North America reconstruction. However, in the China and the Central Europe reconstructions late 20th century warming exceeds the medieval one, although this is not clear from the proxy reconstructions themselves but only from the instrumental temperature data spliced to the reconstructions.

The trends visible in palaeotemperature proxy records with low (multi-decadal to centennial) resolution, which thus cannot be robustly calibrated to the instrumental record, are generally in good agreement with the trends of the calibrated reconstructions from the same regions. This implies that the calibrated regional reconstructions preserve a good degree of the low-frequency variability although the amplitude of the reconstructed changes probably is too small due to the statistical problems of calibrating noisy data to instrumental data (see, e.g., von Storch et al., 2004; Lee et al., 2008).

Although, from the regional reconstructions presented here, it seems doubtful if the late 20th century warming exceeded the medieval one in the Northern Hemisphere, much more data are needed to draw firm conclusions in this matter. In order to truly assess the spatio-temporal pattern of past temperature variability we also need to be able to make reconstructions from other regions as Africa, the Middle East and Central Asia. Much more research in the field is therefore needed to develop new proxy records from different natural recorders of climate variability, such as fossil pollen, ice-cores, lake and marine sediments, speleothems, and tree-ring width and density, as well as from historical records. It is also very important that all new palaeotemperature records are archived, and thus easily accessible, in public databases as the World Data Center for Paleoclimatology (http://www.ncdc.noaa.gov/paleo/paleo.html).

Although it is outside the scope of this article, we can briefly discuss the possible influence of variations in solar and volcanic forcing on the reconstructed temperatures. All the six regional temperature reconstructions show some agreement with the assumed low-frequency variability in solar forcing of the last 12 centuries (Bard et al., 2000). The medieval period, with high temperatures, had a general high solar activity, whereas the cold LIA was dominated by lower solar activity (Ammann et al., 2007). The warming in the 20th century coincides with an increase in solar forcing, although the warming trend has probably also been amplified in the last decades by anthropogenic greenhouse gas emissions (IPCC, 2007).

4. Conclusion

The presently available palaeotemperature proxy data records do not support the assumption that late 20th century temperatures exceeded those of the MWP in most regions, although it is clear that the temperatures of the last few decades exceed those of any multi-decadal period in the last 700–800 years. Previous conclusions (e.g., IPCC, 2007) in the opposite direction have either been based on too few proxy records or been based on instrumental temperatures spliced to the proxy reconstructions. It is also clear that temperature changes, on centennial time-scales, occurred rather coherently in all the investigated regions - Scandinavia, Siberia, Greenland, Central Europe, China, and North America - with data coverage to enable regional reconstructions. Large-scale patterns as the MWP, the LIA and the 20th century warming occur quite coherently in all the regional reconstructions presented here but both their relative and absolute amplitude are not always the same. Exceptional warming in the 10th century is seen in all six regional reconstructions. Assumptions that, in particular, the MWP was restricted to the North Atlantic region can be rejected. Generally, temperature changes during the past 12 centuries in the high latitudes are larger than those in the lower latitudes and changes in annual temperatures also seem to be larger than those of warm season temperatures. In order to truly assess the possible global or hemispheric significance of the observed pattern, we need much more data. The unevenly distributed palaeotemperature data coverage still seriously restricts our possibility to set the observed 20th century warming in a global long-term perspective and investigate the relative importance of natural and anthropogenic forcings behind the modern warming.

5. References

Alley, R.B., 2000: The Younger Dryas cold interval as viewed from Central Greenland. Quaternary Science Reviews, 19: 213–226.

Ammann, C.M. and Wahl, E.R., 2007. The importance of the geophysical context in statistical evaluations of climate reconstruction procedures. Climatic Change, 85: 71–88.

Ammann, C.M., Joos, F., Schimel, D.S., Otto-Bliesner, B.L., and Tomas, R.A., 2007: Solar influence on climate during the past millennium: Results from transient simulations with the NCAR Climate System Model. Proceedings of the National Academy of Sciences, USA, 104, 3713–3718.

Andersen, K.K., Ditlevsen, P.D., Rasmussen, S.O., Clausen, H.B., Vinther, B.M., Johnsen, S.J. and Steffensen, J.P., 2006: Retrieving a common accumulation record from Greenland ice cores for the past 1800 years. Journal of Geophysical Research, 111: D15106, doi:10.1029/2005JD006765.

Andreev, A. A., and Klimanov, V.A., 2000: Quantitative Holocene climatic reconstruction from Arctic Russia. Journal of Paleolimnology, 24: 81–91.

Andreev, A.A., Klimanov, V.A., and Sulerzhitsky, L.D., 2001: Vegetation and climate history of the Yana River lowland, Russia, during the last 6400 yr. Quaternary Science Reviews, 20: 259-266.

Andreev, A.A., Tarasov, P.E., Siegert, C., Ebel, T., Klimanov, V.A., Melles, M., Bobrov, A., Dereviagin, A.Y., Lubinski, D., and Hubberten, H.-W., 2003: Late Pleistocene vegetation and climate on the northern Taymyr Peninsula, Arctic Russia. Boreas, 32: 484-505.

Andreev, A.A., Tarasov, P.E., Klimanov, V.A., Melles, M., Lisitsyna, O.M., and Hubberten, H.-W., 2004: Vegetation, climate changes around Lama Lake, Taymyr Peninsula, Russia, during the Late Pleistocene and Holocene. Quaternatery International, 122: 69-84.

Andreev, A.A., Tarasov, P.E., Ilyashuk, B.P., Ilyashuk, E.A, Cremer, H., Hermichen, W.-D., Wisher, F., and Hubberten, H.-W., 2005: Holocene environmental history recorded in Lake Lyadhej-To sediments, Polar Urals, Russia, Palaeogeography, Palaeoclimatology, Palaeoecology, 223: 181-203.

Auer, I., Böhm, R., Jurkovic, A., Lipa, W., Orlik, A., Potzmann, R., Schöner, W., Ungersböck, M., Matulla, C., Briffa, K., Jones, P.D., Efthymiadis, D., Brunetti, M., Nanni, T., Maugeri, M., Mercalli, L., Mestre, O., Moisselin, J.-M., Begert, M., Müller-Westermeier, G., Kveton, V., Bochnicek, O., Stastny, P., Lapin, M., Szalai, S., Szentimrey, T., Cegnar, T., Dolinar, M., Gajic-Capka, M., Zaninovic, K., Majstorovic, Z., and Nieplova, E., 2007: HISTALP - Historical instrumental climatological surface time series of the greater Alpine region 1760-2003. Intentional Journal of Climatology 27: 17-46.

Barclay, D.J., Wiles, G.C., and Calkin, P.E. 2009. Tree-ring crossdates for a first millennium AD advance of Tebenkof Glacier, southern Alaska. Quaternary Research, 71: 22-26.

Bard, E., Raisbeck, G., Yiou, F., and Jouzel, J., 2000: Solar irradiance during the last 1200 years based on cosmogenic nuclides. Tellus, 52B: 985-992.

Bjune, A.E., Seppä, H., and Birks, H.J.B., 2009: Quantitative summer-temperature reconstructions for the last 2000 years based on pollen-stratigraphical data from northern Fennoscandia. Journal of Paleolimnology, 41: 43-56.

Böhm, R., Jones, P.D., Hiebl, J., Frank, D., Brunetti, M., and Maugeri, M., 2010: The early instrumental warm-bias: a solution for long Central European temperature series, 1760-2007. Climatic Change: in press.

Bradley, R.S., Briffa, K.R., Crowley, T.J., Hughes, M.K., Jones, P.D. and Mann, M.E., 2001: The scope of medieval warming. Science, 292: 2011-2012.

Bradley, R.S., Hughes, M.K. and Diaz, H.F., 2003: Climate in medieval time. Science, 302: 404-405.

Briffa, K.R., 2000: Annual climate variability in the Holocene: interpreting the message of ancient trees. Quaternary Science Reviews, 19: 87-105.

Broecker, W.S., 2001: Was the Medieval Warm Period global?. Science, 291: 1497-1499.

Brohan, P., Kennedy, J., Haris, I., Tett, S.F.B., and Jones, P.D., 2006: Uncertainty estimates in regional and global observed temperature changes: a new dataset from 1850. Journal of Geophysical Research, 111: D12106.

Chylek, P., Dubey, M.K., Lesins, G., 2006: Greenland warming of 1920-1930 and 1995-2005. Geophysical Research Letters, 33: 10.1029/2006GL026510.

Cook, E.R., Esper, J. and D'Arrigo, R.D., 2004: Extra-tropical Northern Hemisphere land temperature variability over the past 1000 years. Quaternary Science Reviews, 23: 2063–2074.

Cook, T.L., Bradley, R.S., Stoner, J.S. and Francus, P., 2009: Five thousand years of sediment transfer in a high arctic watershed recorded in annually laminated sediments from Lower Murray Lake, Ellesmere Island, Nunavut, Canada. Journal of Paleolimnology, 41: 77–94.

Cremer, H., Wagner, B., Melles, M., and Hubberten, H.-W., 2001: The postglacial environmental development of Raffles Sø, East Greenland: inferences from a 10,000 year diatom record. Journal of Paleolimnology, 26: 67–87.

Cronin, T. M., Dwyer, G.S., Kamiya, T., Schwede, S., and Willard, D.A., 2003: Medieval Warm Period, Little Ice Age and 20th century temperature variability from Chesapeake Bay. Global and Planetary Change, 36: 17–29.

Crowley, T.J., 2000: Causes of climate change over the past 1000 years. Science, 289: 270–277.

Crowley, T.J. and Lowery, T., 2000: How warm was the Medieval Warm Period? A comment on "man-made versus natural climate change". Ambio, 29: 51–54.

Crowley, T.J., Baum, S.K., Kim, K.-Y., Hegerl, G.C. and Hyde, W.T., 2003: Modeling ocean heat content changes during the last millennium. Geophysical Research Letters, 30: 1932, doi:10.1029/2003GL017801.

Dahl-Jensen, D., Mosegaard, K., Gundestrup, N., Clow, G.D., Johnsen, S.J., Hansen, A.W., and Balling, N., 1998: Past temperatures directly from the Greenland Ice Sheet. Science, 282: 268–271.

D'Arrigo, R., Jacoby, G., Frank, D., Pederson, N., Cook, E., Buckley, B., Nachin, B., Mijiddorj, R., and Dugarjav, C., 2001: 1738 years of Mongolian temperature variability inferred from a tree-ring width chronology of Siberian pine. Geophysical Research Letters, 28: 543–546.

D'Arrigo, R., Wilson, R. and Jacoby, G., 2006: On the long-term context for late 20th century warming. Journal of Geophysical Research, 111: D3, D03103.

Dansgaard, W., Johnsen S.J., Reeh N., Gundestrup, N., Clausen, H.B., and Hammer, C.U., 1975: Climatic changes, Norsemen and modern man. Nature, 255: 24–28.

Esper, J., Cook, E.R. and Schweingruber, F.H., 2002a: Low-frequency signals in long tree-ring chronologies for reconstructing past temperature variability. Science, 295: 2250–2253.

Esper, J., Schweingruber, F.H. and Winiger, M., 2002b: 1300 years of climatic history for Western Central Asia inferred from tree-rings. The Holocene, 12: 267–277.

Esper, J., Frank, D.C., Wilson, R.J.S. and Briffa, K.R., 2005a: Effect of scaling and regression on reconstructed temperature amplitude for the past millennium. Geophysical Research Letters, 32: L07711.

Esper, J., Wilson, R.J.S., Frank, D.C., Moberg, A., Wanner, H. and Luterbacher, J., 2005b: Climate: past ranges and future changes. Quaternary Science Reviews, 24: 2164–2166.

Esper, J. and Frank, D.C., 2009: IPCC on heterogeneous Medieval Warm Period. Climatic Change, 94: 267–273.

Filippi, M.L., Lambert, P., Hunziker, J., Kubler, B., and Bernasconi, S., 1999: Climatic and anthropogenic influence on the stable isotope record from bulk carbonates and ostracodes in Lake Neuchatel, Switzerland, during the last two millennia. Journal of Paleolimnology, 21: 19-34.

Fisher, D.A., Koerner, R.M., Paterson, W.S.B., Dansgaard, W., Gundestrup, N. and Reeh, N., 1983: Effect of wind scouring on climatic records from icecore oxygen isotope profiles. Nature, 301: 205-209.

Fricke, H.C., O'Neil, J.R., and Lynnerup, N., 1995: Oxygen isotope composition of human tooth enamel from medieval Greenland: Linking climate and society. Geology, 23: 869-872.

Gagen, M., McCarrol, D., and Hicks, S., 2006: The Millennium project: European climate of the last. PAGES News, 14: 4.

Ge, Q., Zheng, J., Fang, X., Man, Z., Zhang, X., Zhang, P., and Wang, W.-C., 2003: Winter half-year temperature reconstruction for the middle and lower reaches of the Yellow River and Yangtze River, China, during the past 2000 years. The Holocene, 13: 933-940.

Ge, Q.S., Zheng, J.-Y., Hao, Z.-X., Shao, X.-M., Wang, W.-C., and Luterbacher, J., 2010: Temperature variation through 2000 years in China: An uncertainty analysis of reconstruction and regional difference. Geophysical Research Letters, 37: 10.1029/2009GL041281.

Grove, J.M., 1988. The Little Ice Age. London, Methuen: 498 pp.

Grudd, H., 2008: Torneträsk tree-ring width and density AD 500-2004: a test of climatic sensitivity and a new 1500-year reconstruction of north Fennoscandian summers. Climate Dynamics, 31: 843-857.

He, Y., Theakstone, W., Zhang, Z., Zhang, D., Yao, T., Chen, T., Shen, Y., and Pang, H., 2004: Asynchronous Holocene climatic change across China. Quaternary Research, 61: 52-63.

Hegerl, G., Crowley, T., Allen, M., Hyde, W., Pollack, H., Smerdon, J. and Zorita, E., 2007: Detection of human influence on a new, validated, 1500 year temperature reconstruction. Journal of Climate, 20: 650-666.

Hu, F.S., Ito, E., Brown, T.A., Curry, B.B., and Engstrom, D.R., 2001: Pronounced climatic variations in Alaska during the last two millennia. Proceedings of the National Academy of Sciences, USA, 98: 10552-10556.

Hu, C., Henderson, G.M., Huang, J., Xie, S., Sun, Y., and Johnson, K.R. 2008: Quantification of Holocene Asian monsoon rainfall from spatially separated cave records. Earth and Planetary Science Letters, 266: 221-232.

Hughes, M.K. and Diaz, H.F., 1994: Was there a 'medieval warm period', and if so, where and when?. Climatic Change, 26, 109-142.

IPCC, 2007: Climate Change 2007: The physical science basis. Contribution of working group I to the fourth assessment report of the Intergovernmental Panel on Climate Change [Solomon, S., Qin, D., Manning, M., Chen, Z., Marquis, M., Averyt, K.B., Tignor, M. and Miller, H.L. (eds.)]. Cambridge and New York: Cambridge University Press: 996 pp.

Jennings, A.E., and Weiner, N.J., 1996: Environmental change in eastern Greenland during the last 1300 years: evidence from foraminifera and lithofacies in Nansen Fjord, 68°N. The Holocene, 6: 179-191.

Jensen, K.G., Kuijpers, A., Koç, N., and Heinemeier, J., 2004: Diatom evidence of hydrografhic changes and ice conditions in Igaliku Fjord, South Greenland, during the past 1500 years. The Holocene, 14: 152-164.

Jones, P.D., Briffa, K.R., Barnett, T.P. and Tett, S.F.B., 1998: High-resolution palaeoclimatic records for the last millennium: interpretation, integration and comparison with General Circulation Model control-run temperatures. The Holocene, 8: 455-471.

Jones, P.D., Osborn, T.J. and Briffa, K.R., 2001: The evolution of climate over the last millennium. Science, 292: 662-667.

Jones, P.D. and Mann, M.E., 2004: Climate over past millennia. Reviews of Geophysics, 42: RG2002.

Jones, P.D., Briffa, K.R., Osborn, T.J., Lough, J.M., van Ommen, T.D., Vinther, B.M., Luterbacher, J., Wahl, E.R., Zwiers, F.W., Mann, M.E., Schmidt, G.A., Ammann, C.M., Buckley, B.M., Cobb, K.M., Esper, J., Goosse, H., Graham, N., Jansen, E., Kiefer, T., Kull, C., Küttel, M., Mosley-Thompson, E., Overpeck, J.T., Riedwyl, N., Schulz, M., Tudhope, A.W., Villalba, R., Wanner, H., Wolff, E. and Xoplaki, E., 2009: High-resolution palaeoclimatology of the last millennium: A review of current status and future prospects. The Holocene, 19: 3-49.

Juckes, M.N., Allen, M.R., Briffa, K.R., Esper, J., Hegerl, G.C., Moberg, A., Osborn, T.J. and Weber, S.L., 2007: Millennial temperature reconstruction intercomparison and evaluation. Climate of the Past, 3: 591-609.

Kaplan, M.R., Wolfe, A.P. and Miller, G.H., 2002: Holocene environmental variability in southern Greenland inferred from lake sediments. Quaternary Research, 58: 149-159.

Kaufman, D.S., Schneider, D.P., McKay, N.P., Ammann, C.M., Bradley, R.S., Briffa K.R., Miller, G.H., Otto-Bliesner, B.L., Overpeck, J.T., Vinther, B.M., Arctic Lakes 2k Project Members (Abbott, M., Axford, Y., Bird, B., Birks, H.J.B., Bjune, A.E., Briner, J., Cook, T., Chipman, M., Francus, P., Gajewski, K., Geirsdóttir, Á., Hu, F.S., Kutchko, B., Lamoureux, S., Loso, M., MacDonald, G., Peros, M., Porinchu, D., Schiff, C., Seppä, H. and Thomas, E.)., 2009. Recent warming reverses long-term Arctic cooling. Science, 325: 1236-1239.

Korhola, A., Weckström, J., Holmström, L., and Erästö, P.A., 2000: A quantitative Holocene climatic record from diatoms in northern Fennoscandia. Quaternary Research, 54: 284-294.

Lamb, H.H., 1977: Climate: Present, past and future 2. Climatic history and the future. London, Methuen: 835 pp.

Larocque, I., Grosjean, M., Heiri, O., Bigler, C., and Blass, A., 2009: Comparison between chironomid-inferred July temperatures and meteorological data AD 1850-2001 from varved Lake Silvaplana, Switzerland. Journal of Paleolimnology, 41: 329-342.

Lee, T.C.K., Zwiers, F.W., and Tsao, M., 2008: Evaluation of proxy-based millennial reconstruction methods. Climate Dynamics, 31: 263-281.

Linderholm, H.W., and Gunnarson, B.E., 2005: Summer temperature variability in central Scandinavia during the last 3600 years. Geografiska Annaler, 87A: 231-241.

Liu, Z., Henderson, A.C.G., and Huang, Y., 2006: Alkenone-based reconstruction of late-Holocene surface temperature and salinity changes in Lake Qinghai, China. Geophysical Research Letters, 33: 10.1029/2006GL026151.

Ljungqvist, F.C., 2009: Temperature proxy records covering the last two millennia: a tabular and visual overview. Geografiska Annaler, 91A: 11–29.

Ljungqvist, F.C., 2010: An improved reconstruction of temperature variability in the extra-tropical Northern Hemisphere during the last two millennia. Geografiska Annaler, 92A: in press.

Loehle, C., 2007: A 2000-year global temperature reconstruction based on non-treering proxies. Energy & Environment, 18: 1049–1058.

Loehle, C., 2009: A mathematical analysis of the divergence problem in dendroclimatology. Climatic Change, 94: 233–245.

Loso, M.G., 2009: Summer temperatures during the Medieval Warm Period and Little Ice Age inferred from varved proglacial lake sediments in southern Alaska. Journal of Paleolimnology, 41: 117–128.

Luckman, B.H., and Wilson, R.J.S., 2005: Summer temperatures in the Canadian Rockies during the last millennium: a revised record. Climate Dynamics, 24: 131–144.

Mangini, A., Spötl, C., and Verdes, P., 2005: Reconstruction of temperature in the Central Alps during the past 2000 yr from a $\delta^{18}O$ stalagmite record. Earth and Planetary Science Letters, 235: 741–751.

Mann, M.E., Bradley, R.S. and Hughes, M.K., 1998: Global-scale temperature patterns and climate forcing over the past six centuries. Nature, 392: 779–787.

Mann, M.E., Bradley, R.S. and Hughes, M.K., 1999: Northern hemisphere temperatures during the past millennium: inferences, uncertainties, and limitations. Geophysical Research Letters, 26: 759–762.

Mann, M.E. and Jones, P.D., 2003: Global surface temperatures over the past two millennia. Geophysical Research Letters, 30: 1820.

Mann, M.E., Cane, M.A., Zebiak, S.E. and Clement, A., 2005: Volcanic and Solar Forcing of the Tropical Pacific over the Past 1000 Years. Journal of Climate, 18: 417–456.

Mann, M.E., Zhang, Z., Hughes, M.K., Bradley, R.S., Miller, S.K., Rutherford, S. and Ni, F., 2008: Proxy-based reconstructions of hemispheric and global surface temperature variations over the past two millennia. Proceedings of the National Academy of Sciences, USA, 105: 13252–13257.

Mann, M.E., Zhang, Z., Rutherford, S., Bradley, R.S., Hughes, M.K., Shindell, D., Ammann, C., Faluvegi, G., and Ni, F., 2009: Global signatures and dynamical origins of the Little Ice Age and Medieval Climate Anomaly. Science, 326: 1256–1260.

Matthews, J.A., and Briffa, K.R., 2005: The 'Little Ice Age': Re-evaluation of an evolving concept. Geografiska Annaler, 87A: 17–36.

Moberg, A., Sonechkln, D.M., Holmgren, K., Datsenko, N.M., and Karlén, W., 2005: Highly variable Northern Hemisphere temperatures reconstructed from low- and high-resolution proxy data. Nature, 433: 613–617.

Moros, M., Jensen, K.G., and Kuijpers, A., 2006: Mid- to late-Holocene hydrological and climatic variability in Disko Bugt, central West Greenland. The Holocene, 16: 357–67.

Møller, H.S., Jensen, K.G., Kuijpers, A., Aagaard-Sørensen, S., Seidenkrantz, M.S., Prins, M., Endler, R., and Mikkelsen, N., 2006: Late-Holocene environment and climatic changes in Ameralik Fjord, southwest Greenland: evidence from the sedimentary record. The Holocene, 16: 685–95.

Naurzbaev, M.M., Vaganov, E.A., Sidorova, O.V. and Schweingruber, F.H., 2002: Summer temperatures in eastern Taimyr inferred from a 2427-year late-Holocene tree-ring chronology and earlier floating series. The Holocene, 12: 727–736.

Neukom, R., Luterbacher, J., Villalba, R., Küttel, M., Frank, D., Jones, P.D., Grosjean, M., Wanner, H., Aravena, J.-C., Black, D.E., Christie, D.A., D'Arrigo, R., Lara, A., Morales, M., Soliz-Gamboa, C., Srur, A., Urrutia, R., and von Gunten, L., 2010: Multiproxy summer and winter surface air temperature field reconstructions for southern South America covering the past centuries. Climate Dynamics: in press.

NRC (National Research Council), 2006: Surface temperature reconstructions for the last 2,000 years. Washington, DC: National Academies Press: 196 pp.

Osborn, T.J. and Briffa, K.R., 2006: The spatial extent of 20th-century warmth in the context of the past 1200 years. Science, 311: 841–844.

Rosén, P., Segerström, U., Eriksson, L., and Renberg I., 2003: Do diatom, chironomid, and pollen records consistently infer Holocene July air temperatures? A comparison using sediment cores from four alpine lakes in Northern Sweden. Arctic, Antarctic and Alpine Research, 35: 279–290.

Seidenkrantz, M.-S., Aagaard-Sørensen, S., Sulsbrück, H., Kuijpers, A., Jensen, K.G., and Kunzendorf, H., 2007: Hydrography and climate of the last 4400 years in a SW Greenland fjord: implications for Labrador Sea palaeoceanography. The Holocene, 17: 387–401.

Solomina, O., and Alverson, K., 2004: High latitude Eurasian paleoenvironments: introduction and synthesis. Palaeogeography, Palaeoclimatology, Palaeoecology, 209: 1–18.

Soon, W., and Baliunas, S., 2003: Proxy climatic and environmental changes of the past 1000 years. Climate Research, 23: 89–110.

von Storch, H., Zorita, E., Jones, J.M., Dimitriev, Y., González-Rouco, F., and Tett, S.F.B., 2004: Reconstructing past climate from noisy proxy data. Science, 306: 679–682.

Sundqvist, H.S., Holmgren, K., Moberg, A., Spötl, C., and Mangini, A., 2010: Stable isotopes in a stalagmite from NW Sweden document environmental changes over the past 4000 years. Boreas, 39: 77–86.

Tan, M., Liu, T.S., Hou, J., Qin, X., Zhang, H., and Li, T., 2003: Cyclic rapid warming on centennial-scale revealed by a 2650-year stalagmite record of warm season temperature. Geophysical Research Letters, 30: 1617, doi:10.1029/2003GL017352.

Yang, B., Braeuning, A., Johnson, K.R., and Yafeng, S., 2002: General characteristics of temperature variation in China during the last two millennia. Geophysical Research Letters, 29: 1324.

Viau, A.E., Gajewski, K., Sawada, M.C., and Fines, P., 2006: Millennial-scale temperature variations in North America during the Holocene. Journal of Geophysical Research, 111: D09102, doi:10.1029/2005JD006031.

Vinther, B.M., Andersen, K.K., Jones, P.D., Briffa, K.R., and Cappelen, J., 2006: Extending Greenland temperature records into the late eighteenth century. Journal of Geophysical Research, 11: D11105.

Wanner, H., Beer, J., Bütikofer, J. Crowley, T., Cubasch, U., Flückiger, J., Goosse, H., Grosjean, M., Joos, F., Kaplan, J.O., Küttel, M., Müller, S., Pentice, C. Solomina, O., Stocker, T., Tarasov, P., Wagner, M., and Widmann, M., 2008: Mid to late Holocene climate change - an overview. Quaternary Science Reviews, 27: 1791–1828.

Wagner, B., and Melles, M., 2001: A Holocene seabird record from Raffles Sø sediments, East Greenland, in response to climatic and oceanic changes. Boreas, 30: 228–39.

Velichko, A.A. (ed.), 1984: Late Quaternary Environments of the Soviet Union. University of Minnesota Press, Minneapolis.

Velichko, A.A., Andrev, A.A., and Klimanov, V.A., 1997: Climate and vegetation dynamics in the tundra and forest zone during the Late-Glacial and Holocene. Quaternary International, 41: 71–96.

Vinther, B.M., Jones, P.D., Briffa, K.R., Clausen, H.B., Andersen, K.K., Dahl-Jensen, D., and Johnsen, S.J., 2010: Climatic signals in multiple highly resolved stable isotope records from Greenland. Quaternary Science Reviews, 29: 522–538.

Zhang, Q.-B., Cheng, G., Yao, T., Kang, X., and Huang, J., 2003: A 2,326-year tree-ring record of climate variability on the northeastern Qinghai-Tibetan Plateau. Geophysical Research Letters, 30: 10.1029/2003GL017425.

Zhang, Q., Gemmer, M., and Chen, J., 2008a. Climate changes and flood/drought risk in the Yangtze Delta, China, during the past millennium. Quaternary International, 176-177: 62–69.

Zhang, P., Cheng, H., Edwards, R.L., Chen, F., Wang, Y., Yang, X., Liu, J., Tan, M., Wang, X., Liu, J., An, C., Dai, Z., Zhou, J., Zhang, D., Jia, J., Jin, L., and Johnson, K.R. 2008b: A test of climate, sun, and culture relationships from an 1810-Year Chinese cave record. Science, 322: 940–942.

2

Multi-months cycles observed in climatic data

Samuel Nicolay, Georges Mabille, Xavier Fettweis and M. Erpicum
University of Liège
Belgium

1. Introduction

Climatic variations happen at all time scales and since the origins of these variations are usually of very complex nature, climatic signals are indeed chaotic data. The identification of the cycles induced by the natural climatic variability is therefore a knotty problem, yet the knowing of these cycles is crucial to better understand and explain the climate (with interests for weather forecasting and climate change projections). Due to the non-stationary nature of the climatic time series, the simplest Fourier-based methods are inefficient for such applications (see e.g. Titchmarsh (1948)). This maybe explains why so few systematic spectral studies have been performed on the numerous datasets allowing to describe some aspects of the climate variability (e.g. climatic indices, temperature data). However, some recent studies (e.g. Matyasovszky (2009); Paluš & Novotná (2006)) show the existence of multi-year cycles in some specific climatic data. This shows that the emergence of new tools issued from signal analysis allows to extract sharper information from time series.
Here, we use a wavelet-based methodology to detect cycles in air-surface temperatures obtained from worldwide weather stations, NCEP/NCAR reanalysis data, climatic indices and some paleoclimatic data. This technique reveals the existence of universal rhythms associated with the periods of 30 and 43 months. However, these cycles do not affect the temperature of the globe uniformly. The regions under the influence of the AO/NAO indices are influenced by a 30 months period cycle, while the areas related to the ENSO index are affected by a 43 months period cycle; as expected, the corresponding indices display the same cycle. We next show that the observed periods are statistically relevant. Finally, we consider some mechanisms that could induce such cycles. This chapter is based on the results obtained in Mabille & Nicolay (2009); Nicolay et al. (2009; 2010).

2. Data

2.1 GISS temperature data

The Goddard Institute for Space Studies (GISS) provides several types of data.
The GISS temperature data (Hansen et al. (1999)) are made of temperatures measured in weather stations coming from several sources: the National Climatic Data Center, the United States Historical Climatology Network and the Scientific Committee on Antarctic Research. These data are then reconstructed and "corrected" to give the GISS temperature data.
The temperatures from the Global Historical Climatology Network are also used to build temperature anomalies on a $2^\circ \times 2^\circ$ grid-box basis. These data are then gathered and "corrected" to obtain hemispherical temperature data (HN-T for the Northern Hemisphere and HS-T for the Southern Hemisphere) and global temperature data (GLB-T).

2.2 CRU global temperature data

The Climate Research Unit of the East Anglia University (CRU) provides several time series related to hemispherical and global temperature data (Jones et al (2001)). All these time series are obtained from a 5° × 5° gridded dataset: CRUTEM3 gives the land air temperature anomalies (CRUTEM3v is a variance-adjusted version of CRUTEM3), HadSST2 gives the sea-surface temperature (SST) anomalies and HadCRUT3 combines land and marine temperature anomalies (a variance-adjusted version of these signals is available as well). For each time series, a Northern Hemispheric mean, a Southern Hemispheric mean and a global mean exist.

2.3 NCEP/NCAR reanalysis

The National Centers for Environmental Prediction (NCEP) and the National Center for Atmospheric Research (NCAR) cooperate to collect climatic data: data obtained from weather stations, buoys, ships, aircrafts, rawinsondes and satellite sounders are used as an input for a model that leads to 2.5° × 2.5° datasets (humidity, windspeed, temperature,...), with 28 vertical levels (Kalnay et al. (1996)). Only the near-surface air temperature data were selected in this study.

2.4 Indices

The Arctic oscillation (AO) is an index obtained from sea-level pressure variations poleward of 20N. Roughly speaking, the AO index is related to the strength of the Westerlies. There are two different, yet similar, definitions of the AO index : the AO CPC (Zhou et al. (2001)) and the AO JISAO.

The North Atlantic Oscillation (NAO) is constructed from pressure differences between the Azores and Iceland (NAO CRU, Hurrel (1995)) or from the 500mb height anomalies over the Northern Hemisphere (NAO CPC, Barnston & Livezey (1987)). This index also characterizes the strength of the Westerlies for the North Atlantic region (Western Europe and Eastern America).

The El Niño/Southern Oscillation (ENSO) is obtained from sea-surface temperature anomalies in the equatorial zone (global-SST ENSO) or is constructed using six different variables, namely the sea-level pressure, the east-west and north-south components of the surface winds, the sea-surface temperature, the surface air temperature and the total amount of cloudiness (Multivariate ENSO Index, MEI, Wolter & Timlin (1993; 1998)). This index is used to explain sea-surface temperature anomalies in the equatorial regions.

The Southern Oscillation Index (SOI, Schwing et al. (2002)) is computed using the difference between the monthly mean sea level pressure anomalies at Tahiti and Darwin.

The extratropical-based Northern Oscillation index (NOI) and the extratropical-based Southern Oscillation index (SOI*) are characterized from sea level pressure anomalies of the North Pacific (NOI) or the South Pacific (SOI*). They reflect the variability in equatorial and extratropical teleconnections (Schwing et al. (2002)).

The Pacific/North American (PNA, Barnston & Livezey (1987)) an North Pacific (NP, Trenberth & Hurrell (1994)) indices reflect the air mass flows over the north pacific. The PNA index is defined over the whole Northern Hemisphere, while the NP index only takes into account the region 30N–65N, 160E–140W.

The Pacific Decadal Oscillation (PDO, Mantua et al (1997)) is derived from the leading principal component of the monthly sea-surface temperature anomalies in the North Pacific Ocean, poleward 20N.

3. Method

3.1 The continuous wavelet transform

The wavelet analysis has been developed (in its final version) by J. Morlet and A. Grossman (see Goupillaud et al. (1984); Kroland-Martinet et al. (1987)) in order to minimize the numerical artifacts observed when processing seismic signals with conventional tools, such as the Fourier transform. It provides a two-dimensional unfolding of a one-dimensional signal by decomposing it into scale (playing the role of the inverse of the frequency) and time coefficients. These coefficients are constructed through a function ψ, called the wavelet, by means of dilatations and translations. For more details about the wavelet transform, the reader is referred to Daubechies (1992); Keller (2004); Mallat (1999); Meyer (1989); Torresani (1995).

Let s be a (square integrable) signal; the continuous wavelet transform is the function W defined as

$$W[s](t,a) = \int s(x)\bar{\psi}(\frac{x-t}{a})\,\frac{dx}{a},$$

where $\bar{\psi}$ denotes the complex conjugate of ψ. The parameter $a > 0$ is the scale (i.e. the dilatation factor) and t the time translation variable. In order to be able to recover s from $W[s]$, the wavelet ψ must be integrable, square integrable and satisfy the admissibility condition

$$\int \frac{|\hat{\psi}(\omega)|^2}{|\omega|}\,d\omega < \infty,$$

where $\hat{\psi}$ denotes the Fourier transform of ψ. In particular, this implies that the mean of ψ is zero,

$$\int \psi(x)\,dx = 0.$$

This explains the denomination of wavelet, since a zero-mean function has to oscillate.

The wavelet transform can be interpreted as a mathematical microscope, for which position and magnification correspond to t and $1/a$ respectively, the performance of the optic being determined by the choice of the lens ψ (see Freysz et al. (1990)).

The continuous wavelet transform has been successfully applied to numerous practical and theoretical problems (see e.g. Arneodo et al. (2002); Keller (2004); Mallat (1999); Nicolay (2006); Ruskai et al. (1992)).

3.2 Wavelets for frequency-based studies

One of the possible applications of the continuous wavelet transform is the investigation of the frequency domain of a function. For more details about wavelet-based tools for frequency analysis, the reader is referred to Mallat (1999); Nicolay (2006); Nicolay et al. (2009); Torresani (1995).

Wavelets for frequency-based studies have to belong to the second complex Hardy space. Such a wavelet is given by the Morlet wavelet ψ_M whose Fourier transform is given by

$$\hat{\psi}_M(\omega) = \exp(-\frac{(\omega-\Omega)^2}{2}) - \exp(-\frac{\omega}{2})\exp(-\frac{\Omega}{2}),$$

where Ω is called the central frequency; one generally chooses $\Omega = \pi\sqrt{2/\log 2}$. For such a wavelet, one directly gets

$$W[\cos(\omega_0 x)](t,a) = \frac{1}{2}\exp(i\omega_0 t)\hat{\psi}_M(a\omega_0).$$

Since the maximum of $\hat{\psi}_M(\cdot\omega_0)$ is reached for $a = \Omega/\omega_0$, if a_0 denotes this maximum, one has $\omega_0 = \Omega/a_0$. The continuous wavelet transform can thus be used in a way similar to the windowed Fourier transform, the role of the frequency being played by the inverse of the scale (times Ω).
There are two main differences between the wavelet transform and the windowed Fourier transform. First, the scale a defines an adaptative window: the numerical support of the function $psi(./a)$ is smaller for higher frequencies. Moreover, if the first m moments of the wavelet vanish, the associated wavelet transform is orthogonal to lower-degree polynomials, i.e. $W[s+P] = W[s]$, where P is a polynomial of degree lower than m. In particular, trends do not affect the wavelet transform.
In this study, we use a slightly modified version of the usual Morlet wavelet with exactly one vanishing moment,

$$\hat{\psi}(\omega) = \sin(\frac{\pi\omega}{2\Omega})\exp(-\frac{\omega-\Omega)^2}{2}).$$

3.3 The scale spectrum

Most of the Fourier spectrum-based tools are rather inefficient when dealing with non-stationary signals (see e.g. Titchmarsh (1948)). The continuous wavelet spectrum provides a method that is relatively stable for signals whose properties do not evolve too quickly: the so-called scale spectrum. Let us recall that we are using a Morlet-like wavelet.
The scale spectrum of a signal s is

$$\Lambda(a) = E|W[s](t,a)|,$$

where E denotes the mean over time t. Let us remark that this spectrum is not defined in terms of density. Nevertheless, such a definition is not devoid of physical meaning (see e.g. Huang et al. (1998)). It can be shown that the scale spectrum is well adapted to detect cycles in a signal, even if it is perturbed with a coloured noise or if it involves "pseudo-frequencies" (see Nicolay et al. (2009)).
As an example, let us consider the function $f = f_1 + f_2 + \epsilon$, where $f_1(x) = 8\cos(2\pi x/12)$,

$$f_2(x) = (0.6 + \frac{\log(x+1)}{16})\cos(\frac{2\pi}{30}x(1+\frac{\log(x+1)}{100}))$$

and (ϵ) is an autoregressive model of the first order (see e.g. Janacek (2001)),

$$\epsilon_n = \alpha\epsilon_{n-1} + \sigma\eta_n,$$

where (η) is a centered Gaussian white noise with unit variance and $\alpha = 0.862$, $\sigma = 2.82$. The parameters α and σ have been chosen in order to simulate the background noise observed in the surface air temperature of the Bierset weather station (see Section 4). The function f (see Fig. 1) has three components: an annual cycle f_1, a background noise (ϵ) and a third component f_2 defined through a cosine function whose phase and amplitude evolve; f_2 is represented in Fig. 2. As we will see, such a component is detected in many climatic time series. As shown in Fig. 3, the scale spectrum of f displays two maxima, associated with the cycles of 12 months and 29.56 months respectively. The components f_1 and f_2 are thus detected, despite the presence of the noise (ϵ). Furthermore, the amplitudes associated with f_1 and f_2 are also recovered.

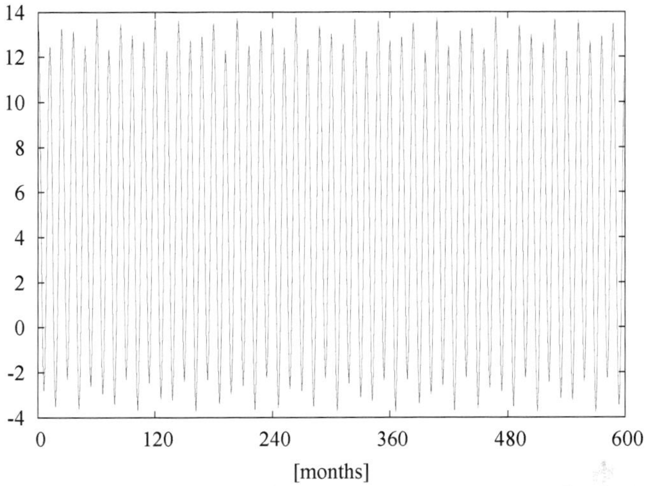

Fig. 1. The function f simulating an air surface temperature time series. The abscissa represent the months.

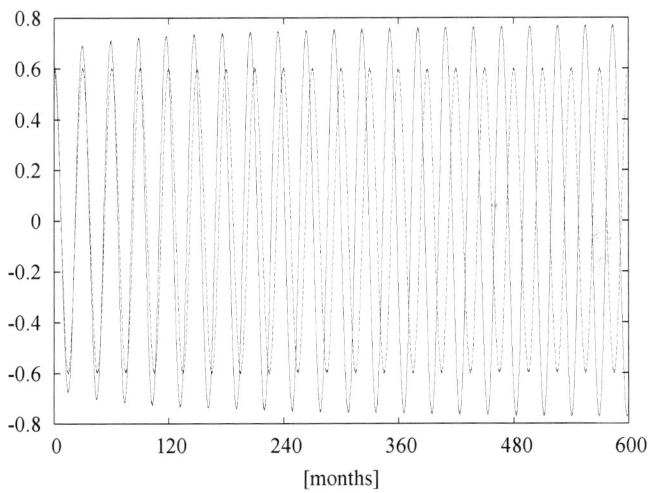

Fig. 2. The component f_2 (solid lines) of the function f, compared with the function $0.6\cos(2\pi x/30)$ (dashed lines). The abscissa represent the months.

Unlike the Fourier transform, which takes into account sine or cosine waves that persisted through the whole time span of the signal, the scale spectrum gives some likelihood for a wave to have appeared locally. This method can thus be used to study non-stationary signals.

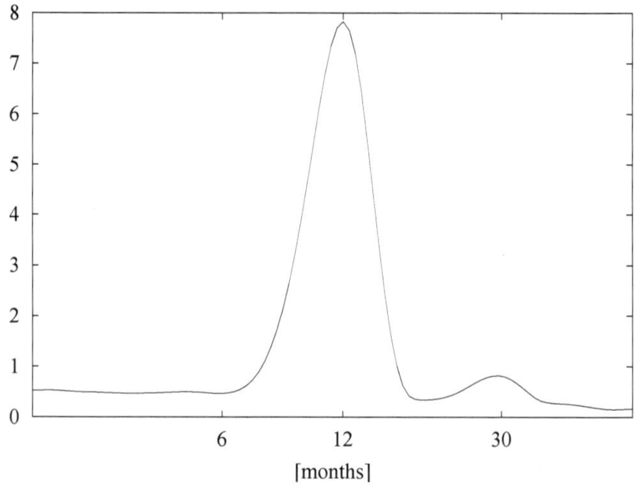

Fig. 3. The scale spectrum Λ of f. The abscissa (logarithmic scale) represent the months.

Let us remark that the scale spectra computed in this work do not take into account values that are subject to border effects.

4. Results

4.1 Scale spectra of global temperature records

The scale spectra of the global temperature data (CRUTEM3gl) display two extrema corresponding to the existence of two cycles $c_1 = 30 \pm 3$ months and $c_2 = 43 \pm 3$ months. The second cycle is also observed in the scale spectra of time series where the SST is taken into account (HadCRUT3, HadCRUT3v, HadSST2 and GLB-T). The existence of c_1 in these data is not so clear. The scale spectra of these series are shown in Fig. 4 and Fig. 5.

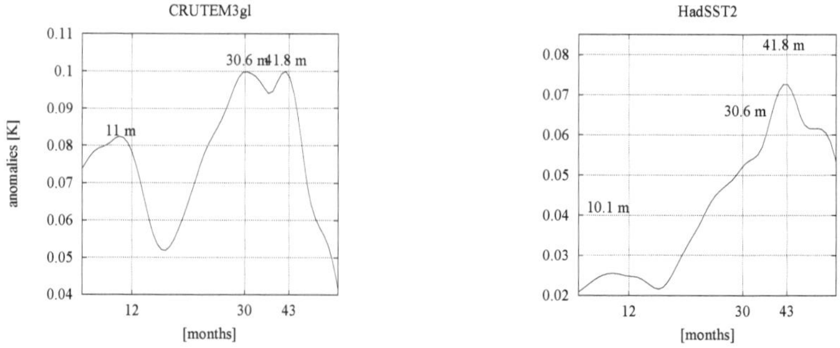

Fig. 4. The scale spectra of global temperature records. Crutem3 (left panel) and HadSST2 (right panel).

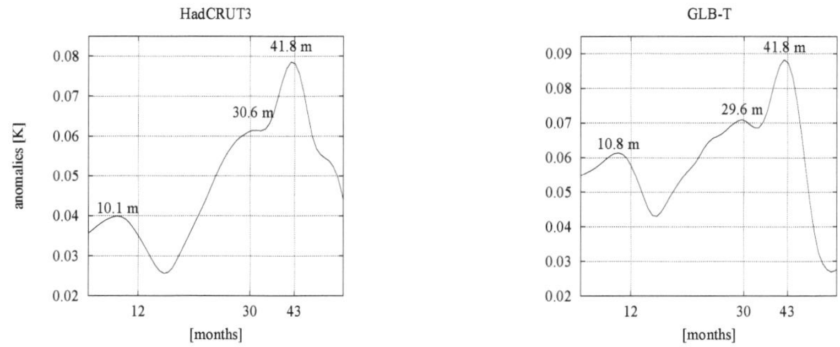

Fig. 5. The scale spectra of global temperature records: HadCRUT3 (left panel) and GLB-T (right panel).

When considering hemispheric data, c_1 and c_2 are still observed. The scale spectra of the global temperature time series in the Northern Hemisphere display a maximum corresponding to c_1. This cycle is more clearly observed in the data where the SST is not taken into account (i.e. with CRUTEM3nh), while c_2 is more distinctly seen in the other time series (NH-T, HadCrut3, HadSST2), as seen in Fig. 6 and Fig. 7. The spectra related to the Southern Hemisphere still display a maximum corresponding to c_2. For the CRU time series (HadCRUT3sh and HadSST2sh), the observed cycle that is the closest to c_1 is about 25 months, while the scale spectrum of the GISS data (SH-T) display a cycle c_1 as marked as the cycle c_2. The scale spectra of these series are shown in Fig. 8 and Fig. 9.

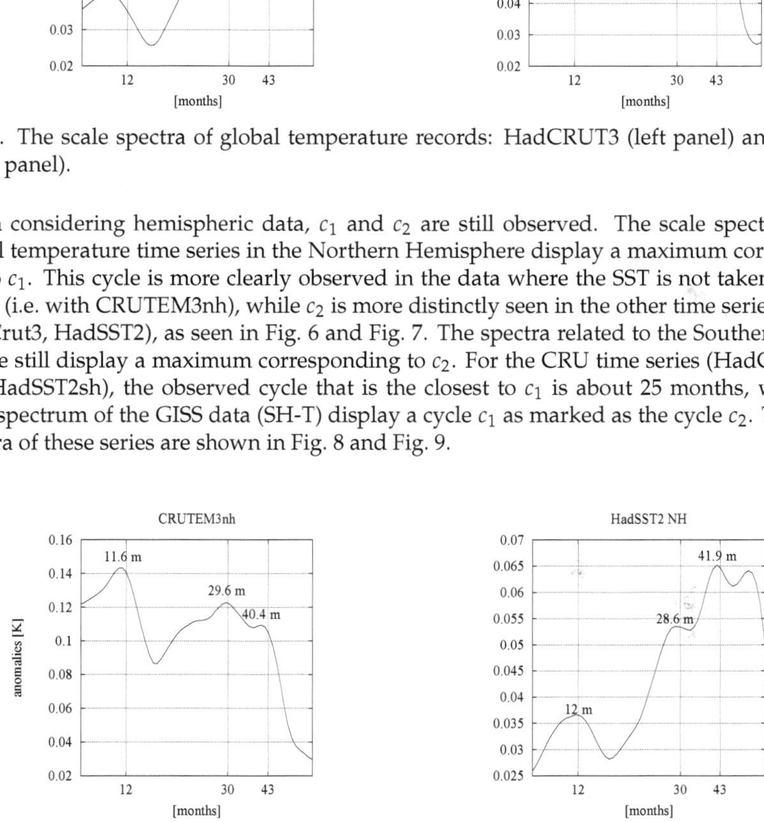

Fig. 6. The scale spectra of Northern Hemisphere temperature records: Crutem3 (left panel) and HadSST2 (right panel).

4.2 Scale spectra of local temperature records

In Nicolay et al. (2009), the scale spectra of a hundred near-surface air temperature time series have been computed using GISS Surface Temperature Analysis data (only the most complete data were chosen). The cycles detected in some weather stations are given by Fig. 10 and Table 1 (the location, the amplitude of the cycles found and the associated class of climate

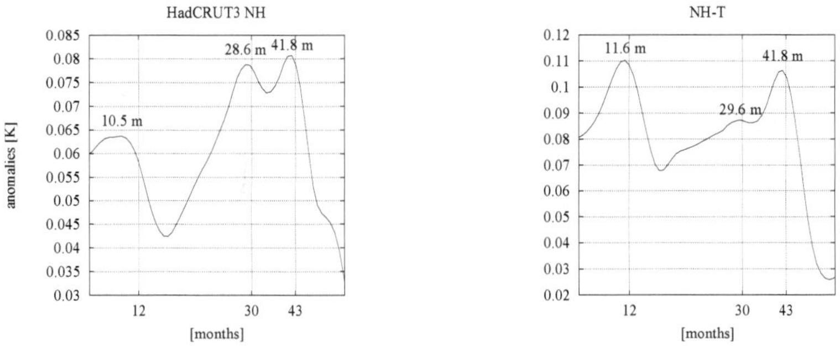

Fig. 7. The scale spectra of Northern Hemisphere temperature records: HadCRUT3 (left panel) and NH-T (right panel).

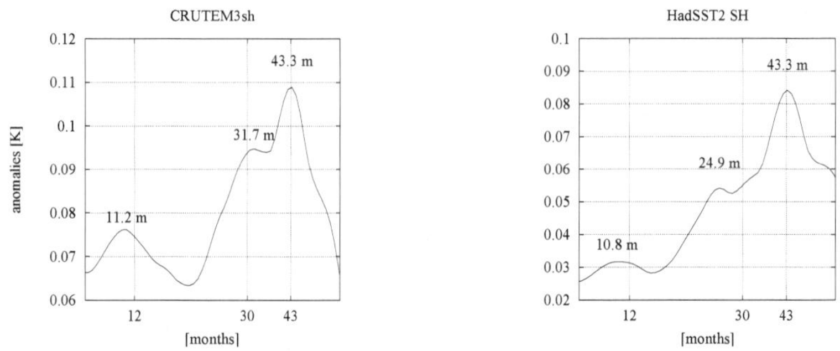

Fig. 8. The scale spectra of Southern Hemisphere temperature records: CRUTEM3 (left panel) and HadSST2 (right panel).

are also presented). These stations were selected in order to cover most of the typical climate areas (see for Rudloff (1981) more details). As expected, the scale spectrum leads to a correct estimation of the annual temperature amplitude (the difference between the mean temperature of the warmest and coldest months). The weather stations located in Europe and Siberia are clearly affected by the cycle c_1, while weather stations in areas such as California, Brazil, Caribbean Sea and Hawaii are influenced by c_2. The North American Weather stations time series analysis shows the presence of both c_1 and c_2. Roughly speaking, the temperature amplitudes induced by the cycles c_1 and c_2 represent about one tenth of the annual amplitude.

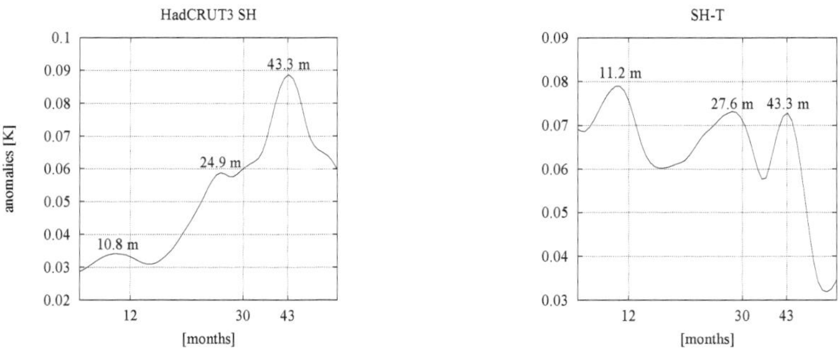

Fig. 9. The scale spectra of Southern Hemisphere temperature records: HadCRUT3 (left panel) and SH-T (right panel).

Weather stations	Lat.	Long.	Cycle (m)	Cycle amp.	An. amp.	Classif.
Uccle (Belgium)	50.8°N	4.3°E	30.4 ± 2.7	0.4 K	15 K	DO
Zaragoza (Spain)	41.6°N	0.9°W	28.4 ± 2.4	0.3 K	18 K	BS
The Pas (Canada)	54.0°N	101.1°W	28.5 ± 2.6	0.6 K	38 K	EC
			44.8 ± 2.4	0.8 K		
Fairbanks (Alaska)	64.8°N	147.9°W	28.5 ± 2.5	0.8 K	40 K	EC
			40.4 ± 2.5	0.8 K		
Verhojansk (Siberia)	67.5°N	133.4°E	31.7 ± 2.5	0.8 K	64 K	EC
Jakutsk (Siberia)	62.0°N	129.7°E	28.6 ± 2.4	0.8 K	60 K	EC
San Francisco (California)	37.6°N	122.4°W	41.8 ± 2.7	0.3 K	8 K	Cs
Lander (Wyoming)	42.8°N	108.7°W	41.8 ± 2.6	0.6 K	28 K	DC
Manaus (Brazil)	3.1°S	60.0°W	43.3 ± 2.4	0.3 K	3 K	Ar
Belo Horizonte (Brazil)	19.9°S	43.9°W	41.8 ± 2.4	0.5 K	4 K	Aw
Tahiti (French Polynesia)	17.6°S	149.6°W	41.8 ± 2.5	0.2 K	3 K	Ar
Lihue (Hawaii)	22.0°N	159.3°W	41.8 ± 2.5	0.3 K	4 K	Ar
Colombo (Sri Lanka)	6.9°N	79.9°E	44.5 ± 2.6	0.2 K	2 K	Ar
Minicoy (India)	8.3°N	73.2°E	41.8 ± 2.6	0.2 K	2 K	Aw

Table 1. Cycles found in some world weather stations (the errors are estimated as in Nicolay et al. (2009)). The stations were selected to represent the main climatic areas. For the class of climates, see Rudloff (1981).

To show, that c_1 and c_2 affect the whole planet, the scale spectrum of each grid point of the NCEP/NCAR reanalysis has been computed. As displayed in Fig. 11 and Fig. 12, 92% of the Earth area is associated to at least one of these cycles. Fig. 11 shows that c_1 is mainly seen in Alaska, Eastern Canada, Europe, Northern Asia and Turkey, while Fig. 12 reveals that c_2 is principally seen in Equatorial Pacific, Northern America and Peru. Roughly speaking, the cycle c_1 is observed in regions associated with the Arctic Oscillation, while c_2 is detected in regions associated to the Southern Oscillation.

4.3 Scale spectra of atmospheric indices

Advection causes the transfer of air masses to neighboring regions, carrying their properties such as air temperature. The climatic indices characterize these air mass movements.

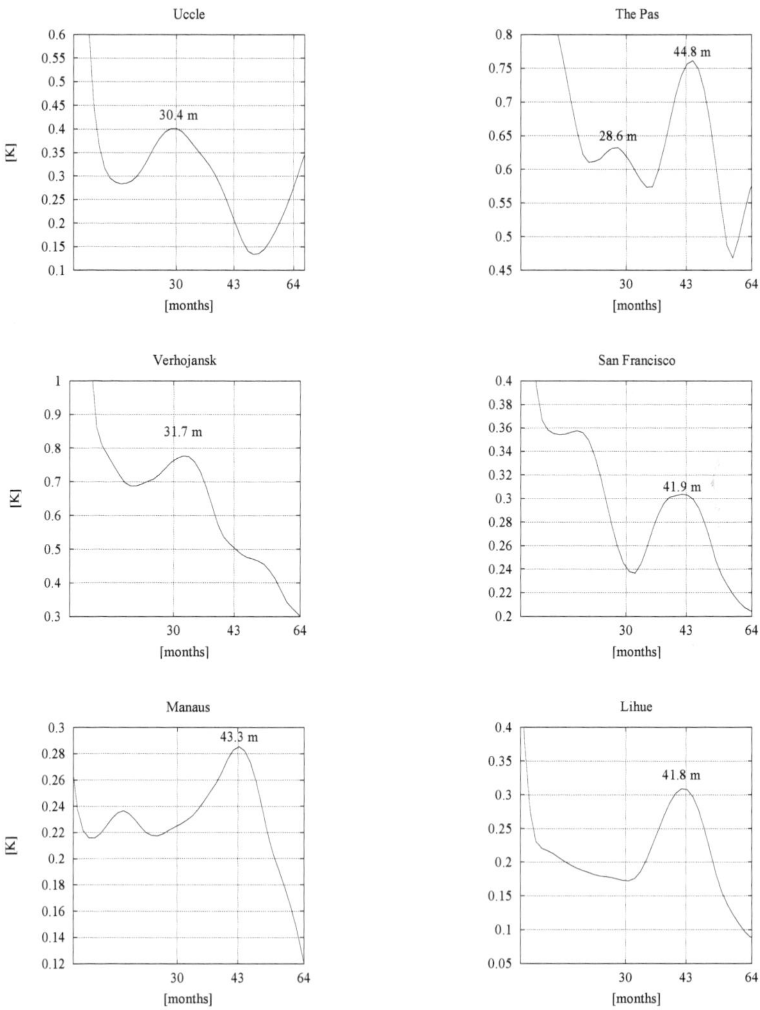

Fig. 10. Scale spectra of near-surface air temperature time series: Uccle (Belgium), The Pas (Canada), Verhojansk (Siberia), San Francisco (California), Manaus (Brazil), Lihue (Hawaii).

The cycles detected in the main climatic indices are reported in Table 2. Almost all these indices display a cycle corresponding to c_1, the notable exceptions being the NP, PNA and global-SST ENSO indices. The cycle c_2 is observed in the AO (CPC), NP, PDO, PNA and SOI* indices, as well as the indices related to the Southern Oscillation (such as the ENSO indices). The scale spectra of these indices are shown in Fig. 13, 14, 15 and 16.

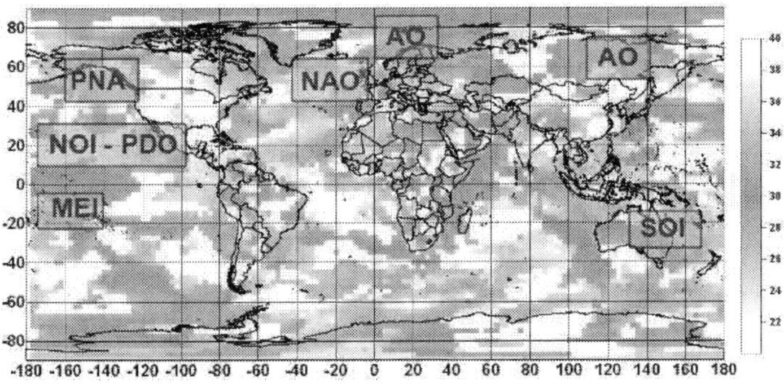

Fig. 11. NCEP/NCAR reanalysis data. The grid points where a cycle corresponding to c_1 has been detected are coloured.

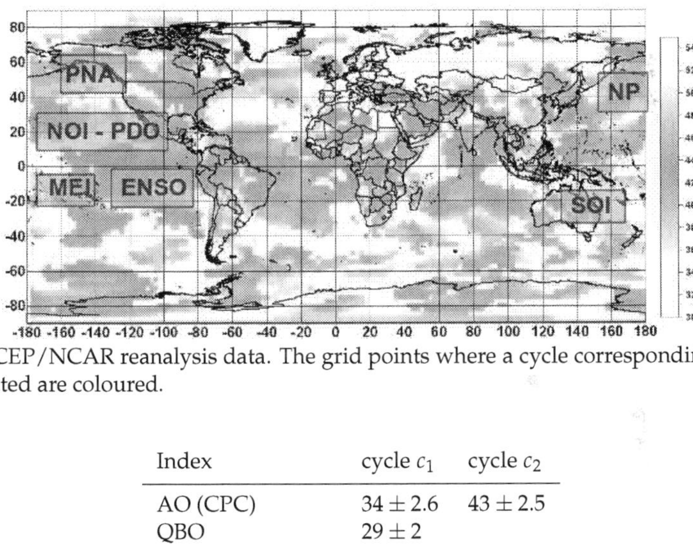

Fig. 12. NCEP/NCAR reanalysis data. The grid points where a cycle corresponding to c_2 has been detected are coloured.

Index	cycle c_1	cycle c_2
AO (CPC)	34 ± 2.6	43 ± 2.5
QBO	29 ± 2	
Global-SST ENSO		45 ± 2.1
MEI ENSO	30 ± 2.1	45 ± 2.1
NAO (CPC)	34 ± 2.1	
NAO (CRU)	34 ± 2.1	
NOI	32 ± 2.3	
NP		43 ± 2.4
PDO	26 ± 2.4	40 ± 2.3
PNA		45 ± 2.4
SOI	30 ± 2.2	
SOI*	30 ± 2.5	44 ± 2.6

Table 2. Cycles found in the main climatic indices (the errors are estimated as in Nicolay et al. (2009)).

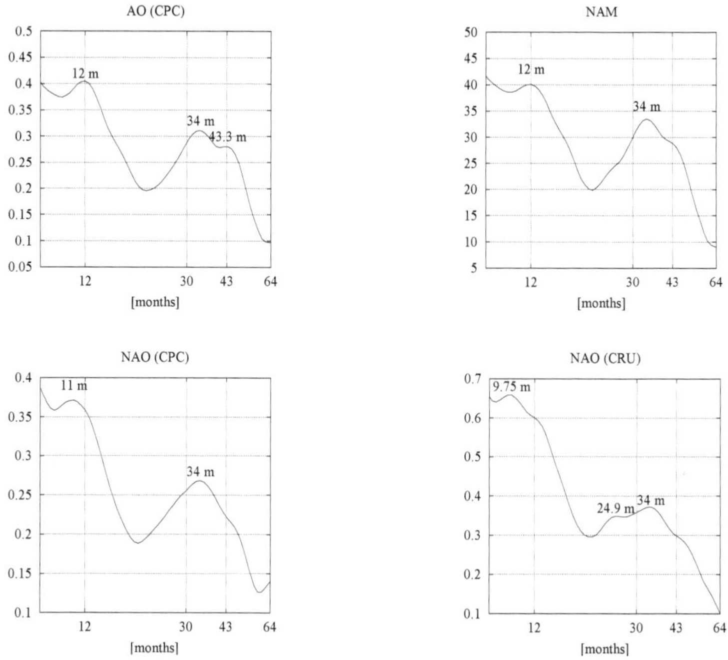

Fig. 13. Scale spectra of the climatic indices related to the Northern Atlantic Oscillation.

4.4 A statistical validation for the observed cycles

Although many evidences attest the validity of the method described above, a question naturally remains: Is there a high probability that the maxima observed in the scale spectra occurred by pure chance?

In Nicolay et al. (2010), to check if the cycles observed in the time series can be trusted, the scale spectra of the NCEP/NCAR reanalysis data have been compared with the spectra of signals made of an autoregressive model of the first order (AR(1) model, see e.g. Janacek (2001)), in which maxima could occur fortuitously. Such processes are observed in many climatic and geophysical data (see e.g. Allen & Robertson (1996); Percival & Walden (1993)) and are well suited for the study of climatic time series (see e.g. Mann & Lees (1996); Mann et al. (2007)).

An artificial signal (y_n) can be associated to the temperature time series (x_n) of a grid point of the NCEP/NCAR reanalysis data as follows:

- One first computes the climatological anomaly time series (δ_n) of (x_n), i.e. for each month, the mean temperature is calculated from the whole signal and the so-obtained monthly-sampled signal (m_n) is subtracted to (x_n), $\delta_n = x_n - m_n$.
- The anomaly time series (δ_n) is fitted with an AR(1) model (ϵ_n),

$$\epsilon_n = \alpha \epsilon_{n-1} + \sigma \eta_n,$$

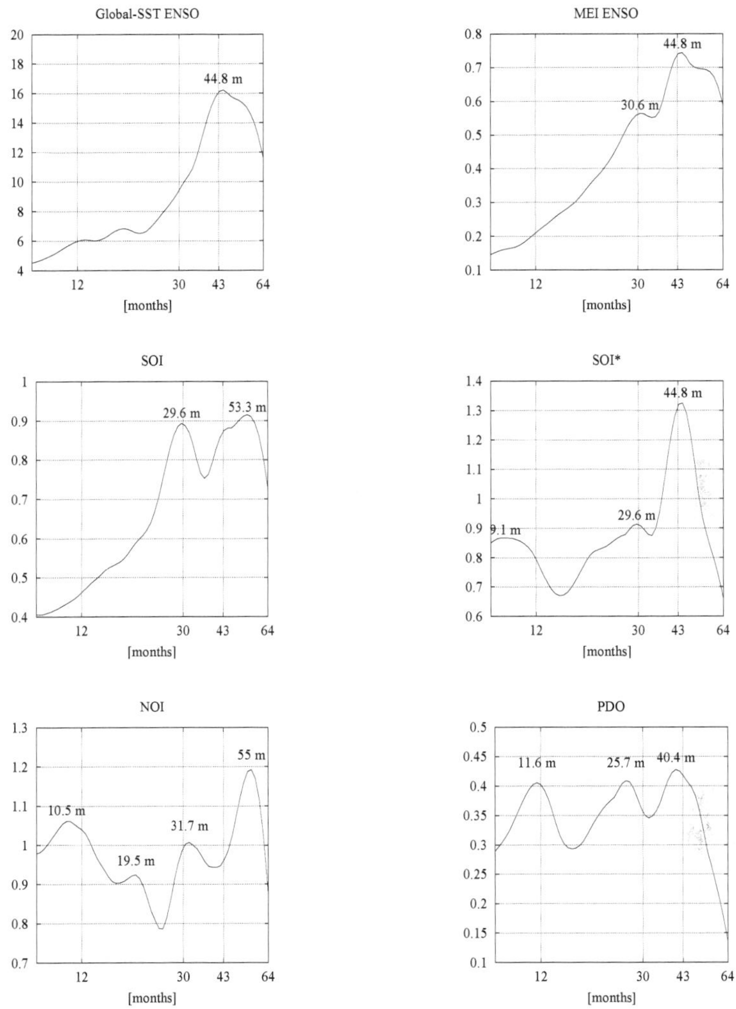

Fig. 14. Scale spectra of the climatic indices related to the Southern Oscillation.

where η_n is a Gaussian white noise with zero mean and unit variance (see e.g. Janacek (2001)).

- The artificial signal (y_n) associated to (x_n) is defined by replacing (δ_n) with (ϵ_n), $y_n = m_n + \epsilon_n$.

Let us remark that (y_n) is indeed a stochastic process; several simulations of the same signal (x_n) will thus yield different realizations.

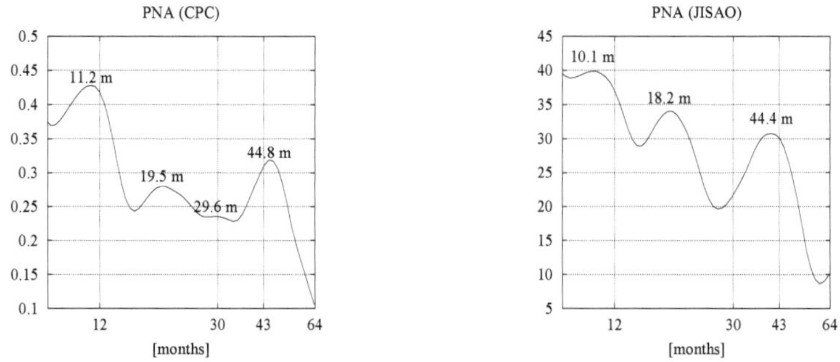

Fig. 15. Scale spectra of the PNA indices.

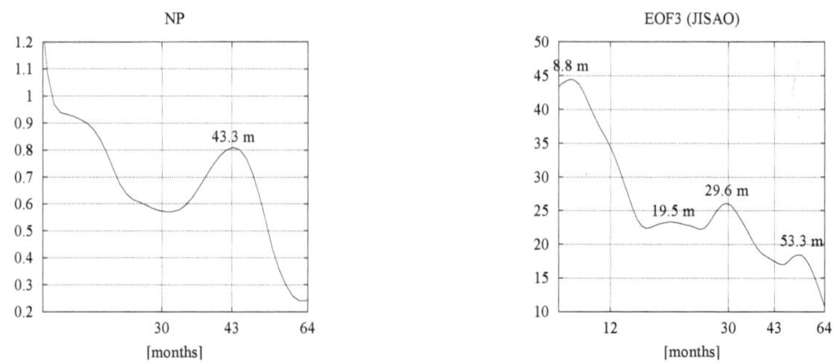

Fig. 16. Scale spectra of other climatic indices.

To check if the cycles c_1 and c_2 appearing in the time series did not occur by pure chance, the subsequent methodology can be applied to each temperature time series (x_n) of the NCEP/NCAR reanalysis data:

- $N = 10,000$ realizations (y_n) of (x_n) are computed.
- The distribution of the highest local maximum y_M of the scale spectrum of the data in the range of 26 to 47 months is estimated from these artificial signals, i.e. one computes the distribution of
$$y_M = \sup_{26 \leq a \leq 47} \tilde{\Lambda}(a),$$
where $\tilde{\Lambda}$ is the scale spectrum of a realization (y_n).
- The probability P to obtain a maximum of higher amplitude than the one corresponding to c_1 or c_2 observed in the scale spectrum of (x_n) is finally computed, using the distribution previously obtained.

It is shown in Nicolay et al. (2010) that such a methodology yields reliable data. The probability values concerning c_1 and c_2 are displayed in Fig. 17 and Fig. 18 respectively. The coloured area correspond to regions where the cycle is significant. These figures show that most of the

cycles associated with c_1 and c_2 can be considered as significant. The cycle observed in the climatic indices are also significant, since one always get $P < 0.1$ (see Mabille & Nicolay (2009); Nicolay et al. (2010)).
Finally, let us remark that c_1 and c_2 can also be detected through the Fourier transform, if the time series are preprocessed in order to free the corresponding spectrum from the dominating cycle corresponding to one year (for more details, see Nicolay et al. (2010)).

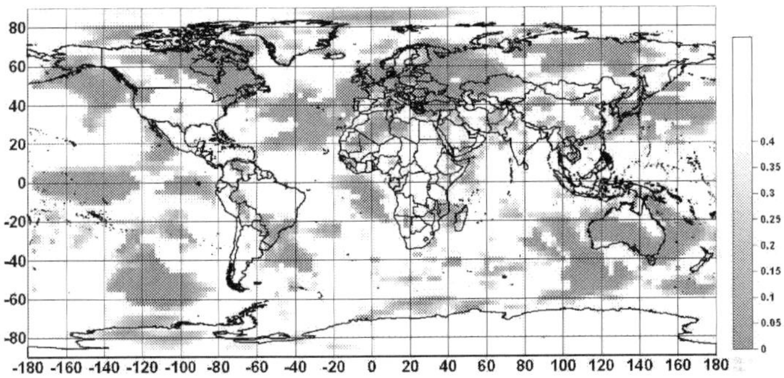

Fig. 17. The probability values associated with c_1 (NCEP/NCAR reanalysis data). The cycles observed in a zone corresponding to the colour white are not significant.

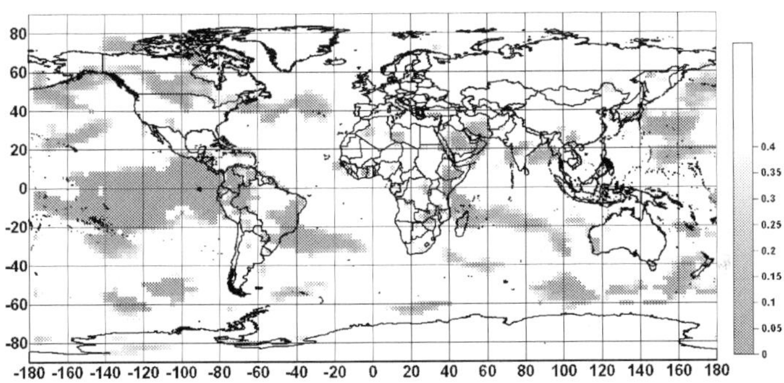

Fig. 18. The probability values associated with c_2 (NCEP/NCAR reanalysis data). The cycles observed in a zone corresponding to the colour white are not significant.

5. Discussion and conclusions

The wavelet-based tool introduced in Sect. 3.1 provides a methodology for detecting cycles in non-stationary signals. Its application to climatic time series has led to the detection of two statistically significant periods of 30 and 43 months respectively.
When looking at the global temperature time series, since most of the lands are situated on the Northern Hemisphere, the cycle c_1 seems to be influenced by the continents, while the cycle c_2 appears to be more influenced by the oceans. However, considering that only a small number

of stations is taken into account in the construction of these records, the above comment has to be taken with circumspection.
Weather station records and NCEP/NCAR reanalysis show that the cycle c_1 is mainly seen in the regions under the influence of the Arctic Oscillation, while the cycle c_2 is observed all over the globe, but more frequently in the regions under the influence of the Southern Oscillation. As a matter of fact, the same cycles are observed in the corresponding indices. In particular c_1 is observed in the spectrum of the AO index and c_2 is detected in the ENSO indices.
As observed in Mabille & Nicolay (2009); Nicolay et al. (2009), the temperature amplitude induced by these cycles always lies between 0.2 and 0.8 K and represents about ten percents of the annual amplitude.
Since the sun is one of the origins of the air mass flows and since the cycles c_1 and c_2 are observed in both the temperature time series and the indices describing the air mass flows, a possible explanation for the existence of these cycles is the solar activity variability. If such hypothesis is true one should find a corresponding cycle in the solar indices such as the solar flux and the sun spot number. Indeed, a cycle corresponding to a period of about 37 months is observed in these data (see Nicolay et al. (2009)). The climate regions could then induce a change of period going from 30 months for continental climates to 43 months for oceanic climates. This cycle corresponding to 37 months detected in the sun is a "flip-flop" type behavior: following Mursula & Hiltula (2004), the solar rotation periodicity undergoes a phase reversal cycle. In Takalo & Mursula (2002), this period is estimated to be about 38 months in the last 40 years, in good agreement with findings based on long series sunspot observations obtained in Berdyugina & Usoskin (2003).

6. References

Allen, M.R. & Robertson, A.W. (1996). *Distinguishing modulated oscillations from coloured noise in multivariate datasets*, Clim. Dyn., 12, 775–784.

Arneodo, A.; Audit, B.; Decoster, N.; Muzy, J.-F. & Vaillant, C. (2002). Climate Disruptions, Market Crashes and Heart Attacks, In: *The Science of Disaster*, A. Bunde and H.J. Sc heelnhuber, (Eds.), pp. 27–102, Springer, Berlin.

Baldwin, M.P. et al. (2001). The quasi-biennial oscillation. *Rev. Geophys.*, Vol. 39, 179–229.

Barnston, A. & Livezey, R. (1987). Classification, seasonality and persistence of low-frequency atmospheric circulation patterns. *Mon. Wea. Rev.*, Vol. 115, 1083–1126.

Berdyugina, S.V. & Usoskin, I.G. (2003). Active longitutes in sunspot activity: Century scale persistence. *Astron. Astrophys.*, Vol. 405, 1121.

Brohan, P.; Kennedy, J.J.; Harris, I.; Tett, S.F.B. & Jones P.D. (2006). Uncertainty estimates in regional and global observed temperature changes: A new dataset from 1850. *J. Geophys. Res.*, Vol. 111, D12,106.

Daubechies, I. (1992). *Ten lectures on wavelets*, SIAM, Philadelphie.

Fedorov, A.V. & Philander, S.G. (2000). Is El Niño changing? *Science*, Vol. 288, 1997–2002.

Freysz, E.; Pouligny, B.; Argoul, F. & Arneodo, A. (1990). Optical wavelet transform of fractal aggregates. *Phys. Rev. Lett.*, Vol. 64, pp. 745–748.

Goupillaud, P.; Grossman, A. & Morlet, J. (1984). Cycle-octave and related transforms in seismic signal analysis. *Geoexploration* , Vol. 23, 85–102.

Hansen J.; Ruedy R.; Glascoe, J. & Sato M. (1999). GISS analysis of surface temperature change. *J. Geophys. Res.*, Vol. 104, 30997–31022.

Huang, N.E. et al. (1998). The empirical mode decomposition and Hilbert spectrum for non-linear and non-stationary time series analysis. *Proc. Roy. Soc. London A*, Vol. 454, pp. 903–995.

Hurrell, J.W. (1995). Decadal trends in the North Atlantic Oscillation: regional temperatures and precipitation. *Science*, 269, 676–679.

Janacek, G. (2001). *Practical time series*, Arnold, London.

Jones, P.D.; Jonsson, T. & Wheeler, D. (1997). Extension to the North Atlantic Oscillation using early instrumental pressure observations from Gibraltar and South-West Iceland. *International Journal of Climatology*, Volume 17, 1433–1550.

Jones, P.D. et al. (2001). Adjusting for sampling density in grid box land and ocean surface temperature time series. *J. Geophys. Res.*, Vol. 106, 3371–3380.

Kalnay, E. et al. (1996). NCEP/NCAR 40-year reanalysis project. *Bull. Amer. Meteor. Soc.*, Vol. 77, 437–471.

Keller, W. (2004). *Wavelets in geodesy and geodynamics*, de Gruyter, Berlin.

Klein Tank, A.M.G. et al. (2002). Daily dataset of 20th-century surface air temperature and precipitation series for the European climate assessment. *International Journal of Climatology*, Vol. 22, 1441–1453.

Klingbjer, P. & Moberg A. (2003). A composite monthly temperature record from Torneladen in Northern Sweden, 1802–2002. *International Journal of Climatology*, Vol. 23, 1465–1494.

Kronland-Martinet, R.; Morlet, J. & Grossmann, A. (1987). Analysis of sound patterns through wavelet transforms. *Int. J. Pattern Recogn. Artific. Intellig.*, Vol. 1, 273–302.

Mabille, G. & Nicolay, S. (2009). Multi-year cycles observed in air temperature data and proxy series. *Eur. Phys. J. Special Topics*, Vol. 174, 135–145.

Mallat, S. (1999). *A wavelet tour of signal processing*, Academic Press, New-York.

Mantua, N.J.; Hare S.R.; Zhang, Y.; Wallace, J.M. &, Francis, R.C. (1997). A pacific interdecadal climate oscillation with impacts on salmon production. *Bulletin of the American Meteorological Society*, Vol. 78, 1069–1079.

Mann, M.E. & Lees, J. (1996). Robust estimation of background noise and signal detection. *Climatic Change*, 33, 409–445.

Mann, M.E.; Rutherford, S.; Wahl, E. & Ammann, C. (2007). Robustness of proxy-based climate field reconstruction methods. *J. Geophys. Res.*, 112, D12109.

Matyasovszky, I. Improving the methodology for spectral analysis of climatic time series. *Theor. Appl. Climatol.*, to appear, DOI: 10.1 007/s00704-009-0212-z.

Meyer, Y. (1989). *Ondelettes et opérateurs*, Hermann, Paris.

Mursula, K. & Hiltula, T. (2004). Systematically asymmetric heliospheric magnetic field: Evidence for a quadrupole mode and non-axis symmetry with polarity flip-flops. *Sol. Phys.*, Vol. 224, 133–143.

Nelder, J. &, Mead R. (1965). A simplex method for function minimization. *Comput. J.*, Vol. 7, 308–313.

Newton, H. & Milsom, A. (1955). Note on the observed differences in spottedness of the Sun's Northern and Southern Hemispheres. *Monthly Not. R. Astron. Soc.*, Vol. 115, 398:404.

Nicolay, S. (2006). *Analyse de séquences ADN par la transformée en ondelettes*, Ph.D. Thesis, University of Liège.

Nicolay, S.; Mabille, G.; Fettweis, X. & Erpicum, M. (2009). 30 and 43 months period cycles found in air temperature time series using the Morlet wavelet method. *Clim. Dyn.*, Vol. 33, pp. 1117–1129.

Nicolay, S.; Mabille, G.; Fettweis, X. & Erpicum, M. (2010). A statistical validation for the cycles found in air temperature data using a Morlet wavelet-based method. *Nonlin. Processes Geophys.*, 17, 269-272.

Paluš, M. & Novotná, D. (2006). Quasi-biennal oscillations extracted from monthly NAO index and temperature records are phase-synchronized. *Nonlin. Processes Geophys.*, Vol. 13, 287–296.

Percival, D.B. & Walden, A.T. (1993). *Spectral analysis for physical applications,* Cambridge University Press, Cambridge.

Rayner, N.A. et al. (2006). Improved analyses of changes and uncertainties in sea surface temperature measured in situ since the mid-nineteenth century: The HadSST2 dataset. *Journal of Climate*, Vol. 19, 446–469.

Rudloff, W. (1981). *World Climates*. Wissenschaftliche Verlagsgesellschaft mbH, Stuttgart.

Ruskai, M.B.; Beylkin, G.; Coifman, R.; Daubechies, I.; Mallat, S.; Meyer, Y. & Raphael, L. (eds.) (1992). *Wavelets and their Applications*, Jones and Bartlett, Boston.

Schwing, F.; Murphree, T. & Green, P. (2002). The northern oscillation index (NOI): A new climate index for the northeast pacific. *Progress in Oceanography*, Vol. 53, 115–139.

Smith, T.M. & Reynolds, R.W. (1997). Extended reconstruction of global sea surface temperature based on COADS data. *J. Climate*, Vol. 16, 1495–1510.

Smith, D.; Cusack, S.; Colman, A.; Folland, C.; Harris, G. & Murphy, J. (2007). Improved surface temperature prediction for the coming decade from global climate model. *Science*, 317, 796–799.

Takalo, J. & Mursula, K. (2002). Annual and solar rotation periodicities in IMF components: Evidence for phase/frequency modulation. *Geophys. Res. Lett.*, Vol. 29, 31–1–31–4.

Titchmarsh, E.C. (1948). *Introduction to the theory of Fourier integrals*, Oxford University Press.

Torresani, B. (1995). *Analyse continue par ondelettes*, CNRS Éditions, Paris.

Trenberth, K. & Hurrell, J.W. (1994). Decadal atmosphere-ocean variations in the pacific. *Climate Dynamics*, Vol. 9, 303–319.

Wolter, K. & Timlin, M.S. (1993). Monitoring ENSO in COADS with a seasonally adjusted principal component index. *Proc of the 17th Climate Diagnostics Workshop*, Norman, OK, NOAA/N MC/CAC, NSSL, Oklahoma Clim. Survey, CIMMS and the School of Meteor, 52–57.

Wolter, K. & Timlin, M.S. (1998). Measuring the strength of ENSO events – how does 1997–98 rank. *Weather*, Vol. 53, 315–324.

Zhang, Y., Wallace, J.M. & Battisti, D.S. (1997). ENSO-like interdecadal variability. *J. Climate*, Vol. 10, 1004–1020.

Zhou, S.; Miller, A.J.; Wang, J. & Angell, J.K. (2001). Trends of NAO and AO and their associations with stratospheric processes. *Geophys. Res. Lett.*, Vol. 28, 4107–4110.

3

Summer-Time Rainfall Variability in the Tropical Atlantic

Guojun Gu
Earth System Science Interdisciplinary Center, University of Maryland, College Park, MD & Laboratory for Atmospheres, NASA Goddard Space Flight Center, Greenbelt, MD U.S.A.

1. Introduction

Convection and rainfall in the tropical Atlantic basin exhibit intense variations on various time scales (e.g., Nobre & Shukla, 1996; Giannini et al., 2001; Chiang et al., 2002). The Atlantic Intertropical Convergence Zone (ITCZ) stays always north of the equator with the trade winds converging into it, representing a prominent climate feature on the seasonal time scale. The ITCZ, though generally over the open ocean, extends to the northeast coast of South America during boreal spring and to the West African continent during boreal summer. Longer-than-seasonal time scale variability is also evident and has been extensively explored in the past through both observational analyses and modeling studies (*e.g.*, Lamb, 1978a, b; Carton & Huang, 1994; Nobre & Shukla, 1996; Sutton et al., 2000). The Atlantic Niño and an interhemispheric sea surface temperature (SST) gradient mode are discovered to be the two major local forcing mechanisms (*e.g.*, Zebiak, 1993; Nobre & Shukla, 1996), in addition to the two other remote large-scale forcings: the El Niño-Southern Oscillation (ENSO) and the North Atlantic Oscillation (NAO) (*e.g.*, Curtis & Hastenrath, 1995; Chiang et al., 2002; Wang, 2002). The most intense year-to-year variability in the tropical Atlantic is usually observed during boreal spring [March-April-May (MAM)], specifically in the western basin and over the northeastern portion of South America, when the marine ITCZ approaches the equator (*e.g.*, Hastenrath & Greischar, 1993; Nobre & Shukla, 1996). Therefore, most of past studies have been primarily focused on this season (*e.g.*, Chiang et al., 2002; Gu & Adler, 2006).

Intense interannual variability has also been seen during boreal summer [June-July-August (JJA)] in the tropical Atlantic (e.g. Sutton et al., 2000; Gu & Adler, 2006, 2009). The Atlantic Niño becomes mature during this season, and the impact of the interhemispheric SST gradient mode and ENSO can still be felt in the equatorial region (*e.g.*, Sutton et al., 2000; Chiang et al., 2002). Particularly, evident interannual variations exist in various distinct severe weather phenomena such as African easterly waves (AEW) and associated convection, and Atlantic hurricane activity (*e.g.*, Thorncroft & Rowell, 1998; Landsea et al., 1999). These severe weather systems frequently appear during boreal summer and fall, and usually propagate westward near the latitudes of the ITCZ (*e.g.*, Chen & Ogura, 1982; Gu & Zhang, 2001).

Hence, based on the two recent studies (Gu & Adler, 2006, 2009), summer-time rainfall variations within the tropical Atlantic basin are further explored here with a focus on the effects of two local SST modes and ENSO. The global monthly precipitation information from the Global Precipitation Climatology Project (GPCP) is applied (Adler et al., 2003). On a global 2.5°×2.5° grid, this satellite-based product is combined from various data sources: the infrared (IR) rainfall estimates from geostationary and polar-orbiting satellites, the microwave estimates from Special Sensor Microwave/Imager (SSM/I), and surface rain gauges from the Global Precipitation Climatological Centre (GPCC) in Germany. The global SST anomalies and three climatic modes are computed using a satellite-based SST product from the National Centers for Environmental Prediction (NCEP) (Reynolds et al., 2002). This product, archived on 1°×1° grids, is built through recons-spatial interpolation, and lasts from January 1982 to present. The results shown here are focused on the time period of January 1982-December 2006. Additionally, monthly surface wind anomalies are derived using the products from the National Centers for Environmental Prediction (NCEP) and National Center for Atmospheric Research (NCAR) reanalysis project (Kalnay et al., 1996). All monthly anomalies, and rainfall and SST indices constructed here are de-trended as we primarily focus on the variations on the interannual time scale.

2. Seasonal-mean rainfall and SST variations during JJA and MAM

Detailed seasonal variations in the tropical Atlantic have been examined in past studies (*e.g.*, Mitchell & Wallace, 1992; Biasutti et al., 2004; Gu & Adler, 2004). Here we just give a brief summary of seasonal-mean features of rainfall and SST, and their corresponding interannual variances during JJA and MAM (Fig. 1). Seasonal-mean oceanic rainfall generally stays over warm SST (≥27°C), featuring the climatological state of the ITCZ. The marine rainfall band is also closely connected to the rainfall zones over the two neighboring continents. During MAM (Fig. 1c), the most intense rainfall zone is located over the coastal region of the northeastern South America; simultaneously the oceanic rainfall band or the marine ITCZ approaches the equator, following the warmest SST in the equatorial region due to the relaxation of trade winds particularly in the eastern basin. During JJA (Fig. 1a), an equatorial cold tongue-ITCZ complex forms with the maritime ITCZ becoming strongest and moving to the north, with two rainfall zones concomitantly being seen over the northern South America and West Africa, respectively.

During MAM (Fig. 1d), the most intense rainfall variability occurs in the western equatorial region, especially along the coastline. In contrast, the maximum SST variances are generally observed in the eastern basin, and are generally concentrated in three areas: tropical north Atlantic (~5°-25°N), the equatorial region, and tropical south Atlantic (~10°-25°S). Furthermore, the most intense SST variability tends to be away from the equator, with thus a much weaker SST variability along the equator. During JJA (Fig. 1b), rainfall variances are mostly found to be over the open ocean along the mean latitudes of the ITCZ. Major SST variability is located in the equatorial region, corresponding to the frequent appearance of the Atlantic Niño (*e.g.*, Zebiak, 1993; Carton & Huang, 1994). Interestingly, rainfall variances within the interior of two continents, particularly over West Africa, are much weaker compared with over ocean and in the coastal zone.

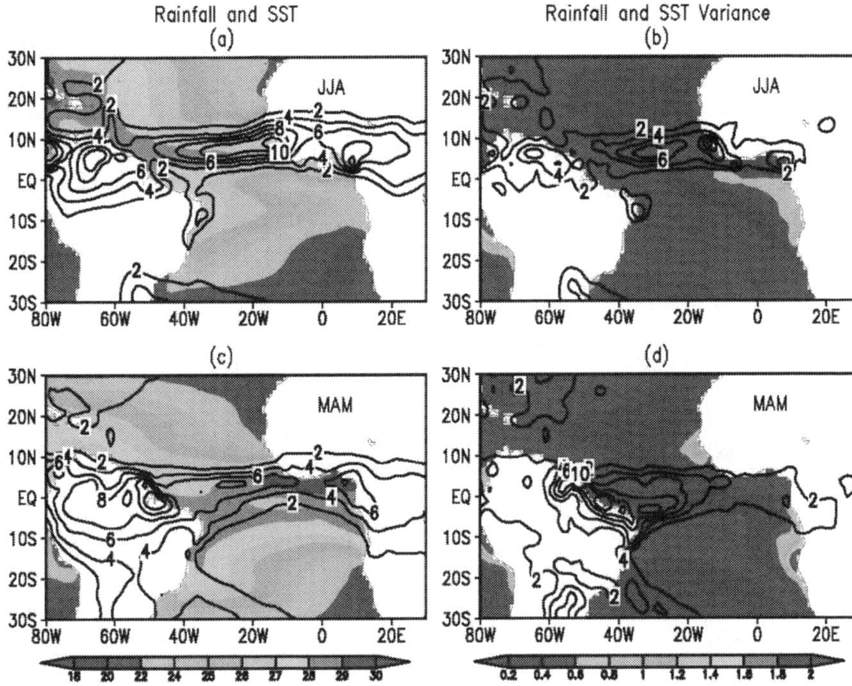

Fig. 1. Seasonal mean rainfall (*mm day*$^{-1}$; contours) and SST (°C; color shades) during (a) June-August (JJA) and (c) March-May (MAM); Seasonal mean variances of rainfall (*mm*^{2} *day*$^{-2}$; contours) and SST [(°C)^2; color shades] during (b) JJA and (d) MAM.

3. Spatial structures of SST anomalies associated with rainfall variations in the tropical Atlantic

To quantitatively investigate rainfall variability, two indices are defined to represent the strength (P_{ITCZ}) and latitude (Lat_{ITCZ}) variations of the marine ITCZ, respectively, with another one denoting the rainfall variability within the entire tropical Atlantic basin (P_{dm}). The tropical (25°S-25°N) meridional rainfall peak ($P_{max}(j)$, here j is an index of latitude) of monthly rainfall averaged along the longitudinal band of 15°-35°W, is determined for each month, including its latitude ($Lat_{max}(j)$). Then, this maximum rainfall value ($P_{max}(j)$) is weighted by the ones at the two neighboring latitudes (P(j-1) and P(j+1)) to yield a relatively smoother value for the ITCZ strength (P_{ITCZ}). The ITCZ latitude (Lat_{ITCZ}) is then estimated based on $Lat_{max}(j)$ and its two neighboring latitudes (Lat(j-1) and Lat(j+1)) weighted by their corresponding rainfall values. The resulting ITCZ latitude (Lat_{ITCZ}) thus depends on all these three rainfall values. The GPCP monthly rainfall product has a coarse (2.5°×2.5°) spatial resolution. This interpolation method can provide a relatively smooth (and reasonable), latitudinal change of the ITCZ north-south migration. The resultant ITCZ latitudes are in general confirmed by derived from both the similar NOAA/NCEP-CMAP satellite-based monthly precipitation product and a merged, short-record (1998-2006) 1°×1° TRMM (3B43) monthly rainfall product (not shown). The basin-mean rainfall is computed over a domain of 15°S-22.5°N, 15°-35°W. Finally, P_{ITCZ}, Lat_{ITCZ}, and P_{dm} are determined by subtracting their corresponding mean seasonal cycles.

Time series for these three indices are depicted in Fig. 2. Rainfall changes during these two seasons are comparable calibrated by either P_{ITCZ} or P_{dm}. However, the ITCZ does not change much its mean latitudes during JJA, in contrast to evident fluctuations during MAM. Thus the major rainfall changes during JJA are related to the variability of the ITCZ strength and/or the basin-mean rainfall. This probably implies a lack of forcing mechanism on the ITCZ location during JJA. Past studies suggested that the Atlantic interhemispheric SST mode, though a dominant factor of the ITCZ position during MAM, becomes secondary during JJA (*e.g.*, Sutton et al., 2000; Gu & Adler, 2006).

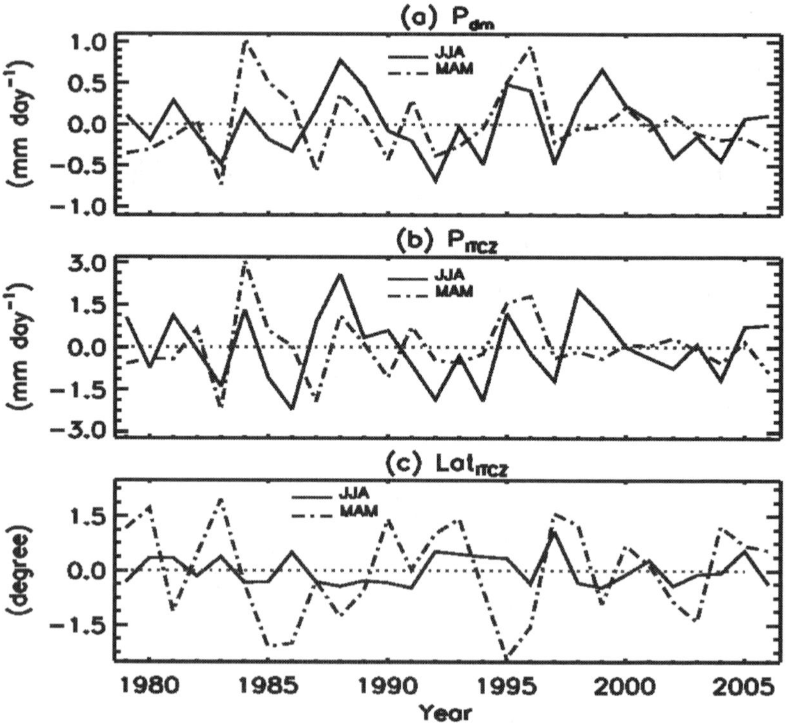

Fig. 2. Time series of (a) the domain-mean rainfall (P_{dm}), (b) the ITCZ strength (P_{ITCZ}), and (c) the ITCZ latitudes (Lat_{ITCZ}) during JJA (solid) and MAM (dash-dot).

Simultaneous correlations between SST anomalies with P_{ITCZ} and Lat_{ITCZ} are estimated during both seasons (Fig. 3). During JJA, the major high-correlation area of SST anomalies with P_{ITCZ} is located west of about 120°W in the tropical central-eastern Pacific, and the correlations between SST anomalies and Lat_{ITCZ} are generally weak in the tropical Pacific. Within the tropical Atlantic, significant, positive correlations with P_{ITCZ} roughly cover the entire basin from 20°S to 20°N. It is of interest to note that the same sign correlation is found both north and south of the equator, suggesting a coherent, local forcing of rainfall variability during JJA. Furthermore, evident negative correlations between SST anomalies and Lat_{ITCZ} are seen within the deep tropics especially along and south of the equator. These confirm the weakening effect of the interhemispheric SST gradient mode during JJA (e.g., Sutton et al., 2000).

During MAM, the ITCZ strength is strongly correlated to SST anomalies in both the equatorial Pacific and Atlantic (e.g., Nobre & Shukla, 1996; Sutton et al., 2000; Chiang et al., 2002). However, the significant negative correlations tend to appear along the equator in the central-eastern equatorial Pacific (east of 180°W) and along the western coast of South America, quite different than during JJA. Roughly similar correlation patterns can also be observed for Lat_{ITCZ} in the tropical Pacific. Within the tropical Atlantic basin, P_{ITCZ} tends to be correlated with SST anomalies along and south of the equator, but the high correlation area shrinks into a much smaller one compared with that during JJA. The lack of high (negative) correlation north of the equator further confirms that the interhemispheric SST mode strongly impacts the ITCZ locations (Fig. 3d), but has a minor effect on the ITCZ strength (*e.g.*, Nobre & Shukla, 1996).

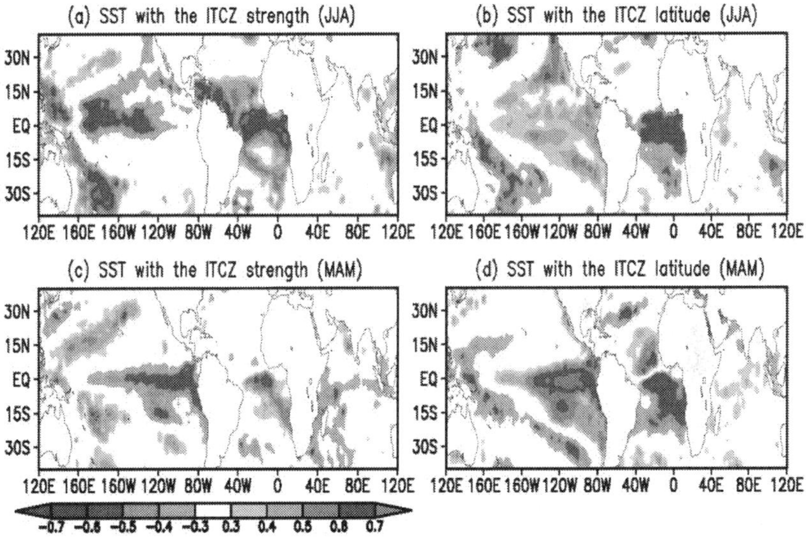

Fig. 3. Correlations of SST anomalies with (a, c) the ITCZ strength (P_{ITCZ}) and (b, d) the ITCZ latitude (Lat_{ITCZ}) during (a, b) MAM and (c, d) JJA. The 5% significance level is ±0.4 based on 23 degrees of freedom (*dofs*).

During these two seasons there are also two major large areas of high correlation for both P_{ITCZ} and Lat_{ITCZ} in the tropical western Pacific, though with different spatial features: One is along the South Pacific Convergence Zone (SPCZ), another is north of 10°N. These two features are probably associated with the ENSO effect and other factors, and not directly related to the changes in the tropical Atlantic, which are supported by weak regressed SST anomalies (not shown).

4. The effects of three major SST modes

To further explore the relationships between rainfall anomalies in the tropical Atlantic and SST variability, particularly during JJA, three major SST indices are constructed. Here, Nino3.4, the mean SST anomalies within a domain of 5°S-5°N, 120°-170°W, is as usual used to denote the interannual variability in the tropical Pacific. As in Gu & Adler (2006), the SST anomalies within 3°S-3°N, 0-20°W are defined as Atl3 to represent the Atlantic Equatorial Oscillation (*e.g.*, Zebiak, 1993; Carton & Huang, 1994). SST variability in the tropical north Atlantic is denoted by SST anomalies averaged within a domain of 5°-25°N, 15°-55°W (TNA). In addition, another index (TNA1) is constructed for comparison by SST anomalies averaged over a slightly smaller domain, 5°-20°N, 15°-55°W. We are not going to focus on the interhemispheric SST mode here because during boreal summer this mode becomes weak and does not impact much on the ITCZ (*e.g.*, Sutton et al., 2000; Gu & Adler, 2006), and the evident variability of the ITCZ is its strength rather than its preferred latitudes (Fig. 2). Same procedures are applied to surface zonal winds in the western basin (5°S-5°N, 25°-45°W) to construct a surface zonal wind index (U_{WAtl}).

As discovered in past studies (*e.g.*, Nobre & Shukla, 1996; Czaja, 2004), evident seasonal preferences exist in these indices (Fig. 4). ENSO usually peaks during boreal winter. The most intense variability in the tropical Atlantic appears during boreal spring and early summer. The maxima of both TNA and TNA1 are in April, about three months later than the strongest ENSO signals (*e.g.*, Curtis & Hastenrath, 1995; Nobre & Shukla, 1996). Surface zonal wind anomaly in the western equatorial region (U_{WAtl}) attains its maximum in May, followed by the most intense equatorial SST oscillation (Atl3) in June. Münnich & Neelin (2005) suggested that there seems a chain reaction during this time period in the equatorial Atlantic region. It is thus further arguable that the tropical western Atlantic (west of 20°W) is a critical region passing and/or inducing climatic anomalies in the equatorial Atlantic basin.

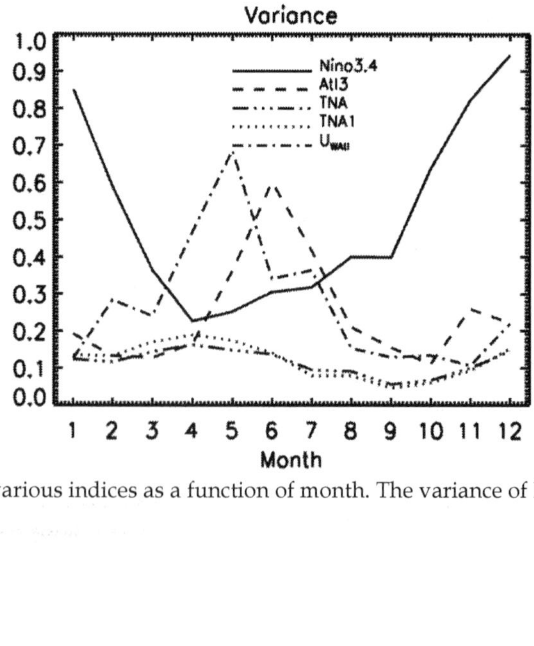

Fig. 4. Variances of various indices as a function of month. The variance of Nino 3.4 is scaled by 2.

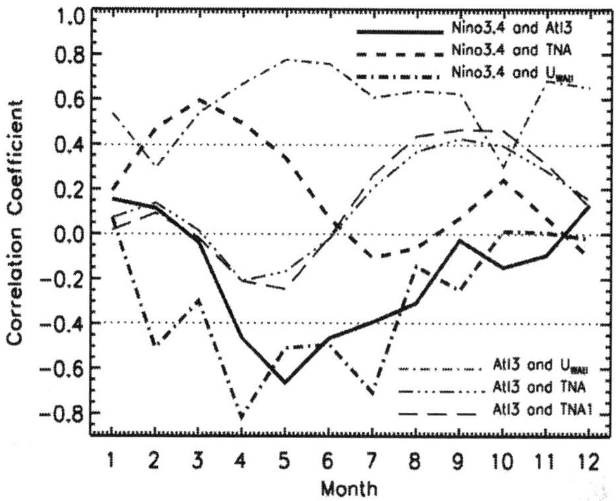

Fig. 5. Correlation coefficients between various indices as function of month. The 5% significance level is ±0.41 based on 21 dofs.

4.1 Relationships between various indices

Simultaneous correlations between SST indices are computed for each month (Fig. 5). The Pacific Niño shows strong impact on the tropical Atlantic indices. Significant correlations are found between Nino3.4 and TNA during February-April with a peak in March. The negative correlation between Nino3.4 and Atl3 becomes statistically significant during April-June, showing the impact of the ENSO on the Atlantic equatorial mode (*e.g.*, Delecluse et al., 1994; Latif & Grötzner, 2000). U_{WAtl} is consistently, negatively correlated with Nino3.4 during April-July except in June when the correlation coefficient is slightly lower than the 5% significance level. Interestingly, there are two peak months (April and July) for the correlation between U_{WAtl} and Nino3.4 as discovered in Münnich & Neelin (2005). High correlations between Atl3 and U_{WAtl} occur during March-July. These correlation relations tend to support that zonal wind anomalies at the surface in the western basin is a critical part of the connection between the equatorial Pacific and the equatorial Atlantic. Münnich & Neelin (2005) even showed a slightly stronger correlation relationship. Atl3 is also significantly correlated to U_{WAtl} in other several months, *i.e.*, January, September, and November, probably corresponding to the occasional appearance of the equatorial oscillation event during boreal fall and winter (*e.g.*, Wang, 2002; Gu & Adler, 2006).

Within the tropical Atlantic basin, the correlations between Atl3 and SST anomalies north of the equator (TNA and TNA1) become positive and strong during late boreal summer, particularly between Atl3 and TNA1 (above the 5% significance level during August-October). As shown in Fig. 4, SST variations north of the equator become weaker during boreal summer. Simultaneously the ITCZ and associated trade wind system move further to the north. It thus seems possible to feel impact in the TNA/TNA1 region from the equatorial region during this time period for surface wind anomalies-driven ocean transport (*e.g.*, Gill, 1982).

Lag-correlations between various SST indices are estimated to further our understanding of the likely, casual relationships among them (Figs. 6, 7, and 8). The base months for SST indices are chosen according to their respective peak months of variances (Fig. 4). The strongest correlation between Atl3 in June and Nino3.4 is found when Nino3.4 leads Atl3 by one month (Fig. 6), further confirming the remote forcing of ENSO on the Atlantic equatorial mode (*e.g.*, Latif & Grötzner, 2000). The 1-3 month leading, significant correlation of U_{WAtl} to Atl3 in June with a peak at one-month leading indicates that the equatorial oscillation is mostly excited by surface zonal wind anomalies in the western basin likely through oceanic dynamics (*e.g.*, Zebiak, 1993; Carton & Huang, 1994; Delecluse et al., 1994; Latif & Grötzner, 2000).

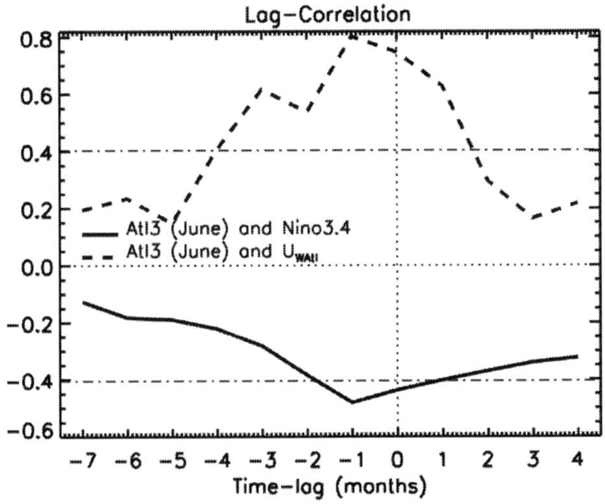

Fig. 6. Lag-correlations between Atl3 in June with Nino3.4 and U_{WAtl}, respectively. Positive (negative) months indicate Atl3 leads (lags) Nino3.4 and U_{WAtl}. The 5% significance level is ±0.42 based on 20 dofs.

The lag-correlation between U_{WAtl} in May and Nino3.4 is depicted in Fig. 7. The highest correlation appears as Nino3.4 leads U_{WAtl} by one-month, suggesting a strong impact from the equatorial Pacific (*e.g.*, Latif & Grötzner, 2000), and this impact probably being through anomalous Walker circulation and not passing through the mid-latitudes.

North of the equator, TNA and TNA1 both peak in April (Fig. 4). Simultaneous correlations between these two and Nino3.4 at the peak month are much weaker than when Nino3.4 leads them by at least one-month (Fig. 8). It is further noticed that the consistent high lag-correlations are seen with Nino3.4 leading by 1-7 months. Significant correlations of TNA and TNA1 in April with Nino3.4 can actually be found as Nino3.4 leads them up to 10 months (not shown). These highly consistent lag-relations suggest that the impact from the equatorial Pacific on the tropical north Atlantic may go through two ways: the Pacific-North-American (PNA) teleconnection and the anomalous Walker circulation (*e.g.*, Nobre & Shukla, 1996; Saravanan & Chang, 2000; Chiang et al., 2002), with the trade wind anomalies being the critical means. Most previous studies generally emphasized the first means being available during boreal winter and spring (*e.g.*, Curtis & Hastenrath, 1995; Nobre & Shukla, 1996).

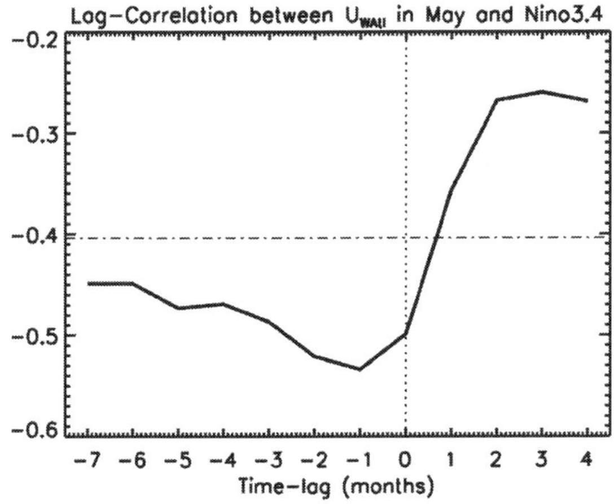

Fig. 7. Lag-correlation between U_{WAtl} in May with Nino3.4. Positive (negative) months indicate U_{WAtl} leads (lags) Nino3.4. The 5% significance level is ±0.42 based on 20 dofs.

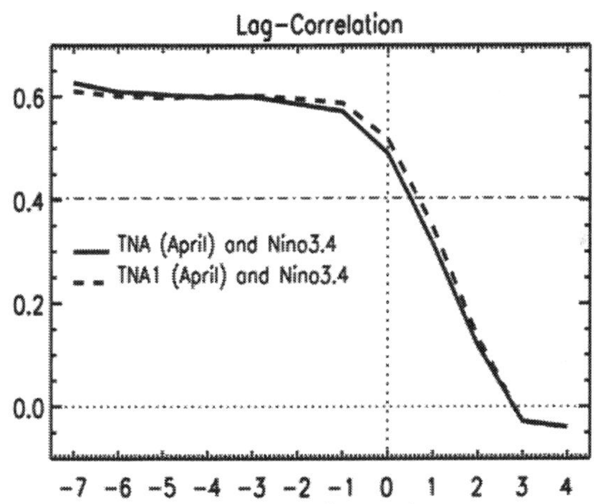

Fig. 8. Lag-correlations between TNA and TNA1 in April with Nino3.4. Positive (negative) months indicate TNA and TNA1 lead (lag) Nino3.4. The 5% significance level is ±0.42 based on 20 dofs.

4.2 Spatial structures of three SST modes related variations

Tables 1 and 2 illustrate the simultaneous correlations between the three SST indices (Atl3, TNA, and Nino3.4) and two rainfall indices (P_{ITCZ} and Lat_{ITCZ}). The ENSO events can effectively impact rainfall variability in the tropical Atlantic (*e.g.*, Nobre & Shukla, 1996;

Enfield & Mayer, 1997; Saravanan & Chang, 2000; Chiang et al., 2002; Giannini et al., 2004). A higher correlation (-0.62) can even be obtained between Nino3.4 and P_{dm}, implying a basin-wide impact in the equatorial region. The correlation between Nino3.4 and Lat_{ITCZ} is relatively weak during JJA, in contrasting to a much stronger impact during MAM.

Significant correlations appear between Atl3, and P_{ITCZ} and Lat_{ITCZ} during JJA and MAM (Tables 1 & 2). Even though the Atlantic equatorial warm/cold events are relatively weak and the ITCZ tends to be located about eight degrees north of the equator during boreal summer, the results suggest that the Atlantic Niño mode could still be a major factor controlling the ITCZ strength.

For the effect of TNA, large seasonal differences exist in its correlations with the rainfall indices (Tables 1 & 2). During JJA, TNA is significantly correlated with P_{ITCZ}. During MAM, however this correlation is much weaker. The correlation coefficient even changes sign between these two seasons. On the other hand, TNA is significantly correlated to Lat_{ITCZ} during MAM, but not during JJA.

γ	Nino3.4	Atl3	TNA
P_{ITCZ}	**-0.51**	**0.68**	**0.51**
Lat_{ITCZ}	0.39	**-0.65**	0.04

Table 1. Correlation coefficients (γ) between P_{ITCZ} (mm day^{-1}) and Lat_{ITCZ} (degree), and various SST indices during JJA. $\gamma=\pm0.40$ is the 5% significance level based on (n-2=) 23 dofs.

γ	Nino3.4	Atl3	TNA
P_{ITCZ}	**-0.50**	**0.56**	-0.18
Lat_{ITCZ}	**0.57**	**-0.67**	**0.41**

Table 2 Correlation coefficients (γ) between P_{ITCZ} (mm day^{-1}) and Lat_{ITCZ} (degree), and three SST indices during MAM. $\gamma=\pm0.40$ is the 5% significance level based on (n-2=) 23 dofs.

The modulations of the three major SST modes on the tropical Atlantic during JJA and MAM are further quantified by computing the regressions based on their seasonal-mean magnitudes normalized by their corresponding standard deviations.

Fig. 9 depicts the SST, surface wind, and precipitation associated with Atl3. During JJA, the spatial patterns generally agree with shown in previous studies that primarily focused on the peak months of the Atlantic equatorial mode (*e.g.*, Ruiz-Barradas et al., 2000; Wang, 2002). Basin-wide warming is seen with the maximum SSTs along the equator and tends to be in the eastern basin (Fig. 9b). Surface wind anomalies in general converge into the maximum, positive SST anomaly zone. Accompanying strong cross-equatorial flows being in the eastern equatorial region, anomalous westerlies are seen in the western basin extending from the equator to about 15°N. These wind anomalies are related to the equatorial warming (Figs. 5, 6), and also might be the major reason for the warming-up in the TNA/TNA1 region. Off the coast of West Africa, there even exist weak southerly anomalies between 10°-15°N. Positive rainfall anomalies are dominant in the entire basin, corresponding to the warm SSTs. It is interesting to note that these rainfall anomalies tend to be over the same area as the seasonal mean rainfall variances (Fig. 1). Particularly, over the open ocean the maximum rainfall anomaly band is roughly sandwiched by the marine ITCZ and the equatorial zone with maximum SST variability (Figs. 1c, 9b, and 9d), confirming the strong modulations of the equatorial mode during this season (Fig. 2). During MAM, positive SST anomalies already appear along the equator (Fig. 9a). However, in addition to the SST anomalies along the equator, the most intense SST variability occurs right off the west coast of Central Africa, reflecting the frequent appearance of the Benguela Niño peaking in March-April (*e.g.*, Florenchie et al., 2004). North of the equator, negative SST anomalies, though very weak, can still be seen off the West African coast. This suggests that the Atlantic Niño may effectively contribute to the interhemispheric SST mode peaking in this season, particularly to its south lobe (Figs. 1b and 9a). Negative-positive rainfall anomalies across the equator forming a dipolar structure are evident, specifically west of 20°W (Fig. 9c). In the Gulf of Guinea, positive rainfall anomalies, though much weaker than in the western basin, can still be observed extending from the open ocean to the west coast of Central Africa, roughly following strong positive SST anomalies.

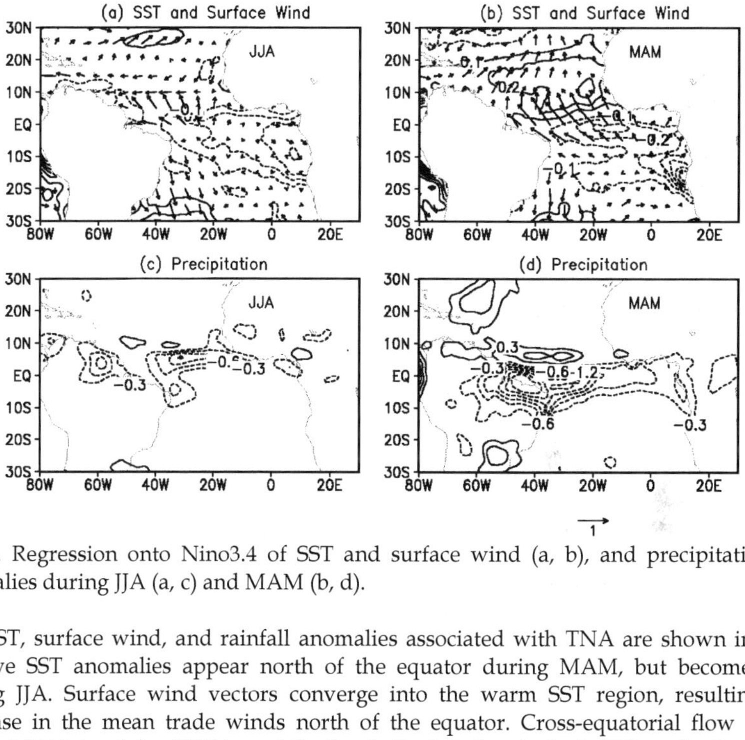

Fig. 9. Regression onto Nino3.4 of SST and surface wind (a, b), and precipitation (c, d) anomalies during JJA (a, c) and MAM (b, d).

The SST, surface wind, and rainfall anomalies associated with TNA are shown in Fig. 10. Positive SST anomalies appear north of the equator during MAM, but become weaker during JJA. Surface wind vectors converge into the warm SST region, resulting in the decrease in the mean trade winds north of the equator. Cross-equatorial flow is strong during MAM, implying TNA's contribution to the interhemispheric SST mode. On the other hand, no evident SST anomalies appear along and south of the equator supporting that the two lobes of the interhemispheric mode are probably not connected (*e.g.*, Enfield et al., 1999). A negative-positive rainfall dipolar feature occurs during MAM with much weaker anomalies east of 20°W, consistent with previous studies (*e.g.*, Nobre & Shukla, 1996; Ruiz-Barradas et al., 2000; Chiang et al., 2002). During JJA, however only appears a single band of positive rainfall anomalies between 5°-20°N, covering the northern portion of the mean rainfall within the ITCZ and its variances (Figs. 1c, 1d, and 10d). Interestingly this band tilts from northwest to southeast, tending to be roughly along the tracks of tropical storms. This may reflect the impact of TNA on the Atlantic hurricane activity (*e.g.*, Xie et al., 2005).

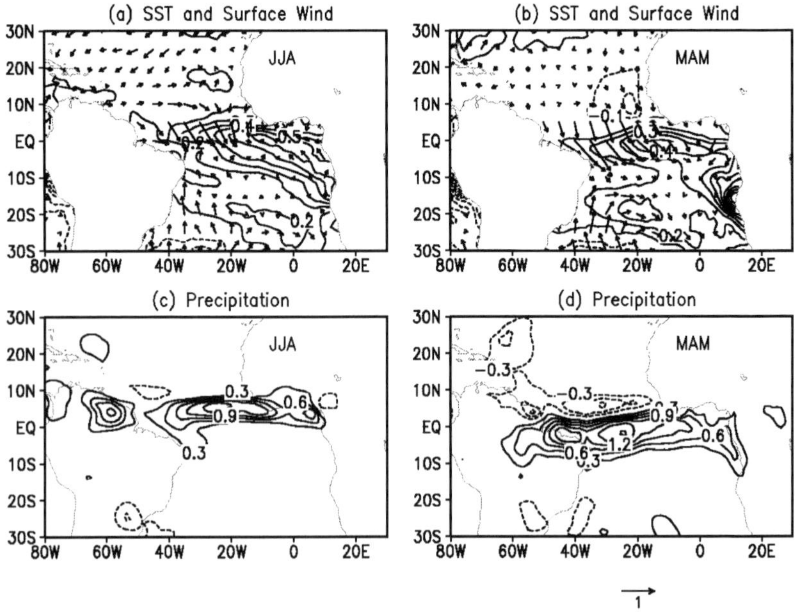

Fig. 10. Regression onto Atl3 of SST and surface wind (a, b), and precipitation (c, d) anomalies during JJA (a, c) and MAM (b, d).

Fig. 11 illustrates the SST, surface wind, and rainfall regressed onto the seasonal mean Nino3.4. During MAM, positive-negative SST anomalies occur in the tropical region, shaping a dipolar structure accompanied by strong cross-equatorial surface winds. Compared with Figs. 9a and 10a, it is likely that ENSO may contribute to both lobes of the interhemispheric SST mode during this season (*e.g.*, Chiang et al., 2002). Rainfall anomalies tend to be in the western basin and manifest as dipolar as well. Compared with Figs. 9a and 9c, it is noticeable that along and south of the equator, ENSO shows a very similar impact feature as the Atlantic equatorial mode except with the opposite sign. This enhances our discussion about their relations (Figs. 5 and 6). During JJA, SST anomalies almost disappear north of the equator. South of the equator, negative SST anomalies can still be seen but become weaker, accompanied by much weaker equatorial wind anomalies. Rainfall anomalies move to the north, as does the ITCZ. The dipolar feature can hardly be discernible. Again, the rainfall anomalies show a very similar pattern as those related to Atl3 (Figs. 9d and 11d), though their signs are opposite. This seems to suggest that during JJA the impact of ENSO on the tropical Atlantic may mostly go through its influence on the Atlantic equatorial mode (Atl3).

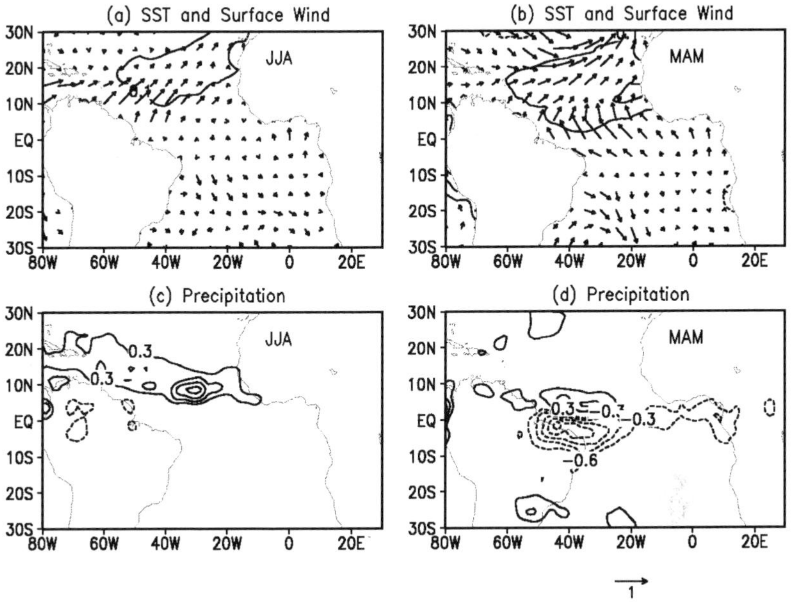

Fig. 11. Regression onto TNA of SST and surface wind (a, b), and precipitation (c, d) anomalies during JJA (a, c) and MAM (b, d).

Therefore, generally consistent with past results (e.g., Nobre & Shukla, 1996; Enfield & Mayer, 1997; Saravanan & Chang, 2000; Chiang et al., 2002; Giannini *et al.* 2004), these three SST modes all seem to influence rainfall variations in the tropical Atlantic, though through differing means. However, strong inter-correlations have been shown above among these SST indices and in past studies (e.g., Münnich & Neelin, 2005; Gu & Adler, 2006). Nino3.4 is significantly correlated with Atl3 during both JJA (-0.46) and MAM (-0.53), and with TNA (0.52) during MAM. Previous studies have demonstrated that the Pacific ENSO can modulate SST in the tropical Atlantic through both mid-latitudes and anomalous Walker circulation (e.g., Horel & Wallace, 1981; Chiang et al., 2002; Chiang & Sobel, 2002). While no significant correlation between Nino3.4 and Atl3 was found in some previous studies (e.g., Enfield & Mayer, 1997), high correlations shown here are generally in agreement with others (e.g., Delecluse et al., 1994; Latif & Grötzner, 2000). Thus, the correlations shown above, particularly the effect of Nino3.4, may be complicated by the inter-correlations among the SST indices. For instance, the high correlation between Nino3.4 and P_{ITCZ} may primarily result from their respective high correlations with Atl3 (Tables 1 & 2), and hence may not actually indicate any effective, direct modulation of convection (P_{ITCZ}) by the ENSO. It is thus necessary to discriminate their effects from each other. Thus, linear correlations and second-order partial correlations are estimated and further compared (Figs. 12 and 13). The second-order correlation here represents the linear correlation between rainfall and one SST index with the effects of other two SST indices removed (or hold constant) (Gu &Adler, 2009). With or without the effects of Nino3.4 and TNA, the spatial structures of correlation with Atl3 do not vary much. With the impact of Nino3.4 and TNA removed, the Atl3 effect only becomes slightly weaker during both JJA and MAM. Given a weak relationship between Atl3 and TNA (0.17 during JJA and -0.16 during MAM; Enfield et al., 1999), this correlation change is in general due to the Pacific ENSO.

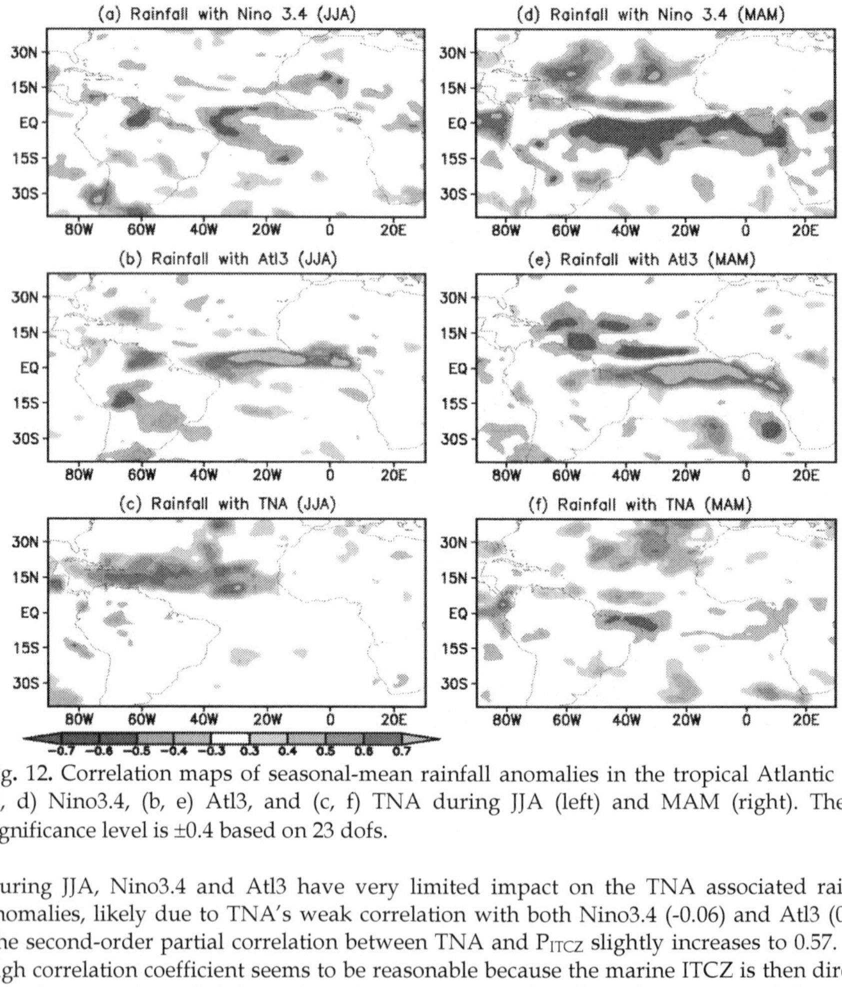

Fig. 12. Correlation maps of seasonal-mean rainfall anomalies in the tropical Atlantic with (a, d) Nino3.4, (b, e) Atl3, and (c, f) TNA during JJA (left) and MAM (right). The 5% significance level is ±0.4 based on 23 dofs.

During JJA, Nino3.4 and Atl3 have very limited impact on the TNA associated rainfall anomalies, likely due to TNA's weak correlation with both Nino3.4 (-0.06) and Atl3 (0.17). The second-order partial correlation between TNA and P_{ITCZ} slightly increases to 0.57. This high correlation coefficient seems to be reasonable because the marine ITCZ is then directly over the tropical North Atlantic (Fig. 1). During MAM, the effect of TNA on rainfall over the tropical open ocean is generally weak. With the effects of Nino3.4 and Atl3 removed, the large area of negative correlation in the western basin near South America shrinks into a much smaller region.

Hence, the direct influence of ENSO through the anomalous Walker circulations could play a role, but in general is confined in the western basin and over the northeastern South American continent where the most intense deep convection and variations are located during MAM (Fig. 1). During JJA, this kind of modulation of deep convection disappears because the ITCZ moves to the north and stays away from the equator. The ENSO impact on rainfall anomalies in the tropical Atlantic may hence primarily go through its effect on the two local SST modes. In particular, its effect on Atl3 seems to be the only means during JJA by means of modulating surface winds in the western basin (Figs. 6 and 7; e.g., Latif & Grötzner, 2000; Münnich & Neelin, 2005). These wind anomalies are essential components for the development of the Atlantic Niño mode (e.g., Zebiak, 1993; Latif & Grötzner, 2000).

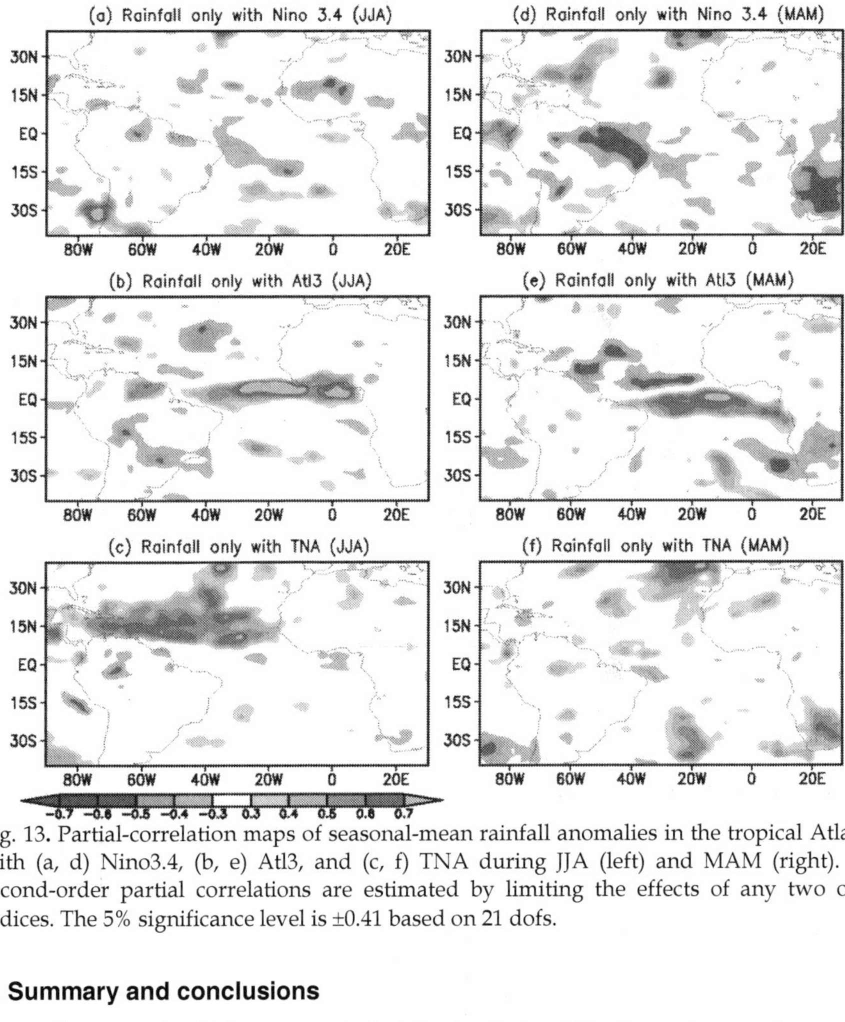

Fig. 13. Partial-correlation maps of seasonal-mean rainfall anomalies in the tropical Atlantic with (a, d) Nino3.4, (b, e) Atl3, and (c, f) TNA during JJA (left) and MAM (right). The second-order partial correlations are estimated by limiting the effects of any two other indices. The 5% significance level is ±0.41 based on 21 dofs.

5. Summary and conclusions

Seasonal-mean rainfall in the tropical Atlantic during JJA shows intense interannual variabilities, which are comparable with during MAM based on both the ITCZ strength and the basin-mean rainfall. The latitudes of the marine ITCZ however do not vary much from-year-to-year during JJA, in contrasting to evident variations occurring during MAM. Hence the summer-time rainfall variability is mostly manifested as the variations in the ITCZ strength and the basin-mean rainfall.

Rainfall variations associated with the two local SST modes and ENSO are further examined. The Atlantic Niño mode can effectively induce rainfall anomalies during JJA through accompanying anomalous surface winds and SST. These rainfall anomalies are generally located over the major area of rainfall variance. TNA can contribute to the rainfall changes as well during this season, but its impact is mostly limited to the northern portion of the ITCZ. The ENSO teleconnection mechanism may still play a role during boreal summer, although it becomes much weaker than during boreal spring. It is noticed that the ENSO-associated spatial patterns tend to be similar to those related to the Atlantic Niño though with an opposite sign. This suggests that the impact of ENSO during JJA may go through its influence on the Atlantic Niño mode.

During MAM, TNA shows an evident impact on rainfall changes specifically in the region near and over the northeastern South America. The correlation/regression patterns are generally consistent with those using the index representing the interhemispheric SST mode (*e.g.*, Ruiz-Barradas et al., 2000), though the TNA-associated SST anomalies are weak and mostly north of the equator. This suggests a strong contribution of TNA to this interhemispheric mode and also its independence from the SST oscillations south of the equator (*e.g.*, Enfield et al., 1999). Atl3 and Nino3.4 can contribute to the interhemispheric SST mode too, in addition to their direct modulations of rainfall change in the basin. Particularly in the western basin (west of 20°W), corresponding to evident oscillations of the ITCZ locations, a dipolar feature of rainfall anomalies occurs in the regression maps for both indices. Simultaneously appear strong surface wind anomalies with evident cross-equatorial components.

To further explore the relationships among the two local SST modes and ENSO, contemporaneous and lag correlations are estimated among various indices. ENSO shows strong impact on the Atlantic equatorial region and the tropical north Atlantic. Significant, simultaneous correlations between Nino3.4 and TNA are seen during February-April. Significant lag-correlation of TNA at its peak month (April) with Nino3.4 one or several months before further confirms that the impact from the tropical Pacific is a major contributor during boreal spring (*e.g.*, Chiang et al., 2000). Nino3.4 is highly correlated with Atl3 during April-June. The correlations between Nino3.4 and zonal wind index in the west basin (U_{Watl}) also become high during April-July. Moreover the maximum correlation between U_{Watl} in May (peak month) and Nino3.4 is seen as Nino3.4 precedes it by one month, indicating the remote modulations of wind anomalies. The Pacific ENSO can effectively modulate convection and surface winds during boreal spring through both ways: the PNA and the anomalous Walker cell (*e.g.*, Nobre & Shukla, 1996; Chiang & Sobel, 2002). Trade wind anomalies are a pathway for the SST oscillations north of the equator (*e. g.*, Curtis & Hastenrath, 1995; Enfield & Mayer, 1997). Along and south of the equator, convective and wind anomalies in the western basin are the critical means for the ENSO impact. During JJA, the pathway from the mid-latitudes becomes impossible due to seasonal changes in the large-scale mean flows, and the ITCZ moves away from the equator. Hence, the ENSO impact on the tropical region is greatly limited. The lag-correlations between Atl3 at the peak month (June) and Nino3.4 and U_{Watl}, respectively, tend to suggest that the equatorial oscillation is excited by the preceding surface wind anomalies in the west basin that are closely related to the ENSO. The lag and simultaneous correlations of Atl3 with U_{Watl} further confirm that it is a coupled mode to a certain extent. It is interesting to further note that high positive correlations can be found between Atl3 and TNA/TNA1 during July-October, implying that during JJA the Atlantic equatorial mode may have a much more comprehensive impact, in addition to its influence on the ITCZ, than expected.

A second-order partial correlation analysis is further applied to discriminate the effects of these three SST modes because of the existence of inter-correlations among them. With the effects of Atl3 and TNA removed, ENSO only has a very limited direct impact on the open ocean in the tropical Atlantic, and its impact is generally confined in the western basin and over the northeastern South America.

Therefore, during JJA, the two local SST modes turn out to be more critical/essential for rainfall variations in the tropical Atlantic. The effect of the Pacific ENSO on the tropical Atlantic is in general through influencing the Atlantic Niño mode, and surface zonal wind anomalies in the western basin are the viable means to realize this effect.

6. References

Adler, R.; & Coauthors (2003). The version 2 Global Precipitation Climatology Project (GPCP) monthly precipitation analysis (1979-present). *J. Hydrometeor*, **4**, 1147-1167.

Biasutti, M.; Battisti, D. & Sarachik, E. (2004). Mechanisms controlling the annual cycle of precipitation in the tropical Atlantic sector in an atmospheric GCM. *J. Climate*, **17**, 4708-4723.

Carton, J. & Huang, B. (1994). Warm events in the tropical Atlantic. *J. Phys. Oceanogr.*, **24**, 888-903.

Chen, Y. & Ogura, Y. (1982). Modulation of convective activity by large-scale flow patterns observed in GATE. *J. Atmos. Sci.*, **39**, 1260-1279.

Chiang, J.; Kushnir, Y. & Zebiak, S. (2000). Interdecadal changes in eastern Pacific ITCZ variability and its influence on the Atlantic ITCZ. *Geophys. Res. Lett.*, **27**, 3687-3690.

Chiang, J.; Kushnir, Y. & Giannini, A. (2002). Reconstructing Atlantic Intertropical Convergence Zone variability: Influence of the local cross-equatorial sea surface temperature gradient and remote forcing from the eastern equatorial Pacific. *J. Geophys. Res.*, **107**(D1), 4004, doi:10.1029/2000JD000307.

Chiang, J. & Sobel, A. (2002). Tropical tropospheric temperature variations caused by ENSO and their influence on the remote tropical climate. *J. Climate*, **15**, 2616-2631.

Curtis, S. & Hastenrath, S. (1995). Forcing of anomalous sea surface temperature evolution in the tropical Atlantic during Pacific warm events. *J. Geophys. Res.*, **100**, 15835-15847.

Czaja, A. (2004). Why is North Tropical Atlantic SST variability stronger in boreal spring? *J. Climate*, **17**, 3017-3025.

Delecluse, P.; Servain, J., Levy, C., Arpe, K. & Bengtsson, L. (1994). On the connection between the 1984 Atlantic warm event and the 1982-1983 ENSO. *Tellus*, **46**A, 448-464.

Enfield, D. & Mayer, D. (1997). Tropical Atlantic sea surface temperature variability and its relation to El Niño-Southern Oscillation. *J. Geophys. Res.*, **102**, 929-945

Enfield, D.; Mestas-Nunez, A., Mayer, D. & and Cid-Serrano, L. (1999). How ubiquitous is the dipole relationship in tropical Atlantic sea surface temperature? *J. Geophys. Res.*, **104**, 7841-7848.

Florenchie, P.; Reason, C., Lutjeharms, J., Rouault, M., Roy, C. & Masson, S. (2004). Evolution of interannual warm and cold events in the southeast Atlantic Ocean. *J. Climate*, **17**, 2318-2334.

Giannini, A.; Chiang, J., Cane, M., Kushnir, Y. & Seager, R. (2001). The ENSO teleconnection to the tropical Atlantic Ocean: Contributions of the remote and local SSTs to rainfall variability in the tropical Americas. *J. Climate*, **14**, 4530-4544.

Giannini, A.; Saravanan, R. & Chang, P. (2004). The preconditioning role of tropical Atlantic variability in the development of the ENSO teleconnection: Implication for the predictability of Nordeste rainfall. *Climate Dyn.*, **22**, 839-855.

Gill, A. (1982). *Atmosphere-Ocean Dynamics*. Academic Press, 662pp.

Gu, G., & Adler, R. (2004). Seasonal evolution and variability associated with the West African monsoon system. *J. Climate*, **17**, 3364-3377.

Gu, G. & Adler, R. (2006). Interannual rainfall variability in the tropical Atlantic region. *J. Geophys. Res.*, **111**, D02106, doi:10.1029/2005JD005944.

Gu, G. & Adler, R. (2009). Interannual Variability of Boreal Summer Rainfall in the Equatorial Atlantic. *Int. J. Climatol.*, **29**, 175-184, doi: 10.1002/joc.1724.

Gu, G. & Zhang, C. (2001). A spectrum analysis of synoptic-scale disturbances in the ITCZ. *J. Climate*, **14**, 2725-2739.

Hastenrath, S. & Greischar, L. (1993). Circulation mechanisms related to northeast Brazil rainfall anomalies. *J. Geophys. Res.*, **98**, 5093-5102.

Kalnay, E.; & Coauthors (1996). The NCEP/NCAR 40-year reanalysis project. *Bull. Amer. Meteor. Soc.*, **77**, 437-471.

Lamb, P. (1978a). Large scale tropical Atlantic surface circulation patterns during recent sub-Saharan weather anomalies. *Tellus*, **30**, 240-251.

Lamb, P. (1978b). Case studies of tropical Atlantic surface circulation patterns during recent sub-Saharan weather anomalies: 1967 and 1978. *Mon. Wea. Rev.*, **106**, 482-491.

Landsea, C.; Pielke Jr., R., Mesta-Nunez, A. & Knaff, J. (1999). Atlantic basin hurricanes: Indices of climate changes. *Climate Change*, **42**, 89-129.

Latif, M. & Grötzner, A. (2000). The equatorial Atlantic oscillation and its response to ENSO. *Climate Dyn.*, **16**, 213-218.

Mitchell, T. & Wallace, J. (1992). The annual cycle in equatorial convection and sea surface temperature. *J. Climate*, **5**, 1140-1156.

Münnich, M. & Neelin, J. (2005). Seasonal influence of ENSO on the Atlantic ITCZ and equatorial South America. *Geophys. Res. Lett.*, **32**, L21709, doi:10.1029/2005GL023900.

Nobre, P. & Shukla, J. (1996). Variations of sea surface temperature, wind stress, and rainfall over the tropical Atlantic and South America. *J. Climate*, **9**, 2464-2479.

Reynolds, R.; Rayner, N., Smith, T., Stokes, D. & and Wang, W. (2002). An improved in situ and satellite SST analysis for climate. *J. Climate*, **15**, 1609-1625.

Ruiz-Barradas, A.; Carton, J. & Nigam, S. (2000). Structure of interannual-to-decadal climate variability in the tropical Atlantic sector. *J. Climate*, **13**, 3285-3297.

Saravanan, R. & Chang, P. (2000). Interaction between tropical Atlantic variability and El Niño-Southern oscillation. *J. Climate*, **13**, 2177-2194.

Sutton, R.; Jewson, S. & Rowell, D. (2000). The elements of climate variability in the tropical Atlantic region. *J. Climate*, **13**, 3261-3284.

Thorncroft, C. & Rowell, D. (1998). Interannual variability of African wave activity in a general circulation model. *Int. J. Climatol.*, **18**, 1306-1323.

Wang, C. (2002). Atlantic climate variability and its associated atmospheric circulation cells. *J. Climate*, **15**, 1516-1536.

Xie, L.; Yan, T. & Pietrafesa, L. (2005). The effect of Atlantic sea surface temperature dipole mode on hurricanes: Implications for the 2004 Atlantic hurricane season. *Geophys. Res. Lett.*, **32**, L03701, doi:10.1029/2004GL021702.

Zebiak, S. (1993). Air-sea interaction in the equatorial Atlantic region. *J. Climate*, **6**, 1567-1586.

4

Tropical cyclones, oceanic circulation and climate

Lingling Liu
*Key Laboratory of Ocean Circulation and Waves,
Institute of Oceanology, Chinese Academy of Sciences, Qingdao,
China*

1. Introduction

Tropical cyclone, also popular known as hurricane or typhoon, is a non-frontal synoptic scale warm-core system characterized by a large low-pressure center. It forms over most of the world's tropical waters between about 5° and 22° latitude in an environment with sufficient sea surface temperature (>26.5°C), moisture instability and weak vertical shear, including North Indian, western North Pacific, eastern North Pacific, North Atlantic, South Indian and western South Pacific (Fig.1). Environmental conditions in the eastern South Pacific and South Atlantic are not favorable for the tropical cyclone's genesis. Thus, so far there has only been one documented tropical cyclone in the South Atlantic basin and it was quite weak. Mostly for the purpose of providing useful warnings, tropical cyclones are categorized according to their maximum wind speed. Tropical cyclones, with maximum winds of 17ms^{-1} or less, are known as tropical depressions; when their wind speeds are in the range of 18 to 32 ms^{-1}, inclusive, they are called tropical storms, whereas tropical cyclones with maximum winds of 33 ms^{-1} or greater are called hurricanes in the western North Atlantic and eastern North Pacific regions, typhoons in the western North Pacific, and severe tropical cyclones elsewhere.

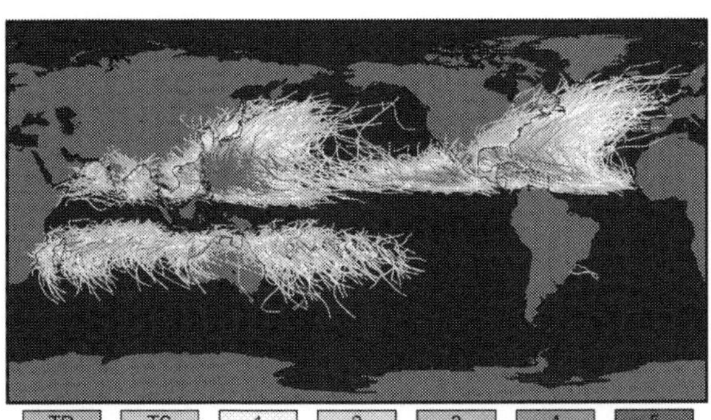

Fig.1. The tracks and intensity of nearly 150 years of tropical cyclones. (http://earthobservatory.nasa.gov/IOTD/view.php?id=7079)

A tropical cyclone is driven principally by heat transfer from the ocean. Thus, the genesis and development of the tropical cyclones and its variability of number and intensity are influenced by the oceans importantly.

Meanwhile, a tropical cyclone can affect the thermal structure and currents of the upper ocean. Beginning with the observations published by Leipper (1967), a number of studies have been made in order to understand the various aspects of the ocean response to tropical cyclones (e.g. Price, 1981; Black, 1983; Greatbatch, 1983; Ginis, 1995; Jacob et al., 2000) and the tropical cyclone-ocean interaction (Chang & Anthes, 1979; Sutyrin & Khain, 1979; Ginis et al., 1989). It became obvious that the tropical cyclones have a profound effect on the uppermost 200-300m of the ocean, deepening the mixed layer by many tens of meters, cooling the surface temperature by as much as 5°C, and causing near-inertial surface currents of 1-2ms^{-1}, detectable at depths up to at least 500m (Withee & Johnson, 1976). However, most of the previous studies focused on the local response of the ocean to the passing tropical cyclone. Relatively little is known about the influence of tropical cyclones on the mean climatology. Emanuel (2001) estimated the oceanic heat transport induced by tropical cyclone activity, comparable the observed peak meridional heat transport by the Meridional Overturning Circulation (MOC) as estimated by Macdonald & Wunsch (1996), suggesting that tropical cyclones may play an important role in driving the thermohaline circulation and thereby in regulating climate.

As we known, the world's oceans is an extremely important part of the Earth's climate control system because the world's oceans carry roughly half of the net equator-to-pole heat flux necessary to balance the meridional distribution of net radiative flux at the top of atmosphere (Macdonald & Wunsch, 1996) and thus play a critical role in setting the global temperature distribution. Furthermore, tropical cyclones threaten lives and property because of their high winds, associated storm surge, excessive rain and flooding, and ability to spawn tornadoes. Of all the natural phenomena that affect our planet, tropical cyclone, which account for the majority of natural catastrophic losses in the developed world, is among the most deadly and destructive. It is therefore of critical importance to understand the mutual influence of the tropical cyclones, oceanic circulation and climate.

Our discussion here focuses on the role of tropical cyclones in regulating the general oceanic circulation and climate, section 2, and the effects of the ocean on tropical cyclones, section 3.

2. The role of tropical cyclones in regulating oceanic circulation and climate

2.1 It's role in ENSO

El Niño/Southern Oscillation (ENSO) is a climate pattern that occurs across the tropical Pacific Ocean on average every four years, but over a period which varies from two to seven years. ENSO is composed of an oceanic component, called El Niño (or La Niña, depending on its phase), which is defined as a warming or cooling of at least 0.5°C (0.9°F) averaged over the east-central tropical Pacific Ocean, and an atmospheric component, Southern Oscillation, which is characterized by changes in surface pressure in the tropical Western Pacific. Measurements from satellite, ships, and buoys reveal El Niño to be a complex phenomenon that affects ocean temperatures across virtually the entire tropical Pacific, also affecting weather in other parts of the world. People are gradually interested in El Niño just because it is usually accompanied with abnormality of global circulation (Horel & Wallace, 1981).

It has been recognized that tropical cyclones, strong nonlinear events in the low and mid-latitudes in the weather system, can influence ENSO greatly. Most tropical cyclones form on the side of the subtropical ridge closer to the equator, then move poleward past the ridge axis before recurving into the main belt of the westerlies. It is well known that surface westerlies on the equator are an essential part of the development of El Niño events. Several studies have pointed out that a single tropical cyclone can also generate significant equatorial westerlies (Harrison & Giese, 1991; kindle & Phoebus, 1995). Gao et al. (1988) proposed a triggering mechanism of the near-equatorial cyclones on El Niño. They pointed out that the near-equatorial tropical cyclones developing equatorward of 10°N can intensify equatorial westerlies and produce Kelvin waves, which propagate to the South American Coasts in about 2-3 months, inducing SST to rise there. According to their result, the near-equatorial cyclones play an essential role in El Niño in its beginning, continuous, and developing period.

Sobel & Camargo (2005) argued that western North Pacific tropical cyclones play an active role in ENSO dynamics, by helping a warm event which is already taking place to persist or strengthen. They proposed that tropical cyclones in the western North Pacific can produce equatorial surface westerly anomalies near the dateline, and an associated SST increase in the central and eastern Pacific. These signals are of the right sign to contribute to the enhancement of a developing El Niño event.

2.2 Tropical cyclone-induced mechanical energy input and its variability

According to the new theory of oceanic general circulation, external sources of mechanical energy are required to maintain the quasi-steady oceanic circulation. Wind stress and tidal dissipation are the primary sources of mechanical energy. However, tropical cyclone, a vitally important component of the atmospheric circulation system at low- and mid-latitudes, may be an important mechanical energy source, which have been ignored because in the commonly used low spatial resolution wind stress data, these strong nonlinear events are smoothed out. Nillson (1995) estimated the energy input to the inertial waves induced by tropical cyclones theoretically as 0.026 TW, while Shay & Jacob (2006) estimated as 0.74TW using the averaged downward vertical energy flux of 2ergs cm^{-2} s^{-1} based on the observational data profile during the passage of the hurricane Gilbert.

Based on a hurricane-ocean coupled model (Schade & Emanuel, 1999), the mechanical energy input to the world's oceans induced by tropical cyclones was estimated (Liu et al., 2008). As shown in Fig.1, tropical cyclones vary greatly in their location and strength and its activity is different each year; thus, for the study of their contribution to the general oceanic circulation and climate, the most objective approach is to estimate the annual mean contribution from these storms. Then the energy input to the ocean induced by over 1500 tropical cyclones from 1984 to 2003 was calculated:

(a) One of the major forms of energy transfer from wind to the ocean is through surface waves. The annual energy input to the surface waves induced by tropical cyclones averaged from 1984 to 2003 is 1.62TW.

(b) The wind energy input to the surface currents, including both the geostrophic and ageostrophic components, by tropical cyclones is 0.1TW.

(c) Tropical cyclones are excellent generators of near-inertial motions, which are the most likely contributor to the subsurface turbulence, internal waves, and the subsurface diapycnal mixing, because of their large wind stress that change on the inertial time scale.

The generation of inertial motions by tropical cyclones has been discussed in previous studies (e.g. Price, 1981, 1983). The energy flux due to wind forcing associated with tropical cyclones averaged from 1984 to 2003 is 0.03TW.

(d) Tropical cyclone-induced cooling in the upper ocean is a striking phenomenon, which has been documented in many studies. Within the vicinity of a tropical cyclone, strong winds blowing across the sea surface drive strong ocean currents in the mixed layer. The vertical shear of the horizontal current at the base of mixed layer induces strong turbulence, driving mixing of warm/old water across the mixed layer base (Emanuel, 2005). As a result, sea surface temperature is cooled down. Most importantly, the warming of water below the mixed layer raises the center of mass, and the gravitational potential energy (GPE) of the water column is increased. According to the calculation, the annual mean GPE increase induced by tropical cyclones averaged from 1984 to 2003 is 0.05TW.

The relationship between the increase of GPE and the energy input to the near-inertial currents and the surface currents for each individual tropical cyclone over the past 20 years are demonstrated in Fig. 2a and 2b, respectively. It is clearly seen that the near-inertial energy input alone cannot account for the increase of the GPE when the hurricanes are strong. The ratio of GPE increase to the wind energy input to the near-inertial currents and the total surface currents versus normalized PDI (power dissipation index: $PDI \equiv \int_0^{T_{life}} v_{\max}^3 dt$, , which indicates the strength of the tropical cyclones) are shown in Fig. 2c and 2d, respectively. For weak tropical cyclones the increase of GPE is limited and it may be dominated by the contribution from the near-inertial energy from the wind.

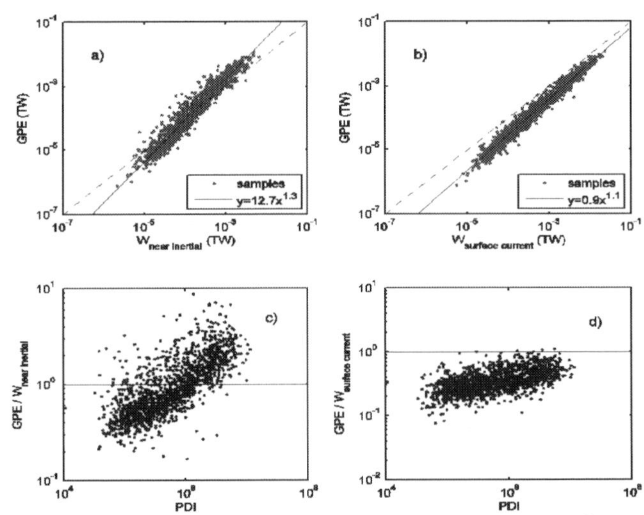

Fig.2. Relationship between the increase of GPE and the energy sources from the hurricanes: (a) GPE vs near-inertial components; (b) GPE vs wind energy input to the surface currents; (c) the ratio of GPE increase to near-inertial energy from the wind vs PDI; and (d) the ratio of GPE increase to energy input from the wind input to the surface currents vs PDI. In the upper panels the solid lines indicate best-fit power laws (Liu et al., 2008).

For hurricanes, however, the near-inertial energy from the wind can only supply a small portion of the energy needed for GPE increase, and the remaining portion of energy should be supplied by subinertial components of the wind energy input to the surface currents. Therefore, when the hurricane is strong, wind energy input to the subinertial motion is not totally dissipated in the mixed layer; instead, it contributes to the increase of GPE. Moreover, the conversion rate of kinetic energy input from the wind to GPE also increases as the strength of the hurricane increases.

The distribution of the energy input to the near-inertial motions from tropical cyclones averaged from 1984 to 2003 is shown in Fig.3. It is readily seen that most of this energy is distributed in the latitudinal band from 10° to 30°N in the western Pacific and in the North Atlantic, with approximately half of the total energy being input into the western North Pacific (The distribution of the other forms of energy generated from tropical cyclones has similar patterns).

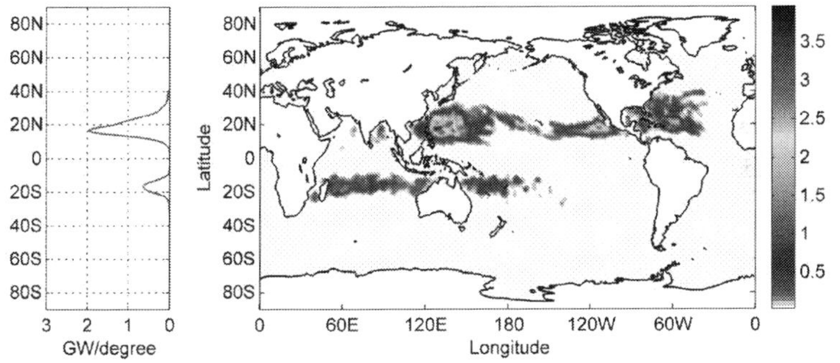

Fig. 3. (right) Energy input to near-inertial motions induced by tropical cyclones averaged from 1984 to 2003 (units: mW m^{-2}); (left) the meridional distribution of the integrated energy. (Liu et al., 2008)

According to the previous studies, the energy input induced by smoothed wind stress to the surface geostrophic currents is estimated as 0.88TW (Wunsch, 1998), the energy input to surface waves is 60TW (Wang & Huang, 2004a) and the energy input to Ekman layer is about 3TW, including 0.5-0.7TW over the near-inertial frequency (Alford, 2003; Watanabe & Hibiya, 2002) and 2.4TW over the subinertial range (Wang & Huang, 2004b). It seems that the energy input by tropical cyclones is much smaller than that from smoothed wind field. However, it may also have a non-ignorable role in the oceanic circulation and climate. Figure 4 shows the distribution of the energy input to surface waves (induced by NCEP-NCAR wind field and tropical cyclones) averaged from 1984 to 2003. The left panel is the meridional distribution of the zonally integrated results, where the blue line is the energy input from the smoothed wind field and the red line is the total energy input. From the meridional distribution, it is readily seen that the energy generated by tropical cyclones greatly enhances the energy input at the midlatitude. In the latitudinal band from 10° to 30°N, tropical cyclones account for 22% increase of the energy, and in the western North Pacific, they account for 57% increase of energy, compared with results calculated from smoothed wind data. Although the total amount of energy input by tropical cyclones is much smaller than that by smoothed wind field, it may be more important for many applications including ecology, fishery, and environmental studies since they occur during a short time period at the midlatitude band where stratification is strong.

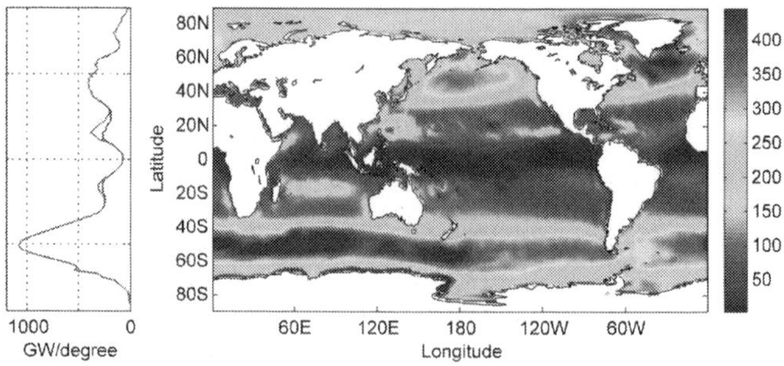

Fig. 4. (right) Distribution of energy input generated from smoothed wind field and tropical cyclones to surface waves averaged from 1984 to 2003 (units: mW m^{-2}) and (left) the meridional distribution of the zonal integrated energy source. The blue line is the energy input generated by smoothed wind stress, and the red line is the total energy input, including contributions due to tropical cyclones. (Liu et al., 2008)

Emanuel (2001) argued that subtropical cyclones are one of the strongest time-varying components in the atmospheric circulation. Accordingly, great changes in energy input to the ocean induced by tropical cyclones are expected. Figure5 shows the decadal variability of the normalized annual mean energy input to the ocean induced by tropical cyclones, the energy input to the ocean based on the NCEP-NCAR wind stress dataset (Huang et al., 2006), the normalized PDI and the normalized number of global tropical cyclones. The energy input from tropical cyclones show strong interannual and decadal variability with an increasing rate of 16% over the past 20 years, which is similar as the variability of the PDI, and the correlation coefficient is 0.92. That is, the energy input induced by tropical cyclones depends upon the strong hurricanes. Moreover, the energy input is also associated with the number of tropical cyclones in each year, and the correlation coefficient is 0.33. In addition, it can be readily seen that the energy input from tropical cyclones varies much more greatly than that from smoothed wind field, which may have an important role in the climate variability.

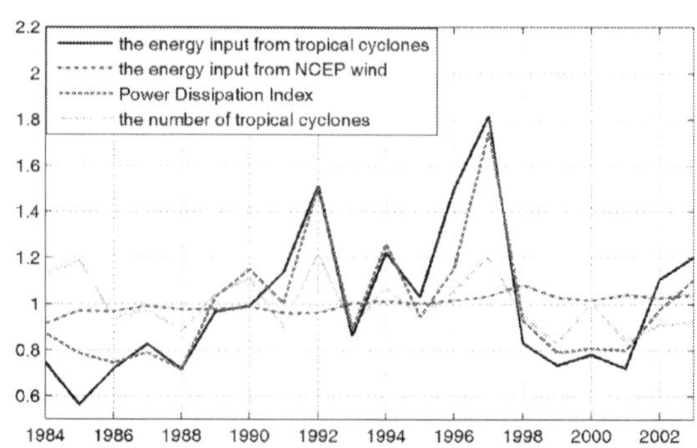

Fig.5. The normalized annual-mean energy input to surface waves from hurricane (black solid line), from the NCEP-NCAR wind stress dataset (blue dash-dot line), the normalized PDI (red dashed line) and the number of tropical cyclones (magenta dotted line).

2.3 The mixed layer deepening induced by tropical cyclones

The strong activity of tropical cyclones can deepen the mixed layer at low- and mid-latitudes. Huang et al. (2007) have shown that the mixed layer deepening at low and middle latitudes can enhance the meridional pressure difference and thus the overturning circulation and poleward heat flux, and at the same time, take less mechanical energy to support to subsurface diapycnal mixing. Then a natural question is, how much do the tropical cyclones contribute to the mixed layer deepening at low and mid-latitudes at the global scale?

Owing to the strong wind associated with tropical cyclones, mixing in the ocean is greatly enhanced, deepening the mixed layer. The mixed layer deepening for an individual tropical cyclone is defined as the difference between the initial mixed layer depth and the maximal mixed layer depth obtained from the model at a given station during the whole process of the passing through of a tropical cyclone. However, there is a possibility that several tropical cyclones passed through the same grid point within one year, each time the mixed layer deepening is denoted as dh_j. If there are N tropical cyclones that passed through this grid in one year, the total mixed layer deepening at this grid is $dH = \sum_{j=1}^{N} dh_j$, and the distribution in the world's oceans is shown in Fig.6. The maximum mixed layer deepening induced by tropical cyclones is on the order of 100m. It is readily seen that the mixed layer deepening induced by tropical cyclones accumulates at low- and mid-latitude, which may be much important for the meridional overturning circulation, and thus the climate.

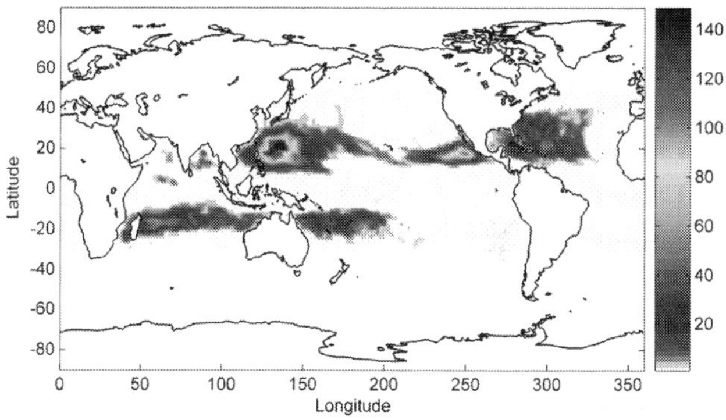

Fig.6. Annual mean (accumulated) mixed layer deepening (m) induced by tropical cyclones averaged from 1984 to 2003. (Liu et al., 2008)

In general, with the passing of a tropical cyclone, the mixed layer depth can be increased remarkably. Furthermore, after the passing through of the tropical cyclone, the mixed layer gradually relaxes back to the initial state. Woods (1985) have demonstrated that the mixed layer deepening/shoaling process can play the most important critical roles in watermass formation. The deepening of the mixed layer enables a mass exchange from the pycnocline, leading to obduction, which indicates the irreversible mass flux from the permanent pycnocline to the mixed layer; on the other hand, mixed layer retreating leaves water mass behind, leading a mass flux from the mixed layer, and thus an enhancement of subduction, which indicates the irreversible mass flux from the mixed layer to the permanet pycnocline.

Thus, the mixed layer deepening/shoaling process induced by tropical cyclones may be an important mechanism for the watermass formation/erosion. The total volume flux of mixed layer deepening induced by tropical cyclones is estimated as 39 Sv (1 Sv=10^6 m^3 s^{-1}), with 22.4 Sv in the North Pacific. Qiu and Huang (1995) discussed subduction and obduction in the oceans; they estimated that the basin-integrated subduction rate is 35.2 Sv and obduction rate 7.8 Sv in the North Pacific (10° off the equator). That is, the total rate of mixed layer deepening induced by tropical cyclones is approximately 50% of the subtropical water mass formation rate through subduction and it is much larger than obduction.

However, the volume flux of mixed layer deepening induced by tropical cyclones cannot be simply regarded as the induced subduction/obduction rate enhancement. In the study of water mass movement, the upper ocean can be divided vertically into four layers: the Ekman layer, the mixed layer, the seasonal pycnocline, and the permanent pycnocline. Water parcels entrained into the mixed layer during the passage of tropical cyclones may be originated from the seasonal pycnocline, rather than from the permanent pycnocline; similarly, water parcels released during the mixed layer retreating period may be re-entrained into the mixed layer downstream. On the other hand, the subduction/obduction rate for the stations near the passage of tropical cyclone can also be affected if the subducted/obducted water parcels pass through the typhoon region during the mixed layer deepening period, which can re-entrain the water parcels in the pycnocline to the mixed layer. However, so far how to estimate the subduction/obduction enhancement induced by tropical cyclones remains unclear. Consequently, the mixed layer deepening and shoaling process induced by tropical cyclones must have a major impact on water mass balance in the regional and global oceans, which needs our further study and water mass balance without taking into consideration of the contribution of tropical cyclones may not be acceptable.

2.4. The vertical diffusivity induced by tropical cyclones

Ocean mixing affects global climate because it is linked to the ocean's ability to store and transport heat (Wunsch & Ferrari, 2004). The winds associated with tropical cyclones are known to lead to localized mixing of the upper ocean (Price, 1981; Jacob et al., 2000; D'Asaro, 2003). Furthermore, they are important mixing agents at the global scale. Sriver & Huber (2007) estimated the vertical diffusivity induced by tropical cyclones based on the temperature data. However, the assumption that all mixing in a given year is achieved during the single largest cooling event may underestimate the vertical mixing rate induced by tropical cyclones.

In our study, the vertical diffusivity averaged over the lifetime of a tropical cyclone can be defined by the following scaling: $k = wdh$, where dh is the mixed layer deepening due to cyclone stirring, and $w = dh / T_{life}$ is the equivalent upwelling velocity averaged over the life cycle of tropical cyclone. However, for the study of oceanic general circulation, it is more appropriate to define the contribution due to each tropical cyclone in terms of the annual mean vertical diffusivity

$$k_j = w_j \cdot dh_j = dh_j^2 / T_{year} \tag{1}$$

The annual meaning (accumulated) vertical diffusivity is defined as

$$k = \sum_{j=1}^{N} k_j = \frac{1}{T_{year}} \sum_{j=1}^{N} dh_j^2 \quad (2)$$

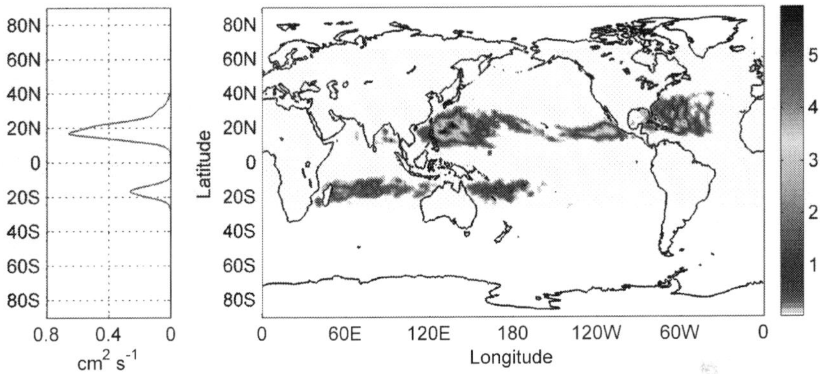

Fig. 7. Annual-mean vertical diffusivity induced by tropical cyclones from 1984 to 2003 (units: $10^{-4}\,m^2\,s^{-1}$): (right) the horizontal distribution and (left) the zonally averaged vertical diffusivity. (Liu et al., 2008)

The horizontal distribution of the tropical cyclone-induced vertical diffusivity is shown in Fig. 7. It indicates that vertical diffusivity induced by tropical cyclones is on the order of $(1\text{-}6)\times10^{-4}m^2s^{-1}$, which is approximately 10 to 100 times larger than the low environmental diffusivity $(0.05\text{-}0.1)\ \times10^{-4}m^2s^{-1}$ observed in the subtropical ocean below the mixed layer. Thus, the winds associated with tropical cyclones generate strong, near-inertial internal waves, making them efficient upper ocean mixers. Although the hurricane-induced mixing takes place in a vertical location in the water column quite different from that of deep mixing induced by tides, they all contribute to the maintenance of the oceanic general circulation and climate. Climate is especially sensitive to mixing variations in the tropical ocean (Bugnion et al., 2006) because this region of strong stratification is the most efficient place for mixing to drive strong heat transport (Scott & Marotzke, 2002; Nof & Gorder, 1999, 2000; McWilliams et al., 1996). The cyclone-induced mixing is a fundamental physical mechanism that may act to stabilize tropical temperature, mix the upper ocean, and cause amplification of climate change (Sriver & Huber, 2007).

3. How the tropical cyclone's behavior is affected by the ocean?

During the past decade or so, many researchers have suggested that because of global warming, the sea surface temperature will likely increase, which will then lead to an increase in both the number and intensity of tropical cyclones (Chan & Liu, 2004). Emanuel (2005) found that the hurricane power dissipation, which indicates the potential destructiveness of hurricanes, is highly correlated with tropical sea surface temperature, reflecting well-documented climate signals. Moreover, the total power of dissipation has increased markedly since the mid-1970s due to both longer storm lifetimes and greater storm intensities. Thus, it was proposed that the future warming may lead to an upward trend in tropical cyclone destructive potential. Sriver and Huber (2006) further proposed that a 0.25°C increase in mean annual tropical sea surface temperature corresponds roughly to a 60% increase in global Power Dissipation of tropical cyclones. Webster et al. (2005) showed that in an environment of increasing sea surface temperature, a large increase was seen in the number and proportion of hurricanes reaching categories 4 and 5 while the number of cyclones and cyclone days has decreased in all basins except the North Atlantic during the past decade. However, in the more recent work published by Emanuel et al. (2008), based on a new technique for deriving hurricane climatologies from global data, he stated that global warming should reduce the global frequency of hurricanes, though their intensity may increase in some locations.

Emanuel (2001) argued that tropical cyclones are one of the strongest time-varying components in the atmospheric circulation. The interannual variations of tropical cyclone activity received much more attention. Among these, most studies focus on the effect of the El Niño/Southern Oscillation (ENSO). ENSO affects tropical cyclones strongly in several basins, though its influence is different in each.

For the western North Pacific, during warm ENSO events, there is the eastward and equatorial shift in genesis location of tropical cyclones and longer lift span (Wang & Chan, 2002). However, due to the differences in data and technique, the results about the influence of ENSO on the frequency of tropical cyclones may be somewhat controversial. Chan (1985, 2000), Wu& Lau (1992), and others reached a common conclusion that the number of tropical storm formation over the western North Pacific is less than normal during El Niño years. However, in more recent work (Chan & Liu, 2004; Camargo & Sobel, 2005), it is shown that the mean annual tropical cyclones in the western North Pacific is higher (lower) during El Niño (La Niña) year.

For the eastern North Pacific, the formation points shifts west, more intense hurricanes are observed, and tropical cyclones track father westerward and maintain a longer lifetime in association with warm ENSO events (Schroeder & Yu, 1995; Irwin & Davis, 1999; Kimberlain, 1999) . Moreover, in the South Pacific, tropical cyclones originated farther east, resulting in more storms in the eastern South Pacific and fewer in the western South Pacific during El Niño years (Revell & Goulber, 1986).

In the North Atlantic, when El Niño is present, there are fewer and/or weaker storms and its genesis father north (Gray et al., 1993; Knaff, 1997).

4. Conclusions

Climate variability and any resulting change in the characteristics of tropical cyclones have become topics of great interest and research. As we discussed above, the climate signals, including ENSO, global warming, can greatly influence the tropical cyclone activity, including its number and intensity. On the other hand, a tropical cyclone can affect the local thermal structure and currents of the upper ocean, which has been discussed much in the previous studies. However, relatively little is known about the influence of tropical cyclone activity on the oceanic circulation and climate. Therefore, in this study we mainly focus on the role of tropical cyclones in regulating the general oceanic circulation and climate.

The tropical cyclones can influence the oceanic circulation in many aspects: 1) Tropical cyclones play an essential role in modulating, even triggering ENSO; 2) Tropical cyclones are an important mechanical energy sources required to maintain the quasi-steady oceanic circulation. 3) Mixed layer deepening induced by tropical cyclones at low and midlatitudes can enhance the meridional pressure difference and thus the overturning circulation and poleward heat flux. Moreover, the mixed layer deepening/shoaling processes induced by tropical cyclones can also affect the water mass formation/erosion, which need the further study. 4) Tropical cyclones can lead to strong localized mixing which affects global climate. Nevertheless, these studies are only the first step toward unraveling the complicated roles of tropical cyclones in the oceanic circulation and climate system. Many important questions remain unclear at the present time. For example, tropical cyclones are active only for a small fraction of the annual cycle. How do the energy input and the increase of vertical diffusivity induced by tropical cyclones over a small fraction of time in a year affect the general oceanic circulation? How do the mixed layer deepening/shoaling processes induced by tropical cyclones affect the formation/erosion, and then the subduction/obduction? To answer these questions further study is clearly needed.

5. References

Alford, M. H. (2003). Improved global maps and 54-year history of wind-work done on the ocean inertial motions. *Geophys. Res. Lett.*, 30, 1424, doi: 10.1029/2002GL016614.

Black, P. G. (1983). Ocean temperature changes induced by tropical cyclones. Ph. D. thesis. The Pennsylvania State University, 278PP.

Bugnion, V., C. Hill & P. H. Stone (2006). An adjoint analysis of the meridional overturning circulation in an ocean model. *J. Clim.*, 19, 3732-3750.

Camargo, S. J. & A. H. Sobel (2005). Western North Pacific tropical cyclone intensity and ENSO . *J. Clim.*, 18, 2996-3006.

Chan, J. C. J. (1985). Tropical cyclone activity in the northwest Pacific in relation to the El Nino/Southern Oscillation phenomenon. *Mon. Wea. Rev.*, 113, 599-606.

Chan, J.C.J. (2000). Tropical cyclone activity over the western North Pacific associated with El Nino and La Nina events. *J. Climate*, 13, 2960-2972.

Chan, J.C.J., K. S. Liu (2004). Global warming and western North Pacific typhoon activity from an observational perspective. *J. Clim.*, 17, 4590-4602.

Chang, S. W. & R. A. Anthes (1979). The mutual response of the tropical cyclone and the ocean. *J. Phys. Oceanogr.*, 9, 128-135.

D'Asaro, E. A. (2003). The ocean boundary below Hurricane Dennis. *J. Phys. Oceanogr.*, 33, 561-579.

Dong, K. (1988). El nino and tropical cyclone frequency in the Australian region and the northwest Pacific. *Aust. Meteor. Mag.*, 36, 219-225.

Emanuel, K. A. (2001). The contribution of tropical cyclones to the oceans meridional heat transport. *J. Geophys. Res.*, 106(D14), 14771-14781.

Emanuel, K. A. (2005). Increasing destrctiveness of tropical cyclones over the past 30 years. *Nature*, 436, 686-688. doi: 10.1038/nature03906.

Emanuel, K. A., R. Sundararjan & J. Williams (2008). Hurricanes and global warming: results from downscaling IPCC AR4 simulations. *Bull. Amer. Meteorol. Soc.*,347-367.

Gao. S., J. Wang & Y. Ding (1988). The triggering effect of near-equatorial cyclones on EL Nino. *Adv. Atmos. Sci.*, 5, 87-95.

Ginis, I. , K. Z. Dikinov & A. P. Khain (1989). A three dimensional model of the atmosphere and the ocean in the zone of a typhoon. *Dikl. Akad. Nauk SSSR*, 307,333-337.

Ginis, I. (1995). Ocean response to the tropical cyclone. Global Perspective on Tropical Cyclones. WMO/TD-NO.693, 198-260.

Gray, W. M., C. W. Landsea, P. W. Mielke, Jr., & K. J. Berry (1993). Predicting Atlantic basin seasonal tropical cyclone activity by 1 August. *Weather and Forecasting*, 8, 73-86.

Greatbatch, R. J. (1983). On the response of the ocean to a moving storm: The nonlinear dynamics. *J. Phys. Oceanogr.*, 13, 357-367.

Harrison, D. E., & B. S. Giese (1991). Episodes of surface westerly winds as observed from islands in the western tropical Pacific. *J. Geophys. Res.*, 96, 3221-3237.

Horel. J. D. & J. M. Wallace (1981). Planetary-scale atmospheric phenomena associated with the Southern Oscillation. *Mon. Wea. Rev.*, 109, 813-829.

Huang, R. X., W. Wang & L. L. Liu (2006). Decadal variability of wind-energy input to the world ocean. *Deep-Sea Res. II.*, 53, 31-41.

Huang, R. X., C. J. Huang & W. Wang (2007). Dynamical roles of mixed layer in regulating the meridional mass/heat fluxes. *J. Geophys. Res.*, 112, C05036, doi: 10.1029/2006JC004046.

Irwin, R. P. & R. E. Davis (1999). The relationship between the Southern Oscillation Index and tropical cyclone tracks in the eastern North Pacific. *Geophys. Res. Lett.*, 20, 2251-2254.

Jacob, S. D., L. K. Shay, A. J. Mariano & P. G. Black (2000). The 3D oceanic mixed layer response to Hurricane Gilbert. *J. Phys. Oceanogr.*, 30, 1407-1429.

Kimberlain, T. B. (1999). The effects of ENSO on North Pacific and North Atlantic tropical cyclone activity. In *preprints of the 23 rd conference on Hurricanes and Tropical Meteorology*, 250-253. Boston: American Meteorological Society.

Kindle, J. C., & P. A. Phoebus (1995). The ocean response to operational westerly wind bursts during the 1991-1992 El Nino. *J. Geophys. Res.*, 100, 4893-4920.

Knaff, J. A. (1997). Implications of summertime sea level pressure anomalies in the tropical Atlantic region. *J. Clim.*, 10, 789-804.

Leipper, D. F. (1967). Observed ocean conditions in Hurricane Hilda. *J. Atmos. Sci.*, 24,

Liu, L. L., W. Wang & R. X. Huang (2008). The mechanical energy input to the ocean induced by tropical cyclones. *J. Phys. Oceanogr.*, 38, 1253-1266.

Macdonald, A. M. & C. Wunsch (1996). The global ocean circulation and heat flux. *Nature*, 382, 436-439.

McWilliams, J. C., G. Danabasoglu & P. R. Gent (1996). Tracer budgets in the warm water sphere. *Tellus A*, 48, 179-192.

Nilsson, J. (1995). Energy flux from traveling hurricanes to the internal wave field. *J. Phys. Oceanogr.*, 25, 558-573.

Nof, D. & Van Gorder, S. (1999). A different perspective on the export of water from the south Atlantic. J. Phys. Oceanogr., 29, 2285-2302.

Nof, D. & Van Gorder, S. (2000). Upwelling into the thermocline of the Pacific Ocean. *Deep-Sea Res. I*, 47, 2317-2340.

Price, J. F. (1981). Upper ocean response to a hurricane. *J. Phys. Oceanogr.*, 11, 153-175.

Price, J. F. (1983). Internal wave wake of a moving storm. Part I: Scales, energy budget and observations. J. Phys. Oceanogr., 13, 949-965.

Qiu, B., & R. X. Huang (1995). Ventilaiton of the North Atlantic and North Pacific: Subduction Versus Obduction. *J. Phys. Oceanogr.*, 25, 2374-2390.

Revell, C. G. & S. W. Goulter (1986). South Pacific tropical cyclones and the Southern Oscillation. *Mon. Wea. Rev.*, 114, 1138-1145.

Schade, L. R. & K. A. Emanuel (1999). The ocean's effect on the intensity of tropical cyclones: Results from a simple coupled Atmosphere-Ocean Model. *J. Atmos. Sci.*, 56, 642-651.

Schroeder, T. A. & Z. P. Yu (1995). Interannual variability of central Pacific tropical cyclones. In *Preprints of the 21st conference on Hurricanes and Tropical Meteorology*, 437-439. Boston: American Meteorological Society.

Scott, J. R. & J. Marotzke (2002). The location of diapycnal mixing and the meridional overturning circulation. *J. Phys. Oceanogr.*, 32, 3578-3595.

Shay, L. K. & S. D. Jacob (2006). Relationship between oceanic energy fluxes and surface winds during tropical cyclone passage. *Atmosphere-Ocean Interactions II: Advances in Fluid Mechanics*, W. Perrie, Ed., WIT Press, Southampton, United Kingdom, 115-142.

Sobel, A. H. & S. J. Camargo (2005). Influence of western North Pacific tropical cyclones on their large-scale environment. *J. Atmos. Sci.*, 62, 3396-3407.

Sriver, R. L. & M. Huber (2006). Low frequency variability in globally integrated tropical cyclone power dissipation. *Geophys. Res. Lett.*, 33, L11705, doi: 10.1029/2006GL026167.

Sriver, R. L. & M. Huber (2007). Observational evidence for an ocean heat pump induced by tropical cyclones. *Nature*, 447, 577-580. doi: 10.1038/nature05785.

Sutyrin, G. G. & A. P. Khain (1979). Interaction of the ocean and atmosphere in the area of moving tropical cyclones. *Dokl. Akad. Nauk SSSR*, 249, 467-470.

Wang, B. & J. C. L. Chan (2002). How strong ENSO events affect tropical storm activity over the western North Pacific. *J. Clim.*, 15, 1643-1658.

Wang, W. & R. X. Huang (2004a). Wind energy input to the surface waves. J. Phys. Oceanogr., 34, 1276-1280.

Wang, W. & R. X. Huang (2004b). Wind energy input to the Ekman layer. J. Phys. Oceanogr., 34, 1267-1275.

Watanabe, M. & T. Hibiya (2002). Global estimate of the wind-induced energy flux to the inertial motion in the surface mixed layer. *Geophys. Res. Lett.*, 29, 1239, doi: 10.1029/2001GL04422.

Webster, P. J., G. J. Holland, J. A. Curry & H. R. Chang (2005). Changes in tropical cyclone number, duration and intensity in a warming environment. *Science*, 309, 1844-1846.

Withee, G. W., & A. Johnson (1976). Data report: buoy observations during Hurricane Eloise (*September 19 to October 11, 1975*), US Dep. Commer., NOAA, NSTL Station, MS

Woods, J. D. (1985). The physics of pycnocline ventilation. Coupled ocean-Atmosphere Models. J.C.J.Nihoul, Ed., Elservier Sci. Pub., 543-590.

Wu, G. & N. C. Lau (1992). A GCM simulation of the relationship between tropical-storm formation and ENSO. *Mon. Wea. Rev.*, 120, 958-977.

Wunsch, C. (1998). The work done by the wind on the oceanic general circulation. *J. Phys. Oceanogr.*, 28, 2332-2340.

Wunsch, C. & Ferrari, R. (2004). Vertical energy flux, and the general circulation of the oceans. *Annu. Rev. Fluid Mech*, 36, 281-314.

5

Possible impacts of global warming on typhoon activity in the vicinity of Taiwan

Chia Chou
Research Center for Environmental Changes, Academia Sinica, Taipei, Taiwan
Department of Atmospheric Sciences, National Taiwan University, Taipei Taiwan
Jien-Yi Tu
Department of Atmospheric Sciences, Chinese Culture University, Taipei Taiwan
Pao-Shin Chu
Department of Meteorology, SOEST, University of Hawaiiat Manoa, Honolulu Hawaii

1. Introduction

Typhoons are one of the most extreme natural events over the western North Pacific–East Asian (WNP–EA sector). Typhoons often affect the spatial distribution of regional precipitation in summer since they are a major source of rainfall over this region. For example, in 2004, 10 typhoons occurred in Japan and brought more than usual precipitation, causing widespread damage, whereas drought occurred in the Philippines and southern China (Kim et al., 2005; Levinson et al., 2005; Wu et al., 2005). Typhoon-related climate studies often focus on the variation of typhoon intensity, frequency, and track in multitemporal scales ranging from intraseasonal to interdecadal (Chan, 1985, 2000; Chia and Ropelewski, 2002; Chu, 2004; Ho et al., 2006; Matsuura et al., 2003; Wang and Chan, 2002). In recent years, the influence of global warming on the intensity of tropical cyclones has received much attention. Emanuel (2005), Hoyos et al. (2006), and Webster et al. (2005) found increasing trends in the western Pacific and Atlantic based on some available best-track datasets. Chan and Liu (2004) and Klotzbach (2006) found small or no trends using alternate analysis techniques. Other studies (e.g., Landsea et al., 2006) have shown opposite trends to those found by Emanuel in the west Pacific by examining other best-track datasets. Besides observations, model simulations also show that intense tropical cyclones will be more frequent in the future warmer climate, while the total number of tropical cyclones tends to decrease (Bender et al., 2010; Emanuel et al., 2008; Zhao et al., 2009).

In addition to the increase in tropical cyclone intensity, one study also found a detectable shift of the typhoon track over the WNP–EA in the past four decades (Wu et al., 2005). Such a change also affects regional precipitation (Ren et al., 2006). For a future climate projection, the typhoon track over the WNP–EA region may potentially be affected by global warming (Wu and Wang, 2004). In our study, we also focus on the variation of typhoon tracks, particularly on the abrupt change of typhoon tracks from a historical perspective. Observational data for Taiwan are useful for studying variations of typhoon tracks over the WNP-EA region because of the island's unique location. Taiwan is located at the turning point of the track for most typhoons in the WNP-EA region (Camargo et al., 2007). Figure 1 shows two major typhoon paths over the WNP-EA region: one is moving westward to the South China Sea directly and the other is turning north to either Japan or Korea. Taiwan is located just between these two major tracks, so the number of typhoon landfalls on Taiwan is sensitive to the shift of the typhoon track. Thus, we use the number of typhoons that passed through the vicinity of Taiwan (21°-26°N, 119°-125°E) as an index to examine the variation of the typhoon track in the WNP-EA region. The results here are mainly from Tu et al. (2009), along with a discussion of global warming impacts on typhoon track. The data and statistical methods used in this study are briefly discussed in section 2. We first identified the abrupt shift of the typhoon track in section 3 and then discussed its association with large-scale environmental changes in section 4. A possible association of this change in typhoon track with global warming was discussed in section 5, followed by a discussion and conclusions.

2. Data, statistical methods, and climate model

2.1 Data

The number of typhoons in the vicinity of Taiwan from 1970 to 2006 is provided by the Central Weather Bureau (CWB) in Taiwan (Chu et al., 2007). Independently, the typhoon-track information for the period of 1970-2009 is obtained from the Regional Specialized Meteorological Center (RSMC) Tokyo—Typhoon Center. Here we defined the maximum surface wind over 34 kt as a typhoon case. To understand the influence of the large-scale environment on typhoon activity, the following two global datasets were analyzed: (1) a monthly optimum interpolation (OI) sea surface temperature (SST) with 1° spatial resolution from January 1982 to December 2009 (Reynolds et al., 2002), and (2) other large-scale variables, such as the geopotential height and wind field, are derived from (1979–2009) the National Centers for Environmental Prediction/Department of Energy (NCEP/DOE) Reanalysis 2 (NCEP-2) data (Kanamitsu et al., 2002), with a horizontal resolution of 2.5° latitude × 2.5° longitude.

2.2 Statistical Methods

To detect abrupt shifts in tropical cyclone records, we use a Bayesian changepoint analysis (Chu and Zhao, 2004; Zhao and Chu, 2006). Because typhoon occurrence in the study domain is regarded as a rare event, a Poisson process is applied to provide a reasonable representation of typhoon frequency. Poisson process is governed by a single parameter: the Poisson intensity. A detail description can be found in Tu et al. (2009).

Besides tropical cyclones, it is also of interest to investigate whether there is any change point in the SST records or typhoon passage frequency series. Since these variables do not follow a Poisson process, we use a different method to detect abrupt shifts in the temperature or passage frequency series: a log-linear regression model in which a step function is expressed as an independent variable is adopted. If the estimated slope is at least twice as large as its standard error, one would reject the null hypothesis (i.e., slope being zero) at the 5% significance level. This model is similar to that used in Elsner et al. (2000) and Chu (2002).

To evaluate the difference in the mean circulation between two samples, we use a classic nonparametric test, known as the Wilcoxon-Mann-Whitney test (Chu, 2002). To perform this test, the two data batches need to be pooled together and ranked. The null hypothesis assumes that the two batches come from the same distribution. The details can also be found in Tu et al. (2009).

To calculate the trends of SST and typhoon activity in 1982-2009 and 1970-2009 respectively, these variables are computed for each typhoon season by a rank regression method, i.e., minimizing a product of usual variable times its rank in a centered ranking system (Hollander and Wofle, 1999; Neelin et al., 2006). A Spearman-rho test is used to examine the statistical significance of the trend.

2.3 Model

A coupled ocean-atmosphere-land model of intermediate complexity (Neelin and Zeng, 2000; Zeng et al., 2000) is used in this study. Based on the analytical solutions derived from the Betts-Miller moist convective adjustment scheme (Betts and Miller, 1993), typical vertical structures of temperature, moisture, and winds for deep convection are used as leading basis functions for a Galerkin expansion (Neelin and Yu, 1994; Yu and Neelin, 1994). The resulting primitive-equation model makes use of constraints on the flow by quasi-equilibrium thermodynamic closures and is referred to as the quasi-equilibrium tropical circulation model with a single vertical structure of temperature and moisture for deep convection (QTCM1). Because the basic functions are based on vertical structures associated with convective regions, these regions are expected to be well represented and similar to a general circulation model (GCM) with the Betts-Miller moist convective adjustment scheme. Instead of coupling a complicated ocean general circulation model, a slab mixed layer ocean model with a fixed mixed layer depth of 50 m is used. By specifying Q flux, which crudely simulates divergence of ocean transport (Hansen et al., 1988, 1997), SST can be determined by the energy balance between surface radiative flux, latent heat flux, sensible heat flux, and Q flux. The Q flux can be obtained from observations or ocean model results (Doney et al., 1998; Keith, 1995; Miller et al., 1983; Russell et al., 1985). In general, the Q flux varies from ocean to ocean as well as from season to season. QTCM version 2.3 is used here, with the solar radiation scheme slightly modified.

3. Abrupt shift of typhoon track

A small area of 21°-26°N, 119°-125°E is defined as the vicinity of Taiwan (Chu et al., 2007). If a typhoon passes through this area, we count it as one that influences Taiwan. Figure 2 shows that a majority (over 90%) of typhoons pass through this region in June-October (JJASO), which is a typical typhoon season over the WNP-EA region (Chu et al., 2007). Thus, we count the number of typhoons only for these months of each year, instead of for the entire 12 months. Figure 3a is the time series of seasonal (JJASO) typhoon numbers in the vicinity of Taiwan from 1970 to 2006. The interannual variation is relatively small before 1982, but becomes much stronger after 1982. An increase of the number of typhoons in recent years is also evident. Figure 3b displays the posterior probability mass function of the changepoint plotted as a function of time (year). High probability on year i implies a more likely change occurring, with year i being the first year of a new epoch, or the so-called changepoint. In Fig. 3b, we note a great likelihood of a changepoint on the typhoon rate in 2000. The average typhoon rate is 3.3 yr^{-1} during the first epoch (1970–99), but increased to 5.7 yr^{-1} during the second epoch (2000–06), thus almost doubling from the first to the second. Because there are two parameters to estimate for each epoch, given a changepoint, the minimum sample size is two. To achieve robust results, however, Chu and Zhao (2004) and Zhao and Chu (2006) suggested the use of 5 yr at each end of the dataset to estimate the prior parameters. Because the data observed after the changepoint year in 2000 are quite consistent, showing higher typhoon counts relative to the first epoch, we feel that our approach and the sample size (n=7) are appropriate to establish a consistent estimation for the "after changepoint" period. Separately, a nonparametric test also shows that the difference of the mean of typhoon numbers (Fig. 3a) between the two epochs is significant at the 5% level. To further substantiate the abrupt shift in typhoon activity, Fig. 3c displays the posterior density function of the rate parameter before and after the changepoint year. The posterior distribution represents a combination of the prior distribution and the likelihood function. Note the very little overlapping areas in the tail areas between two posterior distributions in Fig. 3c, supporting the notion of a rate increase from the first epoch to the second.

To understand the association of the variation of typhoon number over the vicinity of Taiwan with the spatial distribution of typhoon activity over a larger area, such as the entire WNP–EA region, the frequency of the typhoon occurrence is counted for each 6-h interval for each grid box of 2.5°× 2.5° (Stowasser et al., 2007; Wu and Wang, 2004). Figure 3d shows the difference in the mean typhoon rate between the two epochs, that is, the period of 2000–06 minus the period of 1970–99. The spatial distribution exhibits a pronounced increase of typhoon frequency north of 20°N and a decrease south of 20 °N over the western part of the WNP–EA region (100°–140°E), implying a northward shift of the typhoon frequency since 2000. The area with a significance level at the 5% in Fig. 3d (northeast of the Taiwan vicinity) is slightly different from the vicinity of Taiwan defined in Fig. 3a. This is because of the way typhoon counts constructed in Fig. 3a are somewhat different from the typhoon frequency in Fig. 3d, which contains mixed information of typhoon numbers and typhoon translation speed. Examining the total number of typhoon formations over the entire WNP, no noticeable changes are found. However, on a regional scale, we did find a reduction of the formation number over the South China Sea and the Philippine Sea, but little change over the vicinity of Taiwan (not shown). Overall, this suggests that the typhoon track over the western part of the WNP–EA region has shifted northward from the South China Sea and the Philippine Sea toward the vicinity of Taiwan and the East China Sea since 2000. Thus, the increased typhoon number over the vicinity of Taiwan after 2000 (Fig. 3a) is not an isolated local feature, but is consistent with the northward shift of the typhoon track over the WNP–EA region.

We further examined the variation of the typhoon frequency in three subregions of the WNP–EA region with large changes of typhoon frequency (Fig. 4): the South China Sea (15°–20°N, 110°–120°E), the Philippine Sea (15°–20°N, 120°–130°E), and the Taiwan–East China Sea region (25°–30°N, 120°–130°E). Over the South China Sea (Fig. 4a), a clear interannual variation and a downward linear trend of typhoon frequency, which is obtained from a simple best-fit (least squares) method, are found during the past 37 yr (1970–2006). The linear trend is consistent with the findings of Wu et al. (2005). This downward trend is related to a reduction in the number of typhoon formations over the South China Sea. Over the Philippine Sea (Fig. 4b), the variation is more likely characterized by a decadal variation with a phase shift around 1975, 1985, and the late 1990s. Over the Taiwan-East China Sea region (Fig. 4c), it shows a similar variation of typhoon frequency to that in Fig. 3a, with a statistically significant shift occurring in 2000 at the 5% level, as confirmed by the classical changepoint analysis described in section 2b. In other words, the abrupt shift of the typhoon frequency in the vicinity of Taiwan is a part of the change over the Taiwan-East China Sea region (Figs. 3a and 4c), not associated with the changes in the South China Sea and the Philippine Sea.

4. The association with the large-scale pattern

To understand possible causes for such a northward shift of the typhoon track, we examine variations of the large-scale environment, such as the westward extension of the Pacific subtropical high ridge between two epochs. Because the SST data start in 1982, the period of 1982–99 is used to represent the first epoch, instead of 1979–99, for consistency in analyzing the spatial distribution of large-scale variables. We note that the difference of the analyzed periods between the typhoon data (e.g., Figs. 3 and 4) and the large-scale variables (Figs. 5 and 6) may cause an inconsistency. Moreover, the small sample size of the second epoch may also create some uncertainties in the composite analysis. However, the composite analysis may yield insight to the possible mechanisms that induce the abrupt shift of the typhoon number in the Taiwan vicinity.

The ridge of the Pacific subtropical high in the WNP–EA region is the main steering flow controlling the typhoon track (Ho et al., 2004; Wu et al., 2005). Figure 5 shows the 500-hPa geopotential height in the first (dotted line) and second epochs (solid line). The 5880-gpm contour is commonly used to represent the variation of the subtropical high ridge over the WNP–EA region (e.g., Chang et al., 2000). However, the 5875-gpm contour is used in this study because it is closer to the vicinity of Taiwan than the 5880-gpm contour. The results discussed below are not sensitive to the contour that we chose. Averaged over the entire typhoon season (JJASO), the subtropical ridge tends to retreat eastward from the first to the second epoch (Fig. 5a). Because of this possible weakening of the subtropical high, the typhoon track during the second epoch tends to move a little more northward than those in the first epoch.

We further examined the subseasonal variation of the subtropical high over this region since the subtropical high over this region experiences a strong subseasonal variation (e.g., LinHo and Wang, 2002). According to LinHo and Wang (2002), the entire typhoon period (JJASO) is dominated by three major natural periods in the WNP–EA region: June (the first period of summer), July–September (JAS; the second period of summer), and October (early fall). In June, the subtropical high during the second epoch tends to retreat eastward relative to the first epoch (Fig. 5b). In this period, most typhoons move westward into the South China Sea because of the strong westward extent of the subtropical high. The area where the height difference between two epochs is statistically significant is observed over the Taiwan Philippine region. When the subtropical high retreats eastward, such as shown in Fig. 5b, the typhoon track tends to move northward. In JAS, on the other hand, the subtropical high moves northward slightly during the second epoch (Fig. 5c). This period is the peak phase of the typhoon season over the vicinity of Taiwan, which has the most typhoons passing through (Fig. 2). In this period, the subtropical high tends to move northward, so most typhoons move straight to Taiwan or turn northward to Japan and Korea. Thus, the possible northward retreat of the subtropical high (Fig. 5c) also favors the typhoon track shifting northward. In October, unlike the other two periods discussed before, the subtropical high actually becomes stronger and extends more westward (Fig. 5d), so it is not favorable for typhoons moving northward. However, the typhoon frequency in this period is lower than those in the other two periods (Chu et al., 2007). Overall, the eastward retreat of the subtropical high in JJASO, which is associated with the northward shift of typhoon track, could be a result of change of the subtropical high in summer, that is, June–September (JJAS). We note that the differences of the 500-hPa geopotential height over the western North Pacific in the peak season (JAS) are not statistically significant, possibly resulting from the small sample size of the second epoch, so the weakening of the subtropical high should be examined in the future when the second epoch becomes long enough.

We also examined the changes of other large-scale variables between two epochs. Figure 6a shows the change of low-level winds at 850 hPa. A cyclonic circulation anomaly is found between 30°N and the equator over the WNP–EA region during the second epoch, which implies a strengthening of the Asian summer monsoon trough, a favorable condition for typhoon activity. In Fig. 6b, positive and statistically significant vorticity anomalies are also found over the abovementioned region, including Taiwan, which supports the notion of an enhanced Asian summer monsoon trough. Thus, the strengthening of the Asian summer monsoon trough is accompanied by the weakening or eastward retreat of the subtropical high shown in Fig. 5a. We note that positive low-level vorticity anomalies may also favor tropical cyclone genesis. However, it is not a dominant factor here because the number of typhoon formations is decreased over the South China Sea and the Philippine Sea. Vertical wind shear is another factor that may affect typhoon activity (Chia and Ropelewski, 2002; Frank and Ritchie, 2001; Gray, 1979; Kurihara and Tuleya, 1981). Figure 6c does show an increase of vertical wind shear in the region of 0°–15°N, 105°–150°E, which is unfavorable for typhoon formation and development, but this increase of vertical wind shear is not statistically significant. On the other hand, to the north of 15°N in the area where the low-level vorticity anomalies are positive and the cyclonic circulation anomaly dominates, the vertical wind shear has not changed appreciably, so this condition is at least not unfavorable for typhoon activity.

Another important environmental condition for tropical cyclone activity is SST. Warmer SST tends to create a favorable thermodynamic condition for tropical cyclones through the air-sea heat flux exchange (Emanuel, 1999). In the North Atlantic, it is well established that SST is one of the factors impacting the number and severity of tropical cyclones (Landsea et al., 1998; Shapiro, 1982; Shapiro and Goldenberg, 1998). Thus, SST is examined here. In Fig. 6d, warm SST anomalies are found over the equatorial region from 120°E to 120°W. However, only the warm SST anomalies west of the date line are statistically significant. Another region with warm SST anomalies is in the North Pacific region from 30° to 50°N, which is associated with a negative phase of the Pacific decadal oscillation (Mantua et al., 1997). Both regions are away from the region with strong typhoon activity. However, SST anomalies can alter large-scale atmospheric circulation, and thus affect the typhoon track.

To understand which regions of the warm SST anomalies are responsible for creating the favorable conditions of typhoon activity over the WNP–EA region, a climate model with an intermediate complexity (Neelin and Zeng, 2000; Zeng et al., 2000) is used. In these simulations, the warm SST anomalies are prescribed separately in the equatorial region (5°S–5°N, 130°–175°E) and the midlatitude region (25°–45°N, 140°E–120°W), while a mixed layer ocean is used outside those two regions. The experiment design is similar to that used in Lau and Nath (2000). The results shown in Fig. 7 are averages of 10 ensemble runs. Given the warm SST anomalies in the equatorial region, low-level cyclonic circulation anomalies are noted over the WNP-EA region (Fig. 7a). This is a typical Rossby wave response to the equatorial warm SST anomalies (e.g., Gill, 1980; Wang et al, 2003). This pattern over the WNP is qualitatively similar to that in Fig. 6a. In contrast, the warm SST anomalies over the midlatitude region have little bearing on the tropical circulation over the WNP (Fig. 7b). We note that the simulated low-level wind anomalies over the midlatitude Pacific (Fig. 7) are different than the observation shown in Fig. 6a. Thus, both the equatorial and midlatitude warm SST anomalies are not responsible for the wind anomalies over midlatitude Pacific, which may be contributed to by some complicated effects. Accordingly, the warm SST anomalies over the equatorial western and central Pacific are postulated as a possible cause for inducing the favorable conditions for the northward shift of the typhoon track. We then examine the temporal variation of the SST over the equatorial region.

Figure 8 shows the time series of monthly SST anomalies, with the annual cycle removed, along the equatorial western and central Pacific (5°S–5°N, 130°–175°E) from 1982 to July 2007. The SST anomalies exhibit an interannual variation, which is roughly related to the three strongest El Niño events of the twentieth century: 1982/83, 1991/92, and 1997/98, a sharp reduction of the SST anomalies from the growing year (the year before the El Niño peak phase, such as 1997) to the decaying year (the year after the El Niño peak phase, such as 1998). We used the regression model discussed in Chu (2002) to analyze the variation of the monthly SST anomalies. A statistically significant shift occurs in August 2000 (Fig. 8). Specifically, before August 2000, the average of the SST anomalies is about -0.1 °C, but it increases to 0.3°C after August 2000. Although this difference seems to be small, it is considerable when viewed in the context of multiyear seasonal means at the equator. In the interannual time scale, the typhoon number over the vicinity of Taiwan is also strongly associated with the SST anomalies over the equatorial western and central Pacific (C.-T. Lee, 2008, personal communication).We also examined the six typhoons that passed through the vicinity of Taiwan in 2000 and found that five out of six typhoons occurred in or after August 2000.After further examining the SST anomalies in 2007, we found that warm SST anomalies still exist over the equatorial western and central Pacific (Fig. 8), which is consistent with the above-normal number of typhoons (five) that we found so far in 2007. Thus, combining results from observations, especially in Figs. 3 and 8, and model simulations, the warm SST anomalies over the equatorial western to central Pacific appear to be main factors that induced the abrupt change of the northward shift of the typhoon track over the WNP-EA region.

5. Impacts of global warming

Global warming has been observed for more than one hundred years (e.g., Trenberth et al., 2007). Under global warming, El Niño-like warm SST anomalies are often found over the equatorial Pacific region (Meehl and Washington, 1996; Meehl et al., 2007; Teng et al., 2006). In this study, we also found that the mean state of the equatorial Pacific SST transitioned from a cold to a warm phase in 2000. The question that arises is as follows: Do the equatorial

SST anomalies that are associated with the abrupt northward shift of the typhoon track over the WNP-EA region result from global warming?

Figure 9 shows the variation of globally averaged SST in 1982-2009. A clear upward trend of SST is found, which is consistent to the global warming found in the previous studies (e.g., Trenberth et al., 2007). The spatial distribution of the SST trend is shown in Figure 10. This clearly resembles the SST anomalies shown in Fig. 6d. In other words, the SST anomalies that induce the abrupt shift of the typhoon track are similar to the warming trend of SST over the past three decades. Thus, it implies that global warming could be at least partially responsible for the abrupt shift of the typhoon track over the western North Pacific. We further examined the trend of typhoon activity in 1970-2009 and found decreasing trends over the South China Sea and slightly increasing trends north of Taiwan (Fig. 11). We note that the decreasing trends are statistically significant, but the increasing trends are not. The decreasing trend is similar to the pattern found in Fig. 3d, while the increasing trend is much weaker than that shown in Fig. 3d even though the tendency is same. It is consistent with the results shown in Fig. 4: a clear decreasing trend of typhoon activity over the South China Sea (Fig. 4a) and an abrupt increase of typhoon activity north of Taiwan (Fig. 4c). Overall, the warming trend of SST does show an influence on the abrupt shift of the typhoon track over the western North Pacific.

6. Discussion and conclusion

Long-term climate variability of typhoon activity over the western North Pacific-East Asian (WNP-EA) sector has been analyzed here. Because Taiwan is located at a unique location along the typhoon path, the number of typhoons that pass through the vicinity of Taiwan is used to study the movement of the typhoon track over this region. Our analysis suggests an abrupt change in typhoon numbers in 2000, which is consistent with the northward shift of the typhoon track over the WNP-EA region and the abrupt increase of typhoon frequency over the Taiwan-East China Sea region. This northward shift of the typhoon track is mainly associated with warm SST anomalies over the equatorial western and central Pacific, which exhibited an abrupt change in August 2000, concurrent with the change of the typhoon track. Both observations and model simulations show that the equatorial warm SST anomalies tend to induce an eastward and northward retreat of the subtropical high, an enhanced low-level vorticity, and a monsoon trough, which all favor the northward shift of the typhoon track.

A further examination shows that the equatorial warm SST anomalies are partially associated with a global warming trend of SST. Is this global warming trend of SST in 1982-2009 induced by anthropogenic forcings, such as the greenhouse effect? This is an interesting question that should be answered in the future.

7. References

Bender, M. A.; Knutson, T. R.; Tuleya, R. E.; Sirutis, J. J.; Vecchi, G. A.; Garner, S. T. & Held, I. M. (2010). Modeled impact of anthropogenic warming on the frequency of intense atlantic hurricanes. *Science*, 327, 454-458

Betts, A. K. & Miller, M. J. (1993). The Betts-Miller scheme. *The Representation of Cumulus Convection in Numerical Models of the Atmosphere, Meteor. Monogr.*, No. 46, Amer. Meteor. Soc., 107-121

Camargo, S. J.; Robertson, A. W.; Gaffney, S. J.; Smyth, P. & Ghil, M. (2007). Cluster analysis of typhoon tracks. Part I: General properties. *J. Climate*, 20, 3635-3653

Carlin, B. P. & Louis, T. A. (2000). *Bayes and Empirical Bayes Methods for Data Analysis.* Chapman & Hall/CRC

Chan, J. C. L. (1985). Tropical cyclone activity in the northwest Pacific in relation to the El Niño/Southern Oscillation phenomenon. *Mon. Wea. Rev.*, 113, 599-606

Chan, J. C. L. (2000). Tropical cyclone activity over the western North Pacific associated with El Niño and La Niña events. *J. Climate*, 13, 2960-2972

Chan, J. C. L. & Liu, K. S. (2004). Global warming and western North Pacific typhoon activity from an observational perspective. *J. Climate*, 17, 4590-4602

Chang, C.-P.; Zhang, Y. & Li, T. (2000). Interannual and interdecadal variations of the East Asian summer monsoon and tropical Pacific SSTs. Part I: Roles of the subtropical ridge. *J. Climate*, 13, 4310-4325

Chia, H.-H. & Ropelewski, C. F. (2002). The interannual variability in the genesis location of tropical cyclones in the northwest Pacific. *J. Climate*, 15, 2934-2944

Chu, P.-S. (2002). Large-scale circulation features associated with decadal variations of tropical cyclone activity over the central North Pacific. *J. Climate*, 15, 2678-2689

Chu, P.-S. (2004). ENSO and tropical cyclone activity. *Hurricanes and Typhoons: Past, Present, and Potential,* R. J. Murnane and K. B. Liu, Eds., Columbia University Press, 297-332

Chu, P.-S. & Zhao, X. (2004). Bayesian change-point analysis of tropical cyclone activity: The central North Pacific case. *J. Climate*, 17, 4893-4901

Chu, P.-S.; Zhao, X.; Lee, C.-T. & Lu, M.-M. (2007). Climate prediction of tropical cyclone activity in the vicinity of Taiwan using the multivariate least absolute deviation regression method. *Terr. Atmos. Oceanic Sci.*, 18, 805-825

Doney, S. C.; Large, W. G. & Bryan, F. O. (1998). Surface ocean fluxes and water-mass transformation rates in the coupled NCAR climate system model. *J. Climate*, 11, 1420-1441

Elsner, J. B. & Jagger, T. (2004). Ahierarchical Bayesian approach to seasonal hurricane modeling. *J. Climate*, 17, 2813-2827

Emanuel, K. A. (1999). Thermodynamic control of hurricane intensity. *Nature*, 401, 665-669

Emanuel, K. A. (2005). Increasing destructiveness of tropical cyclones over the past 30 years. *Nature*, 436, 686-688

Emanuel, K. A.; Sundararajan, S. & Williams, J. (2008). Hurricanes and global warming: Results from downscaling IPCC AR4 simulations. *Bull. Amer. Meteor. Soc.*, 89, 347-367

Frank, W. M. & Ritchie E. A. (2001). Effects of vertical wind shear on the intensity and structure of numerically simulated hurricanes. *Mon. Wea. Rev.*, 129, 2249-2269

Gill, A. E. (1980). Some simple solutions for heat induced tropical circulation. *Quart. J. Roy. Meteor. Soc.*, 106, 447-462

Gray, W. M. (1979). Hurricanes: Their formation, structure and likely role in the general circulation. *Meteorology over the Tropical Oceans.* D. B. Shaw, Ed., Royal Meteorological Society, 155-218

Hansen, J.; Fung, I.; Lacis, A.; Rind, D.; Lenedeff, S.; Ruedy R. & Russell, G. (1988). Global climate changes as forecast by Goddard Institute for Space Studies three-dimensional model. *J. Geophys. Res.*, 93D, 9341–9364

Hansen, J.; Sato, M. & Ruedy, R. (1997). Radiative forcing and climate response. *J. Geophys. Res.*, 102, 6831–6864

Ho, C.-H.; Baik, J.-J.; Kim, J.-H.; Gong, D.-Y. & Sui, C. H. (2004). Interdecadal changes in summertime typhoon tracks. *J. Climate*, 17, 1767–1776

Ho, C.-H.; Kim, J.-H.; Jeong, J.-H.; Kim, H.-S. & Chen, D. (2006). Variation of tropical cyclone activity in the south Indian Ocean: El Niño–Southern Oscillation and Madden-Julian Oscillation effects. *J. Geophys. Res.*, 111, D22101, doi:10.1029/2006JD007289

Hollander, M. & Wolfe, D. A. (1999). *Nonparametric Statistical Methods*, 2nd ed., John Wiley, New York

Hoyos, C. D.; Agudelo, P. A.; Webster, P. J. & Curry, J. A. (2006). Deconvolution of the factors contributing to the increasing global hurricane intensity. *Science*, 312, 94–97

Kanamitsu, M.; Ebisuzaki, W.; Woollen, J.; Yang, S.-K.; Hnilo, J. J.; Fiorino, M. & Potter, G. L. (2002). NCEP-DOE AMIP-II Reanalysis (R-2). *Bull. Amer. Meteor. Soc.*, 83, 1631–1643

Keith, D. W. (1995). Meridional energy transport: Uncertainty in zonal means. *Tellus*, 47A, 30–44

Kim, J.-H.; Ho, C.-H., & Sui, C.-H. (2005). Circulation features associated with the record-breaking typhoon landfall on Japan in 2004. *Geophys. Res. Lett.*, 32, L1472, doi:10.1029/2005GL022494

Klotzbach, P. J. (2006). Trends in global tropical cyclone activity over the past twenty years (1986–2005). *Geophys. Res. Lett.*, 33, L10805, doi:10.1029/2006GL025881

Kurihara, Y. & Tuleya, R. E. (1981). A numerical-simulation study on the genesis of a tropical storm. *Mon. Wea. Rev.*, 109, 1629–1653

Landsea, C. W.; Bell, G. D.; Gray, W. M. & Goldenberg, S. B. (1998). The extremely active 1995 Atlantic hurricane season: Environment condition and verification of seasonal forecasts. *Mon. Wea. Rev.*, 126, 1174–1193

Landsea, C. W.; Harper, B. A.; Hoarau, K. & Knaff J. A. (2006). Can we detect trends in extreme tropical cyclones? *Science*, 313, 452–454

Lau, N.-C. & Nath, M. J. (2000). Impact of ENSO on the variability of the Asian-Australian monsoons as simulated in GCM experiments. *J. Climate*, 13, 4287–4309

Levinson, D. H. & Coauthors (2005). State of the climate in 2004. *Bull. Amer. Meteor. Soc., 86*, S1-S86

LinHo & Wang B. (2002). The time–space structure of the Asian–Pacific summer monsoon: A fast annual cycle view. *J. Climate*, 15, 2001–2019

Mantua, N. J.; Hare, S. R.; Zhang, Y.; Wallace, J. M. & Francis R. C. (1997). A Pacific interdecadal climate oscillation with impacts on salmon production. *Bull. Amer. Meteor. Soc.*, 78, 1069–1079

Matsuura, T.; Yumoto, M. & Iizuka, S. (2003). A mechanism of interdecadal variability of tropical cyclone activity over the western North Pacific. *Climate Dyn.*, 21, 105–117

Meehl, G. A. & Washington, W. M. (1996). El Niño-like climate change in a model with increased atmospheric CO2 concentrations. *Nature*, 382, 56–60

Meehl, G. A. & Coauthors (2007). Global climate projection. *Climate Change 2007: The Physical Science Basis*, S. Solomon et al. Eds., Cambridge University Press, 747–845

Miller, J. R.; Russell, G. L. & Tsang, L.-C. (1983). Annual oceanic heat transports computed from an atmospheric model. *Dyn. Atmos. Oceans*, 7, 95–109

Neelin, J. D. & Yu, J.-Y. (1994). Modes of tropical variability under convective adjustment and the Madden-Julian oscillation. Part I: Analytical theory. *J. Atmos. Sci.*, 51, 1876–1894

Neelin, J. D. & Zeng, N. (2000). A quasi-equilibrium tropical circulation model—Formulation. *J. Atmos. Sci.*, 57, 1741–1766

Neelin, J. D.; Münnich, M.; Su, H.; Meyerson, J. E. & Holoway, C. E. (2006). Tropical drying trends in global warming models and observations. *Proc. Natl. Acad. Sci. U. S. A.*, 103, 6110-6115

Ren, F.; Wu, G.; Dong, W.; Wang, X.; Wang, Y.; Ai, W. & Li, W. (2006). Changes in tropical cyclone precipitation over China. *Geophys. Res. Lett.*, 33, L20702, doi:10.1029/2006GL027951

Reynolds, R. W.; Rayner, N. A.; Smith, T. M.; Stokes, D. C. & W. Wang, W. (2002). An improved in situ and satellite SST analysis for climate. *J. Climate*, 15, 1609–1625

Russell, G. L. ; Miller, J. R. & Tsang, L.-C. (1985). Seasonal oceanic heat transports computed from an atmospheric model. *Dyn. Atmos. Oceans*, 9, 253–271

Shapiro, L. J. (1982). Hurricane climate fluctuations. Part II: Relation to large-scale circulation. *Mon. Wea. Rev.*, 110, 1014–1040

Shapiro, L. J. & Goldenberg, S. B. (1998). Atlantic sea surface temperature and tropical cyclone formation. *J. Climate*, 11, 578–590

Stowasser, M.; Wang, Y. & Hamilton, K. (2007). Tropical cyclone changes in the western North Pacific in a global warming scenario. *J. Climate*, 20, 2373–2396

Teng, H.; Buja, L. E. & Meehl, G. A. (2006). Twenty-first-century climate change commitment from a multi-model ensemble. *Geophys. Res. Lett.*, 33, L07706, doi:10.1029/2005GL024766

Trenberth, K. E. & Coauthor (2007). Observations: Surface and atmospheric climate change. *Climate Change 2007: The Physical Science Basis*, S. Soloman et al., Eds., Cambridge University Press, 235-336

Tu, J.-Y.; Chou, C. & P.-S. Chu (2009). The abrupt shift of typhoon activity in the vicinity of Taiwan and its association with western North Pacific-East Asian climate change. J. Climate, 22, 3617-3628

Wang, B. & Chan, J. C. L. (2002). How ENSO regulates tropical storm activity over the western North Pacific. *J. Climate*, 15, 1643–1658

Wang, B.; Wu, R. & Li, T. (2003). Atmosphere-warm ocean interaction and its impacts on Asian-Australian monsoon variation. *J. Climate*, 16, 1195–1211

Webster, P. J.; Holland, G. J.; Curry, J. A. & Chang, H. R. (2005). Changes in tropical cyclone number, duration and intensity in a warm environment. *Science*, 309, 1844–1846

Wu, L. & Wang, B. (2004). Assessing impacts of global warming on tropical cyclone tracks. *J. Climate*, 17, 1686–1698

Wu, L.; Wang, B. & Geng, S. (2005). Growing typhoon influence on east Asia. *Geophys. Res. Lett.*, 32, L18703, doi:10.1029/2005GL022937

Yu, J.-Y. & Neelin, J. D. (1994). Modes of tropical variability under convective adjustment and the Madden-Julian oscillation. Part II: Numerical results. *J. Atmos. Sci.*, 51, 1895–1914

Zeng, N.; Neelin, J. D. & Chou, C. (2000). The first quasiequilibrium tropical circulation model-implementation and simulation. *J. Atmos. Sci.*, 57, 1767-1796

Zhao, M.; Held, I. M.; Lin, S.-J. & Vecchi G. A. (2009). Simulations of global hurricane climatology, interannual variability, and response to global warming using a 50km resolution GCM, J. Climate, 22, 6653-6678

Zhao, Z. & Chu, P.-S. (2006). Bayesian multiple change point analysis of hurricane activity in the eastern North Pacific: A Markov chain Monte Carlo approach. *J. Climate*, 19, 564-578

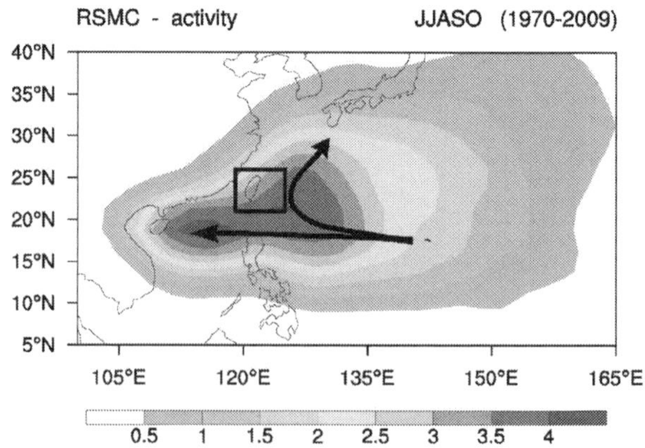

Fig. 1. June-October (JJASO) typhoon frequency climatology averaged over the period of 1970-2009. The contour interval is 0.5 per season (JJASO) per grid box (2.5°×2.5°). The bold arrows represent the majority of typhoon paths in the Western North Pacific-East Asian region. The box represents the area in the vicinity of Taiwan.

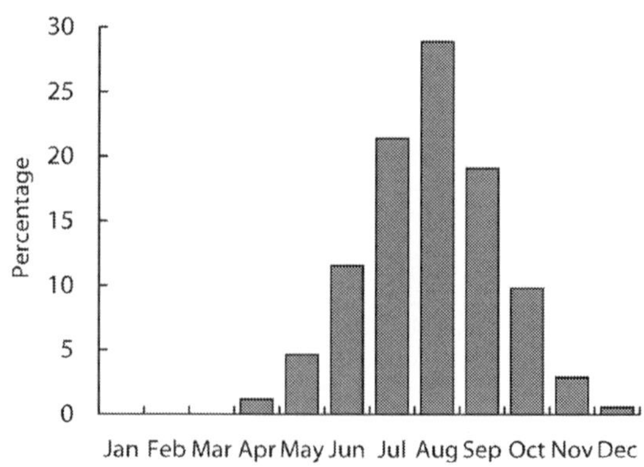

Fig. 2. Monthly percentage of typhoons impacting Taiwan averaged over the period of 1970-2009.

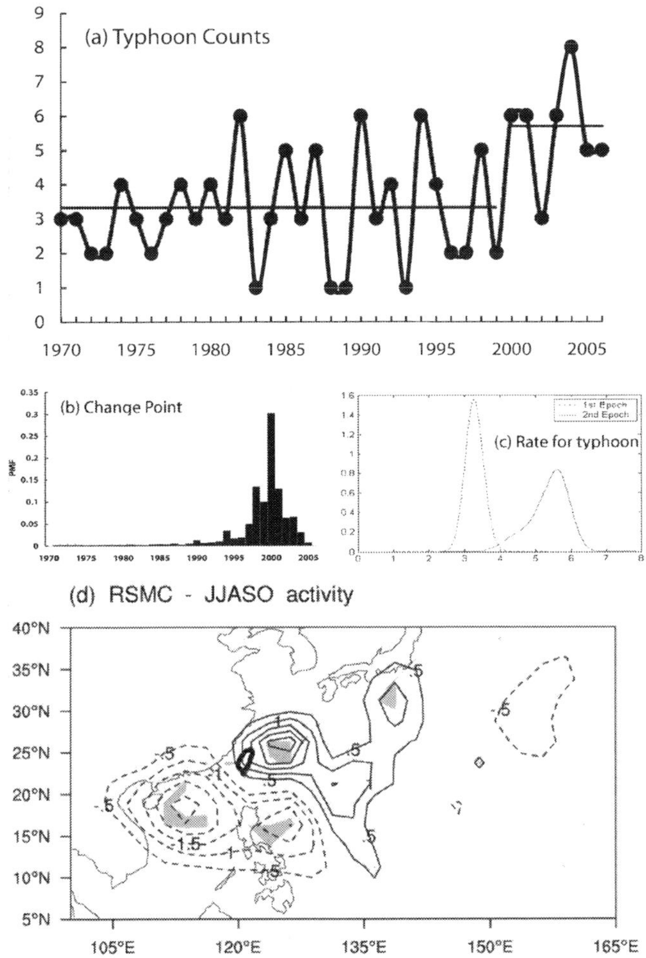

Fig. 3. (a) Time series of seasonal (JJASO) typhoon numbers passing the vicinity of Taiwan from 1970 to 2006 as compiled by the Central Weather Bureau. The vicinity was defined as 21°N-26°N and 119°E-125°E. (b) The conditional posterior probability mass function of change-points is plotted as a function of time. (c) Posterior density function of seasonal typhoon rate before (dashed line) and after (solid) the shift, with the change-point year being set in 2000. (d) June-October typhoon frequency differences for the period of 2000-2006 minus the period of 1970-1999 from the Regional Specialized Meteorological Center, Tokyo. The contour interval is 0.5; shading denotes that the difference between the mean of the two epochs is statistically significant at the 5% level. Adapted from Tu et al., (2009).

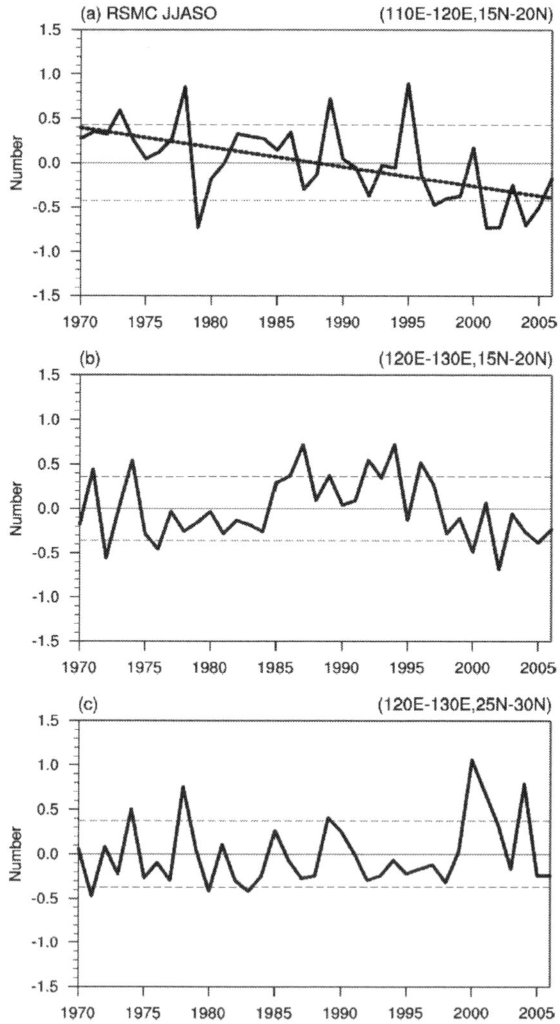

Fig. 4. Time series of seasonal (JJASO) typhoon frequency departure from 1970 to 2006 for three sub regions of the western North Pacific: (a) the South China Sea, (b) the Philippine Sea and (c) the Taiwan and East China Sea region. The thicker dashed line in the upper panel is a best-fit least square linear trend and the thinner dashed lines denote one standard deviation for each area. The unit in the y-axis is the typhoon number per season (JJASO) per grid box (2.5°×2.5°). Adapted from Tu et al. (2009).

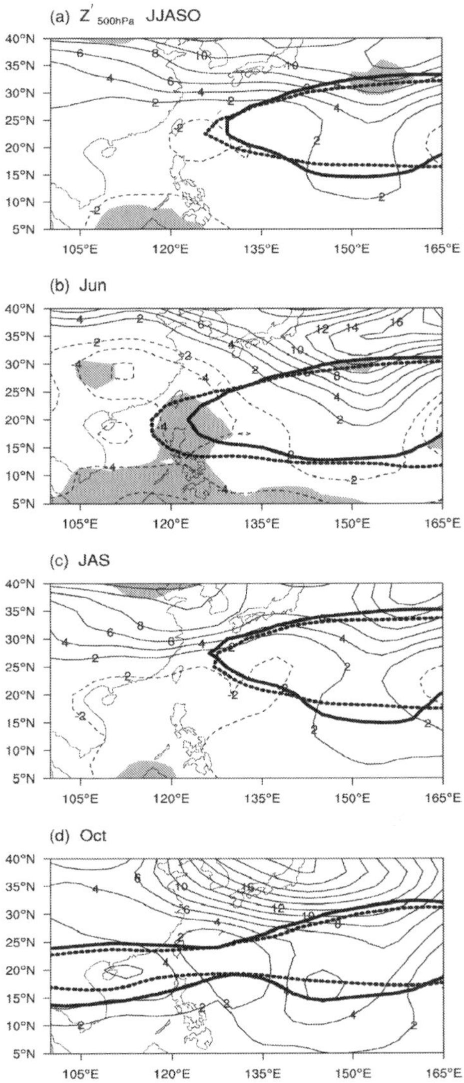

Fig. 5. The 5875 gpm contour of 500hPa geopotential height for the period of 1982-1999 (thick dotted line) and 2000-2006 (thick solid line) in (a) June-October (JJASO), (b) June, (c) July-September (JAS) and (d) October. The contours are the 500hPa geopotential height differences of the second minus first epoch, shaded by the 10% significance level. Adapted from Tu et al. (2009).

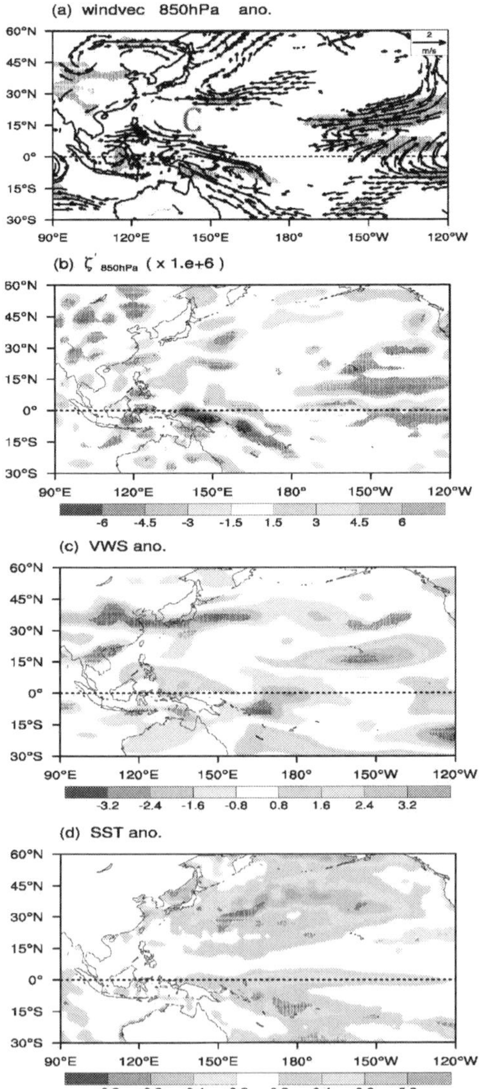

Fig. 6. (a) 850 hPa wind difference between 2000-2006 and 1982-1999 for JJASO; (b) same as (a) but for 850 hPa relative vorticity; (c) same as (a) but for vertical wind shear (200hPa-850hPa); and (d) same as (a) but for sea surface temperature (SST). The contour interval for 850 hPa relative vorticity is 1.5E+6 (s^{-1}), for vertical wind shear is 0.8 (m s^{-1}), and for SST anomalies is 0.2°C. Dotted areas indicate regions where the difference in the mean between two epochs is significant at the 5% level. In (b), (c), and (d), negative values are dashed. Adapted from Tu et al. (2009).

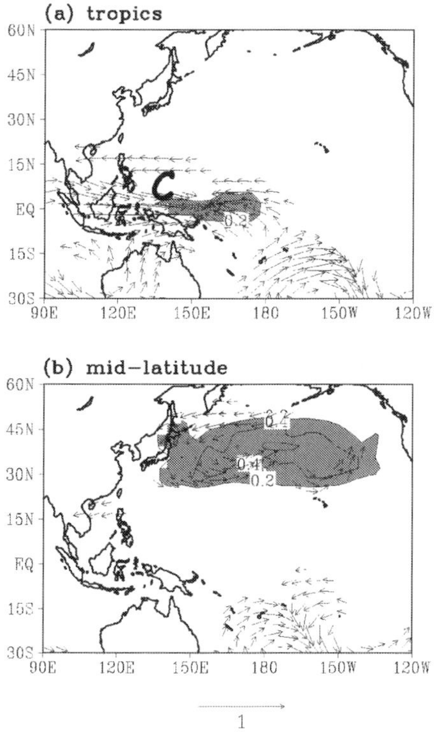

Fig. 7. SST anomalies (contour) and 850 hPa wind anomalies from the model simulations with the prescribed SST anomalies over (a) the equatorial region (130°E-175°E, 5°S-5°N) and (b) mid-latitudes (140°E-120°W, 25°N-45°N). Adapted from Tu et al. (2009).

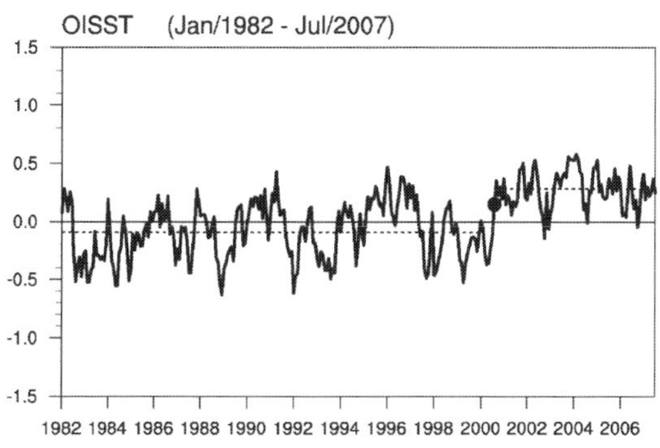

Fig. 8. Variation of monthly SST anomalies averaged over the area of 130°E-175°E and 5°S-5°N from January 1982 to July 2007. The short dashed lines are the means averaged over 1982-1999 (-0.1°C) and 2001-2006 (0.3°C) respectively. Adapted from Tu et al. (2009).

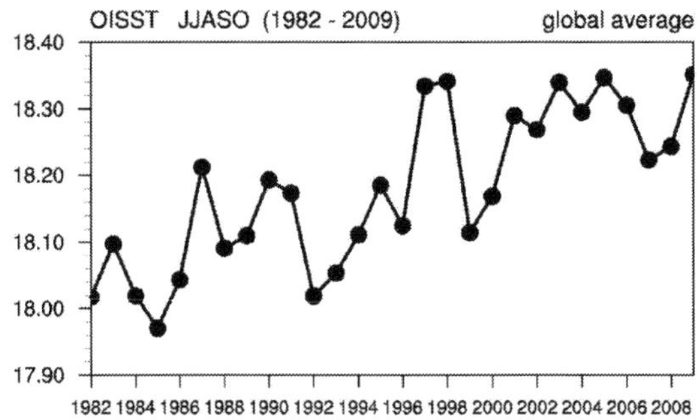

Fig. 9. Globally averaged SST in JJASO for the period of 1982-2009.

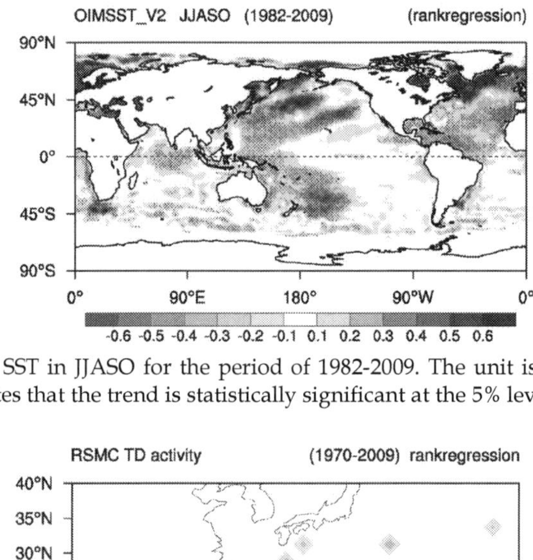

Fig. 10. Trend of SST in JJASO for the period of 1982-2009. The unit is °C per decade. The dotted area denotes that the trend is statistically significant at the 5% level.

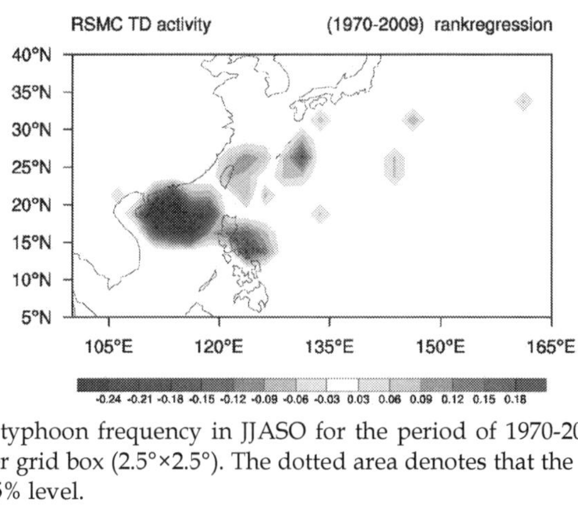

Fig. 11. Trend of typhoon frequency in JJASO for the period of 1970-2009. The unit is per season (JJASO) per grid box (2.5°×2.5°). The dotted area denotes that the trend is statistically significant at the 5% level.

6

Influence of climate variability on reactive nitrogen deposition in temperate and Arctic climate

Lars R. Hole
Norwegian Meteorological Institute (met.no)
Norway

1. Introduction

Depending on wetness of the climate, a large fraction of reactive nitrogen deposited from the atmosphere is deposited as wet deposition, ranging from 10 to 90%. The remaining fraction is deposited as dry deposition (gas and particles) (Delwiche, 1970; Galloway et al., 2004; Wesely & Hicks, 2000). Deposition of long-range transported reactive nitrogen (Nr) has been an issue of concern Europe and North America for a long time. In 1983 the Convention on Long-Range Transboundary Air Pollution entered into force, while the Protocol concerning the Control of Nitrogen Oxides or their Transboundary Fluxes was signed in 1988. While measures to reduce sulphur (S) emissions have been quite successful, nitrogen (N) emissions have proven more difficult to reduce (www.emep.int). Effects of N deposition on terrestrial ecosystems include surface water acidification (Stoddard, 1994) and reductions in biodiversity (Bobbink et al., 1998) while forest growth effects are more difficult to substantiate (Tietema et al., 1998; Emmett et al., 1998). Retention of N in many boreal and temperate ecosystems is usually high, which leads to soil N enrichment which in turn may lead to 'N saturation' of soils and increased leaching of N to surface waters, leading to water acidification (Stoddard, 1994). Recent studies indicate that climate change may affect the biogeochemical Nr cycle profoundly. Evidence is accumulating that interactions between N deposition and terrestrial processes are influenced by climate warming (De Wit et al., 2008).

There are few studies on the linkage between Nr deposition and climate variability in Northern Europe. By coupling of a regional climate model and the Mesoscale Chemical Transport (CTM) Model MATCH, Langner et al. (2005) showed that changes in the precipitation pattern in Europe have a substantial potential impact on deposition of oxidised nitrogen, with a global warming of 2.6 K reached in 2050-2070. Air mass trajectories have been shown to be affected by climate warming and this may potentially lead to changes in N deposition. Fowler et al (2005) were not able to establish a clear connection between Nr wet deposition in the UK and the North Atlantic Oscillation Index (NAOI), suggesting that a much more detailed approach with analysis of individual precipitation events and trajectory studies would have to be used in order to establish relationships between Nr deposition trends and climate variation.

In Norway, Hole and Tørseth (2002) reported the total sulphur and nitrate deposition in five-year periods from 1978-1982 to 1997-2001 by interpolating national and EMEP (European Monitoring and Evaluation Programme) station measurements to the EMEP 50x50 km grid. They found that the total (wet+dry) Nr deposition in the last period had been reduced with 16% compared to the first period although the total precipitation had increased with 10% (Fig 1). However the decline in deposition since the early 1980s is not steady since EMEP area NOx emissions reached a peak around 1990 and the period 1988-1992 was the wettest in Norway of the periods studied. Grid cell total deposition for NOx in the last period varied from 0.04 to 1.2 g N m^{-2} yr^{-1} while corresponding numbers for NHy was 0.06 to 0.9 g N m^{-2} yr^{-1}.

According to Hanssen-Bauer (2005) mean annual precipitation in Norway has increased in 9 of 13 climate regions into which Norway is divided (Fig. 1), with a 15-20% increase in northwestern regions (between Bergen and Trondheim) in the last century.

2. Trend analysis of nitrogen deposition and relation to climate variability

2.1 Measurement network studied

In the following, we explore relations between climate variability and wet N deposition at 7 locations in south Norway, including a range in annual precipitation and atmospheric Nr deposition. We have tested whether various climate indices are significantly correlated with i) bulk concentrations of Nr in precipitation ii) monthly precipitation iii) Nr deposition during summer and winter. Our main focus is deposition. We have separated summer and winter data to test whether there are seasonal differences in the correlations. More details on the measurement network can be found in Hole et al. (2008).

2.2 Climate indices

Different climate indices have been tested for correlation with Nr deposition, precipitation and Nr concentration in precipitation. In addition to the North Atlantic Oscillation Index (NAOI) we have tested for the Arctic Oscillation Index (AOI), the European Blocking Index (EUI), the Scandinavian blocking Index (ScandI) and the East Atlantic Index (EAtlI).

The Arctic oscillation (AO) is the dominant pattern of non-seasonal sea-level pressure (SLP) variations north of 20N, and it is characterized by SLP anomalies of one sign in the Arctic and anomalies of opposite sign centered about 37-45N. The North Atlantic oscillation (NAOI) is a climatic phenomenon in the North Atlantic Ocean of fluctuations in the difference of sea-level pressure between Iceland and the Azores. It controls the strength and direction of westerly winds and storm tracks across the North Atlantic and is a close relative of the AO (www.cpc.noaa.gov).

The European blocking index is based on observations of pentad (5-day average) wind over the region 15W to 25E and 35n to 55N. If the pentad zonal wind equals the climatological value for that time period, the index is zero. If the pentad zonal wind is less than average the index is positive (a blocking high pressure persist over central Europe), while the opposite is true if the index is negative. Similarly, positive ScandI and EatlI are associated with blocking anticyclones over Scandinavia and the East Atlantic, respectively. Jet stream intensity and orientation at the storm trackexit, and in the vicinity of Norway in particular, vary with the phase of these climate patterns. (Orsolini and Doblas-Reyes, 2003).

The winter of 1990 (which was warm and wet with prevailing westerlies in S Norway) was a strong positive event in NAOI whilst the dry and cold winter of 1996 was a prolonged negative event. It also appears that the NAOI and AOI behave similarly and they are also correlated, particularly in winter ($R_{summer} = 0.55$, $R_{winter} = 0.81$).

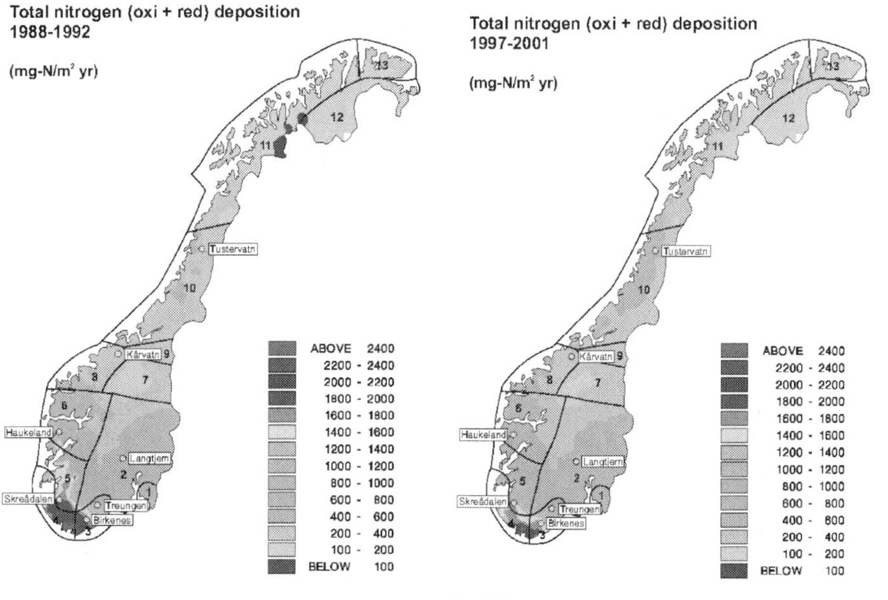

Fig. 1. Total deposition of nitrogen (oxidized + reduced) 1988-92 (maximum total Nr deposition in the monitoring period) and 1997-2001 (minimum total Nr deposition in the monitoring period) in mainland Norway. The unit is mg N/m2 year. From Hole and Tørseth (2002). Precipitation zones from Hanssen-Bauer (2005) are also indicated.

2.3 Statistical method

Precipitation data from seven monitoring stations are presented here as monthly values in winter (December-February) and summer (June-August). In this way we can see seasonal differences since strong anticyclones in the Atlantic with westerlies are particularly common in winter during negative NAOI events. Precipitation concentrations were weighted according to precipitation amount. Existence of a monotonic increasing or decreasing trend in the time series 1980-2005 and 1990-2005 was tested with the nonparametric Mann-Kendall test at the 10% significance level as a two-tailed test (Gilbert, 1987). Some of the stations opened in the 1970s, but we choose to test for the same periods at all stations to be able to compare trends. An estimate for the slope of a linear trend was calculated with the nonparametric Sen's method (Sen, 1968). The Sen's method is not greatly affected by data outliers, and it can be used when data are missing (Salmi et al., 2002).

It is likely that significant trends in deposition are partly a result of changes in emissions. However, it is not obvious which emission areas contribute to deposition in Norway, even though a sector analysis has been carried out for parts of the period studied (Tørseth et al,

2001). The relative contribution could also vary from year to year depending on transport climate. Here, we have tested whether removing significant trends in the data have any influence on the correlations we observe.

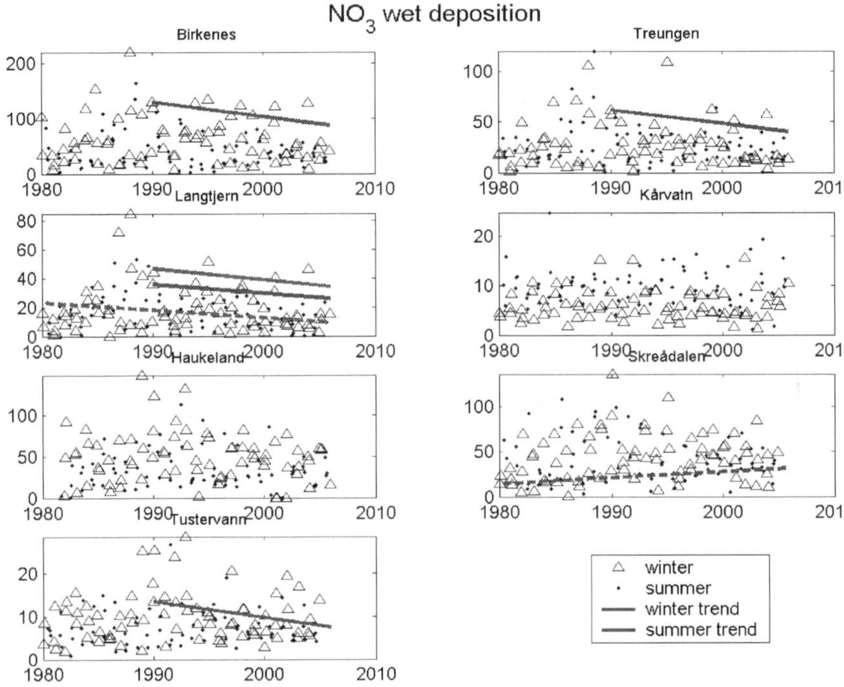

Fig. 2. Monthly average NO_3 wet deposition summer and winter (mg/m^2). Solid lines are 1990-2005 trends, dashed lines are 1980-2005 trends.

2.4. Observed trends

Significant Sen slopes (10% level) in nitrate and ammonia deposition for 1980-2005 and 1990-2005 are shown in Figures 2-3. Trends in nitrate concentrations since 1980 corresponds to a reduction of up to 50% at Kårvatn in summer (Aas et al, 2006) and less at the other stations. For the longest period, there are negative trends (summer, winter or both) in nitrate wet deposition at five out of seven sites. For the shortest period there are negative trends in nitrate wet deposition at four of seven sites, including the most coastal site (Haukeland), where there is also a very strong increase in summer precipitation (32 mm/decade). For the longest period there are few sites with significant trends in nitrate wet deposition and this could be caused by increasing precipitation in the period, although the data analysed here show significant increase in precipitation at only three sites. For 1990- 2005 decreasing nitrate concentration in precipitation is accompanied by decreasing nitrate wet deposition only at the driest site (Langtjern). The positive trend in ammonia wet deposition at Tustervatn could be caused by changes in local farming activity. We should keep in mind that the 25 year studied here is a very short time to detect climatic trends, since there is much variability on decadal scale (Hanssen-Bauer, 2005).

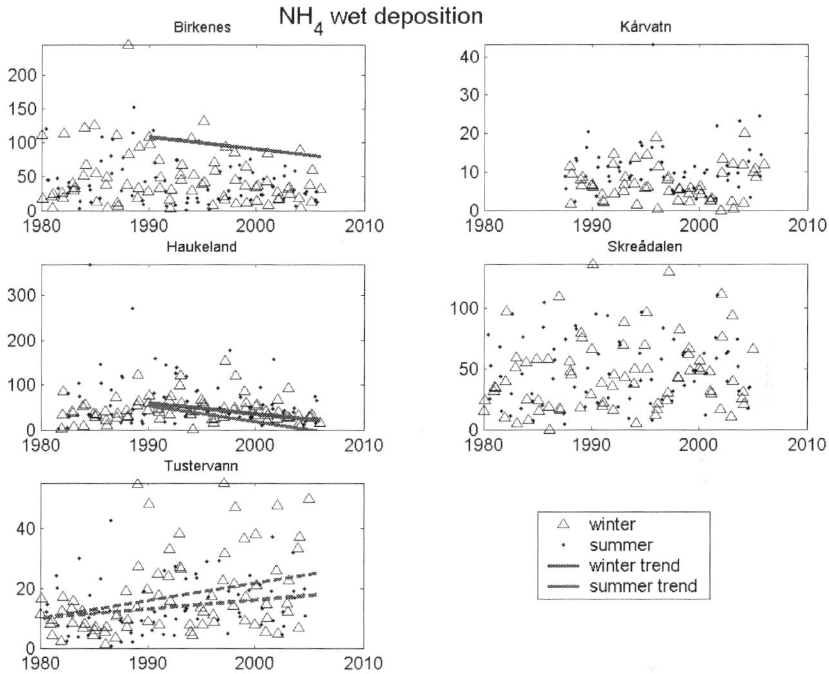

Fig. 3. Monthly average NH4 wet deposition summer and winter (mg/m2). Solid lines are 1990-2005 trends, dashed lines are 1980-2005 trends.

2.5 Climate indices and connection to concentrations, precipitation and deposition

First, we test correlations between Nr concentrations and climate indices. For most stations there was no correlation. The strongest correlation found was R=-0.45 for nitrate concentration and NAOI at Haukeland in winter. Nitrate wet deposition at the western sites (Haukeland and Skreådalen) are well correlated with NAOI and strongest in winter (R=0.60 at Skreådalen) (Table 1). A cluster analysis where the western sites are combined gives R=0.56 for the western sites in winter, and a much lower correlation (R=0.22) for the southern sites (Birkenes and Treungen). For precipitation the corresponding correlations coefficients are 0.75 and 0.38 respectively. Interestingly AOI has a similar regional correlation pattern, but it has a higher correlation at the northern site Tustervatn (R = 0.47 in winter). This regional pattern reflexes the correlation with precipitation in which again corresponds well with Hanssen-Bauer (2005). High correlations with NAOI and AOI in winter is not surprising since strong cyclonic systems in the Atlantic leads to high precipitation at the west coast. Local air temperature is also strongly correlated with winter nitrate wet

Station name	NAOI	AOI	European blocking	Scandinavian blocking	East Atlantic blocking	
Birkenes	0.15	-0.01		-0.06	0.31	Summer
Treungen	0.09	0		0.01	0.24	
Langtjern	0.10	-0.03		-0.05	0.11	
Kårvatn	0.20	0.21		-0.20	0.08	
Haukeland	0.46	0.30		-0.18	0.13	
Skreådalen	0.38	0.21		-0.19	0.37	
Tustervatn	0.11	0.14		0.19	-0.01	
Birkenes	0.24	0.16	-0.45	0.25	0.24	Winter
Treungen	0.25	0.13	-0.47	0.25	0.23	
Langtjern	0.21	0.06	-0.46	0.23	0.32	
Kårvatn	0.04	0.16	0.14	-0.27	-0.15	
Haukeland	0.53	0.60	0.13	-0.20	0.20	
Skreådalen	0.60	0.57	-0.20	-0.22	0.39	
Tustervatn	0.28	0.47	0.24	-0.12	0.22	

Table 1. Correlation coefficients, R, for nitrate deposition vs climate indices 1980-2005.

deposition at the coastal sites (R=0.84), suggesting that mild, humid winter weather with strong transport from west and south-west (positive NAOI) brings high deposition, mostly as rain, and transport from the UK. For the other sites R<0.2. The European blocking index is strongest (and negatively) correlated with winter deposition at the drier, eastern site, Langtjern, (Table 1). This suggests that a certain orientation of the isobars brings in precipitation from the south at these sites. The other blocking indices do not show very high correlation with nitrate wet deposition. However, ScandI shows high correlation (R = -0.49) with winter precipitation at Skreådalen, although much lower than NAOI (R=0.77) and AOI (R=0.73). The pattern for ammonia wet deposition is similar and will not be discussed here.

2.6 Discussion of trend analysis and climate variability

Reductions in nitrate wet deposition are probably a consequence of emission reductions in the EMEP area (EMEP, 2006). There has been a steady decrease in oxidised nitrogen (NOx) emissions in most of Europe since 1990 and looking at the trend 1980-2004 the decrease has been particularly strong in Eastern Europe. Ammonia emission estimates are highly uncertain since agriculture is the main source. Emissions seem to be rather steady in most areas, except in Eastern Europe where reductions have been up to 50% in the 1990s. Sutton et al., (2003) studied trends in reduced nitrogen in different parts of Central Europe and the UK to assess the effectiveness of ammonia abatement. For a range of countries it was shown that atmospheric interactions complicate the expected changes, particularly since sulphur emissions have decreased steadily in the last two decades.

Precipitation is better correlated than deposition with NAOI and AO. This is an indication that deposition is depending more on precipitation amount than on transport sector. NAOI seems to also partly control the variation in atmospheric nitrate concentrations (R = -0.45 at the coastal sites), i.e. westerly wind brings lower concentrations. It is already established that precipitation amounts, particularly on the west coast, are well correlated with NAOI

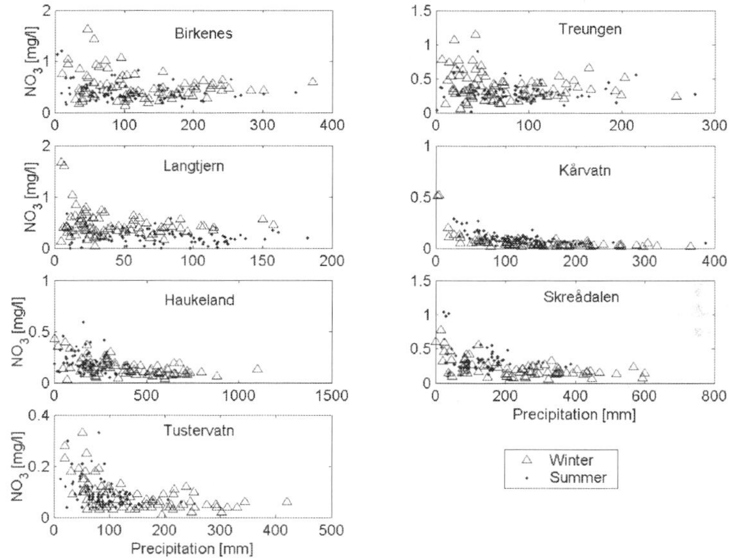

Fig. 4. Monthly average NO_3 concentration in precipitation (mg/l) vs monthly precipitation (mm) 1980-2005.

(Hanssen-Bauer, 2005). On the other hand, it has been shown that transport from continental Europe in south and east is likely to result in higher concentration levels than transport from the Atlantic in west and north (Tørseth et al., 2001). Probably since emissions trends for nitrate are relatively weak and continuous (28% reduction from maximum in 1989 to 2003) it was not possible to establish a correlation between emissions in the EMEP area and wet deposition here. For nitrate concentration in precipitation (Fig. 4) it is clear that the driest months bring the highest concentrations at all sites. The negative correlation between nitrate wet deposition and precipitation amount is weakest at the driest sites (Treungen and Langtjern). In Norway high precipitation events are associated with weather systems with a S component, generally SW wind on the W coast and SE wind in E Norway. We would also expect that these directions with transport from UK and E Europe would give the highest concentrations. Figure 4 suggests a dilution effect in rainy months. Modelling results in Hole and Enghardt (2008) also show that the severe increase in precipitation in W Norway expected in the coming decades (in the order of 50%) will indeed result in lower concentrations. Because 1990 was the warmest (and consequently one of the wettest) year on record in Norway, there are no significant trends in precipitation in 1990-2005 except for a strong increase in winter precipitation at Kårvatn. However, there are significant reductions in nitrate concentration in precipitation at several stations (Hole et al., 2008).

3. Trends in concentrations of sulphur and nitrogen compounds in the Arctic

3.1 Long range transport of air pollution to the Arctic

Arctic acidification in areas with both sensitive ecology and levels of acid deposition elevated to a point that exceeds the system's acid neutralizing capacity. Sulphur is the most important acidifying substance in the Arctic, with nitrogen of secondary importance (Kämäri et al., 1998). Significant anthropogenic sources of sulphur emissions, and to a lesser extent nitrogen emissions, exist within the Arctic region. In addition, long-range transported air pollutants contribute to acidification and Arctic haze. Emissions from natural sources within the Arctic (volcanoes, marine algae, and forest fires) are difficult to quantify and project (Kämäri et al., 1998).

Based on firn core analysis from the Canadian high Arctic, Barrie et al. (1985) suggested that in the first-half of the 20th century the level of winter-time air pollution remained roughly constant, consistent with a pattern of little change in European sulphur dioxide (SO_2) emissions. However, between 1956 and 1977 there was a 75% increase of Arctic air pollution which seems to be associated with a marked increase in SO_2 and total NO_x emissions in the industrialized world Barrie et al. (1985). Weiler et al. (2005) analysed an ice core from a North Siberian ice cap and found that maximum sulphate and nitrate concentrations in the ice could be related to maximum SO_2 and NO_x anthropogenic emissions in the 1970s, probably caused by the nickel- and copper-producing industries in Norilsk and on the Kola peninsula or by industrial combustion processes occurring in the Siberian Arctic. In addition, they found that during recent decades, sulphate (SO_4^{2-}) and nitrate (NO_3^-) concentrations declined by 80% and 60%, respectively, reflecting a decrease in anthropogenic pollution of the Arctic basin.

Kämäri et al., (1998) concluded that there were no trends in atmospheric concentrations of acidifying compounds in Canada and Alaska during the 1980s, but that there were decreasing trends on Svalbard. Background data from Russia were not presented. It was considered that about 75% of the deposition could be dry deposition, but that there was a lack of observations and knowledge at this point. Model output for SO_2 and SO_4^{2-} compared well with time series observations series at one station (Nord, Greenland) and for long term averages at a number of EMEP stations.

Although atmospheric lifetimes of SO_2 , NO_x and their oxidation products are of the order of a some days at temperate latitudes (Schwarz, 1979; Levine and Schwarz, 1982, Logan, 1983), the atmospheric half-life of SO_4^{2-} have been reported to reach even two weeks or more in the high Arctic during winter (Barrie, 1986). The transport distances range from hundreds to thousands of kilometres (Seinfeld and Pandis, 1998). Thus, many factors, besides the primary emissions, affect the observed concentrations and trends of the compounds involved in the acid deposition process, including their relative concentrations in the atmosphere, the reversible nature of some of the reactions and the meteorological situation.

Trend analysis for several indicators of Arctic haze has been performed for the spring months by Quinn et al. (2007). The monthly average SO_4^{2-} concentration in air in March and April has decreased in the Canadian, Norwegian and Finnish Arctic by 30-70% from the early 1990s to early 2000. NO_3^-concentration in air has increased by 50% in Alert, Canada during the same period.

3.2 Monitoring Arctic air pollution

There is a lack of long time series of background concentrations in main atmospheric compounds in the high Arctic. Also there are few stations with co-located air and precipitation sampling. The AMAP[1] atmospheric monitoring network consist of a number of stations spread across the Arctic. Most of these are EMEP stations that also report to the AMAP database. In addition, a few national stations report data. Some stations have been reporting data since the mid 1970s. As of 2002, 24 stations reported data to AMAP relevant for acidification and eutrophication (Hole et al,. 2006a). Most stations are located in the European sector. The nitrogen compounds in air are measured at the EMEP stations as a sum of particulate nitrate and gas phase nitric acid and, respectively, a sum of particulate ammonium and gas phase ammonia. They are referred later in the text as total nitrate and total ammonium in air. The station Alert measure particulate nitrate and ammonium.

The Russian national network for monitoring of precipitation chemical composition and acidity consists of 110 monitoring stations. Precipitation samples collected at these stations are then analysed in regional analytical laboratories for the main atmospheric compounds. The coordinating and analytical centre for the precipitation chemistry monitoring network is the Voieykov Main Geophysical Observatory, Roshydromet whose data are mainly used for this article. In addition to these stations, there are 105 monitoring sites where only pH value is analysed. Stations are unevenly distributed over the territory of Russia. Less than 40% of the stations are situated in the vast Siberian region. The period of observations reaches up to 40 years for some stations. For analysis of the acid precipitation and acidity we have used nine background monitoring stations situated in the Russian Arctic. For these stations, average summer (June-August) and winter (December-February) values were reported. Except for the two EMEP-stations reported here (Janiskoski and Pinega), there are no background air concentrations monitoring sites situated in the Arctic region of Russia.

As pointed out by MacDonald et al. (2005), detection of recent trends in the Arctic is difficult due to the combination of short or incomplete data records at some sites and interference from natural variations on seasonal, annual and decadal timescales (Quinn et al., 2007). In order to remove seasonal variability from the trend analyses, we focus here on monthly concentrations for winter (December-February) and summer (June-August) separately.

It is likely that significant trends in deposition are partly a result of changes in emissions. However, it is not obvious which emission areas contribute to deposition. The relative contributions of different regions could also vary from year to year depending on atmospheric transport paths.

3.3 Description of Danish Eulerian Hemispherical Model

The Danish Eulerian Hemispheric Model (DEHM) system consists of a weather forecast model, the PSU/NCAR Mesoscale Model version 5 (MM5) modelling subsystem (see Grell et al, 1994), which is driven by meteorological data from ECMWF, and a 3-D atmospheric transport model, the DEHM model. The model has a horizontal resolution of 150 km x 150 km and 20 irregularly spaced vertical layers up to 16 km. The coverage is close to hemispheric from nearly 10 degrees N at the corners and 25 degrees N at midpoints of the model domain boundaries.

The original version of the DEHM model was developed for studying the long-range

1 Arctic Monitoring and Assessment Programme, www.amap.no.

transport of SO_2, SO_4^{2-} and Pb to the Arctic (Christensen, 1997) and has been used since 1991. The sulphur version has been used in the first and the second phase of the AMAP program (see Kämäri et al., 1998, Hole et al., 2006a, 2006b) and the Pb version was used in the last AMAP heavy metal assessment. It has been further developed to study transport, transformation and deposition of reactive and elemental mercury, and this version was also used in the heavy metal assessment, see also (Heidam et al. (2004). Other versions calculate the concentrations and depositions of various pollutants (Frohn et al., 2002) through the inclusion of the extensive chemistry scheme, and transport and exchange of atmospheric carbon dioxide (Geels et al., 2004) and Persistant Organic Pollutants (Hansen et al, 2004).

In this work we are using the extensive chemical version which includes 63 species of which 4 relate to primary particulates (PM25, PM10, TSP and sea salt), other species are SO_x, NO_x, total reduced nitrogen (NH_y), VOC's and secondary inorganic particulates (Frohn et al, 2003). The chemical scheme was based on a scheme with 51 species presented in Flatøy and Hov, 1996, which were an ozone chemistry scheme with most of the important inorganic species and as well the most abundant hydrocarbons (explicit treatment of alkanes up to C4, longer alkanes lumped, explicit treatment alkenes up to C3, longer alkenes lumped, xylene, toluene and isoprene). There were added reactions to extend the chemistry to eutrophication issues by using ammonium chemistry based on the old EMEP acidification model and adding reactions in order extend to acidification issues by using aqueous chemistry based on Jonson et al. (2000). The scheme contains 120 chemical reactions where 17 are photolysis reactions calculated by the Phodis routine (Kylling et al, 1998) depending on sun-angle, altitude, Dobson unit and 3-d cloud cover. The used chemical scheme is quite similar to the EMEP scheme described in Simpson et al, 2006.

The dry deposition module used in the DEHM model is based on the resistance method and is very similar to the dry deposition module of the EMEP model (for details and documentation see Simpson et al., 2006). This module calculates deposition of both gaseous species and particulates to 16 different land-use categories based on Olson World Ecosystem Classes, version 1.4D. The dry depositions of gaseous species to water surfaces are depending on the wind speed (surface roughness) and on solubility of the chemical species (see Hertel et al., 1995).

Wet deposition is parameterized by a scavenging ratio formulation, where the scavenging is divided into two contributions. The first contribution is the in-cloud scavenging, which represents the uptake in droplets inside a cloud. The second contribution originates from precipitation events and is uptake in droplets below the cloud base. The scavenging coefficients are also very similar to the EMEP model. Further information about the model run and emission data applied can be found in Hole et al., 2009.

3.4 Trends in concentrations in air and precipitation, 1980-2005 and 1990-2005

Figure 5 shows summer and winter trends after 1990 for non sea salt SO_4^{2-} and NH_4^+ in precipitation. For the SO_4^{2-} concentration, the values are usually higher during summer months than during winter months. Low concentrations are measured at the Oulanka, Pinega and Snare Rapids stations.

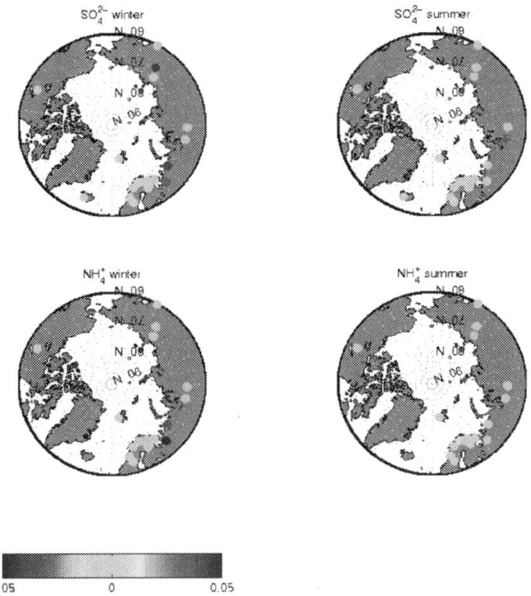

Fig. 5. Significant trends (within 10%) in SO_4^{2-} and NH_4^+ in precipitation after 1990 for winter (December-February) and summer (June-August). No trend is shown as green. $NO3^-$ is not shown because of few significant trends. Units are mg S l^{-1} and mg N l-1.

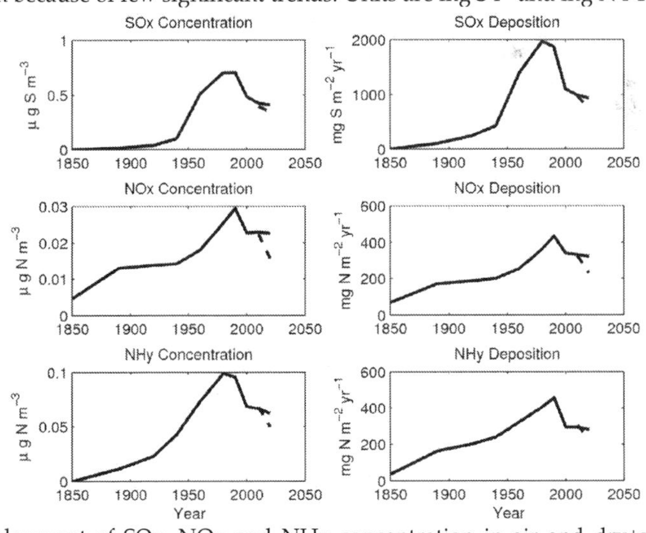

Fig. 6. Development of SOx, NOx and NHy concentration in air and dry+wet deposition north of the Arctic circle from industrialization to 2020. Solid lines: Business as Usual scenario (BAU). Dashed lines: Most Feasible Reduction scenario (MFR).

The level of the monthly SO_4^{2-} concentration in the beginning of the monitoring period is higher than at the end of the period but there is not a significant trend at all of the stations. For the NO_3^- concentration, values are on the contrary higher during the winter months than during the summer months (Hole et al. (2009)). The inter annual variation in the NO_3^- concentration is larger than in the sulphate concentration. The level of the nitrate concentration at the end of the monitoring period is lower than in the beginning at only the Pinega station. At the Jäniskoski station, the concentration has increased during the winter months. There are increasing trends in sulphate in precipitation at Ust-Moma in east Siberia in winter but at this station background concentrations are very low. This could be due to changes in Norilsk (NE Siberia, 69°21′ N 88°12′ E) emission or variability in transport pattern (Hole et al., 2006b). However, Norilsk emissions are not well quantified, so no clear conclusions can be drawn.

SO_4^{2-} concentrations measured in air at monitoring stations in the High Arctic (Alert, Canada; and Zeppelin, Svalbard) and at several monitoring stations in subarctic areas of Fennoscandia and northwestern Russia show decreasing trends since the 1990s, which corresponds well with Quinn et al. (2007). At many stations there are significant downward trends for SO_4^{2-} and SO_2 in air, both summer and winter. There are significant reductions of SO_2 in Svanvik probably because emissions in the area are strongly reduced. For the air concentration of the nitrogen compounds there is no clear pattern, but it is interesting to see a positive trend in summer total NO_3^- concentration at 3 stations. Total ammonium in air also has both positive and negative trends in summer.

3.5 Historical and expected trends 2000-2030 with "constant" climate

The DEHM model with extensive chemistry has been run with two different emissions scenarios: The "Business As Usual" (BAU) and the "Maximum technically Feasible Reduction" (MFR), as described in in Hole et al. (2006b). For each emission scenario the DEHM model has been run for the same meteorological input for the period 1991-1993 in order to reduce the meteorological variations of the model results. The pollution penetrates further north in the eastern Arctic compared to the western Arctic. This is in accordance with Stohl (2006) and Iversen and Jordanger (1985) and is a result of differences in circulation patterns and higher temperatures in the Barents sea region which allows air masses from temperate regions to move to higher latitudes without being lifted.

In Fig. 6 we present the overall development of concentration and deposition of SO_x and NOx and NHy in the Arctic since 1860, based on DEHM model runs and emission climate data as described earlier. The patterns for NHy and NOx are very similar to each other. It is not clear why concentrations and deposition do not have exactly the same development, but changes in temperature and precipitation patterns will influence the historical deposition development. This development with an accelarating depositon during the 19th century and a decline after about 1980, corresponds well with ice core observations such as Weiler et al., 2005.

4. Climate change impact on future atmospheric nitrogen deposition in a temperate climate

4.1 Background

Climate change, with increased air temperatures and changed precipitation patterns, is likely to affect the biogeochemical nitrogen (N) cycle in northwestern Europe significantly (deWit et al., 2008). The >40 years of historical weather data (ERA40) and dynamically downscaled climate scenarios for Europe to the year 2100 have been used to assess the linkage between climate variability and N deposition by means of the MATCH (Multi-scale Atmospheric Transport and Chemistry) model (Hole & Enghardt, 2008).

Total nitrate (NO_3)and total ammonium (NH_4) concentrations in precipitation decreased significantly at the Swedish EMEP stations from the mid 1980s to 2000 (Lövblad et al., 2004). During the same period the pH of precipitation increased from ~4.2 to 4.6. Data from the national throughfall network (Nettelblad et al., 2005) measurements of air- and precipitation chemistry at around 100 sites across Sweden confirm the downward trend in concentrations of NO_3 and NH_4 in rain. The trend was particularly pronounced in southern Sweden. Due to increasing precipitation amounts during the same period, however, the total deposition of reactive nitrogen (NO_3 and NH_4) has not decreased; instead it has remained roughly unchanged.

Increasing precipitation in a region will obviously result in increasing wet deposition if atmospheric N concentrations are unchanged. Altered precipitation patterns and temperatures are also likely to affect mobilisation of N pools in the soil and runoff to rivers, lakes and fjords (de Wit et al., 2008). Since many aquatic ecosystems in Scandinavia are N limited, increasing N fertilization will disturb the natural biological activity.

In the following we focus on future N deposition in northern Europe (Fennoscandia and the Baltic countries) as a result of future climate change. There are substantial regional differences in factors such as topography, annual mean temperature and precipitation in this area, and hence a regional discussion is required. Our purposes are to examine (1) regional and seasonal differences in climate change effects on nitrogen deposition, (2) whether changes in wet deposition are proportional to changes in precipitation, and (3) the distribution between dry and wet deposition. The MATCH model and the experimental set-up applied is described in Hole & Enghardt (2008) and references therein.

4.2 Deposition in future climate – comparison with current climate

Figures 7 and 8 show the calculated relative change in annual mean deposition of NO_y and NH_x over northern Europe. The figures display the difference of the 30-year mean of annually accumulated deposition during a future 30-year period minus the 30-year period labelled "current climate" normalised by the "current climate".

The Norwegian coast will experience a large increase in total N deposition due to increased precipitation projected by the present climate change scenario (ECHAM4/OPYC3–RCA3, SRES A2). The changes are most likely connected to the projected changes in precipitation in northern Europe. On an annual basis the whole of Fennoscandia is expected to receive more precipitation in 2071-2100 compared to "current climate".

The deposition of NO_y and NH_x display similar increasing trends along the coast of Norway. In northern Fennoscandia and in parts of southeast Sweden NH_x decreases, while NO_y is projected to increase. East and south of the Baltic Sea, the increase in NH_x deposition is much smaller than the increase in NO_y deposition. This is mostly because scavenging of NH_x is more effective in source areas than scavenging of NO_y.

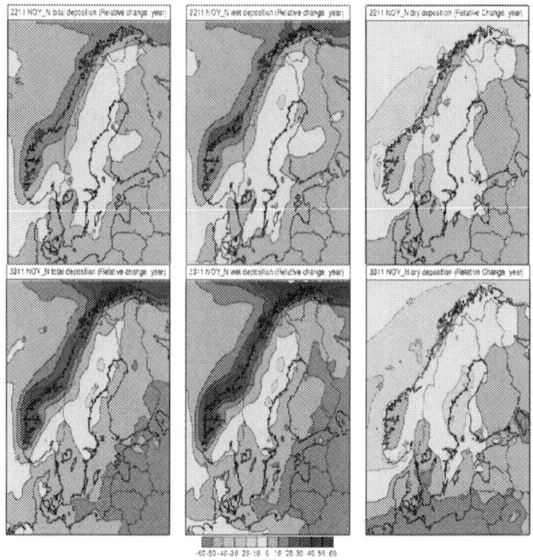

Fig. 7. Relative change in annually accumulated deposition of oxidised nitrogen (NO_y) from the period 1961-1990 to 2021-2050 (top row) and from 1961-1990 to 2071-2100 (bottom row). Left panel is total deposition, middle panel is wet deposition, right panel is dry deposition.

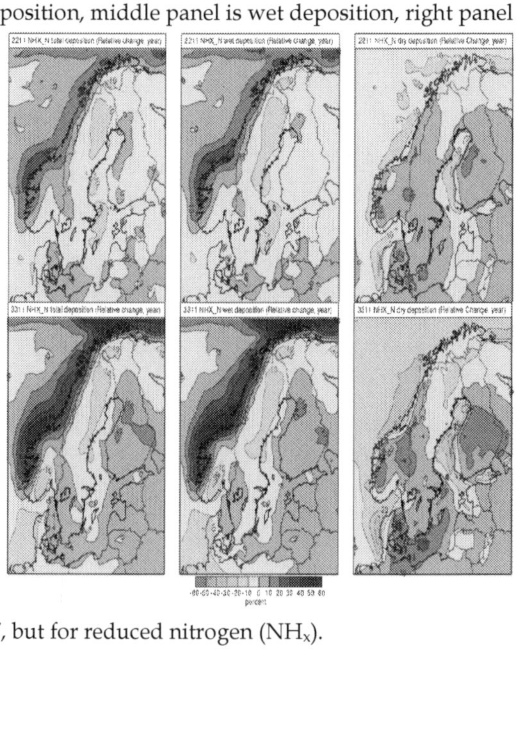

Fig. 8. Same as Fig. 7, but for reduced nitrogen (NH_x).

The total deposition of NO_y over Norway is expected to increase from 96 Gg N year^{-1} during current climate to 107 Gg N year^{-1} by the year 2100 due only to changes in climate (Hole & Enghardt, 2008). The corresponding values for Sweden are more modest, 137 Gg N year^{-1} to 139 Gg N year^{-1}. Finland, the Baltic countries, Poland and Denmark will also experience increases in total NO_y deposition. A large part of the increase in total NO_y deposition south and east of the Baltic is due to increased dry deposition. Reduced precipitation and increased atmospheric lifetimes of NO_y results in higher surface concentrations here, which drive up the dry deposition. In Norway and Sweden the change in annual dry deposition from current to future climate is only minor and virtually all change in total NO_y deposition emanates from changes in wet deposition.

The total deposition of NH_x decreases marginally in many countries around the Baltic Sea. Decreasing wet deposition of NH_x causes the decrease in total deposition in Sweden, Poland and Denmark. Norway will experience a moderate increase in total NH_x deposition in both during 2021-2050 and 2071-2100 compared to "current climate" (52 Gg N year^{-1} and 53 Gg N year^{-1} compared to 50 Gg N year^{-1}).

Trends in deposition pattern for the two compounds are not identical because primary emissions occur in different parts of Europe and because their deposition pathways differ. NH_x generally has a shorter atmospheric lifetime than NO_y; the increased scavenging over the coast of Norway will leave very little NH_x to be deposited in northern Finland and the Kola Peninsula, where NH_x emissions are minor.

The relative increase in deposition is slightly smaller than the predicted increase in precipitation. In Fig. 9 this dilution effect for NO_y is apparent along the Norwegian coast (where precipitation will increase most), but further north and east it is stronger because much of the NO_y is scavenged out before it reaches these areas.

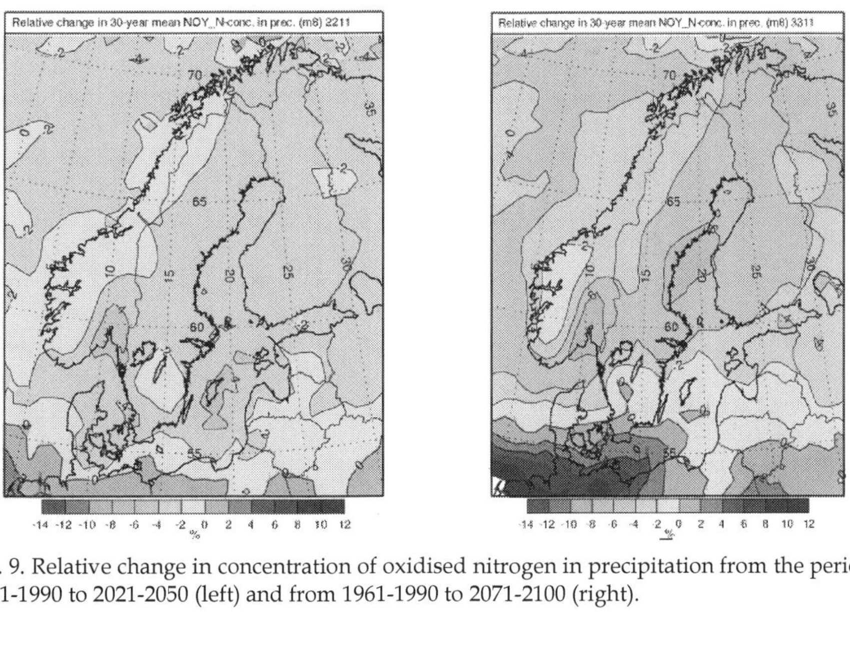

Fig. 9. Relative change in concentration of oxidised nitrogen in precipitation from the period 1961-1990 to 2021-2050 (left) and from 1961-1990 to 2071-2100 (right).

4.3 What can we say from these model results?

The accuracy of our results is determined by the accuracy of the utilised models and the input to the models. MATCH has been used in a number of previous studies and has proven capable to realistically simulate most species of interest. The model has, however, always had limitations in its capability to simulate NH_x species. This we have attributed to relatively larger uncertainties in the emission inventory of NH_3 and to the fact that subgrid emission/deposition processes not fully resolved in the system.

The model (RCA3) used to create the meteorological data in the present study has been evaluated in Kjellström et al. (2005). Using observed meteorology (ERA40 from ECMWF; "perfect boundary condition") on the boundaries they compare the model output with observations from a number of different sources. The increase in resolution from ERA40 produces precipitation fields more in line with observations although many topographical and coastal effects are still not resolved. This could explain the underestimation of precipitation at the sites located in western Norway. The precipitation in northern Europe is also generally overestimated in RCA3 when ECHAM4/OPYC3 is used on its boundaries. The degree of certainty we can attribute to RCA3's predictions of future climate is not only dependent on the climate model's ability to describe "current climate" and how the regional climate will respond to the increased greenhouse gas forcing. The RCA3 results are to a large degree forced by the boundary data from the global climate model. The EU project PRUDENCE and BALTEX presented a wide range of possible down-scaled scenarios for northwestern Europe showing, for example, that winter precipitation can increase by 20 to 60% in Scandinavia (see (Christensen et al., 2007) and references therein). These uncertainties are thus of the same order of magnitude as the projected changes in N deposition.

Estimates of precursor (NO_X, VOCs, CO etc.) emission strengths comprise a large uncertainty when assessing future N deposition. In order to only study the impact that possible climate changes may have on the deposition of N species we have kept emissions at their 2000-levels. This is a simplification and future N loading in north-western Europe will also be affected by changes in Europe as well as America and Asia. This study has focussed on the change in N deposition due to climate change and not evaluated the relative importance of altered precursor emissions or changed inter-hemispheric transport. The change in deposition over an area may not always be the result of changes in the driving meteorology over that area. It can of course also be due to changes in atmospheric transport pathways or deposition *en route* to the area under consideration.

5. Discussion and conclusions

In section 2 we studied observations of N deposition and its relation to climate variability. We showed that 36 % of the variation in winter nitrate wet deposition is described by the North Atlantic Oscillation Index in coastal stations, while deposition at the inland station Langtjern seems to be more controlled by the European blocking index. The Arctic Oscillation Index gives good correlation at the northernmost station in addition to the coastal (western) stations. Local air temperature is highly correlated (R=0.84) with winter nitrate deposition at the western stations, suggesting that warm, humid winter weather results in high wet deposition. For concentrations the best correlation was found for the coastal station Haukeland in winter (R=-0.45). In addition, there was a tendency in the data that high precipitation resulted in lower Nr concentrations. Removing trends in the data did not have significant influence on the correlations observed. However, a careful sector analysis for each month and for each station could improve the understanding of the separate effects of emission variability and climate variability on the deposition.

For the Business as Usual (BAU) emission scenarios, northern hemisphere sulphur emissions will only decline from 52.3 mt to 51.3 mt from 2000 to 2020 (section 3). For the Most Feasible Reduction (MFR) scenario 2020 emissions will be only 20.2 mt. However, the two different scenarios show much smaller differences in concentration and deposition of sulphur in the Arctic. This is because the largest potential for improvement in SO_2 emissions is in China and SE Asia. These regions have little influence on Arctic pollution according to Stohl (2006) and others. For oxidized and reduced nitrogen compounds there is more reduction in the emissions in Russia and Europe in the MFR scenario, and hence the potential for improvement in the Arctic is larger.

SO_4^{2-} concentrations are decreasing significantly at many Arctic stations. For NO_3^- and NH_4^+ the pattern is unclear (some positive and some negative trends). There are few signs of significant trends in precipitation for the period studied here (last 3 decades). However, expected future occurrence of rain events in both summer and winter can result in increasing wet deposition in the Arctic (ACIA, 2004, www.amap.no/acia).

There is relatively good monitoring data coverage in Fennoscandia and on Kola peninsula in Russia, but there are otherwise few stations for background air and precipitation concentration measurements in the Arctic. In our observations there are few differences between summer and winter observations, although NO_3^- wet deposition is higher in winter in some stations in NW Russia and Fennoscandia (Pinega, Oulanka, Bredkal and Karasjok). The explanation for this is not clear, but in Hole et al (2006b) seasonal exposure differences for SO_2 at Oulanka are revealed which can indicate that transport path differences are part of the explanation for the seasonal pattern.

Because of new technologies and climate change, future emissions and deposition are particularly uncertain due to the expected increase in human activities in the polar and sub-polar regions. Increased extraction of natural resources and increased sea traffic can be expected. Climate change is also likely to influence transport and deposition patterns (ACIA, 2004, www.amap.no/acia). There is a need for a deeper insight in plans and consequences with respect to the Arctic. Modelling results presented here seem to rule out SE Asia as an important contributor to pollution close to the surface in the Arctic atmosphere. This is in accordance with earlier studies (e.g. Iversen and Jordanger, 1985, Stohl, 2006) giving thermodynamic arguments why SE Asian emissions will have minor influence in the Arctic.

As for the relation between future Nr deposition and climate scenarios in temperate climate (section 4), our results suggest that prediction of future Nr deposition for different climate scenarios most of all need good predictions of precipitation amount and precipitation distribution in space and time. Climate indices can be a tool to understand this connection.

Regional differences in the expected changes are large. This is due to expected large increase in precipitation along the Norwegian coast, while other areas can expect much smaller changes. Country-averaged changes are moderate. Wet deposition will increase relatively less than precipitation because of dilution. In Norway the contribution from dry deposition will be relatively reduced because most of the N will be effectively removed by wet deposition. In the Baltic countries both wet and dry deposition will increase. Dry deposition will increase here probably because of increased occurrence of wet surfaces.

According to our model results, northwestern Europe will generally experience small changes in N deposition as a consequence of climate change. The exception is the west coast of Norway, which will experience an increase in N deposition of 10-20% in the period 2021-2050 and 20-40% in 2071-2100 (compared to current climate). Although Norway as a whole will only experience a moderate increase in N deposition of about 10%, there are large regional differences. RCA3/MATCH forced by ECHAM4/OPYC3 (SRES A2) prescribes that a large part of the Norwegian coast is expected to receive at least 50% increase of the precipitation during the period 2071-2100 compared to period 1961-1990, which is in line with other regional climate scenarios. This region has already experienced increasing precipitation in recent decades. The total effect on soil and watercourse chemistry of the dramatic change in these regions remains to be thoroughly understood.

Our studies shows that expected reduction in future N deposition (as a consequence of emission reductions in Europe) could be partly offset due to increasing precipitation in some regions in the coming century. Future long term N emissions in Europe are difficult to predict, however, since they depend on highly uncertain factors such as the future use of fossil fuels and farming technology. The same uncertainty obviously also applies to the greenhouse gas emission scenarios.

6. References

Aas, W.; Solberg, S.; Berg, T.; Manø, S. & Yttri, K. E. (2006). Monitoring of long range transported pollution in Norway. Atmospheric transport, 2005. (In Norwegian). *Norwegian Pollution Control Authority*. Rapport 955/2006. TA-2180/2006. NILU OR 36/2006. www.nilu.no.

Barrie L.A., 1986. Arctic air pollution: An overview of current knowledge. Atm. Env. 20, 643-663.

Barrie, L.A.; Fisher, D. & Koerner, R.M. (2005). Twentieth century trends in Arctic air pollution revealed by conductivity and acidity observations in snow and ice in the Canadian High Arctic. *Atmospheric Environment*, 19 (12), 2055-2063.

Bobbink, R.; Hornung, M. & Roelofs, J.G.M. (1998). The effects of air-borne nitrogen pollutants on species diversity in natural and semi-natural European vegetation. *Journal Of Ecology* 86(5): 717-738.

Christensen, J. (1997). The Danish Eulerian Hemispheric Model - A Three Dimensional Air Pollution Model Used for the Arctic. *Atm. Env*, 31, 4169-4191.

Christensen, J.H.; Carter, T.R.; Rummukainen M. & Amanatidis, G. (2007). Evaluating the performance and utility of climate models: the PRUDENCE project. *Climatic Change*, Vol 81. doi:10.1007/s10584-006-9211-6.

de Wit, H.A.; Hindar, A. & Hole, L. (2008). Winter climate affects long-term trends in streamwater nitrate in acid-sensitive catchments in southern Norway. *Hydrology and Earth System Sciences*, **12**, 393-403.

Delwiche, C. C. (1970). The nitrogen cycle. Sci. Am. **223**: 137-146, 1970.

EMEP (2006). Transboundary acidification, eutrophication and ground level ozone in Europe since 1990 to 2004. *EMEP Status Report1/2006*. The Norwegian Meteorological Institute, Oslo, EMEP/MSC-W Report 1/97..

Flatøy, F. & Hov, Ø. (1996). Three-dimensional model studies of the effect of NOx emissions from aircrafts on ozone in the upper troposphere over Europe and the North Atlantic. *J. Geophys. Res.*, 101, 1401-1422.

Fowler, D.; Smith, R. I.; Muller, J. B. A.; Hayman, G. & Vincent, K. J. (2006). Changes in the atmospheric deposition of acidifying compounds in the UK between 1986 and 2001. *Env. Poll.*, **137**(1): 15-25.

Frohn, L.M.; Christensen, J. H.; Brandt, J.; Geels, C. & Hansen, K. (2003). Validation of a 3-D hemispheric nested air pollution model. *Atmospheric Chemistry and Physics*, 3,3543-3588

Frohn, L.M.; Christensen, J. H. & Brandt, J., (2002). Development and testing of numerical methods for two-way nested air pollution modelling. *Physics and Chemistry of the Earth*, Parts A/B/C, **27** (35), P. 1487-1494

Galloway, J. N.; Dentener, F. J.; Capone, D. G.; Boyer, E. W.; Howarth, R. W.; Seitzinger, S. P.; Asner, G. P.; Cleveland, C.; Green, P.; Holland, E.; Karl, D. M.; Michaels, A. F.; Porter, J. H. Townsend, A. & Vörösmarty, C. (2004). Nitrogen Cycles: Past, Present and Future. *Biogeochemistry* **70**: 153-226.

Geels, C.; Doney, S.C.; Dargaville, R. J. Brandt, J.; Christensen, J.H. (2004). Investigating the sources of synoptic variability in atmospheric CO2 measurements over the Northern Hemisphere continents: a regional model study. *Tellus* B 56 (1), 35–50. doi:10.1111/j.1600-0889.2004.00084.x

Gilbert, R. O.: *Statistical methods for environmental pollution monitoring*. Van Nostrand Reinhold , New York, 1987.

Grell, G.; J. Dudhia, and Stauffer, D. (1994). A description of the Fifth-Generation Penn State/NCAR Mesoscale Model (MM5), NCAR Tech. Note TN-398, Natl. Cent. for Atmos. Res., Boulder, Colo..

Hansen, K.M.; Christensen, J.H.; Brandt, J.; Frohn, L.M.; & Geels, C.(2004). Modelling atmospheric transport of α-hexachlorocyclohexane in the Northern Hemispherewith a 3-D dynamical model: DEHM-POP, *Atmos. Chem. Phys.*, 4, 1125-1137.

Hanssen-Bauer, I. (2005). Regional temperature and precipitation series for Norway: Analyses of time-series updated to 2004. Met.no report 15/2005.

Heidam, N.Z.; Christensen, J.; Wåhlin, P. & Skov, H. (2004). Arctic atmospheric contaminants in NE Greenland: levels, variations, origins, transport, transformations and trends 1990–2001 *Science of The Total Environment*, 331 (1-3). Pages 5-28.

Hertel, O.; Christensen, J.; Runge, E.H.; Asman, W.A.H.; Berkowicz, R.& Hovmand, M.F. (1995). Development and Testing of a new Variable Scale Air Pollution Model - ACDEP. *Atmospheric Environment*, 29 1267-1290.

Hole, L. R. & Tørseth, K. (2002). Deposition of major inorganic compounds in Norway 1978-1982 and 1997-2001: status and trends. Naturens tålegrenser. Norwegian Pollution Control Authority. Report 115. NILU OR 61/2002, ISBN: 82-425-1410-0. www.nilu.no , 2002.

Hole, L.R, Christensen, J.; Ruoho-Airola, T.; Wilson, S.; Ginzburg, V. A.; Vasilenko, V.N.; Polishok, A.I. & Stohl, A.I. (2006). *Acidifying pollutants, Arctic Haze and Acidification in the Arctic*. AMAP assessment report 2006, ch. 3, pp 11-31.

Hole, L.R. & Engardt, M.; (2008) . Climate change impact on atmospheric nitrogen deposition in northwestern Europe - a model study. *AMBIO* 37 (1), 9-17.

Hole, L.R.; Brunner, S.H.; J.E. Hansen & L. Zhang, (2008). Low cost measurements of nitrogen and sulphur dry deposition velocities at a semi-alpine site: Gradient measurements and a comparison with deposition model estimates. *Env. Poll.*, **154**, 473-481. Special issue on biosphere-atmosphere fluxes, .

Hole, L.R.; Christensen, J. Forsius, M.; Nyman, M.; Stohl, A. & Wilson, S. (2006b). Sources of acidifying pollutants and Arctic haze precursors. AMAP assessment report , chapter 2.

Hole, L.R.; de Wit, H.; & Aas, W. (2008). Trends in N deposition in Norway: A regional perspective. *Hydrology and Earth System Sciences* 12, 405-414.

Iversen, T. & Jordanger, E. (2008). Arctic air pollution and large scale atmospheric flows, *Atm. Env.*, 19, 2099-2108.

Jonson, J.E. , Kylling, A. , Berntsen, T. , Isaksen, I.S.A. , Zerefos, C.S. , & Kourtidis, K. (2000), Chemical effects of UV fluctuations inferred from total ozone and tropospheric aerosol variations, *J. Geophys. Res.*, 105, 14561-14574.

Kämäri, J. & Joki-Heiskala, P., (eds), (1998). AMAP assessment report ch. 9, 621-658. Acidifying Pollutants, Arctic haze, and Acidification in the Arctic. *Arctic Monitoring and Assessment Programme*, www.amap.no.

Kjellström, E.; Bärring, L.; Gollvik, S.; Hansson, U.; Jones, C.; Samuelsson, P.; Rummukainen, M.; Ullerstig, A.; Willén, U. & Wyser, K. (2005). *A 140-year simulation of the European climate with the new version of the Rossby Centre regional atmospheric climate model (RCA3).* SMHI Reports Meteorology and Climatology No. 108, SMHI, SE-60176 Norrköping, Sweden 54 pp.

Kylling, A. , Bais, A.F. , Blumthaler, M. , Schreder, J. , Zerefos, C. S. , & Kosmidis, E. , (1998), The effect of aerosols on solar UV irradiances during the Photochemical Activity and Solar Radiation campaign, *J. Geophys. Res.*, 103, 21051-26060

Langner, J.; Bergström, R. & Foltescu, V. (2005). Impact of climate change on surface ozone and deposition of sulphur and nitrogen in Europe. *Atm. Env.*, **39** (6), 1129-1141.

Levine S.Z. & Schwarz S.E.; (1982). In-cloud and below-cloud scavenging of nitric acid vapor. *Atm. Env.* 16, 1725-1734.

Logan J.A.; (1983). Nitrogen oxides in the troposphere; global and regional budgets. *J. Geophys. Res.* 88, 10785-10807.

Lövblad, G.; Henningsson, E.; Sjöberg, K.; Brorström-Lundén, E.; Lindskog, A. & Munthe, J. (2004). *Trends in Swedish background air 1980-2000. In: EMEP Assessment part II National Contributions.* (. pp. 211-220. Oslo ISBN-82-7144-032-2.

MacDonald, R.W.; Harner, T. and Fyfe, J. (2005). Recent climate change in the Arctic and its impact on contaminant pathways and interpretation of temporal trend data. *Sci. Tot. Environ.* 342, 5–86.

Nettelblad, A.; Westling, O.; Akselsson, C.; Svensson, A. & Hellsten, S. (2006). *Air pollution at forest sites – results until September 2005.* (In Swedish). IVL Rapport B 1682. 50 pp. (In Swedish).

Orsolini, Y. J. & Doblas-Reyes, F. J. (2002) Ozone signatures of climate patterns over the Euro-Atlantic sector in the spring, *Q. J. R. Meteorol. Soc.*, 129, 3251-3263.

Quinn PK, Shaw G, Andrews E, Dutton EG, Ruoho-Airola T, & Gong SL. (2007) Arctic haze: current trends and knowledge gaps *Tellus* B 59 (1): 99-114.

Salmi, T.; Määttä, A.; Anttila, P.; Ruoho-Airola, T. & Amnell, T. (2002). Detecting trends of annual values of atmospheric pollutants by the Mann-Kendall test and Sen's slope estimates - the Excel template application MAKESENS, *Publications on Air Quality*, no. 31, FMI-AQ-31, FMI, Helsinki, Finland.

Schwarz S.E. (1979). Residence times in reservoirs under non-steady-state conditions: application to atmospheric SO2 and aerosol sulphate. *Tellus* 31, 520-547.

Seinfeld J.H. & Pandis S.N. (1998). *Atmospheric Chemistry and Physics: From Air Pollution to Climate Change*, John Wiley & Sons, Inc., New York.

Sen P. K. (1968). Estimates of the regression coefficient based on Kendall's tau. *J. of the American Statistical Association*, 63, 1379-1389.

Simpson, D.; Fagerli, H.; Hellsten, S.; Knulst, K.; Westling, O. (2006). Comparison of modelled and monitored deposition fluxes of sulphur and nitrogen to ICP-forest sites in Europe. Biogeosciences 3, 337–355.

Stoddard, J. L. Long-Term Changes In Watershed Retention Of Nitrogen - Its Causes And Aquatic Consequences (1994). *Environmental Chemistry Of Lakes And Reservoirs*. 237: 223-284.

Stohl, A. (2006). Characteristics of atmospheric transport into the Arctic troposphere. *J. Geophys. Res.* 111, D11306, doi:10.1029/2005JD006888.

Sutton, M. A.; Asman, W. A. H.; Ellermann, T.; van Jaarsveld, J. A.; Acker, K.; Aneja, V.; Duyzer, J.; Horvath, L.; Paramonov, S.; Mitosinkova, M.; Tang, Y. S.; Achtermann, B.; Gauger, T.; Bartniki, J.; Neftel, A. and Erisma, J.W. (2003). Establishing the link between ammonia emission control and measurements of reduced nitrogen concentrations and deposition. *Environ Monit. Asessm.* 82:149-85.

Tietema, A.; A.W. Boxman, A.W.; Bredemeier M.; Emmett, B.A.; Moldan F.; Gundersen P.; Schleppi P. & Wright R.F.: Nitrogen saturation experiments (NITREX) in coniferous forest ecosystems in Europe: a summary of results. *Environmental Pollution* 102: 433-437, 1998

Tørseth, K.; Aas, W. & Solberg, S. (2001). Trends in airborne sulphur and nitrogen compounds in Norway during 1985-1996 in relation to airmass origin. *Water, Air and Soil. Poll.* **130**, 1493-1498..

Weiler, K.; Fischer, H.; Fritzsche, Ruth, U.; Wilhelms, F. & Miller H. (2005). Glaciochemical reconnaissance of a new ice core from Severnaya Zemlya, Eurasian Arctic. *J. Glaciology*, Vol. 51, No. 172, 64-74.

Wesely M.L. & Hicks B.B. (2000). A review of the current status of knowledge on dry deposition. *Atm. Env.* 34, 2261-2282.

7

Climate change: impacts on fisheries and aquaculture

Bimal P Mohanty[1], Sasmita Mohanty[2],
Jyanendra K Sahoo[3] and Anil P Sharma[1]
[1]Central Inland Fisheries Research Institute, Barrackpore, Kolkata 700120;
[2]School of Biotechnology, KIIT University, Bhubaneswar 751024,
[3]Orissa University of Agriculture & Technology, College of Fisheries, Berhampur760007; India.

Climate change has been recognized as the foremost environmental problem of the twenty-first century and has been a subject of considerable debate and controversy. It is predicted to lead to adverse, irreversible impacts on the earth and the ecosystem as a whole. Although it is difficult to connect specific weather events to climate change, increases in global temperature has been predicted to cause broader changes, including glacial retreat, arctic shrinkage and worldwide sea level rise. Climate change has been implicated in mass mortalities of several aquatic species including plants, fish, corals and mammals. The present chapter has been divided in to two parts; the first part discusses the causes and general concerns of global climate change and the second part deals, specifically, on the impacts of climate change on fisheries and aquaculture, possible mitigation options and development of suitable monitoring tools.

1. Global Climate change: Causes and concerns

Climate change is the variation in the earth's global climate or in regional climates over time and it involves changes in the variability or average state of the atmosphere over durations ranging from decades to millions of years. The United Nations Framework Convention on Climate Change (UNFCCC) uses the term 'climate change' for human-caused change and 'climate variability' for other changes. In last 100 years, ending in 2005, the average global air temperature near the earth's surface has been estimated to increase at the rate of 0.74 +/- 0.18 °C (1.33 +/- 0.32 °F) (IPCC 2007). In recent usage, especially in the context of environmental policy, the term 'climate change' often refers to changes in the modern climate.

2. Causes of climate change

There are both natural processes and anthropogenic activities affecting the earth's temperature and the resultant climate change. The steep increases in the global anthropogenic greenhouse gas (GHG) emissions over the decades are major contributors to the global warming.

2.1. Natural processes affecting the earth's temperature

Sun is the primary source of energy on earth. Though the sun's output is nearly constant, small changes over an extended period of time can lead to climate change. The earth's climate changes are in response to many natural processes like orbital forcing (variations in its orbit around the Sun), volcanic eruptions, and atmospheric greenhouse gas concentrations. Changes in atmospheric concentrations of greenhouse gases and aerosols, land-cover and solar radiation alter the energy balance of the climate system and causes warming or cooling of the earth's atmosphere. Volcanic eruptions emit many gases and one of the most important of these is sulfur dioxide (SO_2) which forms sulfate aerosol (SO_4) in the atmosphere.

2.2 Greenhouse gases

Greenhouse gases (GHGs) are those gaseous constituents of the atmosphere, both natural and anthropogenic, that are responsible for the greenhouse effect, leading to an increase in the amount of infrared or thermal radiation near the surface. While water vapor (H_2O), carbon dioxide (CO_2), nitrous oxide (N_2O), methane (CH_4), and ozone (O_3) are the primary greenhouse gases in the Earth's atmosphere, there are a number of entirely human-made greenhouse gases in the atmosphere, such as the halocarbons and other chlorine- and bromine-containing substances. Halocarbons such as CFCs (chlorofluorocarbons) are completely artificial (man-made), and are produced from the chemical industry in which they are used as coolants and in foam blowing.

Increases in CO_2 are the single largest factor contributing more than 60% of human-enhanced increases and more than 90% of rapid increase in past decade. Most CO_2 emissions are from the burning of fossil fuels such as coal, oil, and gas. Rising CO_2 is also related to deforestation, which eliminates an important carbon sink of the terrestrial biosphere (www.ncdc.noaa.gov/oa/climate/globalwarming.html; Shea et al., 2007). Currently, the atmosphere contains about 370 ppm of CO_2, which is the highest concentration in 420000 years and perhaps as long as 2 million years. Estimates of CO_2 concentrations at the end of the 21st century range from 490 to 1260 ppm, or a 75% to 350% increase above preindustrial concentrations (WMO World Data Centre for Greenhouse Gases. Greenhouse gas bulletin, 2006; Shea KM and the Committee on Environmental Health, 2007).

3. Impacts of climate change

Although it is difficult to connect specific weather events to global warming, an increase in global temperatures may in turn cause broader changes, including glacial retreat, arctic shrinkage, and worldwide sea level rise. Changes in the amount and pattern of precipitation may result in flooding and drought. Other effects may include changes in agricultural yields, addition of new trade routes, reduced summer stream flows, species extinctions, and increases in the range of disease vectors (Understanding and responding to Climate Change. 2008: http://www.national-academies.org).

Most models on Global climate change indicate that snow pack is likely to decline on many mountain ranges in the west, which would bring adverse impact on fish populations, hydropower, water recreation and water availability for agricultural, industrial and residential use. Partial loss of ice sheets on polar land could imply meters of sea level rise, major changes in coastlines and inundation of low-lying areas, with greatest effects in river deltas and low-lying islands. Such changes are projected to occur over millennial time scales, but more rapid sea level rise on century time scales cannot be excluded. Current models of climate change predict a rise in sea surface temperatures of between 2 °C and 5 °C by the year 2100 (IPCC Third Assessment Report, 2001: Done et al., 2003).

Climate change will affect ecosystems and human systems like agricultural, transportation and health infrastructure. The regions that will be most severely affected are often the regions that are the least able to adept. Bangladesh is projected to lose 17.5 % of its land if sea level rises about 1 meter (39 inches), displacing millions of people. Several islands in the South Pacific and Indian oceans may disappear. Many other coastal regions will be at increased risk of flooding, especially during storm surges, threatening animals, plants and human infrastructure such as roads, bridges and water supplies.

There are many ways in which climate change might affect human health, including heat stress, heat (sun) stroke, increased air pollution, and food scarcities due to drought and other agricultural stresses. Because many disease pathogens and carriers are strongly influenced by temperature, humidity and other climate variables, climate change may also influence the spread of infectious diseases or the intensity of disease outbreaks. During the last 100 years, anthropogenic activities related to burning fossil fuel, deforestation and agriculture has led to a 35% increase in the CO_2 levels in the temperature and this has resulted in increased trapping of heat and the resultant increase in the earth's atmosphere. Most of the observed increase in globally-averaged temperatures has been attributed to the greenhouse gas concentrations. The globally averaged surface temperature rise has been projected to be 1.1-6.4 °C by end of the 21st century (2090-2099) which is mainly due to thermal expansion of the ocean (www.searo.who.int/en/Section260/Section2468_14335.htm, 2008). The global average sea level rose at an average rate of 1.8 mm per year from 1961 to 2003 and the total rise during the 20th century was estimated to be 0.17 m (The Fourth Assessment Report of IPCC, 2007). Due to such surface warming it is predicted that heat waves and heavy precipitations will continue to become more frequent with more intense and devastating tropical cyclones (typhoons and hurricanes). Due to the resultant disruption in ecosystem's services to support human health and livelihood, there will be strong negative impact on the health system. IPCC has projected an increase in malnutrition and consequent disorders, with implications for child growth and development. Increased burden of diarrheal diseases and infectious disease vectors are expected due to the erratic rainfall patterns.

Climate change is likely to lead to some irreversible impacts. Approximately 20- 30 % of species assessed so far are likely to be at increased risk of extinction if increases in global average warming exceed 1.5-2.5 °C (relative to 1980-1999). As global average temperature increase exceeds about 3.5 °C, model projections suggest significant extinctions (40-70 % of species assessed) around the globe. Some projected regional impacts of Climate change have been systematically listed in the IPCC Fourth Assessment Report, 2007.

4. Impacts of Climate Change on Fisheries and Aquaculture

Fish has been an important part of the human diet in almost all countries of the world. It is highly nutritious; it can provide vital nutrients absent in typical starchy staples which dominate poor people's diets (FAO, 2005a; FAO, 2007a). Fish provides about 20 % of animal protein intake (Thorpe et al., 2006) and is one of the cheapest sources of animal proteins as far as availability and affordability is concerned. While it serves as a health food for the affluent world owing to the fish oils rich in polyunsaturated fatty acids (PUFAs), for the people in the other extreme of the nutrition scale, fish is a health food owing to its proteins, oils, vitamins and minerals and the benefits associated with the consumption of small indigenous fishes (Mohanty et al., 2010a).

Although aquaculture has been contributing an increasingly significant proportion of fish over recent decades, approximately two-thirds of fish are still caught in capture fisheries. The number of people directly employed in fisheries and aquaculture is estimated at 43.5 million, of which over 90 % are small -scale fishers (FAO, 2005a). In addition to those directly employed in fishing, over 200 million people are thought to be dependent on small-scale fishing in developing countries, in terms of other economic activities generated by the supply of fish (trade, processing, transport, retail, etc.) and supporting activities (boat building, net making, engine manufacture and repair, supply of services to fisherman and fuel to fishing boats etc.) in addition to millions for whom fisheries provide a supplemental income (FAO, 2005a). Fisheries are often available in remote and rural areas where other economic activities are limited and can thus be important sources for economic growth and livelihoods in rural areas with few other economic activities (FAO, 2005a)

4.1 Potential impacts of climate change on fisheries

Climate change is projected to impact broadly across ecosystems, societies and economics, increasing pressure on all livelihoods and food supplies. The major chunk of earth is encompassed by water that harbors vast majority of marine and freshwater fishery resources and thus likely to be affected to a greater extent by vagaries of climate change. Capture fisheries has unique features of natural resource harvesting linked with global ecosystem processes and thus is more prone to such problems. Aquaculture complements and increasingly adds to the supply chain and has important links with capture fisheries and is likely to be affected when the capture fisheries is affected.

The ecological systems which support fisheries are already known to be sensitive to climate variability. For example, in 2007, the International Panel on Climate Change (IPCC) highlighted various risks to aquatic systems from climate change, including loss of coastal wetlands, coral bleaching and changes in the distribution and timing of fresh water flows, and acknowledged the uncertain effect of acidification of oceanic water which is predicted to have profound impacts on marine ecosystems (Orr et al., 2005). Similarly, fishing communities and related industries are concentrated in coastal or low lying zones which are increasingly at risk from sea level rise, extreme weather events and wide range of human pressures (Nicholls et al., 2007a). While poverty in fishing communities or other forms of marginalization reduces their ability to adapt and respond to change, increasingly globalized fish markets are creating new vulnerabilities to market disruptions which may result from climate change.

Fisheries and fisher folk may have the impact in a wide range of ways due to climate change. The distribution or productivity of marine and fresh water fish stocks might be affected owing to the processes such as ocean acidification, habitat damage, changes in oceanography, disruption to precipitation and freshwater availability (Daw et al., 2009).

Climate change, in particular, rising temperatures, can have both direct and indirect effects on global fish production. With increased global temperature, the spatial distribution of fish stocks might change due to the migration of fishes from one region to another in search of suitable conditions. Climate change will have major consequences for population dynamics of marine biota via changes in transport processes that influence dispersals and recruitment (Barange and Perry, 2009). These impacts will differ in magnitude and direction for populations within individual marine species whose geographical ranges span large gradients in latitude and temperature, as experimented by Mantzouni and Mackenzie (2010) in cod recruitment throughout the north Atlantic. The effects of increasing temperature on marine and freshwater ecosystems are already evident, with rapid pole ward shifts in distributions of fish and plankton in regions such as North East Atlantic, where temperature change has been rapid (Brander, 2007). Climate change has been implicated in mass mortalities of many aquatic species, including plants, fish, corals, and mammals (Harvell et al., 1999; Battin et al., 2007).

Climate change will have impact on global biodiversity; alien species would expand into regions in which they previously could not survive and reproduce (Walther et al., 2009). Climate driven changes in species composition and abundance will alter species diversity and it is also likely to affect the ecosystems and the availability, accessibility, and quality of resources upon which human populations rely, both directly and indirectly through food web processes. Extreme weather events could result in escape of farmed stock and contribute to reduction in genetic diversity of wild stock affecting biodiversity.

Climate variability and change is projected to have significant effects on the physical, chemical, and biological components of northern Canadian marine, terrestrial, and freshwater systems. According to a study conducted by Prowse et al. (2009), the northward migration of species and the disruption and competition from invading species are already occurring and will continue to affect marine, terrestrial, and freshwater communities. This will have implications for the protection and management of wildlife, fish, and fisheries resources; protected areas; and forests. Shifting environmental conditions will likely introduce new animal-transmitted diseases and redistribute some existing diseases, affecting key economic resources and some human populations. Stress on populations of iconic wildlife species, such as the polar bear, ringed seals, and whales, will continue as a result of changes in critical sea-ice habitat interactions. Where these stresses affect economically and culturally important species, they will have significant effects on people and regional economies. Further integrated, field-based monitoring and research programs, and the development of predictive models are required to allow for more detailed and comprehensive projections of change to be made, and to inform the development and implementation of appropriate adaptation, wildlife, and habitat conservation and protection strategies.

Fisheries will also be exposed to a diverse range of direct and indirect climate impacts, including displacement and migration of human populations; impacts on coastal communities and infrastructure due to sea level rise; and changes in the frequency, distribution or intensity of tropical storms. Inland fisheries ecology is profoundly affected by changes in precipitation and run-off which may occur due to climate change. Lake fisheries in Southern Africa for example, will likely be heavily impacted by reduced lake levels and catches. The variety of different impact mechanisms, complex interactions between social, ecological and economic systems and the possibility of sudden and surprising changes make future effects of climate change on fisheries difficult to predict. In fact, understanding the ecological impacts of climate change is a crucial challenge of the twenty-first century. There is a clear lack of general rules regarding the impacts of global warming on biota. A study conducted by Daufresne et al. (2009) provided evidence that reduced body size is the third universal ecological response to global warming in aquatic systems besides the shift of species ranges toward higher altitudes and latitudes and the seasonal shifts in life cycle events.

Apart from fisheries, global primary production (planktonic primary production) which is related to global fisheries catches at the scale of Large Marine Ecosystems appears to be declining, in some part due to climate variability and change, with consequences for the near future fisheries catches (Chassot et al., 2010).

Other climatic change impacts on fisheries include surface winds, high CO_2 levels and variability in precipitations. While surface wind would alter both the delivery of nutrients in to the photic zone and strength and distribution of ocean currents, higher CO_2 levels can change the ocean acidity and variability in precipitation would affect sea levels. Global average sea level is rising at an average rate of 1.8 mm per year since 1961 and there is evidence of increased variability in sea level in recent decades. It is recently reported that ocean temperature and associated sea level increases between 1961 and 2003 were 50% larger than estimated in the 2007 IPCC Report. All coastal ecosystems are vulnerable to sea level rise and more direct anthropogenic impacts. Sea level rise may reduce intertidal habitat areas in ecologically important regions thus affecting fish and fisheries.

4.2 Impact of climate change on the parasites and infectious diseases of aquatic animals

The potential trends of climate change on aquatic organisms and in turn in fisheries and aquaculture are less well documented and have primarily concentrated on coral bleaching and associated changes. An increase in the incidence of disease outbreaks in corals and marine mammals together with the incidence of new diseases has been reported. It was suggested that both the climate and human activities may have accelerated the global transport of species, bringing together of pathogens and previously unexposed populations (Harvell et al., 1999; De Silva and Sato, 2009).

Climate changes could affect productivity of aquaculture systems and increase the vulnerability of cultured fish to diseases. All aquatic ecosystems, including freshwater lakes and rivers, coastal estuarine habitats and marine waters, are influenced by climate change (Parry et al., 2007; Scavia et al., 2002; Schindler, 2001). Relatively small temperature changes alter fish metabolism and physiology, with consequences for growth, fecundity, feeding behavior, distribution, migration and abundance (Marcogliese, 2008). The general effects of increased temperature on parasites include, rapid growth and maturation, earlier onset of spring maturation, increased parasite mortality, increased number of generations per year, increased rates of parasitism and disease, earlier and prolonged transmission, the possibility of continuous, year-round transmission (Marcogliese, 2001).

Many diseases display greater virulence at higher temperatures that might be the result of reduced resistance of the host due to stress or increased expression of virulence factors/ increased transmission of the vectors. Some examples have been summarized in table 1.

Host	**Disease /Parasite**	**Response to high temperature**	**Reference**
Largemouth bass (*Micropterus salmoides*)	Red sore disease /bacterium *Aeromonas hydrophila*	Susceptibility to the disease increases	Esch and Hazen (1980)
Mosquitofish (*Gambusia affinis*)	Asian fish tapeworm (*Bothriocephalus acheilognathi*)	-do-	Granath and Esch (1983)
Trout (*Onchorhynchus spp.*)	Whirling disease / Myxozoan *Myxobolus cerebralis*	-do-	Hiner and Moffitt (2001)
Juvenile coho salmon (*O. kisutch*)	Blackspot disease/ trematode larvae (metacercariae)	Virulence is directly correlated with daily maximum temperature	Cairns et al., 2005
A variety of reef fish	Ciguatera fish poisoning (CFP) caused by bioaccumulation of algal toxins	Increased incidence of CFP due to increased temperature	Tester et al., 2010
Rainbow trout, *Oncorhynchus mykiss*	Infected with *Ichthyophonus* sp.	More rapid onset of disease, higher parasite load, more severe host tissue reaction and reduced mean-day-to-death at higher temperature	Kocan et al., 2009
Freshwater bryozoans infected with myxozoan, *Tetracapsuloides bryosalmonae*	Spores released from sacs produced by the parasite during infection of freshwater bryozoans are infective to salmonid fish, causing the devastating Proliferative Kidney Disease (PKD)	Exacerbate PKD outbreaks and increase the geographic range of PKD as a result of the combined responses of T. bryosalmonae and its bryozoan hosts to higher temperatures.	Tops et al., 2009

Table 1. Impact of climate change on parasitic and other diseases of aquatic animals.

As the emergence of disease is linked directly to changes in the ecology of hosts or pathogens, or both (Harvell et al., 1999), climate change will have a profound impact on the spread of parasites and disease in aquatic ecosystems (Harvell et al., 1999; Marcogliese, 2001; Harvell et al., 2002). Climate change will affect parasite species directly resulting from the extension of the geographical range of pathogens (Harvell et al., 2002). In addition, increased temperature may cause thermal stress in aquatic animals, leading to reduced growth, sub- optimal behaviors and reduced immunocompetence (Harvell et al., 1999; Harvell et al., 2002; Roessig et al., 2004) resulting in changes in the distribution and abundance of their hosts (Marcogliese, 2001). In the oceans, diseases are shown to increase in corals, sea urchins, molluscs, sea turtles and marine mammals, although not all can be linked unequivocally to climate alone (Lafferty et al., 2004). However, it was recently suggested that diseases may not increase with climate change, although distributions of parasites and pathogens will undoubtedly shift (Lafferty, 2009). Other factors may dominate over climate in controlling the distribution and abundance of pathogens, including: habitat alteration, invasive species, agricultural practices and human activities.

Effects on parasites	**Effects on hosts**	**Effects on transmission**
Faster embryonic development and hatching	Altered feeding	Earlier reproduction in spring
Faster rates of development and maturation	Altered behavior	More generations per year
Decreased longevity of larvae and adults	Altered range	Prolonged transmission in the fall
Increased mortality of all stages	Altered ecology Reduced host resistance	Potential transmission year round

Table 2. General effects of increased temperature on parasite life cycles, their hosts and transmission processes (Marcogliese, 2008)

Outbreaks of numerous water- borne diseases in both humans and aquatic organisms are linked to climatic events, although it is often difficult to disentangle climatic from other anthropogenic effects. In some cases, these outbreaks occur in foundation or keystone species, with consequences throughout whole ecosystems. There is much evidence to suggest that parasite and disease transmission, and possibly virulence, will increase with global warming. However, the effects of climate change will be superimposed on a multitude of other anthropogenic environmental changes. Climate change itself may exacerbate these anthropogenic effects. Moreover, parasitism and disease may act synergistically with these anthropogenic stressors to further increase the detrimental effects of global warming on animal and human populations, with debilitating social economic ramifications (Marcogliese, 2008).

The repercussions of climate change are not limited solely to temperature effects on hosts and their parasites, but also have other possible effects such as: alteration in water levels and flow regimes, eutrophication, stratification, changes in acidification, reduced ice cover, changes in ocean currents, increased ultra- violet (UV) light penetration, run off, weather extremes (Cochrane et al., 2009).

5. Anticipated impacts in next few decades

In addition to incremental changes of existing trends, complex social and ecological systems such as coastal zones and fisheries, may exhibit sudden qualitative shifts in behaviour when forcing variables past certain thresholds (Daw et al., 2009). For example, IPCC originally estimated that the Greenland ice sheet would take more than 1000 years to melt, but recent observations suggest that the process is already happening faster owing to mechanisms for ice collapse that were not incorporated into the projections (Lenton et al., 2008). The infamous collapse of the Northwest Atlantic northern cod fishery provides a non-climate-related example where chronic over fishing led to a sudden, unexpected and irreversible loss in production from this fishery. Thus, existing observations of linear trends cannot be used to reliably predict impacts within the next 50 years (Daw et al., 2009).

A study by Veron et al. (2009) also emphasizes impact of increasing atmospheric CO_2 levels due to global warming on mass coral bleaching world-wide. According to this group, temperature-induced mass coral bleaching causing mortality on a wide geographic scale started when atmospheric CO_2 levels exceeded approximately 320 ppm. At today's level of approximately 387 ppm, allowing a lag-time of 10 years for sea temperatures to respond, most reefs world-wide are committed to an irreversible decline. Mass bleaching will in future become annual, departing from the 4 to 7 years return-time of El Niño events. Bleaching will be exacerbated by the effects of degraded water-quality and increased severe weather events. In addition, the progressive onset of ocean acidification will cause reduction of coral growth and retardation of the growth of high magnesium calcite-secreting coralline algae. If CO_2 levels are allowed to reach 450 ppm (due to occur by 2030-2040 at the current rates), reefs will be in rapid and terminal decline world-wide from multiple synergies arising from mass bleaching, ocean acidification, and other environmental impacts. Damage to shallow reef communities will become extensive with consequent reduction of biodiversity followed by extinctions. Reefs will cease to be large-scale nursery grounds for fish and will cease to have most of their current value to humanity. There will be knock-on effects to ecosystems associated with reefs, and to other pelagic and benthic ecosystems. This is likely to have been the path of great mass extinctions of the past, adding to the case that anthropogenic CO_2 emissions could trigger the Earth's sixth mass extinction (Veron et al., 2009).

6. Climate change impacts on inland fisheries - the Indian scenario

In recent years the climate is showing perceptible changes in the Indian subcontinent, where the average temperature is on the rise over the last few decades. In India, observed changes include an increase in air temperature, regional monsoon variation, frequent droughts and regional increase in severe storm incidences in coastal states and Himalayan glacier recession (Vass et al., 2009). In some states like West Bengal, the average minimum and maximum temperatures has increased in the range of 0.1 - 0.9 °C throughout the state. The average rainfall has decreased and monsoon is also delayed; consequently, the climate change impact is being felt on the temperature of the inland water bodies and on the breeding behavior of fishes. It is well known that temperature is an important factor which strongly influence the reproductive cycle in fishes. Temperature, along with rainfall and photoperiod, stimulate the endocrine glands of fishes which help in the maturation of the gonads. In India, the inland aquaculture is centered on the Indian major carps, *Catla catla,*

Labeo rohita and Cirrhinus mrigala and their spawning occurs during the monsoon (June-July) and extend till September. In recent years the phenomenon of IMC maturing and spawning as early as March is observed, making it possible to breed them twice a year. Thus, there is an extended breeding activity as compared to a couple of decades ago (Dey et al., 2007), which appears to be a positive impact of the climate change regime.

Fig. 1. Course of the River Ganga showing different stretches (http://www.gits4u.com/water/ganga1.gif)

The mighty river Ganga forms the largest river system in India and not only millions of people depend on its water but it provides livelihood to a large group of fishermen also. The entire length of the river, with a span of 2,525 km from source to mouth is divided into three main stretches consisting of upper (Tehri to Kanauji), middle (Kanpur to Patna) and lower (Sultanpur to Katwa) (Figure 1). From analysis of 30 years' time series data on river Ganga and water bodies in the plains, Vass et al. (2009) reported an increase in annual mean minimum water temperature in the upper cold-water stretch of the river (Haridwar) by 1.5 °C (from 13 °C during 1970-86 to 14.5 °C during 1987-2003) and by 0.2- 1.6 °C in the aquaculture farms in the lower stretches in the Gangetic plains. This change in temperature clime has resulted in a perceptible biogeographically distribution of the Gangetic fish fauna. A number of fish species which were never reported in the upper stretch of the river and were predominantly available in the lower and middle stretches in the 1950s (Menon, 1954) have now been recorded from the upper cold-water region. Among them, *Mastocembelus armatus* has been reported to be available at Tehri-Rishikesh and *Glossogobius gurius* is available in the Haridwar stretch (Sinha et al., 1998) and *Xenentodon cancila* has also been reported in the cold-water stretch (Vass et al., 2009). The predator-prey ratio in the middle stretch of the river has been reported to be declined from 1:4.2 to 1:1.4 in the last three decades. Fish production has been shown to have a distinct change in the last two decades where the contribution from IMCs has decreased from 41.4% to 8.3% and that from catfishes and miscellaneous species increased (Vass et al., 2009).

7. Adaptation and mitigation options

Adaptation to climate change is defined in the climate change literature as an adjustment in ecological, social or economic systems, in response to observed or expected changes in climatic stimuli and their effects and impacts in order to alleviate adverse impacts of change, or take advantage of new opportunities. Adaptation is an active set of strategies and actions taken by peoples in response to, or in anticipation to the change in order to enhance or maintain their well being. Hence adaptation is a continuous stream of activities, actions, decisions and attitudes that informs decisions about all aspects of life and that reflects existing social norms and processes (Daw et al., 2009).

Many capture fisheries and their supporting ecosystems have been poorly managed, and the economic losses due to overfishing, pollution and habitat loss are estimated to exceed $50 billion per year (World Bank & FAO, 2008). The capacity to adapt to climate change is determined partly by material resources and also by networks, technologies and appropriate governance structures. Improved governance, innovative technologies and more responsible practices can generate increased and sustainable benefits from fisheries.

There is a wide range of potential adaptation options for fisheries. To build resilience to the effects of climate change and derive sustainable benefits, fisheries and aquaculture managers need to adopt and adhere to best practices such as those described in the FAO 'Code of Conduct for Responsible Fisheries', reducing overfishing and rebuilding fish stocks. These practices need to be integrated more effectively with the management of river basins, watersheds and coastal zones. Fisheries and aquaculture need to be blended into National Climate Change Adaptation Strategies. In absence of careful planning, aquatic ecosystems, fisheries and aquaculture can potentially suffer as a result of adaptation measures applied by other sectors such as increased use of dams and hydro power in catchments with high rainfall, or the construction of artificial coastal defenses or marine wind farms (ftp://ftp.fao.org/FI/brochure/climate_change/policy_brief.pdf).

Mitigation solutions reducing the carbon footprint of Fisheries and Aquaculture will require innovative approaches. One example is the recent inclusion of Mangrove conservation as eligible for reducing emissions from deforestation and forest degradation in developing countries, which demonstrates the potential for catchment forest protection. Other approaches to explore include finding innovative but environmentally safe ways to sequester carbon in aquatic ecosystems, and developing low-carbon aquaculture production systems (ftp://ftp.fao.org/FI/brochure/climate_change/policy_brief.pdf).

There is mounting interest in exploiting the importance of herbivorous fishes as a tool to help ecosystems recover from climate change impacts. Aquaculture of herbivorous species can provide nutritious food with a small carbon footprint. This approach might be particularly suitable for recovery of coral reefs, which are acutely threatened by climate change. Surveys of ten sites inside and outside a Bahamian marine reserve over a 2.5-year period demonstrated that increases in coral cover, including adjustments for the initial size-distribution of corals, were significantly higher at reserve sites than those in non-reserve sites: macroalgal cover was significantly negatively correlated with the change in total coral cover over time. Reducing herbivore exploitation as part of an ecosystem-based management strategy for coral reefs appears to be justified (Mumby and Harborne, 2010). Furthermore, farming of shellfish, such as oysters and mussels, is not only good business, but also helps clean coastal water, while culturing aquatic plants help to remove waste from polluted water. In contrast to the potential declines in agricultural yields in many areas of the world, climate change opens new opportunities for aquaculture as increasing numbers of species are cultured (ftp://ftp.fao.org/FI/brochure/climate_change/policy_brief.pdf).

Marine fish is one of the most important sources of animal protein for human use, especially in developing countries with coastlines. Marine fishery is also an important industry in many countries. The depletion of fishery resources is happening mainly due to anthropogenic factors such as overfishing, habitat destruction, pollution, invasive species introduction, and climate change. The most effective ways to reverse this downward trend and restore fishery resources are to promote fishery conservation, establish marine-protected areas, adopt ecosystem-based management, and implement a "precautionary principle." Additionally, enhancing public awareness of marine conservation, which includes eco-labeling, fishery ban or enclosure, slow fishing, and MPA (marine protected areas) enforcement is important and effective (Shao, 2009).

The assessment report of the 4th International Panel on Climate Change confirms that global warming is strongly affecting biological systems and that 20-30% of species risk extinction from projected future increases in temperature. One of the widespread management strategies taken to conserve individual species and their constituent populations against climate-mediated declines has been the release of captive bred animals to wild in order to augment wild populations for many species. Using a regression model based on a 37-year study of wild and sea ranched Atlantic salmon (*Salmo salar*) spawning together in the wild, McGinnity et al. (2009) showed that the escape of captive bred animals into the wild can substantially depress recruitment and more specifically disrupt the capacity of natural populations to adapt to higher winter water temperatures associated with climate variability, thus increasing the risk of extinction for the studied population within 20 generations. According to them, positive outcomes to climate change are possible if captive bred animals are prevented from breeding in the wild. Rather than imposing an additional genetic load on wild populations by releasing maladapted captive bred animals, they propose that conservation efforts should focus on optimizing conditions for adaptation to occur by reducing exploitation and protecting critical habitats.

8. Monitoring stress in aquatic animals and HSP70 as a possible monitoring tool

Temperature above the normal optimum are sensed as heat stress by all organisms, Heat stress (HS) disturbs cellular homeostasis and can lead to severe retardation in growth and development and even death. Heat shock (stress) proteins (HSP) are a class of functionally related proteins whose expression is increased when cells are exposed to elevated temperatures or other stress. The dramatic up regulation of the HSPs is a key part of heat shock (stress) response (HSR). The accumulation of HSPs under the control of heat shock (stress) transcription factors (HSFs) play a central role in the heat stress response (HSR) and acquired thermo tolerance. HSPs are highly conserved and ubiquitous and occur in all organisms from bacteria to yeast to humans. Cells from virtually all organisms respond to different stress by rapidly synthesizing the HSPs and therefore, HSPs are widely used as biomarkers for stress response (Jolly and Marimoto, 2000). HSPs have multiple housekeeping functions, such as activation of specific regulatory proteins and folding and translocation of newly synthesized proteins. HSPs are usually produced in large amounts (induction) in response to distinct stressors such as ischemia, hypoxia, chemical/toxic insult, heavy metals, oxidative stress, inflammation and altered temperature or heat shock (Marimoto, 1998).

Out of different HSPs, the HSP70 is unique in many ways; it acts as molecular chaperone in both unstressed and stressed cells. HSC70, the constitutive HSP70 is crucial for the chaperoning functions of unstressed cells, where as the inducible HSP70 is important for allowing cells to cope with acute stress, especially those affecting the protein machinery. HSP70 in marine mussels are widely used as a potential biomarker for stress response and aquatic environmental monitoring of the marine ecosystem (Li et al., 2000).

The success of any organism depends not only on niche adaptation but also the ability to survive environmental perturbation from homeostasis, a situation generally described as stress (Clark et al., 2008a). Although species-specific mechanisms to combat stress have been described, the production of heat shock proteins (HSPs), such as HSP70, is universally described across all taxa. We have studied expression profile of the HSP70 proteins, in different tissues of the large riverine catfish *Sperata seenghala* (Mohanty et al., 2008), freshwater catfish *Rita rita* (Mohanty et al., 2010b), Indian catfish *Clarias batrachus*, Indian major carps *Labeo rohita, Catla catla, Cirrhinus mrigala,* exotic carp *Cyprinus carpio var. communis* and the murrel *Channa striatus,* the climbing perch *Anabas testudineus* (CIFRI, 2009; Mohanty et al., 2009). Out of these, the IMCs are the major aquaculture species and therefore are of much economic significance. Similarly, Anabas and Channa fetch good market value and their demand is increasing owing to their perceived therapeutic value (Mohanty et al., 2010a). The large riverine catfish *S. seenghala* comprises the major fisheries in majority of rivers and reservoirs and the freshwater catfish *Rita rita* has a good market demand and these two comprise a major share of the capture fisheries in India.

Monoclonal anti-HSP70 antibody (H5147, Sigma), developed in mouse against purified bovine brain HSP70, in immunoblotting localizes both the constitutive (HSP73) and inducible (HSP72) forms of HSP70. The antibody recognizes brain HSP70 of bovine, human, rat, rabbit, chicken, and guinea pig. We observed immunoreactivity of this antibody with HSP70 proteins in different organs and tissues of a variety of fish species (Table 3). The strong immunoreactivity indicates that the HSP70 proteins of bovine and this riverine catfish *Rita rita* share strong homology although fish belong to a clade phylogenetically distant from the bovines. Persistent, high level of expression of HSP70 was observed in muscle tissues of *Rita rita* and for this reason, we have used and recommend use of white muscle tissue of *Rita rita* as a suitable positive control in analysis of HSP70 expression in tissues of other organisms (Mohanty et al., 2010b).

Early studies on heat shock response in Antarctic marine ectoderms had led to the conclusion that both microorganisms and fish lack the classical heat shock response, i.e. there is no increase in HSP70 expression when warmed (Carratti et al., 1998; Hofmann et al., 2000). However, later it was reported that other Antarctic animals, show an inducible heat shock response, at a level probably set during their temperate evolutionary past (Clark et al., 2008 a, b); the bivalve (clam) *Laternula elliptica* and gastropod (limpet) *Nacella concinna* show an inducible heat shock response at 8 °C and 15 °C, respectively and these are temperatures in excess of that which is currently experienced by these animals, which can be attributed to the global warming (Waller et al., 2006). Permanent expression of the inducible HSP70 genes, species-specific high expression of HSC70 (*N. concinna*) and permanent expression of GRP78 (*N concinna* and *L. elliptica*) indicates that, as for fish, chaperone proteins form an essential part of the adaptation of the biochemical machinery of these animals to low but stable temperatures. High constitutive levels of HSP gene family member expression may be a compensatory mechanism for coping with elevated protein damage at low temperature analogous to the permanent expression of HSP70 in the Antarctic notothenoids (Clark et al., 2008 a). Such studies clearly indicate that both genetics and environment play important role in spatio-temporal gene expression.

Fish species	Liver	Muscle	Kidney	Gill	Remarks
Labeo rohita	-	++	++	++	Mohanty et al. 2009
Cirrhinus mrigala	++	-	-	++	CIFRI 2009; Mohanty et al. 2009
Cyprinous carpio var communis	++	++	++	-	-do-
Anabas testudineus	++	-	-	++	-do-
Channa punctatus	-	-	++		-do-
Sperrata seenghala	++	++	++	+	Mohanty et al. 2008
Rita rita	++	++	++	+	Mohanty et al. 2010b

Table 3. HSP70 expression profile in different tissues of some freshwater fishes, both aquacultured and wild stock.

There is need to standardize tools suitable for monitoring stress resulting from global warming and climate change impacts, in the aquatic animals from both aqua culture and capture fisheries systems. As HSP70 expression has been reported in many fish species (Table 3) it might serve as a suitable tool for monitoring impact of thermal stress/global warming; however, as HSP70 proteins are expressed under other conditions also, it is necessary to identify the heat shock (stress) transcription factors (HSFs) that can be specifically attributed to global warming (thermal stress) and climate change. It is also necessary to distinguish the constitutive and induced forms of the transcripts/proteins by qPCR/proteomic analysis so that specific HSP70 forms suitable for monitoring performance of the farmed fishes can be monitored for better management of aquacultured animals.

IPCC have predicted an average global warming between +2 and +6 °C, depending on the scenarios, within the next 90 years (IPCC 2007). The consequences of this increase in temperature are now well documented on both the abundance and geographic distribution of numerous taxa i.e. at population or community levels; in contrast, studies at the cellular level are still scarce. The study of the physiological or metabolic effects of such small increases in temperature is difficult because they are below the amplitude of the daily or seasonal thermal variations occurring in most environments. The underground water organisms are highly thermally buffered and thus are well suited for characterization of cellular responses of global warming. Colson-Proch et al. (2010) studied the genes encoding HSP70 family chaperones in amphipod crustaceans belonging to the ubiquitous sub-terranean genus *Niphargus* and HSP 70 sequence in 8 populations of 2 complexes of species of this genus (*Niphargus rhenorhodanensis* and *Niphargus virei* complexes). Expression profiles of HSP70 were determined for one of these populations by reverse transcription and quantitative polymerase chain reaction, confirming the inducible nature of this gene. An increase of 2 °C seem to be without any effect on *N. rhenorhodanensis* physiology whereas a heat shock of + 6 °C represented an important thermal stress for these individuals. Thus this study showed that although *Niphargus* individuals do not undergo any daily or seasonal thermal variations in underground water, they display an inducible HSP70 heat shock response (Colson-Proch et al., 2010).

9. Epilogue

There are opposing viewpoints on the predicted impacts of 'global warming' also. Scientists warn against overselling climate change. Some experts feel that the data produced by models used to project weather changes, risk being over-interpreted by governments, organizations and individuals keen to make plans for a changing climate, with dangerous results. The point made is that the Global Climate Models (GCMs) help us understand pieces of the climate system, but that does not mean we can predict the details. Thus, indications of changes in the earth's future climate must be treated with the utmost seriousness and with the precautionary principle uppermost in our minds. Extensive climate change may alter and threaten the living conditions of much of mankind. They may induce large-scale migration and lead to greater competition for the earth's resources. Such changes will place particularly heavy burdens on the world's most vulnerable countries. There may be increased danger of violent conflicts and wars, within and between states. A wide array of adaptation options is available, but more extensive adaptation than is currently occurring is required to reduce vulnerability to climate change.

Although the understanding of climate change has advanced significantly during the past few decades, many questions remain unanswered. The task of mitigating and adapting to the impacts of climate change will require worldwide collaborative input from a wide range of experts from various fields. The common man's contribution will play a major role in reducing the impacts of climate change and protecting the earth from climate change-related hazards. The impacts of climate change to freshwater aquaculture in tropical and subtropical region is difficult to predict as marine and freshwater populations are affected by synergistic effects of multiple climate and noncelibate stressors. If such noncelibate factors are identified and understood then it may be possible for local predictions of climate change impacts to be made with high confidence (De Silva and Soto, 2009).

Coastal communities, fishers and fish farmers are profoundly affected by climate change. Climate change is modifying the distribution and productivity of marine and freshwater species and is already affecting biological processes and altering food webs, thus making the consequences for sustainability of aquatic ecosystems for fisheries and aquaculture, and for the people dependent on them, uncertain. Fisheries, aquaculture and fish habitats are at risk. Deltas and estuaries are in the fore front and thus, most vulnerable to climate change. Mitigation measures are urgently needed to neutralize and alleviate these growing threats, to adapt to their impacts and also to build our knowledge base on Complex Ocean and aquatic processes. The prime need is to reduce the global emissions of GHGs, which is the primary anthropogenic factor responsible for climate change (ProAct Network, 2008).

Healthy aquatic ecosystems contribute greatly to food security and livelihoods. They are critical for production of wild fish and for some of the seed and much of the feed (trash fish) for aquaculture. Coastal ecosystems provide food, habitats and nursery grounds for fish. Estuaries, coral reefs, mangroves and sea grass beds are particularly important. Mangroves create barriers to destructive waves from storms and hold sediments in place with their extensive root systems thereby reducing coastal erosion. Healthy coral reefs, sea grass beds and wetlands provide similar benefits. Thus, these natural systems not only support fisheries, but help protect communities from the terrible impacts of natural hazards and disasters also (ProAct Network, 2008). In freshwater systems, ecosystem health and productivity is linked to water quality and flow and the health of wetlands. Ecosystem-based approaches to fisheries and coastal zone management are highly beneficial as such approaches recognize the need for people to use the ecosystem for their food security and livelihoods while enabling these valuable natural assets to adapt to the effects of climate change, and to reduce the threats from other environmental stresses (Hoegh-Guldberg et al., 2007).

Fish and shellfish provide essential nutrition for 3 billion people and about 50% of animal protein and micronutrients to 400 million people in the poorest countries of the world. Fish is one of the cheapest sources of animal proteins and play important role in preventing protein-calorie malnutrition. The health benefits of eating fish are being increasingly understood by the consumers. Over 500 million people in the developing countries depend on fisheries and aquaculture for their livelihoods. Aquaculture is the world's fastest growing food production system, growing at 7% annually. Fish products are among the most widely traded foods internationally (ftp://ftp.fao.org/FI/brochure/climate_change/policy_brief.pdf).

Implementing adaptation and mitigation pathways for communities dependent on fisheries, aquaculture and aquatic ecosystems will need increased attention from policy-makers and planners. Sustainable and resilient aquatic ecosystems will benefit the fishers as well as the coastal communities and will provide good and services at national and global levels. Fisheries and aquaculture need specific adaptation and mitigation measures like: improving the management of fisheries and aquaculture as well as the integrity and resilience of aquatic ecosystems; responding to the opportunities for and threats to food and livelihood security due to climate change impacts; and helping the fisheries and aquaculture sector reduce GHG emissions. To conclude, the present generation is already facing the harmful effects of the climate change; however, the future generations will suffer most of the harmful effects of global climate change. So, the present generation need to decide, whether to aggressively reduce the chances of future harm at the cost of sacrificing some luxuries or to let our descendants largely fend for themselves (Broome, 2008). Thus, how we handle the issue of Climate Change is more of an ethical question and the global community must act sensibly and responsibly.

10. References

Barange, M., & Perry, R.I. (2009) Physical and ecological impacts of climate change relevant to marine and inland capture fisheries and aquaculture In: *Climate change implications for fisheries and aquaculture overview of current scientific Knowledge,* Cochrane, K., Young, C. De, Soto, D., & Bahri, T. (Eds). FAO Fisheries and Aquaculture Technical paper: No. 530, pp. 7-106, FAO, Rome.

Battin, J., Wiley, M. W., Ruckelshaus, M. H., Palmer, R. N,. Korb, E., Bartz, K. K., & Imaki, H. (2007) Projected impacts of climate change on salmon habitat restoration, *Proc. Natl. Acad. Sci, USA,* 104, 6720-6725.

Brander, K. M. (2007) Global fish production and climate change, *Proc. Natl. Acad. Sci., USA*, 104, 19709-19714.

Broome, J. (2008) The ethics of climate change, *Sci. Am.*, 298, 96-100.

Cairns, M. A., Ebersole, J. L., Baker, J. P., Wigngton, P. J. Jr., Lavigne, H. R., & Davis, S. M. (2005) Influence of summer stream temperatures on black spot infestation of juvenile coho salmon in the Oregon Coast Range, *Trans. Am. Fish. Soc.*, 134, 1471-1479.

Carrattù, L., Gracey, A. Y, B.uono, S., & Maresca, B. (1998) Do Antarctic fish respond to heat shock? In: *Fishes of Antarctica. A Biological Overview*. di Prisco, G., Pisano, E., Clarke, A. (Eds) Springer, Italy.

Chassot, E., Bonhommeau, S., Dulvy NK, Mélin F, Watson R, Gascuel D, Le Pape O. (2010) Global marine primary production constrains fisheries catches. *Ecol Lett.*, Feb 5. [Epub ahead of print]

CIFRI (2009) Annual Report. Central Inland Fisheries Research Institute, Barrackpore, Kolkata, India. ISSN 0970 6267.

Clark, M. S., Fraser, K. P. P., & Peck, L. S. (2008a) Antarctic marine molluscs do have an HSP70 heat shock response, *Cell Stress Chaperon.*, 13, 39-49.

Clark, M. S., Geissler, P., Waller, C., Fraser, K. P. P., Barnes, D. K. A., & Peck, L. S. (2008b) Low heat shock thresholds in wild Antarctic inter-tidal limpets (*Nacella concinna*). *Cell Stress Chaperon.*, 13, 51-58.

Cochrane, K., Young, C. De, Soto, D., & Bahri, T. (2009) *Climate change implications for fisheries and aquaculture: overview of current scientific knowledge*. FAO Fisheries and Aquaculture Technical paper: No. 530,FAO, Rome.

Colson-Proch, C., Morales, A., Hervant, F., Konecny, L., Moulin, C., & Douady, C. J. (2010) First cellular approach of the effects of global warming on groundwater organisms: a study of the HSP70 gene expression. Cell Stress Chaperon., 15, 3, 259-270.

Daufresne, M., Lengfellner, K., & Sommer, U. (2009) Global warming benefits the small in aquatic ecosystems. *Proc Natl Acad Sci USA.*, 106, 31, 12788-12793.

Daw, T., Adger, W. N., Brown, K., & Badjeck, M.-C. (2009) Climate change and capture fisheries: potential impacts, adaptation and mitigation. In: *Climate change implications for fisheries and aquaculture overview of current scientific Knowledge*, Cochrane, K., Young, C. De, Soto, D., & Bahri, T. (Eds). FAO Fisheries and Aquaculture Technical paper: No. 530, pp.107-150, FAO, Rome.

De Silva, S. S. and Soto, D. 2009, Climate change and aquaculture: potential impacts, adaptation and mitigation In: *Climate change implications for fisheries and aquaculture overview of current scientific Knowledge*, Cochrane, K., Young, C. De, Soto, D., & Bahri, T. (Eds). FAO Fisheries and Aquaculture Technical paper: No. 530, pp. 151-212, FAO, Rome.

Dey, S., Srivastava, P. K., Maji, S., Das, M. K., Mukhopadhyay, M. K., & Saha, P. K. (2007) Impact of climate change on the breeding of Indian major carps in West Bengal. *J. Inland Fish. Soc. India*, 39, 1, 26-34.

Done, T., Whetton, P., Jones, R. *et al.* (2003) Global climate change and coral bleaching on the Great Barrier Reef. Final report to the State of Queensland Greenhouse taskforce through the Department of Natural Resources and Mines, Queensland,.

Esch, G. W., & Hazen, T. C. (1980) Stress and body condition in a population of largemouth bass: implications for red-sore disease, *Trans. Am. Fish. Soc.*, 109, 532-536.

FAO (2005) *Increasing the contribution of small-scale fisheries to poverty alleviation and food security.* FAO Technical Guidelines for Responsible Fisheries. No. 10, 79 p., FAO, Rome.

FAO (2007) *The state of world fisheries and aquaculture – 2006*, 162 p., FAO, Rome.

Granath, W. O. Jr., & Esch, G. W. (1983) Survivorship and parasite- induced host mortality among mosquio fish in a predator- free, North Carolina cooling reservoir, *Am. Midland Naturalist*, 110, 314-323.

Harvell, C. D., Kim, K., Burkholder, J. M., Colwell, R. R., Epstein, P. R., Grimes, D. J., Hofmann, E. E., Lipp, E. K., & Osterhaus, A. D. Overstreet RM et al. (1999) Emerging marine diseases- climate links and anthropogenic factors, *Science*, 285, 1505-1510.

Harvell, C. D., Mitchell, C. E., Ward, J. R., Altizer, S., Dobson, A. P., Ostfeld, R. S & Samuel, M.D. (2002) Climate warming and disease risks for terrestrial and marine biota, *Science*, 296, 5576, 2158-2162.

Hiner, M., & Moffitt, C. M. (2001) Variation in infections of *Myxobolus cerebralis* in field-exposed cutthroat and rainbow trout in Idaho, *J. aquat. Anim. Hlth*, 13, 124-132.

Hoegh-Goldberg, O. et al. (2007) Coral reefs under rapid climate change and ocean acidification. Science, 318, 1737-1742.

Hofmann, G. E., Buckley, B. A., Airaksine, S., Keen, J. E., & Somero, G. N. (2000) Heat-shock protein expression is absent in the Antarctic fish *Trematomus bernacchii* family Nototheniidae. *J Exp Biol.*, 203, 2331-2339.

IPCC (2007) Fourth Assessment Report - Climate Change 2007: Synthesis Report, 2007.

IPCC (2001) Climate Change 2001. IPCC Third Assessment Report, 2001.

Jolly, C., & Marimoto, R. I. (2000) Role of the heat shock response and molecular chaperones in oncogenesis and cell death, *J. Natl Cancer Inst.* 92, 1564-1572.

Kocan, R., Hershberger, P., Sanders, G., & Winton, J. (2009) Effects of temperature on disease progression and swimming stamina in Ichthyophonus-infected rainbow trout, *Oncorhynchus mykiss* (Walbaum), *J Fish Dis.*, 32, 10, 835-43.

Lafferty, K. D. (2009) The ecology of climate change and infectious diseases. *Ecology, 90, 888-900.*

Lafferty, K. D., Porter, J. W & Ford, S. E. (2004) Are diseases increasing in the ocean? *Ann. Rev. Ecol. Evol. Syst.*, 35, 31-54.

Lenton, T. M., Held, H., Kriegler, E., Hall, J. W., Lucht, W., Rahmstorf, S., & Schellnhuber, H. J. (2008) Tipping elements in the earth's climate system, *Proc Natl Acad Sci, USA*, 105, 6, 1786-1793.

Li, C. Y., Lee, J. S., Ko, Y. G., Kim, J. I. & Seo, J. S. (2000) Heat shock protein 70 inhibits apoptosis downstream of cytochrome c release and upstream of caspase-3 activation. *J. Biol. Chem.*, 275, 25665-25671.

Ling, S. D., Johnson, C. R., Frusher, S. D., & Ridgway, K. R. (2009) Overfishing reduces resilience of kelp beds to climate-driven catastrophic phase shift, *Proc Natl Acad Sci USA.*, 106, 52, 22341-22345.

Mantzouni, I., & Mackenzie, B. R. (2010) Productivity responses of a widespread marine piscivore, *Gadus morhua*, to oceanic thermal extremes and trends. *Proc Biol Sci.* Feb 10. [Epub ahead of print].

Marcogliese, D. J. (2008) The impact of climate change on the parasites and infectious diseases of aquatic animals, *Rev. sci. tech. Off. int. Epiz.*, 27, 2, 467-484.

Marimoto, R.J. (1998) Regulation of the heat shock transcriptional response: cross talk between a family of heat shock factors, molecular chaperones and negative regulators. *Genes Dev.*, 12, 3788-3796.

McGinnity, P., Jennings, E., DeEyto, E., Allott, N., Samuelsson, P., Rogan, G., Whelan, K., & Cross, T. (2009) Impact of naturally spawning captive-bred Atlantic salmon on wild populations: depressed recruitment and increased risk of climate-mediated extinction, *Proc Biol Sci.* 276, 1673, 3601-3610.

Menon, A. G. K. (1954) *Fish geography of the Himalayas.* Zoological Survey of India, Calcutta. 11, 4, 467-493.

Mohanty, B. P., Mondal, K., Bhattacharjee, S., & Vass, K. K. (2008) HSP 70 expression profile in tissues of the large riverine catfish *Aorichthys seenghala* (Sykes). P-GNB-58, p.153. 8th Indian Fisheries Forum 22-26 Nov 2008, Kolkata, India; jointly organized by CIFRI, Inland Fisheries Society of India and Indian Fisheries Forum. ISBN-81-85482-14-4.

Mohanty, B. P., Bhattacharjee, S, Mondal, K., & Das, M. K. (2009) HSP 70 expression in different tissues of some important tropical freshwater fishes. 96th Indian Science Congress, 3-7 January 2009, organized by NEHU, Shillong, India.

Mohanty, S., & Mohanty, B. P. (2009) Global climate change: a cause of concern, *Natl Acad Sci Lett,* 32, 5 & 6, 149-156.

Mohanty, B. P., Behera, B. K., & Sharma, A. P. (2010a) Nutritional significance of small indigenous fishes in human health. Bulletin No. 162, Central Inland Fisheries Research Institute, Barrackpore, Kolkata, India. ISSN 0970-616X.

Mohanty, B. P., Bhattacharjee, S., Mondal, K., & Das, M. K. (2010b) HSP70 expression profiles in white muscles of riverine catfish *Rita rita* show promise as biomarker for pollution monitoring in tropical rivers. *Natl Acad Sci Lett.,* 33, 5 & 6, 177-182.

Mumby, P. J., & Harborne, A. R. (2010) Marine reserves enhance the recovery of corals on Caribbean reefs, *PLoS One* 5, 1, e8657.

National Climate Data Centre, National Oceanic and Atmospheric Administration. Global warming: frequently asked questions. Available at: www.ncdc.noaa.gov/oa/climate/globalwarming.html. Accessed December 9, 2008.

Nicholls, R. J., Wong, P. P., Burkett, V. R., Codignotto, J. O., Hay, J. E., McLean, R. F., Ragoonaden, S., & Woodroffe, C. D. (2007) Coastal systems and low-lying areas. In: *Climate Change 2007: impacts, adaptation and vulnerability,* Parry, M. L., Canziani, O. F., Palutikof, J. P., Linden, V. D. & Hanson, C. E., (Eds.), pp. 315-356. Contribution of working group II to the Fourth Assessment Report of the Intergovernmental Panel on Climate Change, Cambridge University Press, Cambridge, UK.

Orr, J. C., Fabry, V. J., Aumont, O., Bopp, L., Doney, S. C., Feely, R. A., Gnanadesikan, A., Gruber, N., Ishida, A., Joos, F., Key, R. M., Lindsay, K., Maier-Reimer, E., Matear, R., Monfray, P., Mouchet, A., Najjar, R. G., Plattner, G-K, Rodgers, K. B., Sabine, C. L., Sarmiento, J. L., Schlitzer, R., slater, R. D., Totterdell, I. J., Weirig, M-F., Yamanaka, Y., & Yool, A. (2005) Anthropogenic ocean acidification over the twenty-first century and its impact on calcifying organisms. *Nature,* 437, 681-686.

ProAct Network (2008) The role of environmental management and eco-engineering in disaster risk reduction and climate change adaptation.

Prowse, T. D., Furgal, C., Wrona, F. J., & Reist, J. D. (2009) Implications of climate change for northern Canada: freshwater, marine, and terrestrial ecosystems, *Ambio,* 38, 5, 282-289.

Regional Framework for action to protect human health from effects of climate change in the South East Asia and Pacific Region. 2007. Available at http://www.searo.who.int/en/Section260/Section2468_14335.htm. Accessed December 9, 2008.

Scott, M. A., Locke, M., & Buck, L. T (2003) Tissue- specific expression of inducible and constitutive Hsp70 isoforms in the western painted turtle, *J Exptl. Biol.*, 206, 303-311.

Shao, K. T. (2009) Marine biodiversity and fishery sustainability. *Asia Pac J Clin Nutr.*, 18, 4, 527-531.

Shea, K. M., & the Committee on Environmental Health. (2007) Global Climate Change and children's health. *Pediatrics,* 120, e1359-e1367.

Sinha, M., De, D. K., & Jha, B. C. (1998) *The Ganga- Environment and Fishery*. Central Inland Fisheries Research Institute, Barrackpore, Kolkata, India.

Tester, P. A., Feldman, R. L., Nau, A. W., Kibler, S. R., & Wayne Litaker, R. (2010) Ciguatera fish poisoning and sea surface temperatures in the Caribbean Sea and the West Indies. Toxicon. Mar 3. [Epub ahead of print]

Thorpe, A., Reid, C., Anrooy, R. V., Brugere, C., & Becker, D. (2006) Poverty reduction strategy papers and the fisheries sector: an opportunity forgone?, J Intl. Dev., 18, 4, 487-517.

Tops, S., Hartikainen, H. L., & Okamura, B. (2009) The effects of infection by *Tetracapsuloides bryosalmonae* (Myxozoa) and temperature on *Fredericella sultana* (Bryozoa). *Int J Parasitol.*, 39, 9, 1003-1010.

Understanding and responding to Climate Change. 2008 Edn. pp. 1-24. The National Academies, USA (http://www.national-academies.org)

Vass, K. K., Das, M. K., Srivastava, P. K. & Dey, S. (2009) Assessing the impact of climate change on inland fisheries in River Ganga and its plains in India. *Aqu Ecosys Health & Management.*, 12, 2, 138-151.

Veron, J. E., Hoegh-Guldberg, O., Lenton, T. M., Lough, J. M., Obura, D. O., Pearce-Kelly, P., Sheppard, C. R., Spalding, M., Stafford-Smith, M. G., & Rogers, A. D. (2009) The coral reef crisis: the critical importance of<350 ppm CO2. *Mar Pollut Bull.*, 58, 10, 1428-1436.

Waller, C., Barnes, D. K. A., & Convey, P. (2006) Ecological contrasts across an Atlantic land-sea interface, *Austral Ecol,* 31, 656-666.

Walther, G. R., Roques, A., Hulme, P. E., Sykes, M. T., Pysek, P., Kühn, I., Zobel, M., Bacher, S., Botta-Dukát, Z., Bugmann, H., Czúcz, B., Dauber, J., Hickler, T., Jarosík, V., Kenis, M., Klotz, S., Minchin, D., Moora, M., Nentwig, W., Ott, J., Panov, V. E., Reineking, B., Robinet, C., Semenchenko, V., Solarz, W., Thuiller, W., Vilà, M., Vohland, K., & Settele, J. (2009) Alien species in a warmer world: risks and opportunities. *Trends Ecol Evol.*, 24, 12, 686-693.

WMO World Data Centre for Greenhouse Gases. Greenhouse gas bulletin: the state of greenhouse gases in the atmosphere using global observations up to December 2004. Vol.1, March 14, 2006.

World Bank & FAO (2008) The sunken billions: the economic justification for fisheries reform. Agriculture and Rural Development Dept. The World Bank: Washington DC. www.worldbank.org.sunkenbillions

8

Community ecological effects of climate change

Csaba Sipkay[1], Ágota Drégelyi-Kiss[2], Levente Horváth[3], Ágnes Garamvölgyi[4], Keve Tihamér Kiss[1] and Levente Hufnagel[3]

[1.] *Hungarian Danube Research Station, Hungarian Academy of Sciences*

[2.] *Bánki Donát Faculty of Mechanical and Safety Engineering, Óbuda University*

[3]*Adaptation to Climate Change Research Group of Hungarian Academy of Sciences*

[4]*Department of Mathematics and Informatics, Corvinus University of Budapest*

Hungary

1. Introduction

The ranges of the species making up the biosphere and the quantitative and species composition of the communities have continuously changed from the beginning of life on earth. Earlier the changing of the species during the history of the earth could be interpreted as a natural process, however, in the changes of the last several thousand years the effects due to human activity have greater and greater importance. One of the most significant anthropogenic effects taken on our environment is the issue of climate change. Climate change has undoubtedly a significant influence on natural ecological systems and thus on social and economic processes. Nowadays it is already an established fact that our economic and social life is based on the limited natural resources and enjoys different benefits of the ecosystems ("ecosystem services"). By reason of this, ecosystems do not only mean one sector among the others but due to the ecosystem services they are in relationship with most of the sectors and global changes influence our life mainly through their changes.

In the last decades direct and indirect effects of the climate change on terrestrial and marine ecosystems can already be observed, on the level of individuals, populations, species, ecosystem composition and function as well. Based on the analysis of data series covering at least twenty years, statistically significant relationship can be revealed between temperature and the change in biological-physical parameters of the given tax on in case of more than 500 taxes. Researchers have shown changes in the phonological, morphological, physiological and behaviour characteristics of the taxes, in the frequency of epidemics and damages, in the ranges of species and other indirect effects.

In our present study we would like to examine closely the effects of climate change on community ecology, throwing light on some methodological questions and possibilities of studying the topic. To understand the effects of climate change it is not enough to collect ecological field observations and experimental approaches yield results only with limited validity as well. Therefore great importance is attached to the presentation of modelling methods and some possibilities of application are described by means of concrete case studies. This chapter describes the so-called strategic model of a theoretical community in detail, with the help of which relevant results can be yielded in relation to ecological issues such as "Intermediate Disturbance Hypothesis" (IDH). Adapting the model to real field data, the so-called tactical model of the phytoplankton community of a great atrophic river (Danube, Hungary) was developed. Thus we show in a hydro biological case study which influence warming can have on the maximum amount of phytoplankton in the examined aquatic habitat. The case studies of the strategic and tactical models are contrasted with other approaches, such as the method of „geographical analogy". The usefulness of the method is demonstrated with the example of Hungarian agro-ecosystems.

2. Literature overview

2.1. Ways of examination of community ecological effects of climate change

In the first half of the 20th century, when community ecology was evolving, two different concepts stood out. The concept of a „super organism" came into existence in North America and was related to Clements (1905). According to his opinion, community composition can be regarded as determined by climatic, geological and soil conditions. In case of disturbance, when the community status changes, the original state will be reached by succession. Practically, the community is characterized by stability or homeostasis. Since the 1910s, the Zürich-Montpellier Phytocoenological School has evolved within this framework with the participation of Braun-Blanquet, and the same tendency can be observed in the field of animal ecology, in the principal work of Elton (1927). The same concept characterizes the Gaia concept of Lovelock (1972, 1990), which is the extension of the above-mentioned approach to biosphere level. Another concept, entitled „individualistic" (Gleason, 1926), stands in contrast with it. It postulates that the observed assembly pattern is generated by the stochastic sum of the populations individually adapted to the environment.

Nowadays, contrasting these concepts seems to be rather superfluous, as it is obvious that one of them describes communities regulated by competition, which are often disturbed, whereas the other one implies coevolved, stable communities, which have been permanent for a long time. However, it is true for both habitat types that community ecological and production biological processes, as well as species composition and biodiversity depend on the existing climate and the seasonal patterns of weather parameters.

According to our central research hypothesis, climate change takes its main ecological effects through the transitions between these two different habitats and ecological states. Testing of the present hypothesis can be realized by simulation models and related case studies, as it is evident that practically; these phenomena cannot be investigated either by field observations or by manipulative experiments.

The important community ecological researches have three main approaches related to methodology considering climate change. Ecologists working in the field observing real natural processes aspire to interfere as little as possible with the processes (Spellerberg, 1991). The aim is to describe the community ecological patterns.

The other school of ecological researches examines hypotheses about natural processes. The basis of these researches is testing different predictions in manipulative trials. The third group of ecologists deals with modelling where a precise mathematical model is made for basic and simple rules of the examined phenomena.

The work of the modelling ecologists consists of two parts. The first one is testing the mathematical model with case studies and the second one is developing (repairing and fitting again) the model. These available models are sometimes far away from the observations of field ecologists because there are different viewpoints. In the course of modelling the purpose is to simplify the phenomena of nature whereas in case of field observations ecosystems appear as complex phenomena.

It is obvious that all the three approaches have advantages and disadvantages. There are two approaches: monitoring- and hypothesis-centred ones. In case of monitoring approaches the main purpose is to discover the relationships and patterns among empirical data. This is a multidimensional problem where the tools of biomathematics and statistics are necessary. Data originate from large monitoring systems (e.g. national light trap network, Long Term Ecological Research (LTER)).

In case of hypothesis-centred approaches known or assumed relationships mean the starting point. There are three types of researches in this case:

- Testing simple hypotheses with laboratory or field experiments (e.g. fitotron plant growth room).
- Analyzing given ecosystems with tactical models (e.g. local case studies, vegetation models, food web models, models of biogeochemical cycles) (Fischlin et al., 2007, Sipkay et al., 2008a, Vadadi et al., 2008).
- Examination of general questions with strategic modelling (e.g. competition and predation models, cellular automata, evolutionary-ecological models).

In the examination of the interactions between climate change, biodiversity and community ecological processes the combined application of these main schools, methodological approaches and viewpoints can yield results.

2.2. Intermediate Disturbance Hypothesis (IDH)

Species richness in tropical forests as well as that of the atolls is unsurpassable, and the question arises why the theory of competitive exclusion does not prevail here. Trees often fall and perish in tropical rainforests due to storms and landslide, and corals often perish as a result of freshwater circulation and predation. It can be said with good reason that disturbances of various quality and intensity appear several times in the life of the above mentioned communities, therefore these communities cannot reach the state of equilibrium. The Intermediate Disturbance Hypothesis (IDH) (Connell, 1978) is based on this observation and states the following:

- In case of no disturbance the number of the surviving species decreases to minimum due to competitive exclusion.
- In case of large disturbance only pioneers are able to grow after the specific disturbance events.
- If the frequency and the intensity of the disturbance are medium, there is a bigger chance to affect the community.

There are some great examples of IDH in case of phytoplankton communities in natural waters (Haffner et al., 1980; Sommer, 1995; Viner & Kemp, 1983; Padisák, 1998; Olrik & Nauwerk, 1993; Fulbright, 1996). Nowadays it is accepted that diversity is the largest in the second and third generations after the disturbance event (Reynolds, 2006).

2.3. Connection between IDH and diversity

The connection between the diversity and the frequency of the disturbance can be described by a parabola (Connell, 1978). If the frequency and the strength of the disturbance are large, species appear which can resist the effects, develop fast and populate the area quickly (r-strategists). In case of a disturbance of low frequency and intensity the principle of competitive exclusion prevails so dominant species, which grow slowly and maximize the use of sources, spread (K-strategists).

Padisák (1998) continuously took samples from different Hungarian lakes (such as Balaton and Lake Fertő) and the abundance, uniformity (in percentage) and Shannon diversity of phytoplankton were examined. In order to be able to generalize, serial numbers of the phytoplankton generations between the single disturbance events are represented on the horizontal axis, and this diagram shows similarity with that of Connell (1978). This graph also shows that the curve doesn't have symmetrical run as the effect of the disturbance is significantly greater in the initial phase than afterwards.

According to Elliott et al. (2001), the relationship between disturbance and diversity cannot be described by a Connell-type parabola (Connell, 1978) because a sudden breakdown occurs on a critically high frequency. This diagram is called a cliff-shaped curve. The model is known as PROTECH (Phytoplankton ResPonses To Environmental CHange); it is a phytoplankton community model and is used to examine the responses given to environmental changes (Reynolds, 2006).

2.4. Expected effects of climate change on fresh-water ecosystems

Rising water temperatures induce direct physiological effects on aquatic organisms through their physiological tolerance. This mostly species-specific effect can be demonstrated with the examples of two fish species, the eurythermal carp (Cyprinids cardio) and the stenothermal Splenius alpines (Ficke et al., 2007). Physiological processes such as growth, reproduction and activity of fish are affected by temperature directly (Schmidt-Nielsen, 1990). Species may react to changed environmental conditions by migration or acclimatization. Endemic species, species of fragmented habitats and systems with east-west orientation are less able to follow the drastic habitat changes due to global warming (Ficke et al., 2007). At the same time, invasive species may spread, which are able to tolerate the changed hydrological conditions to a greater extent (Baltz & Moyle, 1993).

What is more, global warming induces further changes in the physical and chemical characteristics of the water bodies. Such indirect effects include decrease in dissolved oxygen content (DO), change in toxicity (mostly increasing levels), tropic status (mostly indicating eutrophication) and thermal stratification.

DO content is related to water temperature. Oxygen gets into water through diffusion (e. g. stirring up mechanism by wind) and photosynthesis. Plant, animal and microbial respiration decrease the content of DO, particularly at night when photosynthesis based oxygen production does not work. When oxygen concentration decreases below 2-3 mg/l, we have to face the hypoxia. There is an inverse relationship between water temperature and oxygen solubility. Increasing temperatures induce decreasing content of DO whereas the biological oxygen demand (BOD) increases (Kalff, 2000), thus posing double negative effect on aquatic organisms in most systems. In the side arms of atrophic rivers, the natural process of phytoplankton production-decomposition has an unfavourable effect as well. Case studies of the side arms in the area of Szigetköz and Gemenc also draw attention to this phenomenon: high biomass of phytoplankton caused oxygen depletion in the deeper layers and oversaturation in the surface (Kiss et al., 2007).

Several experiments were run on the effects of temperature on toxicity. In general, temperature dependent toxicity decreases in time (Nussey et al., 1996). On the other hand, toxicity of pollutants increases with rising temperatures (Murty, 1986.b), moreover there is a positive correlation between rising temperatures and the rate at which toxic pollutants are taken up (Murty, 1986.a). Metabolism of poikilothermal organisms such as fish increases with increasing temperatures, which enhances the disposal of toxic elements indirectly (MacLeod & Pessah, 1993). Nevertheless, the accumulation of toxic elements is enhanced in aquatic organisms with rising temperatures (Köck et al., 1996). All things considered, rising temperatures because increasing toxicity of pollutants.

Particularly in lentil waters, global warming has an essential effect on tropic state and primary production of inland waters through increasing the water temperature and changing the stratification patterns (Lofgren, 2002). Bacterial metabolism, rate of nutrient cycle and algal abundance increase with rising temperatures (Klapper, 1991). Generally, climate change related to pollution of human origin enhances eutrophication processes (Klapper, 1991; Adrian et al., 1995). On the other hand, there is a reverse effect of climate change inasmuch as enhancement of stratification (in time as well) may result in concentration of nutrients into the hypolimnion, where they are no longer available for primary production (Magnuson, 2002). The latter phenomenon is only valid for deep, stratified lakes with distinct aphetic and tropholitic layers.

According to the predictions of global circulation models climate change is more than rise in temperatures purely. The seasonal patterns of precipitation and related flooding will also change. Frequency of extreme weather conditions may intensify in water systems as well (Magnuson, 2002). Populations of aquatic organisms are susceptible to the frequency, duration and timing of extreme precipitation events including also extreme dry or wet episodes. Drought and elongation of arid periods may cause changes in species composition and harm several populations (Matthews & Marsh-Matthews, 2003). Seasonal changes in melting of the snow influence the physical behaviour of rivers resulting in changed reproduction periods of several aquatic organisms (Poff et al., 2002). Due to melting of ice rising sea levels may affect communities of river estuaries in a negative way causing increased erosion (Wood et al., 2002). What is more, sea-water flow into rivers may increase because of rising sea levels; also drought contributes to this process causing decreased current velocities in the river.

Climate change may enhance UV radiation. UV-B radiation can influence the survival of primary producers and the biological availability of dissolved organic carbon (DOC). The interaction between acidification and pollution, UV-B penetration and eutrophication has been little studied and is expected to have significant impacts on lake systems (Magnuson, 2002; Allan et al., 2002).

2.5. Feedback mechanisms in the climate-ecosystem complex

The latest IPCC report (Fischlin et al., 2007) points out that a rise of 1.5-2.5 ^{0}C in global average temperature causes important changes in the structure and functioning of ecosystems, primarily with negative consequences for the biodiversity and goods and services of the ecological systems.

Ecosystems can control the climate (precipitation, temperature) in a way that an increase in an atmosphere component (e.g. CO_2 concentration) induces the processes in biosphere to decrease the amount of that component through biogeochemical cycles. Pale climatic researches proved this control mechanism existing for more than 100,000 years. The surplus CO_2 content has most likely been absorbed by the ocean, thus controlling the temperature of the Earth through the green house effect. This feedback is negative therefore the equilibrium is stable.

During the climate control there may be not only negative but positive feedbacks as well. One of the most important factors affecting the temperature of the Earth is the albino of the poles. While the average temperature on the Earth is increasing, the amount of the arctic ice is decreasing. Therefore the amount of the sunlight reflected back decreases, which warms the surface of the Earth with increasing intensity. This is not the only positive feedback during the control; another good example is the melting of frozen methane hydrate in the tundra.

The environment, the local and the global climate are affected by the ecosystems through the climate-ecosystem feedbacks. There is a great amount of carbon in the living vegetation and the soil as organic substance which could be formed to atmospheric CO_2 or methane hereby affecting the climate. CO_2 is taken up by terrestrial ecosystems during the photosynthesis and is lost during the respiration process, but carbon could be emitted as methane, volatile organic compound and solved carbon. The feedback of the climate-carbon cycle is difficult to determine because of the difficulties of the biological processes (Drégelyi-Kiss & Hufnagel, 2008).

The biological simplification is essential during the modelling of vegetation processes. It is important to consider several feedbacks to the climate system to decrease the uncertainty of the estimations.

3. Strategic modelling of the climate-ecosystem complex based on the example of a theoretical community

3.1. TEGM model (Theoretical Ecosystem Growth Model)

An algae community consisting of 33 species in a freshwater ecosystem was modelled (Drégelyi-Kiss & Hufnagel, 2009). During the examinations the behaviour of a theoretical ecosystem was studied by changing the temperature variously.

Theoretical algae species are characterized by the temperature interval in which they are able to reproduce. The simulation was made in Excel with simple mathematical background. There are four types of species based on their temperature sensitivity: super-generalists, generalists, transitional species and specialists. The temperature optimum curve originates from the normal (Gaussian) distribution, where the expected value is the temperature optimum. The dispersion depends on the niche overlap among the species. The overlap is set in a way that the results correspond with the niche overlap of the lizard species studied by Pianka (1974) where the average of the total niche overlap decreases with the number of the lizard species. 33 algae species with various temperature sensitivity can be seen in Figure 1. The daily reproductive rate of the species can be seen on the vertical axis, which means by how many times the number of specimens can increase at a given temperature. This corresponds to the reproductive ability of freshwater algae in the temperate zone (Felföldy, 1981). Since the reproductive ability is given, the daily number of specimens related to the daily average temperature is definitely determinable.

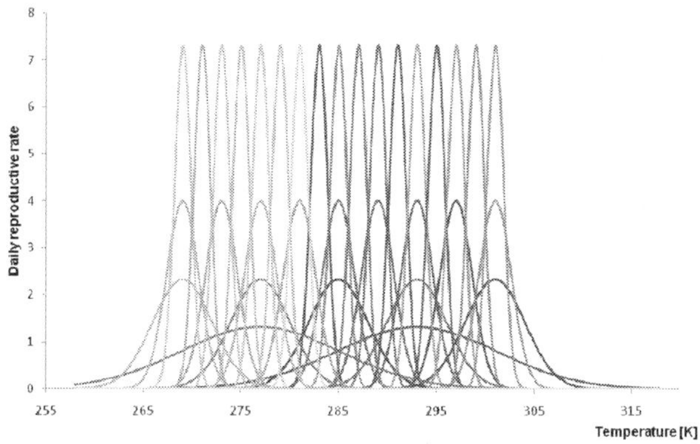

Fig. 1. Reproductive temperature pattern of 33 algae species

The 33 species are described by the Gaussian distribution with the following parameters:

- 2 super-generalists (μ_{SG1}=277 K; μ_{SG2}=293 K; σ_{SG}=8.1)
- 5 generalists (μ_{G1}=269 K; μ_{G2}=277 K; μ_{G3}=285 K; μ_{G4}=293 K; μ_{G5}=301 K; σ_G=3.1)
- 9 transitional species (μ_{T1}=269 K; μ_{T2}=273 K; μ_{T3}=277 K; μ_{T4}=281 K; μ_{T5}=285 K; μ_{T6}=289 K; μ_{T7}=293 K; μ_{T8}=297 K; μ_{T9}=301K; σ_T=1.66)
- 17 specialists (μ_{S1}=269 K; μ_{S2}=271 K; μ_{S3}=273 K; μ_{S4}=275 K; μ_{S5}=277 K; μ_{S6}=279 K; μ_{S7}=281 K; μ_{S8}=283 K; μ_{S9}=285 K; μ_{S10}=287 K; μ_{S11}=289 K; μ_{S12}=291 K; μ_{S13}=293 K; μ_{S14}=295 K; μ_{S15}=297 K; μ_{S16}=299 K; μ_{S17}=301 K; σ_S=0.85).

We suppose 0.01 specimens for every species as a starting value and the following minimum function describes the change in the number of specimens.

$$N(X_i)_j = N(X_i)_{j-1} \cdot Min\left\{\left(RR(X_i)_j\right)^r; (RF_j)^r\right\} + 0.01 \tag{1}$$

where i denotes the species, i=1,2,...,33; j is the number of the days (usually j=1, 2,..., 3655);
$RR(X_i)_j$ is the reproduction rate of the X_i species on the jth day;
RF_j is the restrictive function related to the accessibility of the sunlight;
r is the velocity parameter (r=1 or 0.1);
the 0.01 constant means the number of the spore in the model which inhibits the extinction of the population.

The temperature-dependent growth rate can be described with the density function of the normal distribution, whereas the light-dependent growth rate includes a term of environmental sustainability, which was defined with a sine curve representing the scale of light availability within a year. The constant values of the restrictive function were set so that the period of the function is 365.25, the maximum place is on 23rd June and the minimum place is on 22nd December. (These are the most and the least sunny days.)

In every temperature interval there are dominant species which win the competition. The output parameters of the experiments are the determination of the dominant species, the largest number of specimens, the first year of the equilibrium and the use of resources. The use of the resources shows how much is utilized from the available resources (in this case from sunlight) during the increase of the ecosystem.

Functions of temperature patterns

1. Simulation experiments were made at constant 293 K, 294 K and 295 K using the two velocity parameters (r=1 and 0.1). The fluctuation was added as ±1...±11 K random numbers.
2. The temperature changes as a sine function over the year (with a period of 365.25 days):

$$T=s_1 \cdot \sin(s_2 \cdot t+s_3)+s_4 \tag{2}$$

   where s_2=0.0172, s_3=-1.4045 since the period of the function is 365.25 and the maximum and the minimum place are given (23th June and 22nd December, these are the most and the least sunny days).
3. Existing climate patterns
   a. Historical daily temperature values in Hungary (Budapest) from 1960 to 1990
   b. Historical daily temperature values from various climate zones (from tropical, dry, temperate, continental and polar climate)
   c. Future temperature patterns in Hungary from 2070-2100
   d. Analogous places related to Hungary by 2100
   It is predicted that the climate in Hungary will become the same by 2100 as the present-day climate on the border of Romania and Bulgaria or near Thessaloniki. According to the worst prediction the climate will be like the current North-African climate (Hufnagel et al., 2008).

The conceptual diagram of the TEGM model summarizes the build-up of the model (Figure 2.).

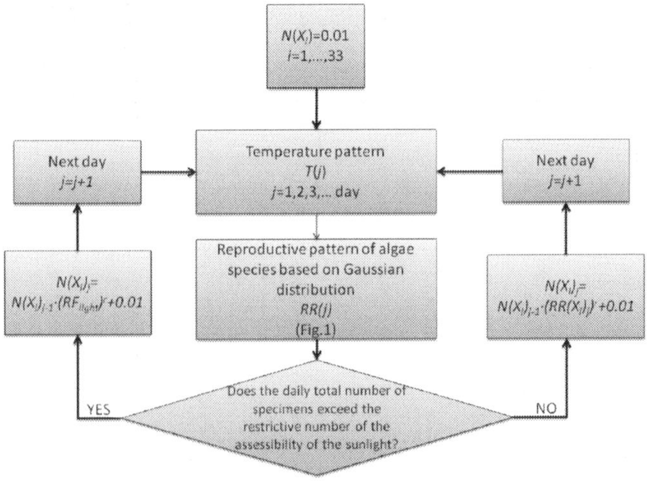

Fig. 2. Conceptual diagram of the TEGM model (*RR*: reproduction rate, *RF*: restriction function related to the accessibility of the sunlight, $N(X_i)$: the number of the ith algae species, *r*: velocity parameter)

3.2. Main observations based on simulation model examinations

Changing climate means not only the increase in the annual average temperature but in variability as well, which is a larger fluctuation among daily temperature data (Fischlin et al., 2007). As a consequence, species with narrow adaptation ability disappear, species with wide adaptation ability become dominant and biodiversity decreases.

In the course of our simulations it has been shown what kind of effects the change in temperature has on the composition of and on the competition in an ecosystem. Specialists reproducing in narrow temperature interval are dominant species in case of constant or slowly changing temperature patterns but these species disappear in case of fluctuation in the temperature (Drégelyi-Kiss & Hufnagel, 2009). The best use of resources occurs in the tropical climate.

Comparing the Hungarian historical data with the regional predictions of huge climate centres (Hadley Centre: HC, Max Planck Institute: MPI) it can be stated that recent estimations (such as HC adhfa, HC adhfd and MPI 3009) show a decrease in the number of specimens in our theoretical ecosystem.

Simulations with historical temperature patterns of analogous places show that our ecosystem works similarly in the less hot Rumanian lowland (Turnu Magurele), while the number of specimens and the use of resources increase using North African temperature data series. In further research it could be interesting to analyze the differences in the radiation regime of the analogous places.

Regarding diversity the annual value of the Shannon index increases in the future (in case of the data series HC adhfa and MPI 3009), but the HC adhfd prognosis shows the same pattern as historical data do (Budapest, 1960-1990). According to the former predictions (such as UKLO, UKHI and UKTR31) the composition of the ecosystem does not change in proportion to the results based on historical data (Drégelyi-Kiss & Hufnagel, 2010).

Further simulations were made in order to answer the following question: what kind of environmental conditions result in larger diversity in an ecosystem related to the velocity of reproduction. The diversity value of the slower process is the half of that of the faster process. Under the various climate conditions the number of specimens decreases earlier in case of the slower reproduction (r=0.1) than in the faster case (r=1), and there are larger changes in diversity values. Generally it can be said that an ecosystem with low number of specimens evolves finally. Using the real climate functions it can be stated that from the predicted analogous places (Turnu Magurele, Romania; Cairo, Egypt (Hufnagel et al., 2008)) Budapest shows similarity with Turnu Magurele in the number of specimens and in diversity values (Hufnagel et al., 2010).

Our strategic model was adapted for tactical modelling, which is described later as "Danubian Phytoplankton Model".

3.3. Manifestation of the Intermediate Disturbance Hypothesis (IDH) in the course of the simulation of a theoretical ecosystem

In the simulation study of a theoretical community made of 33 hypothetical algae species the temperature was varied and it was observed that the species richness showed a pattern in accordance with the intermediate disturbance hypothesis (IDH).

In case of constant temperature pattern the results of the simulation study can be seen in Fig. 3, which is the part of the examinations where random fluctuations were changed by up to ± 11K. The number of specimens in the community is permanent and maximum until daily random fluctuation values are between 0 and ±2K. Significant decrease in the number of specimens depends on the velocity factor of the ecosystem. There is a sudden decrease in case of a fluctuation of ± 3K in the slower processes while the faster ecosystems react in case of a random fluctuation of about ± 6K.

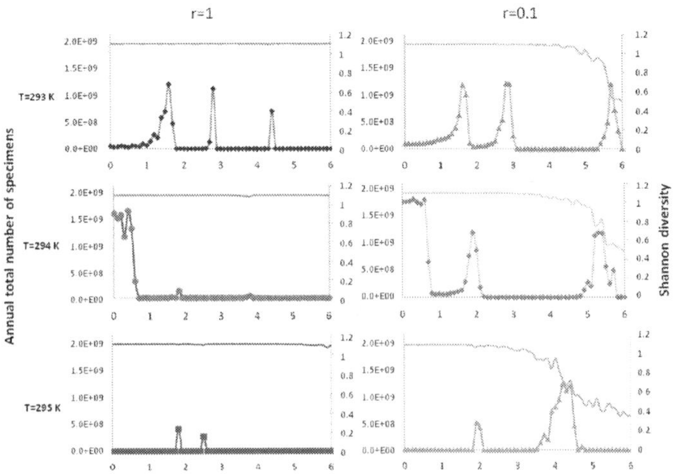

Fig. 3. Annual total number of specimens and diversity values versus the daily random fluctuation in constant temperature environment (The signed plots show the diversity values.)

There are some local maximums in the diversity function. In case of low fluctuation the diversity values are low; the largest diversity can be observed in case of medium daily variation in temperature; in case of large fluctuations, just like in case of the low ones, the diversity value is quite low. The diversity of the ecosystem which has faster reproductive ability shows lower local maximum values than that of the slower system in the experiments.

The degree of the diversity is greater in case of r=0.1 velocity factor than in case of the faster system. If there is no disturbance, the largest diversity can be observed at 294 K in case of both speed values. If the fluctuation is between ± 6K and ± 9K, the diversity values are nearly equally low. In case of the largest variation (± 11K) the degree of the diversity increases strongly.

In case of constant temperature pattern the Intermediate Disturbance Hypothesis can be seen well (Fig. 3.). In case of r=1 and T=293 K the specialist (S13) wins the competition when the random daily fluctuation has rather low values (up to ±1.5K). Then, increasing the random fluctuation the generalist (T7) is the winner and the transition between the exchanges of the two type genres shows the local maximum value in case of disturbance, which is related to IDH. The following competition is between the species T7 and G4 in case of a fluctuation of about ±2.8K, then between G4 and the super generalist (SG1) in case of about ±4.5K. These are similar fluctuation values where the IDH can be observed as it can be seen in Fig. 3.

The shapes of the IDH local maximum curves show similarity in all cases. The maximum curves increase slowly and decrease steeply. The main reason of this pattern is the competition between the various species. If the environmental conditions are better for a genre, the existing genre disappears faster, which explains the steep decrease in the diversity values after the competition. There are controversies regarding the shape of the local maximum curves in diversity values versus the random daily fluctuation (Connell, 1978; Elliott et al., 2001).

In case of sine temperature pattern the parameter s_1 was changed during the simulations. The results of the experiments can be seen in Fig.4. The initial low diversity value increases as the value of the parameter s_1 grows then decreases again.

There are two peaks in diversity when increasing the amplitude of the annual sine temperature function (s_1) in case of low values. The annual total number of specimens is permanent when s_1=0…3.5 in case of both velocity parameters, only the diversity value changes. In case of annual fluctuation (i.e. sine temperature pattern) the Intermediate Disturbance Hypothesis could be observed as well, and there are two local peaks similarly to the case of daily fluctuation.

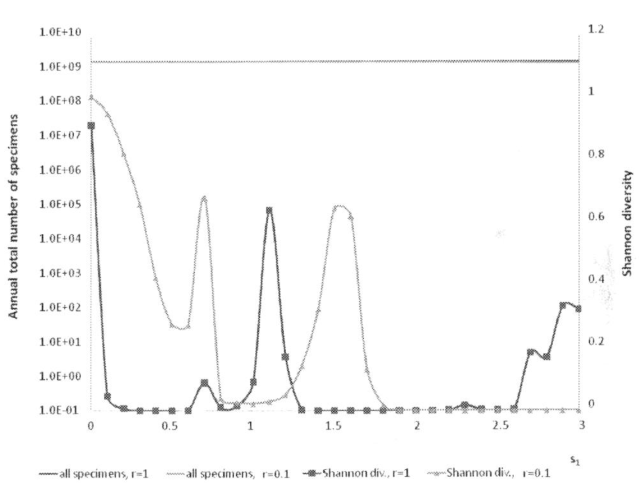

Fig. 4. Annual total numbers of specimens and Shannon diversity values plotted against the parameter s_1 in case of sine temperature pattern

3.4. Future research

Ecosystems have an important role in the biosphere in development and maintenance of the equilibrium. Regarding the temperature patterns it is not only the climate environment which affects the composition of ecosystems but plants also provides a feedback to their environment through the photosynthesis and respiration in the global carbon cycle.

The specimens of the ecosystems do not only suffer the change in climate but they can affect the equilibrium of the biosphere and the composition of the air through the biogeochemical cycles. There is an opportunity to examine the controlling ability of temperature and climate with the theoretical ecosystem.

In our further research we would like to examine the feedback of the ecosystem to the climate. These temperature feedbacks are very important related to DGVM models with large computation needs (Friedlingstein et al., 2006), but the feedbacks are not estimated directly. We would like to examine the process of the feedback with PC calculations in order to answer easy questions.

4. Tactical modelling case study using the example of the phytoplankton community of a large river (Hungarian stetch of River Danube)

The present subchapter describes the seasonal dynamics of the phytoplankton by means of a discrete-deterministic model on the basis of the data gathered in the Danube River at Göd (Hungary). The strategic model, so-called "TEGM" was adapted to field data (tactical model). The "tactical model" is a simulation model fitted to the observed temperature data set (Sipkay et al. 2009). The tactical models could be beneficial if the general functioning of ecosystems is in the focus (Hufnagel & Gaál 2005; Sipkay et al. 2008a, 2008b; Sipkay et al. 2009; Vadadi et al. 2009).

4.1. Materials and methods

Long-term series of phytoplankton data are available on the river Danube at Göd (1669 rkm) owing to the continuous record of the Hungarian Danube Research Station of the Hungarian Academy of Sciences collecting quantitative samples of weekly frequency between 1979 and 2002 (Kiss, 1994). Phytoplankton was sampled from the streamline near the surface and after processing of samples biomass was calculated (mg l^{-1}).

The relatively intensive sampling makes our data capable of being used in simulation models, which are functions of weather conditions. We assume that temperature is of major importance when discussing the seasonal dynamics of phytoplankton. What is more, the reaction curve describing the temperature dependency may be the sum of optimum curves, because the temperature optimum curves of species or units of phytoplankton and of biological phenomena determining growth rate are expected to be summed. On the other hand, the availability of light has also a major influence on the seasonal variation of phytoplankton abundance; therefore it was taken into account as well. Further biotic and biotic effects appear within the above-mentioned or hidden.

First, a strategic model, the so-called TEGM (Theoretical Ecosystem Growth Model) (Drégelyi & Hufnagel, 2009) was used, which involves the temperature optimum curves of 33 theoretical species covering the possible spectrum of temperature. The strategic model of the theoretical algal community was adapted to field data derived from the river Danube (tactical model), with respect to the fact that the degree of nutrient oversupply varied regularly during the study period (Horváth & Tevanné Bartalis, 1999). Assuming that nutrient oversupply of high magnitude represents a specific environment for phytoplankton, two sub models were developed, one for the period 1979-1990 with nutrient oversupply of great magnitude (sub model „A") and a second one for the period 1991-2002 with lower oversupply (sub model „B"). Either sub model can be described as the linear combination of 20 theoretical species. These sub models vary slightly in the parameters of the temperature reaction curves. Biomass (mg l^{-1}) of a certain theoretical species is the function of its biomass measured the day before and the temperature or light coefficient. So as to define whether temperature or light is the driving force, a minimum function was applied. Temperature-dependent growth rate can be described with the density function of normal distribution, whereas light-dependent growth rate includes a term of environmental sustainability, which was defined with a sine curve representing the scale of light availability within a year.

The model was run with the data series of climate change scenarios as input parameters after being fitted (with the Solver optimization program of MS Excel) to the data series of daily temperatures supplied by the Hungarian Meteorological Service. Data base of the PRUDENCE EU project (Christensen, 2005) was used, that is, A2 and B2 scenarios proposed by the IPCC (2007), the daily temperatures of which are specified for the period 2070-2100. Three data series were used including the A2 and B2 scenarios of the HadCM3 model developed by the Hadley Centre (HC) and the A2 scenario of the Max Planck Institute (MPI). Each scenario covers 31 replicates of which we selected 24 so as to compare to measured data of 24 years between 1979 and 2002. In addition, the effect of linear temperature rise was tested as follows: each value of the measured temperatures between 1979 and 2002 was increased by 0.5, 1, 1.5 and 2 °C, and then the model was run with these data.

The outcomes were analyzed with statistical methods using the Past software (Hammer et al., 2001). Yearly total phytoplankton biomass was defined as an indicator; however, it was calculated as the sum of the monthly average biomass in order to avoid the „side-effect" of extreme values. One-way ANOVA was applied to demonstrate possible differences between model outcomes. In order to point out which groups do differ from each other, the post-hoc Turkey test was used, homogeneity of variance was tested with Levene's test and standard deviations were compared with Welch test.

4.2. Results

On the basis of field and simulated data of phytoplankton abundance (Fig. 5), it can be said that the model fits to the observed values quite well. Yearly total biomass measured in the field and calculated as the sum of monthly average biomass correlated with the simulated values ($r=0.74$).

Phytoplankton biomass varied significantly within outcomes for scenarios and real data (one-way ANOVA, $p<0.001$), however, variances did not prove to be homogeneous (Levene's test, $p<0.001$), resulting from the significant differences of standard deviations (Welch test, $p<0.001$). Turkey's pair wise comparisons implied significant differences between outcomes of the scenario A2 (of MPI) and the others in sub model „A" only ($p<0.05$).

Examining the effect of linear temperature rise there were also significant differences between outputs (one-way ANOVA, $p<0.001$), similarly, variances were not homogeneous (Levene's test, $p<0.001$), and again, this was interpreted by the significant differences of standard deviations (Welch test, $p<0.001$). Turkey's pair wise comparisons pointed out that there are significant differences between the outcomes for the period 1979-2002 and outcomes at a temperature rise of 2 °C in case of sub model „A", furthermore, rises in temperature of 0.5, 1 and 1.5 °C in sub model „A" implied significant differences from the outcomes of sub model „B" ($p<0.05$). However, outcomes within sub model „B" showed large similarity ($p=1$).

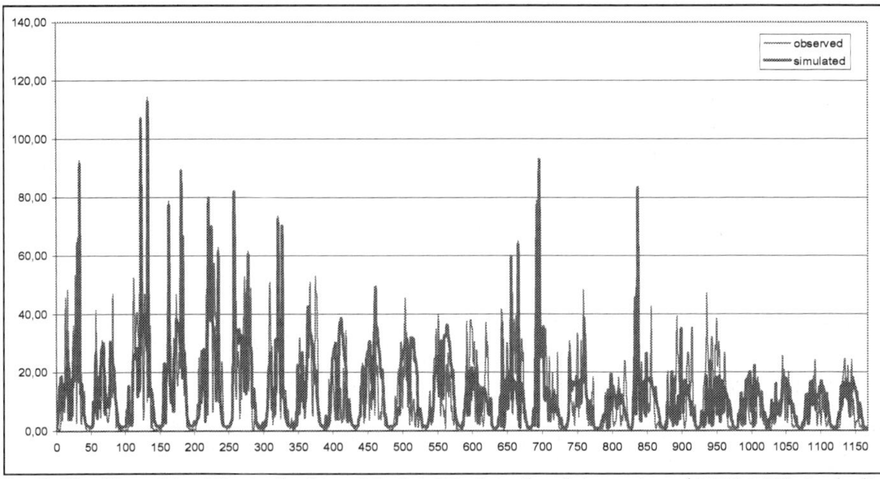

Fig. 5. The Danubian Phytoplankton Model fitted to the data series of 1979-1990, including sub model "A" for the period 1979-1990 and sub model "B" for the period 1991-2002.

Fig. 6. shows outputs of means and standard deviations of yearly total biomass of algae on the basis of monthly averages, from which it is evident that biomass increased largely in case of scenario A2 (MPI), however, biomass output for the period 1960-1990 increased notably as well (sub model „A"). Sub model „B" showed minor variations in biomass compared to model outputs for the period 1979-2002. Nevertheless, standard deviation showed major increase in each case. Temperature rise of 0.5 °C (Fig. 7.) obviously implied remarkable increase in biomass in case of sub model „A" and negligible changes in case of sub model „B".

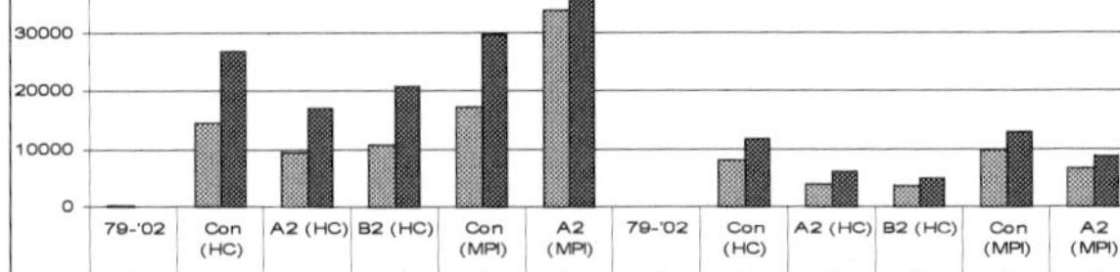

Fig. 6. Monthly means and standard deviations of yearly total algal biomass (mg l^{-1}) in case of the model outputs for the period 1979-2002 (based on measured data of temperatures) compared with outputs of the model run with data series of the climate change scenarios. Sub models „A" and „B" are presented separately. HC=Hadley Centre; MPI=Max Planck Institute; A2 and B2 scenarios; Con=control

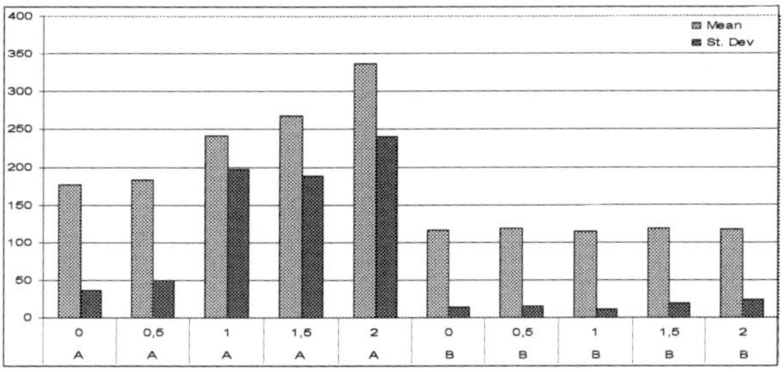

Fig. 7. Monthly means and standard deviations of yearly total algal biomass (mg l^{-1}) in case of the model outputs for the period 1979-2002 (based on measured data of temperatures) compared with outputs of the model run with data series of linear temperature rises of 0.5 °C. Sub models „A" and „B" are presented separately. 0; 0.5; 1; 1.5; 2: degree of linear temperature rise (°C).

4.3. Discussion

Adapting the TEGM model we managed to develop a model that fits to the measured data quite well. Beyond the indicator of yearly total biomass introduced in the paper, further indicators can be defined and used in order to get a better understanding of the possible effects of climate change concerning the phytoplankton in the Danube River.

Interpretation of model outcomes for climate change scenarios is rather difficult. Comparing the model outcomes for the control period 1960-1990 with the outcomes for the case of applying the observed temperature data of 1979-2002 as input parameters, there are remarkable differences. Major variation within data series of certain climate change scenarios may account for the increased standard deviations observed in the model outputs. Sub model „B", assuming minor nutrient oversupply, implied minor increase in biomass and variation within scenarios was of minor importance.

In case of linear temperature rise a clear answer is received: the more drastic warming results in greater abundance of phytoplankton if only the environment rich in nutrients is assumed (as it was experienced between 1979 and 1990). Moderate nutrient supply does not favour algae even if temperatures increase by 2 °C, thus biomass is not expected to increase notably. All these draw attention to the increased hazard of nutrient loading in rivers: global warming brings more drastic changes to nutrient rich environments.

Global warming can influence the tropic state and primary productivity of inland waters in a very fundamental way (Lofgren, 2002). Bacterial metabolism, rate of nutrient cycling and algal production all increase with rising temperatures (Klapper, 1991). Generally, climate change together with anthropogenic pollution enhances eutrophication (Klapper, 1991; Adrian et al., 1995). Several studies derived their findings from models forecasting increased phytoplankton abundance mostly through rising trophy (Mooij et al., 2007; Elliot et al., 2005; Komatsu et al., 2007). These are in line with our results of linear temperature rise, but only in case of high nutrient load. Global warming does not have so severe effects on phytoplankton biomass under conditions of lower nutrient load.

Looking at the variation in phytoplankton biomass only is not relevant enough in order to explore the effects of climate change. Freshwater food webs show rather characteristic seasonal dynamics, thus the effect of climate is the function of the season (Straile, 2005). If we are interested in the phenomena within a year, further indicators can be introduced into the model. Such indicators include those showing the day of population peak (blooms) within a year or those representing 50% of the yearly total biomass. Some day the model will be able to provide us with an insight into the effects of global warming from a broader perspective by taking advantage of new indicators.

5. Spatial climatic analogy (SCA)

With the method of the spatial climatic analogy we search for regions which have the same climate at present as the scenarios indicate for the future. With the spatial analogy we can make the climate scenarios easier to understand. In this research the analogue regions for Debrecen, which is an agriculturally important region in Hungary, were examined.
According to the result we can say, in line with international results, that our climate will be similar to that of regions south of Hungary. This shifting will be 250-450 km in the next decades (2011-2040) and it could be 450-650 km in the middle of the century and maybe there will be no spatial analogues in Europe by the end of the century. Different methods were used to calculate the analogues, but they indicate the same regions. These analogue regions are North-Serbia, the region Vojvodina, South Romania and North Bulgaria in the next decades, and South Bulgaria and North Greece in the middle of the century. We developed the inverse analogy method, with the help of which we can search for regions which will have the same climate in future as there is in Debrecen at present. If we accept the results of spatial analogy, we can identify the analogue regions and compare their data.
Data were collected from the databases EUROSTATs and CORINE. We collected all the data on land use, crops and natural vegetation. These data show that land use may become more diverse, which is advantageous in adaptation to climate change. The ratio of forests and pastures may become higher. In the next decade's maize and wheat will be more important because climatic conditions will become better for them (because of the "corn belt").
The tendency of a potential global climate change is still not obvious, but the most accepted models predict warming and increase in extreme weather events. As the climate change has an overall impact on human health, natural systems and agricultural production and also has socio-economic impacts, it is very important to predict potential changes to have enough time for appropriate decision-making. Analogue scenarios involve the use of past warm climates as scenarios of future climate (temporal analogue scenario) and the use of the current climate in another (usually warmer) location as a scenario of future climate in the study area (spatial analogue scenario). Our aim was to find spatial analogues to describe the potential future climate of Hungary. However, we must note that climate depends also on other effects, especially on elevation, topography and storm track conditions, which cannot be considered in this kind of analysis.

5.1. Materials and methods of spatial climatic analogy

Climate scenarios can be defined as relevant and adequate pictures of how the climate may look like in the future. Our work is based on General Circulation Models (GCMs) downscaled to Debrecen, an important centre of agricultural production in Hungary and we used the method of geographical analogies to explain the results. We used different GCM scenarios (HadCM3 A1FI, B2), the IPCC CRU Global Climate dataset and the Hungarian meteorological database for 30 years (1961-1990). To find the analogues, monthly data were used. Monthly temperature averages and precipitation sums were used for 4 different time periods, for the base period 1961-1990 and for the future periods 2010-2019, 2020-2029, 2020-2039 and 2040-2069. To calculate and find the analogue regions we used the SCA method, which was improved by us.

$$T_{dj} = \frac{1}{12} \cdot \sum_{i=1}^{12} \left| TEMP_{ji} - T_i \right| \tag{3}$$

$$P_{dj} = \frac{1}{12} \cdot \sum_{i=1}^{12} \frac{\left| PREC_{ji} - P_i \right|}{1 + a \cdot (PREC_{ji} + P_i)} \tag{4}$$

$$I_{Tj} = e^{-\lambda \cdot k_T \cdot T_{dj}} \tag{5}$$

$$I_{Pj} = e^{-(1-\lambda) \cdot k_P \cdot P_{dj}} \tag{6}$$

$$CMI_j = I_{Tj} \cdot I_{Pj} \tag{7}$$

Where:

- j: grid point identity number (j=1-31143)
- i: month (i=1-12)
- $TEMP_{ji}$: monthly mean temperature of the grid point j for the base period
- T_i: monthly mean temperature of the grid point j for the scenario
- $PREC_{ji}$: monthly precipitation of the grid point j for the base period
- P_i: monthly precipitation of the grid point j for the scenario
- T_{dj}: average of temperature differences
- P_{dj}: average of precipitation differences
- I_{Tj}: similarity of the climate for the scenario in temperature
- I_{Pj}: similarity of the climate for the scenario in precipitation
- CMI_j: „Composite Match Index", if CMI>90%, we can call the grid point the analogue for the scenario

5.2. Analogue regions

First we looked for the analogue regions for the base period. We found that we got back our regions after looking for the analogues in the future. We found that the analogue regions are south of Debrecen. This climate shifting was the same in case of different scenarios because in the first decades they do not differ very much but we can see more differences in the middle of the century. Finally, we defined the analogue regions for the scenarios and time periods. We found that the climatic shifting would be 250-450 km in the next decades and 450- 650 km by the middle of the century. Unfortunately we could not find any similar regions for the end of the century, but some analogues can be found in North Africa.

TIME	Base period	
1961-1990	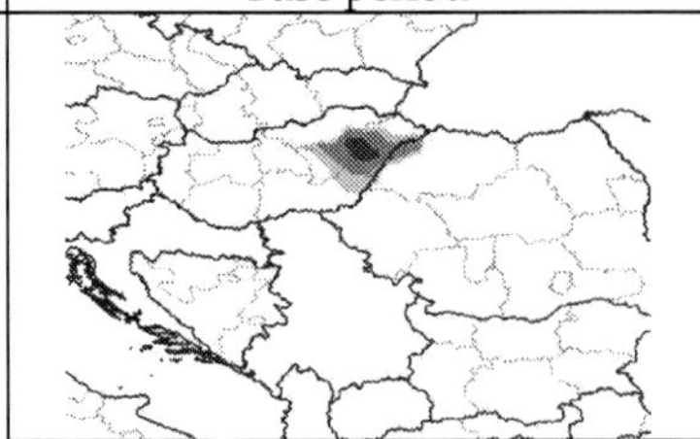	For validating the method, analogues for the observed climate were calculated. As a result we got back the region of Debrecen.
	A1F1	B2
2010-2019	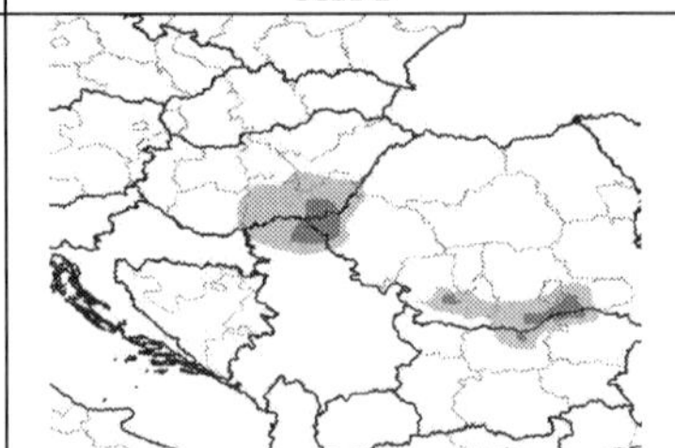	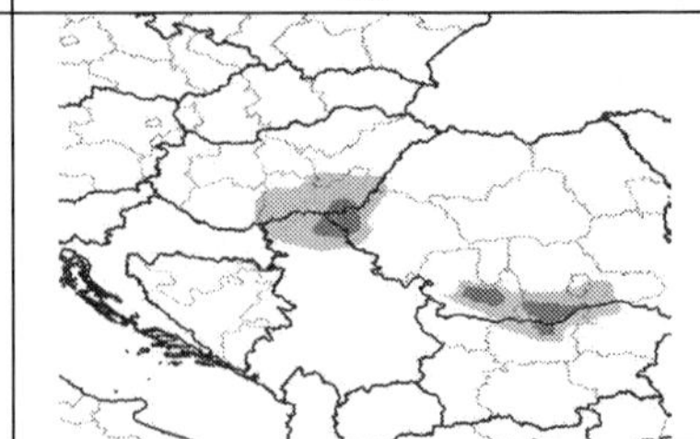
2020-2029	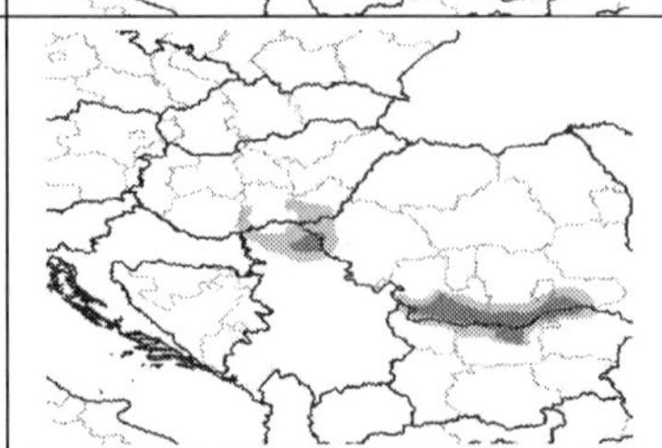	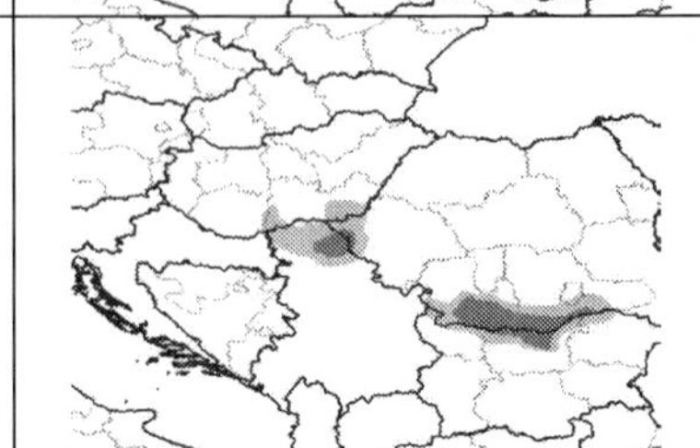
2030-2039	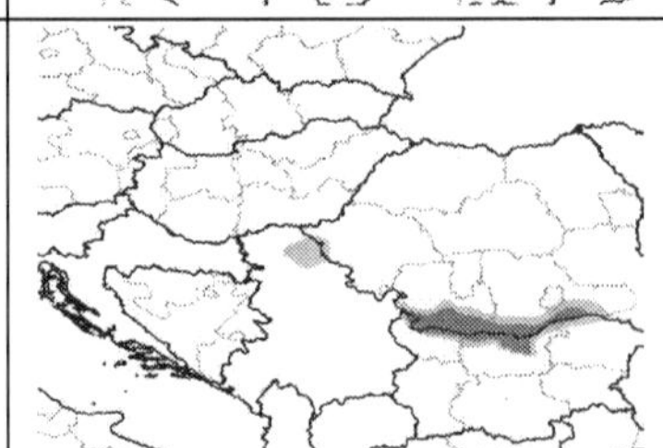	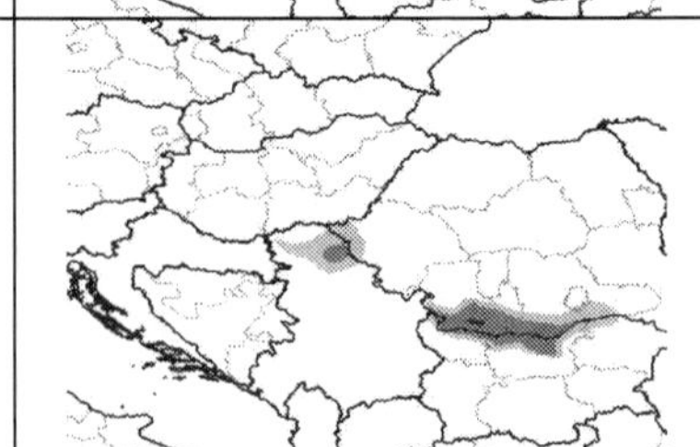
2040-2069	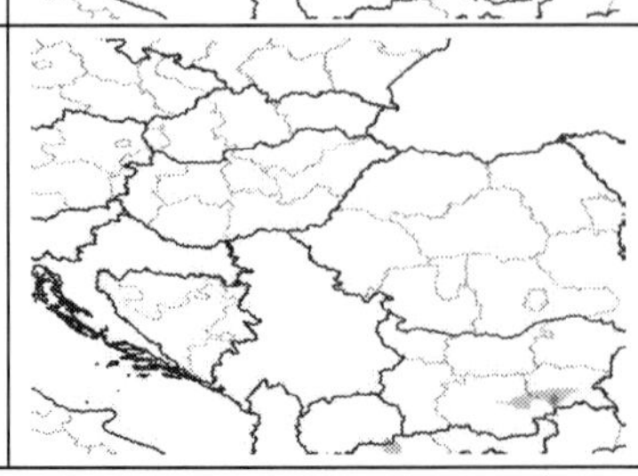	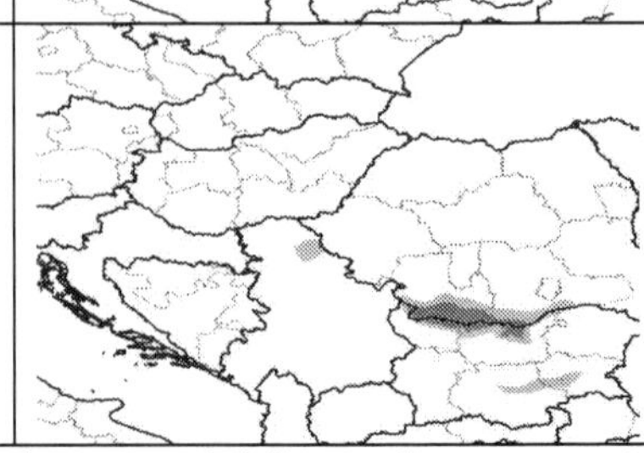

Fig. 8. Analogue regions in the next decades and in case of different climate scenarios for Debrecen

We developed a new method to find inverse analogue regions; these are the regions the climate of which will be similar to that of our study area in the future. We found a same shifting to the north. These analogue regions are in Poland (Fig. 9.) and were defined for the scenario A2.

Similarity (%)	2011-2040	2041-2070
-90 90,01 - 91 91,01 - 92 92,01 - 93 93,01 - 94 94,01 - 95 95,01 - 96 96,01 - 97 97,01 - 98 98,01 - 99 99,01 - 100	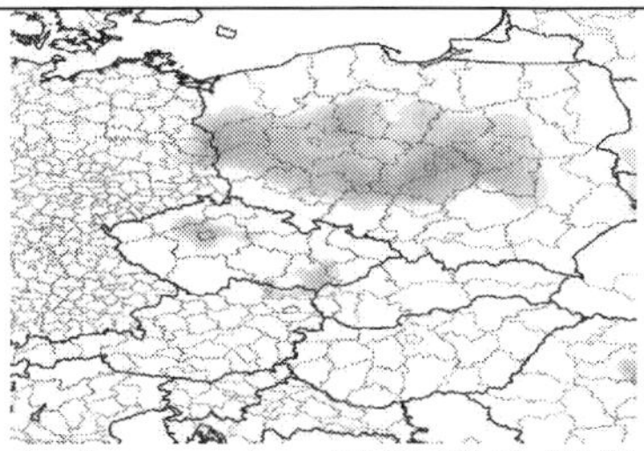 Analogue regions: PL11, PL43, PL41	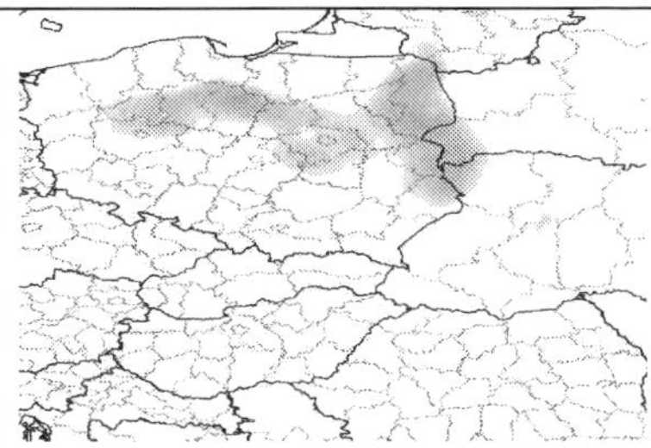 Analogue regions: PL61, PL12, PL31

Fig. 9. Analogue regions in the next decades and in case of different climate scenarios for Debrecen

It can be seen that analogue regions are south-east of Debrecen about 250-450 km away but later this distance is larger. The analogue regions are Vojvodina in Serbia as well as the RO04 (Sud-Vest) and the RO03 (Sud) NUTS regions in Romania. For further analyses only these regions were taken into consideration (Fig. 10.). We calculated the diversity of croplands and the land use and we found opposite changes. While the diversity of croplands is lower than in Hungary, the diversity of land use is higher. It is mostly because of the main crops. The ratio of wheat and maize is higher in the South, just because the climatic conditions are better for them. Meanwhile the yield is lower; it is more economical to use these crops because the better conditions necessitate less agronomic techniques.

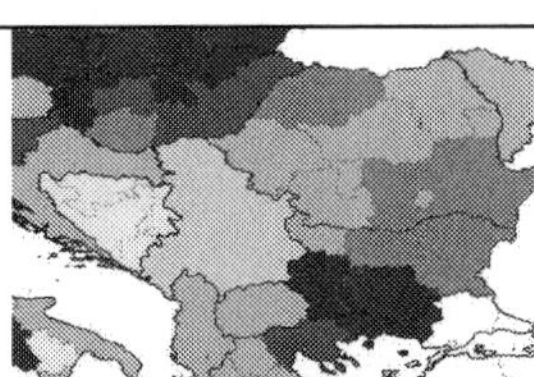	Diversity of the croplands. In the analogue regions the diversity is lower, so the structure of the croplands in Hungary will probably change, too.
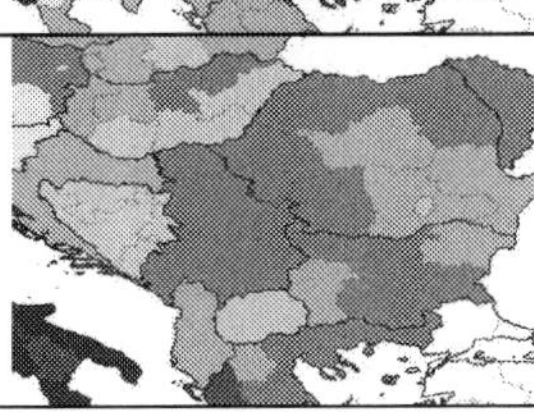	Diversity of the land use types is higher in the analogue regions, but it could be caused by the topography as well.

<table>
<tr><td>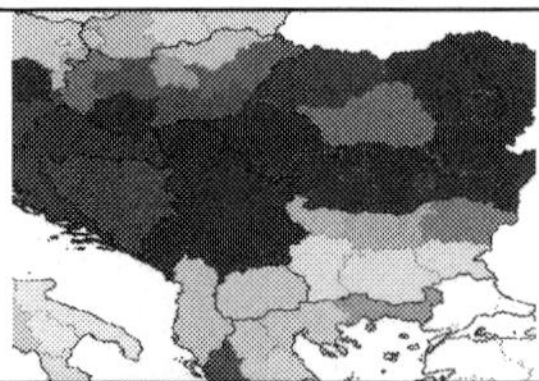</td><td>Ratio of the maize fields in the arable land is higher in the analogue regions than in Hungary. Climate change could have a positive effect on the Hungarian maize production, too.</td></tr>
<tr><td>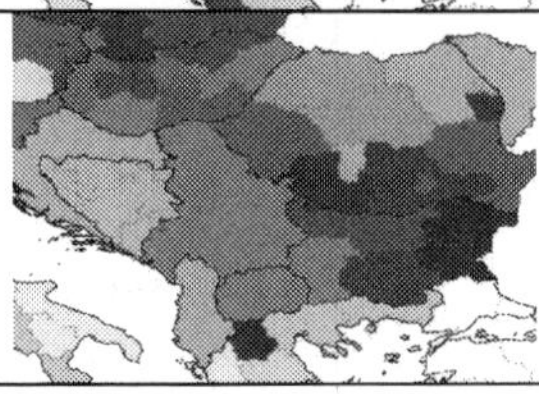</td><td>The ratio of the wheat fields in the analogue regions is similar to the present situation in Hungary. Climate change may not have an effect on it.</td></tr>
</table>

Fig. 10. Land use and cropland ratio of the analogue regions

5.3. Discussion

Debrecen, the basic object of our calculations, is an important centre of agricultural production in Hungary, so we would like to interpret the results in this aspect.

Climate, especially temperature and precipitation, basically determines agricultural production. Results show that in Hungary we have to count on an increase in temperature and decrease in precipitation. The possible future climate predicted by the scenarios will be similar to the present climate of South-Southeast Europe. Of course climate depends also on other effects, especially on elevation, topography and storm-track conditions, which could not be considered in this kind of analysis. However, the method of spatial analogies seems to be a good tool to understand and interpret the results of the GCM scenarios and the effects of climate change, so we want to go ahead in this research. This method and additional data on the analogue regions can provide information on the impacts of climate change on ecosystems or on agricultural production, such as the changes in land use, cropping system or yields and on the possibilities for disappearing or introducing new crops or weeds and pests into an area.

Increase in mean annual temperatures in our region, if limited to two or three degrees, can generally be expected to extend the growing season. In case of crops (or animals), where phonological phases depend on accumulated heat units, the phenophases can become shorter. Whether crops respond to higher temperatures with an increase or decrease in yield depends on whether their yield is currently strongly limited by insufficient warmth or the temperature is near or little above the optimum. In Central Europe, where temperatures are near the optimum under current climatic conditions, increases in temperature would probably lead to decreased yields in case of several crops. Increased temperature could be favourable for example for pepper and grapes; however, it is unfavourable for green peas and potato. Decrease in precipitation could be a great limiting factor in agriculture.

If we accept the results of the GCMs, according to the A1FI scenario for the period 2011-2040, the analogue regions of Debrecen will be the Vojvodina region in Serbia and South Romania. It means a shifting of about 250-450 km south, which corresponds to other international results.

The detailed analyses of the analogue regions can help us to adapt to the changing climate. From the analogue regions we should collect all kind of available ecological, agricultural,

economic, social and public sanitation data. We can study what kind of problems there are, and what the solutions are. We can learn from there how to solve the possible problems and develop strategies. This would be a good base for further research and an important base for decision makers.
With the method of spatial analogy we can build a new way of knowledge transfer from where we can learn adaptation techniques and to where we can transfer our knowledge.

6. References

Adrian, R., Deneke, R., Mischke, U., Stellmacher, R., Lederer, P. (1995). A long-term study of the Heilingensee (1975–1992). Evidence for effects of climatic change on the dynamics of eutrophied lake ecosystems. *Archiv fur Hydrobiol* 133: 315–337.

Allan, J.D., Palmer, M., Poff, N.L. (2005). Climate change and freshwater ecosystems. In: *Lovejoy, T.E., Hannah, L. (eds): Climate change and biodiversity.* Yale University Press, New Haven, CT, pp. 274–295.

Baltz, D.M., Moyle, P.B. (1993). Invasion resistance to introduced species by a native assemblage of California stream fishes. *Ecol Appl* 3: 246–255.

Christensen, J. H. (2005). Prediction of Regional scenarios and Uncertainties for Defining European Climate change risks and Effects. *Final Report.* DMI, Copenhagen.

Clements, F. E. (1905). *Research methods in ecology.* University Publishing. Lincoln, Nebraska, USA.

Connell, J. H. (1978). Diversity in tropical rain forests and coral reefs. *Science,* 199., 1302 - 1310, 0036-8075

Drégelyi-Kiss, Á. & Hufnagel, L. (2010). Effects of temperature-climate patterns on the production of some competitive species on grounds of modelling. *Environmental Modeling & Assessment,* Doi :10.1007/s10666-009-9216-4, 1573-2967

Drégelyi-Kiss, Á. & Hufnagel, L. (2009). Simulations of Theoretical Ecosystem Growth Model (TEGM) during various climate conditions. *Applied Ecology and Environmental Research,* 7(1)., 71-78, 1785-0037

Drégelyi-Kiss, Á., Drégelyi-Kiss, G. & Hufnagel, L. (2008). Ecosystems as climate controllers - biotic feedbacks (a review). *Applied Ecology and Environmental Research,* 6(2)., 111-135, 1785-0037

Elliott, J. A., Irish, A. E. & Reynolds, C. S. (2001). The effects of vertical mixing on a phytoplankton community: a modelling approach to the intermediate disturbance hypothesis. *Freshwater Biology,* 46., 1291–1297, 0046-5070

Elliott, J. A., Thackeray, S. J., Huntingford, C. & Jones, R.G. (2005). Combining a regional climate model with a phytoplankton community model to predict future changes in phytoplankton in lakes. *Freshwater Biology* 50: 1404-1411.

Elton, C.S. (1927). *Animal ecology.* Sidgwick and Jackson, London, GB 207 pp.

Felföldy, L (1981). *Water environmental sciences.* Mezőgazdasági Kiadó, Budapest [in Hungarian]

Ficke, A.D., Myrick, C.A., Hansen, L.J. (2007). Potential impacts of global climate change on freshwater fisheries. *Rev. Fish. Biol. Fisheries* 17: 581-613.

Fischlin, A., G.F. Midgley, J.T. Price, R. Leemans, B. Gopal, C. Turley, M.D.A. Rounsevell, O.P. Dube, J. Tarazona, A.A. Velichko (2007). Ecosystems, their properties, goods, and services. In: *Climate Change. (2007). Impacts, Adaptation and Vulnerability. Contribution of Working Group II to the Fourth Assessment Report of the Intergovernmental Panel on Climate Change,* M.L. Parry, O.F. Canziani, J.P. Palutikof, P.J. van der Linden and C.E. Hanson, Eds., Cambridge University Press, Cambridge, pp. 211 - 272

Friedlingstein, P., Cox, P. M., Betts, R. A., Bopp, L., von Bloh, W., Brovkin, V., Cadule, P., Doney, S., Eby, M., Fung, I., Bala, G., John, J., Jones, C. D., Joos, F., Kato, T., Kawamiya, M., Knorr, W., Lindsay, K., Matthews, H. D., Raddatz, T., Rayner, P., Reick, C., Roeckner, E., Schnitzler, K. G., Schnur, R., Strassmann, K., Weaver, A. J., Yoshikawa, C. & Zeng, N. (2006). Climate-Carbon Cycle feedback analysis: Results from the C4MIP model incomparison, *Journal of Climate, 19.*, 3337-3353, 0894-8755

Fulbright, T. E. (1996). Viewpoint: a theoretical basis for planning woody plant control to maintain species diversity. *Journal of Range Management,* 49., 554- 559, 0022-409X

Gleason, Henry A. (1926). The Individualistic Concept of the Plant Association. *Bulletin of the Torrey Botanical Club* 53: 7-26

Haffner, G. D., Harris, G. P. & Jarai, M. K. (1980). Physical variability and phytoplankton communities. III. Vertical structure in phytoplankton populations. *Archiv für Hydrobiologie,* 89., 363 - 381, 0003-9136

Hammer, O., Harper, D. A. T. & Ryan, P. D. (2001). PAST: Paleontological statistics software package for education and data nalysis. *Paleontologia Electronica,* 4(1), 9.

Horváth, L. & Tevanné Bartalis, É. (1999). A vízkémiai viszonyok jellemzése a Duna Rajka-Szob közötti szakaszán. *Vízügyi Közlemények.* 81: 54-85.(in Hungarian)

Hufnagel L., Sipkay Cs., Drégelyi-Kiss Á., Farkas E., Türei D. , Gergócs V., Petrányi G., Baksa A., Gimesi L., Eppich B., Dede L., Horváth L. (2008). Interactions between the processes of climate change, bio-diversity and community ecology. *In Climate Change: Environment, Risk, Society. Harnos Zs., Csete L. (Eds.), 227-264.* Szaktudás Kiadó Ház, ISBN 978-963-9736-87-0, Budapest (in Hungarian)

Hufnagel, L. & Gaál, M. (2005): Seasonal dynamic pattern analysis service of climate change research. *Applied Ecology and Environmental Research* 3(1): 79–132.

Hufnagel, L., Drégelyi-Kiss, G. & Drégelyi-Kiss, Á. (2010). The effect of the reproductivity's velocity on the biodiversity of a theoretical ecosystem, *Applied Ecology and Environmental Research,* in press, 1785-0037

IPCC (2007). *The Fourth Assessment Report "Climate Change 2007"* Cambridge University Press 2008 ISBN-13:9780521705974

Kalff, J. (2000). *Limnology.* Prentice Hall, Upper Saddle River, New Jersey.

Kiss, K. T. 1994. Trophic level and eutrophication of the River Danube in Hungary. *Verh.Internat.Verein.Limnol.* 25: 1688-1691.

Klapper, H. (1991). *Control of eutrophication in Inland waters.* Ellis Horwood Ltd., West Sussex, UK.

Komatsu, E., Fukushima, T. & Harasawa, H. (2007). A modeling approach to forecast the effect of long-term climate change on lake water quality. *Ecological Modelling* 209: 351-366.

Köck, G., Triendl, M., Hofer, R. (1996). Seasonal patterns of metal accumulation in Arctic char (Salvelinus alpinus) from an oligotrophic Alpine lake related to temperature. *Can J Fisher Aqua Sci* 53: 780–786.

Lofgren, B.M. (2002). Global warming influences on water levels, ice, and chemical and biological cycles in lakes: some examples. *In: McGinn NA (ed) Fisheries in a changing climate.* American Fisheries Society, Bethesda, MD, pp. 15–22.

Lovelock J.E (1972). "Gaia as seen through the atmosphere". *Atmospheric Environment* 6 (8): 579–580.

Lovelock J.E (1990). "Hands up for the Gaia hypothesis". *Nature* 344 (6262): 100–2.

MacLeod, J.C., Pessah, E. (1973). Temperature effects on mercury accumulation, toxicity, and metabolic rate in rainbow trout (Salmo gairdneri). J *Fisher Res Board Can* 30: 485–492.

Magnuson, J.J. (2002.a). Future of adapting to climate change and variability. *In: McGinn NA (ed) Fisheries in a changing climate.* American Fisheries Society, Bethesda, MD, pp. 283–287.

Matthews, W.J., Marsh-Matthews, E. (2003). Effects of drought on fish across axes of space, time and ecological complexity. *Freshwater Biol* 48: 1232–1253.

Mitchell T.D., Carter T.R., Jones P.D., Hulme,M., New, M., (2003). A comprehensive set of high-resolution grids of monthly climate for Europe and the globe: the observed record (1901-2000) and 16 scenarios (2001-2100). *Journal of Climate*

Mooij, W.M., Janse, J.H., Domis, L.N., Hülsmann, S., Ibelings, B.W. (2007): Predicting the effect of climate change on temperate shallow lakes with the ecosystem model PCLake. *Hydrobiologia* 584:443–454.

Murty, A.S. (1986.a). *Toxicity of pesticides to Fish. I.* CRC Press, Boca Raton, FL.

Murty, A.S. (1986.b). *Toxicity of pesticides to Fish. II.* CRC Press, Boca Raton, FL.

Nussey, G., van Vuren, J.H.J., du Preez, H.H. (1996). Acute toxicity of copper on juvenile Mozambique tilapia, Oreochromis mossambicus (Cichlidae), at different temperatures. *S Afr J Wildlife Res* 26: 47–55.

Olrik, K. & Nauwerck, A. (1993). Stress and disturbance in the phytoplankton community of a shallow, hypertrophic lake. *Hydrobiologia,* 249., 15 - 24, 0018-8158

Padisák, J. (1998). Sudden and gradual responses of phytoplankton to global climate change: case studies from two large, shallow lakes (Balaton, Hungary and the Neusiedlersee, Austria/Hungary). *In Management of Lakes and Reservoirs during Global Change. George, D. G, Jones, J. G, Puncochar, P., Reynolds, C. S. & Sutcliffe, D. W. (Eds.),* 111-125, Kluwer Academic Publishers, ISBN-0-7923-5055-3, Dordrecht, Boston, London.

Pianka, E. R. (1974). Niche overlap and diffuse competition, *Proceedings of the National Academy of Sciences of the United States of America,* 71., 2141-2145, 0027-8424

Poff, N.L., Brinson, M.M., Day, J.W. (2002). *Aquatic ecosystem & Climate change. Potential impacts on inland freshwater and coastal wetland ecosystems in the United States.* Pew Center on Globa Climate Change. pp. 44.

Reynolds, C. S. (2006). *The ecology of phytoplankton.* Cambridge University Press, ISBN-13 978-0-521-84413-0, Cambridge

Sipkay, Cs., Kiss, K. T., Vadadi-Fülöp, Cs. & Hufnagel, L. (2009). Trends in research on the possible effects of climate change concerning aquatic ecosystems with special emphasis on the modelling approach. *Applied Ecology and Environmental Research* 7(2): 171-198.

Sipkay, Cs., Horváth, L., Nosek, J., Oertel, N., Vadadi-Fülöp, Cs., Farkas, E., Drégelyi-Kiss, Á. & Hufnagel, L. (2008a). Analysis of climate change scenarios based on modelling of the seasonal dynamics of a Danubian copepod species. *Applied Ecology and Environmental Research* 6(4): 101-108.

Sipkay, Cs., Hufnagel, L. Révés,z A. & Petrányi, G. (2008b). Seasonal dynamics of an aquatic macroinvertebrate assembly (Hydrobiological case study of Lake Balaton, №. 2) *Applied Ecology and Environmental Research* 5(2): 63-78.

Solymosi N., Kern, A. Horváth L., Maróti-Agócs Á., Erdélyi K. (2008). TETYN: An easy to use tool for extracting climatic parameters from Tyndall datasets, *Environmental Modelling and Software* 948-949 p.

Sommer, U. (1995). An experimental test of the intermediate disturbance hypothesis using cultures of marine phytoplankton. *Limnology and Oceanography*, 40, 1271 - 1277, 0024-3590

Spellerberg, I.F. (1991). *Monitoring ecological change*, Cambridge University Press, Cambridge

Straile, D. (2005). Food webs in lakes - seasonal dynamics and the impacts of climate variability. *In: Belgrano, A, Scharler, U.M, Dunne, J & Ulanowicz, R.E.*

Szenteleki K. (2007). A Kötnyezet-Kockázat- Társadalom (KLIMAKKT) Klímakutatás adatbázis kezelő rendszerei, *Klíma21 füzetek* vol51. 89-115 p. (in Hungarian)

Vadadi-Fülöp Cs., D. Türei, Cs. Sipkay, Cs. Verasztó, Á. Drégelyi-Kiss, L. Hufnagel (2009). Comparative Assessment of Climate Change Scenarios Based on Aquatic Food Web Modeling. *Environmental Modeling and Assessment* 14(5) : 563-576

Viner, B. & Kemp, L. (1983). The effect of vertical mixing on the phytoplankton of Lake Rotongaio (July 1979 -January 1981). *New Zealand Journal of Marine and Freshwater Research*, 17, 407 – 422, 0028-8330

Wood, R.J., Boesch, D.F., Kennedy, V.S. (2002): Future consequences of climate change for the Chesapeake Bay ecosystem and its fisheries. *In: McGinn, N.A. (ed): Fisheries in a changing climate*. American Fisheries Society, Bethesda, MD, pp. 171–184.

9

Pelagic ecosystem response to climate variability in the Pacific Ocean off Baja California

Gilberto Gaxiola-Castro[1], Bertha E. Lavaniegos[1], Antonio Martínez[2], Rubén Castro[2], T. Leticia Espinosa-Carreón[3]

[1]Centro de Investigación Científica y Educación Superior de Ensenada. Departamento de Oceanografía Biológica. Carretera Ensenada-Tijuana No. 328. Fracc. Zona Playitas. Ensenada, Baja California, México. C.P. 22860. [2]Universidad Autónoma de Baja California. Facultad de Ciencias Marinas. Kilómetro 107 carretera Tijuana-Ensenada. Ensenada, Baja California, México. C.P. 22810. [3]Centro Interdisciplinario de Investigación para el Desarrollo Integral Regional (CIIDIR) Unidad Sinaloa-IPN. Blvd. Juan de Dios Bátiz Paredes #250. Col. San Joachin. C.P. 81101. Guasave, Sinaloa, México.

1. Introduction

The oceans enclose 72 percent of the Earth's surface, controlling the global deliveries of heat and freshwater which drive our climate and weather. In turn, the enormous and varied oceanic ecosystems are affected by the ocean climate, generating changes mainly in the upper part of the water column which can be detected as response to this variability. In particular, the Pacific Ocean environment is affected by changes in the world climate, responding to seasonal, interannual, and interdecadal variability, as well as El Niño-La Niña cycles.

The California Current System (CCS) located in the northeastern Pacific Ocean, is one of the largest marine ecosystems of the world, with physical and biological mechanism forced by regional (coastal upwelling, eddies, wind driven currents), and large scale processes (e.g. El Niño-La Niña cycles). One example of these large scale processes affecting the CCS region was the anomalous intrusion of Subartic fresher water into during the 2002-2006 periods, which disturb the California Current (CC) ecosystems from Canada to Mexico (Bograd & Lynn, 2003; Durazo et al., 2005; Durazo, 2009; Freeland et al., 2003; Gaxiola-Castro et al., 2008; Lavaniegos, 2009). The ocean climate response of the CC region off Baja California is particularly significant because this is a transitional ocean basin, highly influenced by the equatorward flow of Subartic waters mainly during spring and summer (Durazo & Baumgartner, 2002), and by subtropical signatures triggered by poleward flows mostly during late summer and autumn (Bograd et al., 2000; Durazo, 2009). Moreover, CCS off Baja California is a useful region to understand large and mesoscale ocean climate effects on physical and biological variability of the marine environment, and for instance of climate change.

In this contribution we examine the associations between large-scale temporal climate physical forcing and the plankton variability off Baja California. The pelagic ecosystem off Baja California has been influenced by large scale processes like the 1997-1998, and 2008-2009 El Niño events, which were characterized by low chlorophyll, high sea surface height, high surface salinity, and high sea surface temperature, with opposite conditions during the 1998-1999 and 2007-2008 La Niña events (Espinosa-Carreon et al., 2004; McClatchie et al., 2009). Seasonal and interannual patterns are also observed off Baja California, with a maximum of phytoplankton Chlorophyll-*a* occurring in spring, as a result of the phytoplankton growth in response of the seasonal maximum of upwelling-favorable winds (Espinosa-Carreon et al. 2004; Perez-Bruinus et al. 2007). Zooplankton biomass is larger in summer and autumn, characterized by abundance of copepods, euphausids, and other minor groups, higher abundance of salps during the 1997-1998 El Niño event, and by jelly fish and ctenophores abundances in April 2009 (Lavaniegos et al., 2010).
The main objectives of this chapter are to describe the general conditions in the CCS off Baja California coast from 1997-2010, reviewing the governing ocean climate and physical forcing at different time and scales, that might have influenced the pelagic ecosystem of the study region.

2. Climate conditions off Baja California

Northeast Pacific Ocean atmospheric circulation is dominated by the North Pacific High Pressure center (NPH). In February, this pressure system is weak and located ~25°N-130°W off Baja California coast. During summer the NPH migrates near of ~38°N-145°W, where it reaches the maximum values. Between spring and summer the strong pressure gradient generate high equatorward upwelling favorable winds between northern California and the Baja California coast (Huyer, 1983; Strub et al., 1987; Strub & James, 2002; Perez-Brunius et al., 2007; Castro & Martinez, 2010). In the vicinity of Baja California coast the wind is highly persistent with a south-southeast direction. Wind speeds are stronger and less variables in spring and summer than during autumn and winter. Annual harmonic of the wind has small amplitude, with low (<30%) explained variance.
In general, high frequencies (order of days) play a dominant role in the region. Interannual variability of the NPH is frequently linked with negative (positive) sea level pressure anomalies during El Niño (La Niña), which produces weak, and/or reversal winds during El Niño and strong equatorward winds during La Niña episodes (Lynn et al., 1998; Schwing et al., 2002; Peterson & Schwing, 2003). Off Baja California coast the influence of the El Niño conditions produces low chlorophyll, high SSH, and high SST, with opposite conditions during La Niña (Espinosa-Carreon et al., 2004; Durazo, 2009). During 2002 a weak El Niño condition was observed on southern California and Baja California waters, at the same time that anomalous influence of subarctic waters in the California Current System was evident (Schwing et al., 2002; Bograd & Lynn, 2003; Durazo et al., 2005).
The roll of the wind stress curl has not received much attention in this region, and their contribution to upwelling process it is not clear. Some general characteristics of the wind field along the Baja California Peninsula are given in Castro & Martinez (2010), where the potential importance of the curl wind stress on the area close to the coast along the Baja California Peninsula is suggested. Next to the shore, the coast affects wind intensity and decay, inducing positive curl (and then Ekman pumping) contributing to the injection of

nutrients to the surface (Cappet et al., 2004). Chelton et al. (2007) discuss the relevance of the wind stress curl and its divergence and their relation to ocean sea surface temperature.

3. Data and Methods

In this chapter we examine satellite observations on the Pacific Ocean off Baja California Peninsula coast, with emphasis on interannual variability patterns. The analysis was made with wind stress data, sea surface height (SSH) anomaly, and sea surface temperature (SST). The wind data were obtained from PODAAC-CCMP product with spatial resolution 0.25°× 0.25° described by Atlas et al. (1996) and Hoffman (1984). Sea surface height was obtained by Ssalto/Duacs and distributed by AVISO (http://www.aviso.oceanobs.com/duacs/). Sea surface temperature data (SST) was obtained from the PODAAC-AVHRR (http://podaac.jpl.nasa.gov). All the satellite data covers the time period between January-1996 to December-2007, and were processed to obtain weekly time series, and low passed filtered with a 400 days cutoff filter. The wind stress (N m^{-2}) was calculated with the equation

$$\tau = \rho C_d |V| V \tag{1}$$

where ρ is the air density (kg m^{-3}), C_d is the drag coefficient (nondimensional and function of the wind speed magnitude), V is the vector velocity and $|V|$ its magnitude. Multivariate ENSO Index (MEI; http://www.esrl.noaa.gov/psd/data/) is compared with standardized anomalies of satellite variables described above.

Weekly remote sensed composite imagery of chlorophyll (CHL) were obtained from September 1997 to February 2010 for the region off Baja California between 22°N-33°N and 112°-120°W (Fig. 1). SeaWiFS (Sea Viewing Wide Field of View Sensor) ocean color data were downloading from http://oceandata.sci.gsfc.nasa.gov/SeaWiFS/Mapped/8Day/chlor/ with 9-km resolution for the last re-processing version of December 2009. Use of the geometric mean and log transformation follow the assumption that oceanic bio-optical data are lognormal distributed. We obtained the fraction of observations available from 540 weekly images. For each of the three selected areas (north, central, and south; see title of figure 1), the temporal mean, as well as the annual and semiannual harmonics were removed. These were calculated by fitting the time series to mean plus annual and semiannual harmonics:

$$F(\bar{x},t) = A_0(\bar{x}) + A_1(\bar{x})\cos(wt-\phi_1) + A_2(\bar{x})\cos(2wt-\phi_2) \tag{2}$$

where A_0, A_1, and A_2 are the temporal mean, annual amplitude, and semiannual amplitude for each time series at each pixel; w=$2\pi/365.25$ is the annual radian frequency; φ_1, φ_2 are the phases of annual and semiannual harmonics respectively; and t is the time (as year-day).

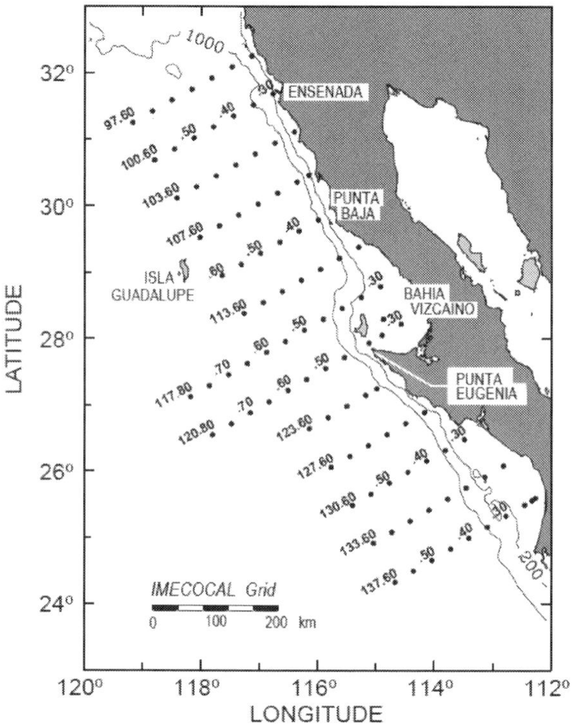

Fig. 1. IMECOCAL station grid off Baja California, México surveyed from September 1997 to April 2010. Line and station number follow the CalCOFI nomenclature. Line 97 was only covered during spring season cruises. Chlorophyll-*a*, and SeaWIFS imagery data analyses were separated in three regions: north (lines 100 to 110), central (lines 113 to 123), and south (lines 127 to 137). Also, zooplankton volume means were obtained of data grouped in regions named as north (lines 100-110), and central (lines 113-137).

We use IMECOCAL (Spanish acronym of *Mexican Oceanographic Investigations of the California Current*) research program data collected from 1997-2010 surveys, together with CalCOFI (*California Cooperative Fisheries Investigations)* 1950-2009 data, to illustrate the temporal links between lower trophic biological levels and large-scale climatic processes in the northeastern Pacific Ocean off Baja California. Quarterly sampling routine of the IMECOCAL cruises conducted on the Mexican RV *Francisco de Ulloa* includes CTD/Rossete casts (conductivity, temperature, depth) to 1000 m, depth permitting; also water samples from the upper 200 m collected in 5-liter Niskin bottles to determine phytoplankton Chlorophyll-*a*, and zooplankton tows were analyzed. The present study includes information from 45 cruises performed between September 1997 and April 2010 through a station grid of 92 stations (Fig. 1).

Water column phytoplankton Chlorophyll-*a* was determined from 1-liter discrete seawater samples, filtered onto Whatman GF/F filters, following the fluorometric method (Yentsch &

Menzel, 1963; Holm Hansen et al., 1965) with modifications of Venrick & Hayward (1984). Integrated 100 m water column Chlorophyll-*a* anomalies were calculated for the period 1998-2010 removing the long-term seasonal mean. Since the IMECOCAL domain is characterized by different hydrographic provinces (Subarctic north of 28°N; Subtropical and tropical south of 28°N), regional averages of Chlorophyll-*a* anomalies for three separated regions were calculated.

During the IMECOCAL surveys zooplankton was collected with a bongo net of 505 µm mesh-width doing double oblique hauls in the upper 210 m or from 10 m above the bottom to the surface at shallow stations. The diameter of the net was 61 cm before October 2001, and then it was replaced by one of 71 cm. The volume of water strained was measured with a flow meter in the mouth of the net. Samples were preserved with 4% formalin and sodium borate. A total of 3044 zooplankton samples were processed for displacement volume determination with a graduate cylinder one month after the end of each cruise. Taxonomic identification of major taxa were done selecting 1215 nighttime samples from oceanic stations (>200 m depth). Here only data of copepods and euphausiids are presented separately, while the rest of the taxa were combined in the category of other zooplankton. They were counted from a fraction (1/8, 1/16, or 1/32) of the original sample. Zooplankton biomass and abundance were log-transformed to normalize data. Biomass and abundance anomalies for the period 1997-2008 were estimated removing the long-term seasonal mean. Time series of abundance anomalies for copepods and euphausiids were subject to linear regression analyses.

Data of daily upwelling indices (UI) from www.pfeg.noaa.gov/products/products.html, were also used to illustrate the variability in coastal upwelling activity. The data were obtained for two sites: off Punta Baja (30°N, 119°W) and off Punta Eugenia (27°N, 116°W; see figure 1). UI monthly means (and standard deviations) were estimated. Further, UI anomalies were calculated removing long-term monthly means from the period 1997-2008. In addition, we used oceanic climate indices (MEI, Multivariate ENSO Index; http://www.esrl.noaa.gov/psd/people/klaus.wolter/MEI/; PDO, Pacific Decadal Oscillation; http://jisao.washington.edu/pdo/PDO.latest; NPGO, North Pacific Gyre Oscillation; http://ocean.eas.gatech.edu/npgo/), which are contrasted with integrated water column Chlorophyll-*a* anomalies, computed from the IMECOCAL hydrographic data.

4. Physical Forcing Analysis

The response to the forcing at interannual variability scales off the Baja California coast is evident from the time series on figure 2. In this figure, the wind stress magnitude, wind stress curl, SST (Sea Surface Temperature) and SSH (Sea Surface Height) anomalies were averaged on the overall IMECOCAL area, and compared with the MEI. Anomalies were computed averaging spatially all data over the area on figure 4, subtracting the temporal mean and dividing by the standard deviation. The correlation coefficient (r) between wind stress (magnitude and curl) and MEI was very low (r~0.1). Negative anomalies of the wind stress occurred during 1997, 2002, and 2004-2005, coinciding roughly with positive values of MEI, which could be associated to El Niño conditions and relatively weak winds on the northeastern Pacific (Peterson & Schwing, 2003).

The strength of the wind anomalies showed poor agreement with MEI; for example in 1997, negative anomalies of the wind stress reached values as low as -2.0, while during the 2004 event were just ~-0.7. Wind positive anomalies were present from 1998 to 2000, matching much of the time with La Niña. At the second half of 2005, high positive wind anomalies were observed between the last part of El Niño 2004-2005 and the transition to negative values of MEI (weak La Niña conditions). Summarizing, the MEI does not represent adequately the wind events off Baja California coast.

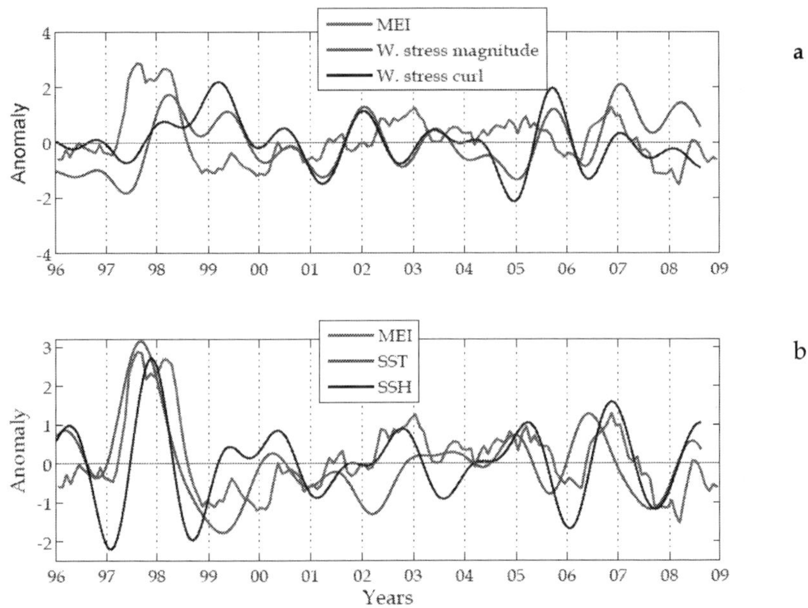

Fig. 2. (a) Anomalies of the wind stress magnitude and wind stress curl; (b) SST (Sea Surface Temperature) and SSH (Sea Surface Height). Multivariate El Niño Southern Oscillation Index (MEI) is included in both figures as reference.

Interannual SST anomalies were better correlated with the MEI than the wind (Fig. 2). The correlation (at 0 lag) between SST and MEI was relatively high (r= 0.66). The most remarkable SST events were during 1997-1998, and 2006 corresponding to El Niño conditions, where both MEI and SST showed positive anomalies (Espinosa-Carreon et al., 2004). Negative SST anomalies occurred during 2002-2003 with positive MEI values, which possibly was related with a strong incursion of subartic waters to the region (Durazo et al., 2005).

SST negatives anomalies were evident during La Niña episodes, between winter 1998 and 2000-2001, and remained until 2003. A shift between anomalies of SST-MEI is evident; with SST anomalies preceding MEI (figures 2, 4a). This lag has been reported by Lynn et al. (1998). SSH anomalies were well correlated with MEI (r= 0.52), except during the La Niña conditions (1999-2000) (Fig. 3). During La Niña 1999-2000 the MEI and SST showed negative values, while SSH was positive, which is not consistent with a thermal expansion.

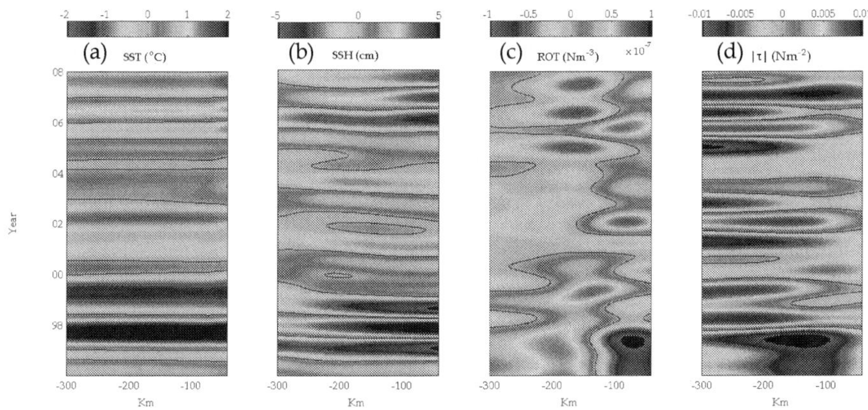

Fig. 3. Hovmuller diagram: (a) SST; (b) SSH; (c) wind stress curl magnitude; (d) wind setress curl anomalies. Each variable has been low passed filtered and averaged on the along shore direction. The horizontal axis represent the distance in km from the coast (at x=0). The vertical axis represents the time beginning in January 1996.

Time response of the Pacific Ocean along the Baja California Peninsula, as observed in figure 3, shows a lagged relation for each variable. In particular, the positive anomalies of SST (Fig. 3a) clearly preceded positive anomalies of SSH (Fig. 3b) during El Niño years. Casey & Adamec (2002) found that the relationship between SST-SSH is consistent with thermal expansion related to ENSO, which was proposed before by Leuliette & Wahr (1999). The lag between SST and SSH is more evident if we relate those variables with some diagnostic index like the MEI. The correlation was made between the MEI and the weekly-unfiltered time series of each variable on every grid point to show the short term lag. SST is somewhat correlated with MEI. The maximum correlation (Fig. 4a) occurs around Punta Eugenia (28°N). In general the correlation is low (but significant at the 95% confidence level). The lag between MEI and SST (black contour line in Fig. 4a) shows that SST precedes MEI by 1 to 1.5 months and shows a northward propagation. By the other side, the correlation between SSH and MEI (Fig. 4b) is much higher for the most part of the area. In contrast, the lag shows that the MEI leads SSH by 1.5 months along the coast, and the lag decreases on the offshore direction to 0.5 months. It is important to mention that if the SST and SSH data are low-passed filtered, the correlation, as expected, will increase significantly, but the short-term lag will be lost.

In addition, figure 3 also shows a rich offshore structure of the wind stress curl (Fig. 3c) and magnitude (Fig. 3d). The wind stress curl is positive most of the time near the coast, and changes sign away from the drop off zone (Castro & Martinez, 2010). The effect of the wind stress curl seems to be related to SST in such way that when the wind stress curl is weak (negative anomaly in figure 3c) SST increases. Almost two years previous to the 1997-1998 El Niño, the wind stress curl was very weak next to the coast (Fig. 3c), as was the wind intensity (Fig. 3d). In contrast, during the 2002 El Niño, the SST anomaly was negative (colder). There was not a response of SSH, with a positive wind stress curl anomaly next to the coast, and the intensification of the wind magnitude. The 2006 El Niño event produced very similar (but moderate) response on SST, SSH, and wind magnitude anomalies to the 1997-1998 event. It is important to notice that the wind magnitude anomalies in figure 3d are more intense off the coast, while the wind stress curl has large values near the coast.

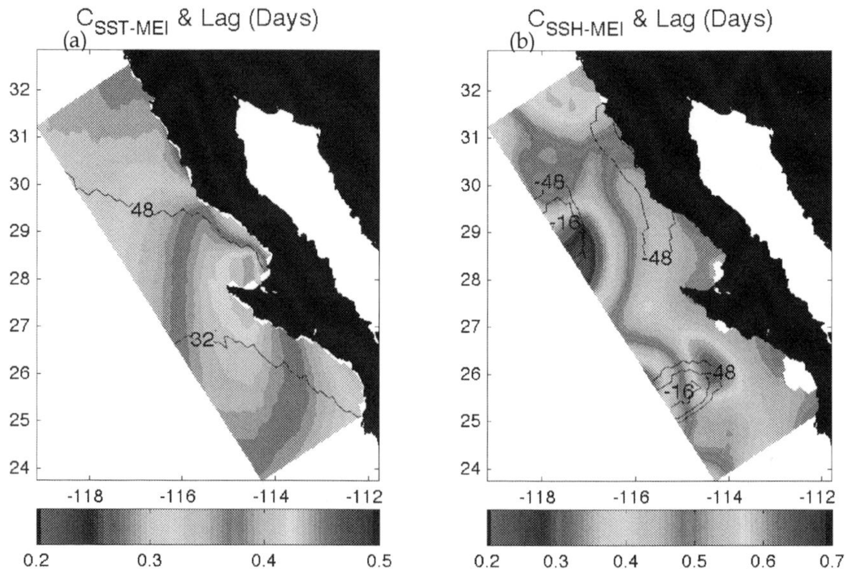

Fig. 4. Correlation between the MEI with SST (a), and MEI with SSH (b) on every grid point. Data cover the period from 1996-2007, without filtered. The black contour-line represents the lag in days.

To investigate a possible dynamical relationship connecting SST and the other analyzed variables, we calculate the zero-lagged time correlation of the low passed filtered data described previosly. The correlations between SST and SSH, wind stress curl ($\nabla x \tau$), wind stress major component (τ_x) and wind stress minor component (τ_y) were calculated for every grid point and then averaged on the along-coast direction. Just the significant correlations were considered for the averaging. The black line in figure 5 shows the correlation between SST and SSH.

Largest correlation was on the coast vicinity (offshore distance $\leq$ 150 km). Beyond this point the correlation decaes gradually. The positive sign shows two possible mechanism involved: the upwelling scenario, where sea level and SST decreases; and the thermal expansion scenario, where as the water temperature increases sea level also increases. The latest scenario does not seems to be probable in this analysis since figure 5 shows the zero lag correlation, namely the short term response. The major component of the wind stress (τ_x) is almost aligned with the shore line. It is well correlated with SST, except near the coast (red line in figure 5). The minor component of the wind stress (τ_y), does not show a clear relation with SST, and the wind stress curl shows the strongest relation with SST at about 120 km off the coast.

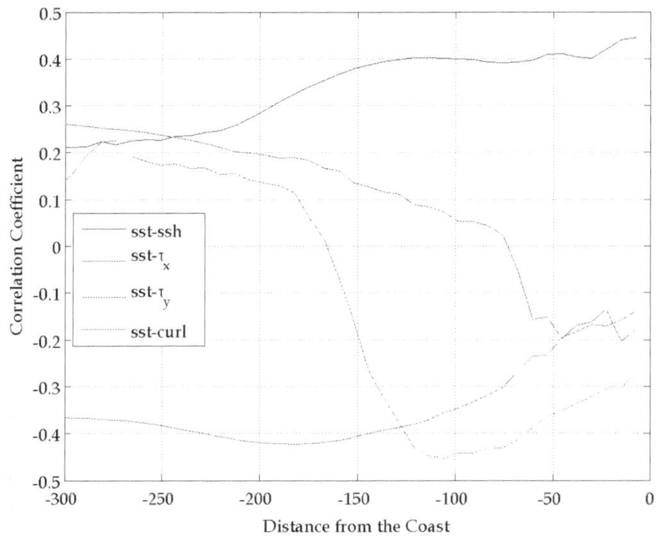

Fig. 5. Time correlation between SST and SSH (black line). Wind stress principal component τ_x (red line) and minor component τ_y (blue), and wind stress curl (magenta). The correlation was done for every point and then averaged on the along shore direction.

Near the coast (~120 km off shore), SST is well related with SSH, being the wind stress curl the second dominant forcing. Beyond this point, the dominant forcing is the component of the wind stress parallel to the coast (τ_x). The negative sign of the SST-τ_x correlation indicates upwelling (since the major component is esentially positive), but from figures 3, and 5, the SST is mainly related with the wind 150 km away from the coast. It is clear that the averaged correlation shown in figure 5, fades the local relation between SST, SSH, and the wind. The relevance of the relationships derived from figure 5, is that the dependence of SST on the wind or SSH, can not be sumarized on a single mechanism. The wind stress curl dominates on the wind intensity near the coast. This result implies that Ekman pumping is also important on the near coast area. Ekman pumping could be a possible alternative mechanism to explain the presence of SST anomalies. Cold (warm) SST during La Niña (El Niño) events seems to be in phase with positive (negative) anomalies of the wind stress curl (Fig. 3a,c). Also the role of SSH is important on the coast neighborhood, as it is well correlated with SST. Offshore, the wind component parallel to the coast dominates.

5. Remote sensed and in situ Phytoplankton Chlorophyll

The fraction of available remote sensed composite imagery of chlorophyll (CHL) SeaWiFS 1997-2010 data for the region off Baja California (percent of current data in the weekly images) show high values (>80%) near coastal zone, diminishing offshore, with lowest values (<50%) in the southwest region.

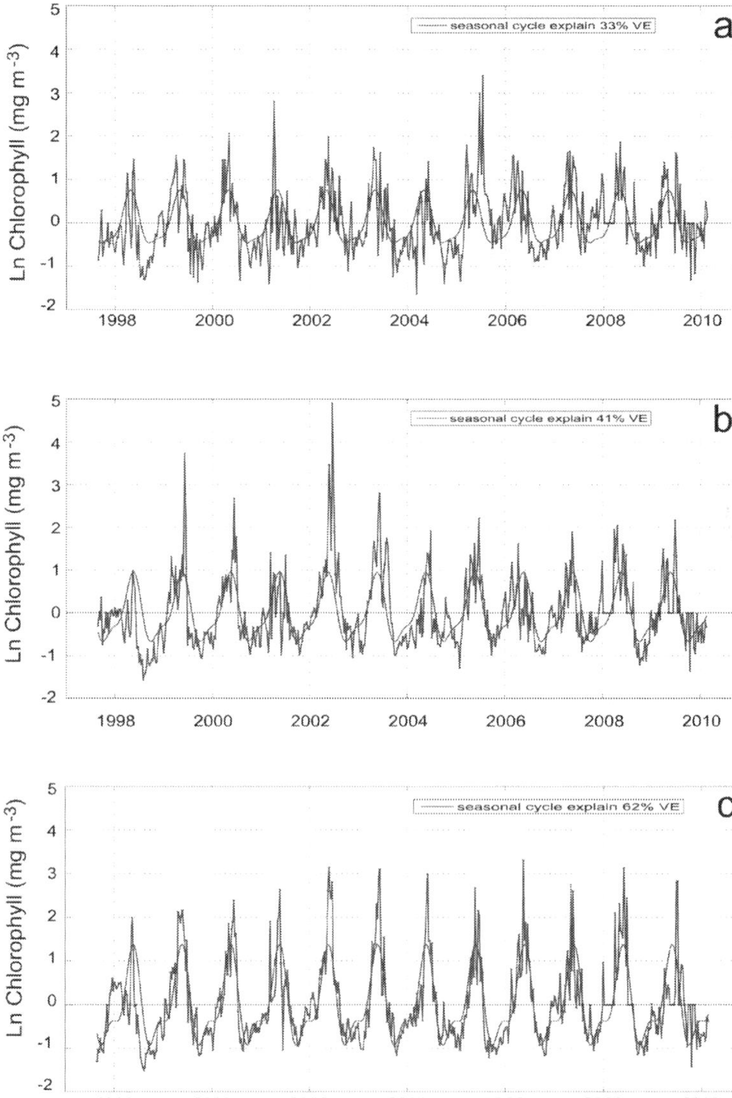

Fig. 6. Time series of log-transformed chlorophyll data (CHL, mg m^{-3}) obtained from SeaWIFS weekly composite imagery for the region off Baja California, México. Red color indicated the adjusted seasonal cycle. In blue color are the CHL time series for the three regions described in figure 1. a) north; b) central; c) south.

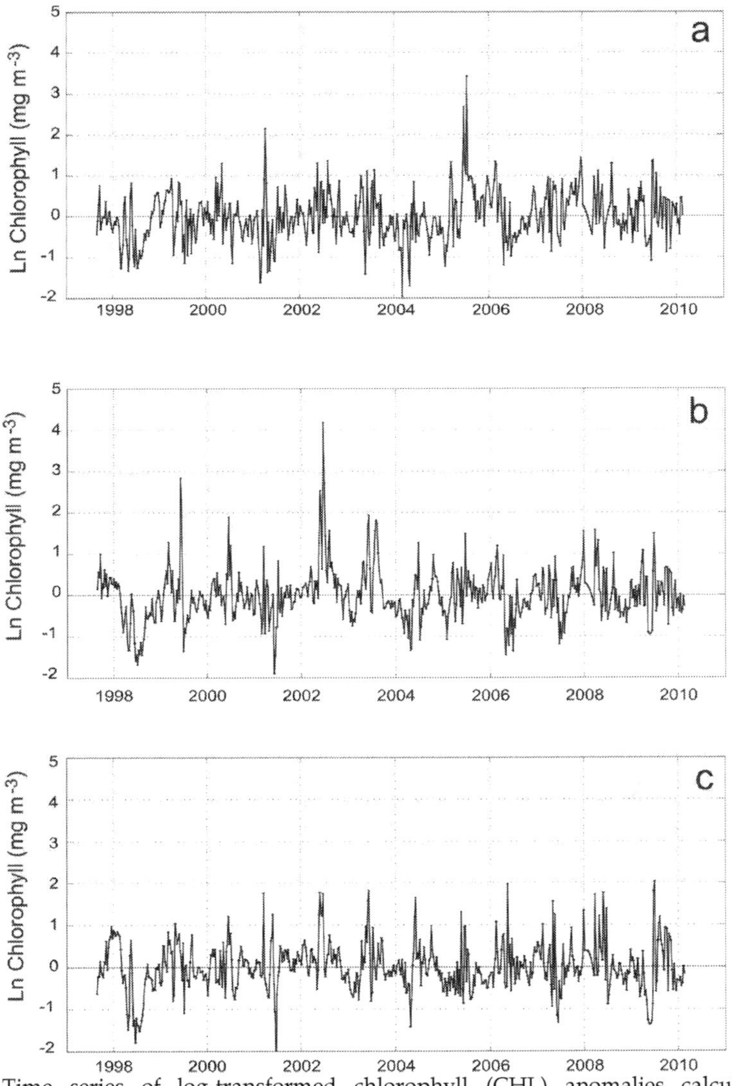

Fig. 7. Time series of log-transformed chlorophyll (CHL) anomalies calculated by subtracting mean seasonal cycles from weekly original SeaWIFS imagery data, collected for the three regions off Baja California described in figure 1. a) north; b) central; c) south.

The variance of CHL explained by the seasonal cycle (%VE) show some differences at coastal and offshore zones. Offshore Baja California appears to be at least three main processes influencing spatial variability: from 29-33°N the 10-40%VE is related with the influence of the California Current flowing southward. In the southwest zone the VE was

10-30%, probably associated with mesoscale processes as was illustrate before by Espinosa-Carreon et al. (2004). The low variance explained (<10%VE) should be associated to other different spatially scales processes (data not shown).

For the entire region off Baja California the variance explained by the seasonal cycle increased from northern to southern regions (see figure 1), with 33% in the north region, 41% in the central region, and 62% in the south region (Fig. 6a, b, c). The maximum increase on phytoplankton Chlorophyll-*a* occurred during spring, as a result of phytoplankton growth in response of seasonal maximum in upwelling-favorable winds (Perez-Brunius et al., 2007; Castro & Martinez, 2010), transporting inorganic nutrients to the ocean surface and enhancing primary production. Seasonal cycle amplitude increased also from north to south regions, with an annual signal dominating. However, the southern region during winter season shows a raise, which represents an unambiguous semiannual signal.

Satellite chlorophyll diminished at northern and central regions during the 1997-98 El Niño event, without any important decreasing at the southern region. This pattern for the lasted region has been discussed before (Kahru & Mitchell, 2000; Yoder & Kennelly, 2003; Espinosa-Carreon et al., 2004), without a comprehensible exposition of the processes implicated. The chlorophyll increase during 2005 at the north region (Fig. 6a), has been associated with coastal upwelling improved through this particular year in the California Current off Baja California (Durazo, 2009; Lavaniegos, 2009).

Baja California central region is a well representative area of the large scale processes influencing this basin (Fig. 6b), generally in the lower trophic pelagic organisms. These effects are mostly apparent during 1998, 2002, and 2003, related with El Niño-La Niña cycles. In the southern region, spring satellite chlorophyll show positive anomalies, except during 1997-1998 (Fig. 6c). In that period with an El Niño influence, CHL time series show an increase on late fall and winter, decreasing in spring season (Fig. 6c).

Interanual time series variability for each region is shown in figure 7 (time series without seasonal cycle). In general, variability of log-transformed chlorophyll shown low values during 1998, without an apparent pattern in the northern region. In 2005 positive CHL anomalies were related with warm and cool conditions respectively (Fig. 7a). In the central region CHL changes from positive to negatives anomalies during 1997-98 El Niño cycles, with positive anomalies associated to the 2002-2003 La Niña cycle (Fig. 7b). The southern region had lower variability, but during 1997-1998 positive anomalies are evident in late fall and winter, and negative during spring and summer. After that, positive CHL anomalies were observed on all spring seasons (Fig. 7c).

Water column integrated phytoplankton Chlorophyll-*a* anomalies estimated for three regions off Baja California from 1998-2010 were related with the MEI (Multivariate ENSO Index), PDO (Pacific Decadal Oscillation), and NPGO (North Pacific Gyre Oscillation) indices (Fig. 8). In general, chlorophyll anomalies for these three selected regions follow a similar tendency that the NPOG index, except during 2003 and in fewer manners 2004, when the index had positive values and the chlorophyll had negative anomalies (Fig. 8c, d, e, f). According with Di Lorenzo et al. (2008) the NPGO index response to changes in the sea surface height variability, and measure alterations of the North Pacific Ocean gyres, related with atmospheric forcing along the coast. A positive NPGO index value would indicate suitable oceanic conditions for wind-driven upwelling and horizontal advection enhancement, increasing surface nutrient concentration and for instance the phytoplankton growth. However, long-term phytoplankton variability in the regions off Baja California during the period from 2003 to 2007 appears to respond more properly to decadal forcing changes (PDO index) with an inverse trend between this ocean climate variability index and the phytoplankton biomass (Gaxiola-Castro et al., 2008).

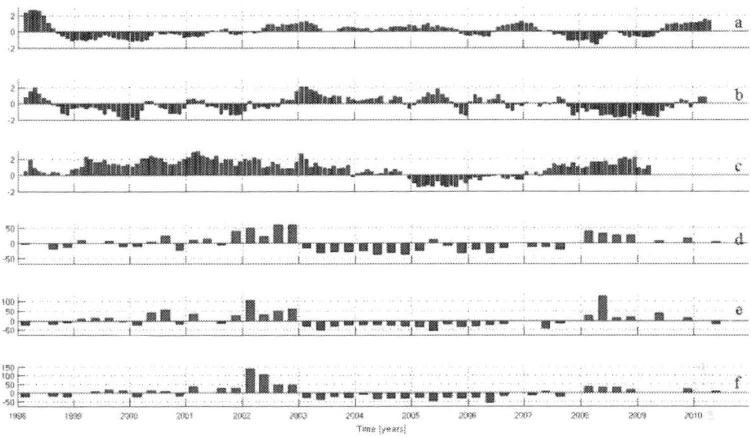

Fig. 8. Pacific Ocean climate variability indices, and water column integrated phytoplankton Chlorophyll-*a* anomalies for three oceanic regions off Baja California described in figure 1. a) MEI (Multivariate ENSO Index); b) PDO (Pacific Decadal Oscillation); c) NPGO (North Pacific Gyre Oscillation). Water column (100 m depth) integrated Chlorophyll-*a* anomalies calculated after removing the long-term seasonal means for the period 1998-2010. d) north region; e) central region; f) south region.

For the period 1998-2010 we calculated a positive linear relationship (correlation coefficient of 0.48, at 95% confidence level) between NPGO index and Chlorophyll-*a* anomalies off Baja California, without any significant differences between the Baja California regions. Interannual relationships between NPGO and chlorophyll have been reported before for the CalCOFI region (Di Lorenzo et al., 2008; http://ocean.eas.gatech.edu/npgo/), located northern Baja California. Linear relationships between water column integrated Chlorophyll-*a* *versus* MEI and PDO index for the three regions were negative, with a low correlation coefficient (~-0.20, at 95% confidence level) for these two indices.

6. Zooplankton variability

The seasonal zooplankton biomass variability off Baja California is less remarkable related to northern sectors of the California Current System (CCS) (USGLOBEC, 1994). Following this latitudinal tendency, Baja California waters exhibits more seasonality in temperature and coastal upwelling activity in the north region (30-32°N) compared to the central region (24-30°N).

Considering the interannual changes, zooplankton biomass anomalies were below the 1997-2008 mean associated with La Niña conditions mainly in the northern region from October 1998 to April 2000 (Fig. 9). Low biomass was also associated to other interannual event known as the subarctic water intrusion, which brought to the CCS cool and low salinity water in the upper sea layer (Freeland et al., 2003; Durazo et al., 2005; Gaxiola-Castro et al.,

2008). Though the subarctic water intrusion started in July 2002, the zooplankton off Baja California response to this strong perturbation was more clearly observed by October 2002, with a dramatic reduction in zooplankton biomass. Sea surface salinity continued a decreasing tendency until February 2004, and then anomalies remained negative but stable until the end of 2006 (Lavaniegos, 2009). Zooplankton biomass during the freshening period decreased in the beginning of subarctic water intrusion, but by 2003 in the north region the zooplankton has progressively increased. In the central region the biomass rebounded until 2005. Notable increase in biomass occurred between 2005 and 2008 particularly in winter and spring, which appeared to be related to enhancement of coastal upwelling (Fig. 9).

Lavaniegos (2009) analyzed the relation of zooplankton biomass with regional (UI) and basin-scale (PDO, NPGO) proxies using multiple regression analysis. This analysis applied to data of 1997-2007 showed a significant influence of the NPGO in the variability of zooplankton biomass at both regions. In the central region UI was also significant indicating similar importance of basin and local processes.

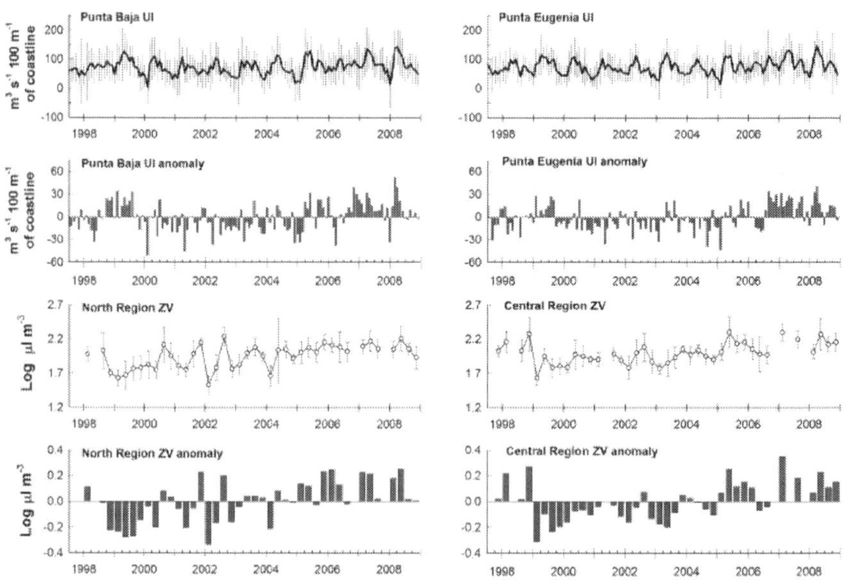

Fig. 9. Upwelling indices (UI) and zooplankton volumes (ZV) off Baja California. The UI monthly means (and standard deviations) were estimated using daily upwelling indices from www.pfeg.noaa.gov/products/products.html, in two locations (Punta Baja and Punta Eugenia). UI anomalies resulted after removing long-term monthly means from the period 1997-2008. The ZV mean for each cruise (and 95% confidence interval) were obtained after logarithmic transformation of data grouped in two regions: north (lines 100-110) and central (lines 113-137). ZV anomalies were calculated removing long-term seasonal means from the period 1997-2008.

Zooplankton abundance diverged between regions in relation to the year season (figure 10 upper panels). By example, during summer both oceanic regions presented a long-term mean of 41 ind m-3. In contrast, in winter the north region has lesser zooplankton than the central region (28 and 39 ind m-3 long-term geometric mean respectively). It is important to bear in mind that shallow stations near the coast were excluded, where zooplankton abundance use to be from 5 to 15 times higher than oceanic waters. These stations were excluded due to high variability. Their abundance must to be estimated separately but the number of nighttime samples is low to present here precise estimations.

Copepods were more abundant in April at the north region with geometric mean of 30 ind m-3. However, they represented a higher percentage of the zooplankton in January (65%) due to the decrease in euphausiids and other taxa. In the central region copepods were also more abundant in April and represented similar percentages in January as in April (55-56%). The opposite tendency was observed for euphausiids which represented 8% of the total zooplankton during January in both regions (Fig. 10). The maximum contribution of euphausiids to the zooplankton community occurred in April, with 11% in the north and 15% in the central regions.

The high frequency variability make difficult to understand interannual tendencies through the period 1997-2008. Removing the seasonal variability, a long-term increase in copepods (figure 10 middle panels) and euphausiids (figure 10 lower panels) was evident in one or both oceanic regions. This increasing tendency could be promoted by increasing upwelling since 2005 to 2008 (Fig. 9).

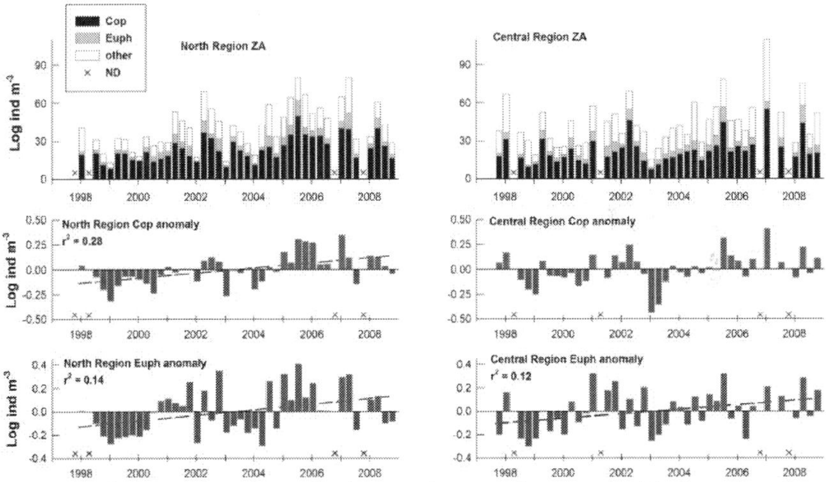

Fig. 10. Zooplankton abundance (ZA), contribution of copepods (Cop) and euphausiids (Euph) for two oceanic regions off Baja California. Stacked bars in the upper panels are geometric means of Cop, Euph, and the rest of the taxa added to obtain total zooplankton abundance. Anomalies are in logarithms and were estimated removing the long-term seasonal mean. Dashed lines are predicted values from linear regression analysis, only shown in significant cases ($p<0.05$). Data used came from nighttime samples. Cruises with no data (ND) are marked with x at the bottom of the graph.

Abundance anomalies estimated for copepods presented a well defined tendency in the north region with an interval of negative anomalies between 1998 and 2004 followed by positive anomalies in 2005-2008. Linear regression analysis was highly significant ($F = 18.0$, $p<0.001$). Euphausiids showed also a significant increase ($F = 8.2$, $p = 0.007$) in the north region, but anomalies shifted above and below zero in bi-or tri-annual intervals, with negative anomalies being associated to La Niña (1998-2000), and the subarctic water intrusion (2003-2004), while the rest of the time series remained mostly positive with the notable exception of January 2002.

In the central Baja California region, the negative perturbation caused in copepod abundance by the subarctic intrusion appeared to be stronger than La Niña and this was the reason for not significant result in regression analysis despite the predominance of positive anomalies in the last three years. In contrast, negative anomalies from euphausiids associated to the subarctic perturbation were lightly weaker to those corresponding to La Niña, therefore the long-term increasing tendency resulted significant ($F = 5.3$, $p = 0.026$). The shifting anomalies entail structural changes in the euphausiid communities (Lavaniegos & Ambriz-Arreola, 2010). Lavaniegos (2009) found a significant relationship between crustacean herbivores (copepods + euphausiids) and the NPGO during 1997-2007 only at northern region. In the central region the relation was significant but with gelatinous herbivorous. This last also were significantly related with UI.

7. Conclusions

Seasonal, and interannual anomalies of wind stress, SST, SSH, CHL, and phytoplankton Chlorophyll-*a* and zooplankton data collected during the IMECOCAL surveys off Baja California from 1997 to 2010 were examined. The analyses help to identify the influence of El Niño-La Niña cycles in the pelagic ecosystem (lower trophic level), and the strong long-term 2002-2006 effect of basin physical processes, diminishing salinity and phytoplankton biomass, and enhancing zooplankton volume.

Wind field variability off Baja California is dominated by the North Pacific high pressure center. The winds are highly persistent with south-southeast direction, favorable to coastal upwelling. During spring and summer winds are stronger and less variable than in autumn-winter, with a cyclonic curl of the wind stress over well-defined areas near the coast during the whole year. In general high wind frequencies (order of few days) play also a dominant role in the region. Wind stress on the ocean off Baja California is manifest in low sea surface temperature, and high phytoplankton production along the coast.

From the relationship of wind field off Baja California with ocean climate temporal variability, we conclude that MEI (Multivariate Enso Index) does not represent adequately the wind events in this particular coastal region, because the spatially mesoscale and local wind stress components are mainly influenced by North Pacific climate conditions, and in a less extent by the tropical and subtropical wind forcing events.

The largest part of the noteworthy sea surface temperature (SST) variability events off Baja California occur during 1997-1998, 2004 and 2006, consequent of El Niño conditions, with both MEI and sea surface temperature (SST) showing positive anomalies. Negative SST anomalies occurred during 2002-2003 related with positive MEI values. In general, SST negatives anomalies were manifest during La Niña phase, mainly between winter 1998 and 2000-2001, and continue until 2003. Sea Surface Height (SSH) anomalies were well connected with MEI, with the exception during La Niña conditions (1999-2000). Throughout La Niña 1999-2000 the MEI and SST showed negative values, while SSH was positive, which is not consistent with a thermal oceanic expansion.

Satellite chlorophyll time series described the influence of El Niño-La Niña cycles off Baja California, with different responses for central and southern regions. Phytoplankton Chlorophyll-*a* positive anomalies were associated with a cold period, characterized by negative anomalies of the Pacific Decadal Oscillation (PDO index), with a warm phase of this index from summer 2002 to at least summer 2006, differentiated by negative Chlorophyll-*a* anomalies off Baja California. After 2006, PDO index shows more neutral or negative values, defining the beginning of a cool phase (2007-2008), with positive anomalies of phytoplankton biomass off Baja California. Following 2007 a shift to a cool ocean climate phase appears to be present in the region, revealed by strong negative values of the PDO index, similar to those obtained by 2001 and early 2002. Nevertheless those changes off Baja California are observed as positive anomalies of salinity and zooplankton biomass (mainly salps), they are not apparent in phytoplankton positive anomalies.

Phytoplankton Chlorophyll-*a* linear relationships were mainly related with NPGO (North Pacific Gyres Oscillation) index, without any significant relationship with MEI and PDO indices. Perhaps, the increase in phytoplankton growth off Baja California during some times of the 1998-2010 events was influenced by changes in surface circulation of the northeast Pacific Ocean, which in turn respond to North Pacific climate forcing. From long-term Pacific Ocean climate index (MEI, PDO, and NPGO) and Chlorophyll-*a* anomalies trends, this is not very apparent that the IMECOCAL subdivided regions response differently to physical forcing, in spite of the domains identified by hydrologic ocean properties. Nevertheless, zooplankton biomass differences are reflected in these provinces (north and central regions) related to the year season, through contrasting responses to large scale events and local coastal upwelling processes. According our data set from 1997-2008, NPGO index showed a significant influence of the variability of zooplankton biomass at both regions (northern and central).

8. Future research

In order to have a better understanding of the pelagic ecosystem response off Baja California to ocean climate variability, analyses of main phytoplankton groups and their relationship with a more accurate and regional ocean index are necessary. Also, these phytoplankton assemblages could be related with the main constituent of zooplankton groups, to understand food web transference in the IMECOCAL region, and their relation with local and large spatially physical processes. For instance, a first step in this direction is to generate a regional ocean climate variability index for Baja California waters.

The effect of the wind stress curl on the pelagic ecosystem is other of the topics to be revisited for the region off Baja California. Because of the wind stress curl and sea surface temperature relationships, this parameter could have a relation with plankton variability in the oceanic waters off Baja California.

There are preliminary indications of a colder tendency in the California Current System, based on long-term sea surface temperature analyses. If this trend is correct, we will expect an increase in the lower trophic organism's biomass off Baja California for the next years. In this chapter, merely some of the zooplankton groups (Copepods and Euphausiids) have been shown a similar tendency, and only for the northern region, mainly influenced by subarctic waters. More research we need to realize in this direction to understand the effect of long-term climate variability on the pelagic marine ecosystems, mainly on plankton abundance and biomass changes.

9. Acknowledgements

Thanks are given to students, technicians, researchers, and crew of *R/V Francisco de Ulloa* working on the IMECOCAL cruises. Special thanks to J.L. Cadena-Ramírez & P. García-García for assistance in zooplankton counting, and to S. Nájera-Martínez & M.E. De la Cruz-Orozco working with chlorophyll analyses. The altimeter products were produced by Ssalto/Duacs and distributed by AVISO, with support from Cnes (http://www.aviso.oceanobs.com/duacs/). Sea Surface Temperature (SST) was obtained from the site ftp://podaac.jpl.nasa.gov/pub/sea_surface_temperature/avhrr/pathfinder. The wind product was obtained from ftp://podaac.jpl.nasa.gov/ocean_wind/ccmp/L3.0/. We would like to thanks the responsible of all satellite products. This research was supported with grants from the Mexican Council of Science and Technology (CONACYT: G0041T, 017Pñ-1297, G35326T, C0125343, C02-42569, 47044, 23947, 48367, SEMARNAT-CONACYT 23804), and by SIP-IPN 20090831. We are grateful with CICESE (Centro de Investigación Científica y de Educación Superior de Ensenada, México) for its IMECOCAL Program support, and the facilities of the *R/V Francisco de Ulloa*, together with funds from special research projects.

10. References

Atlas, R.; Hoffman, R.N., Bloom, S.C., Jusem, J.C. & Ardizzone, J. (1996). A multiyear global surface wind velocity data set using SSM/I wind observations. *Bulletin American Meteorology Society*, 77, 869–882.

Bograd, S.J.; DiGiacomo, P.M., Durazo, R., Hayward, T.L., Hyrenbach, K.D., Lynn, R.J., Mantyla, A.W., Schwing, F.B., Sydemna, W.J., Baumgartner, T., Lavaniegos, B. & Moore, C.S. (2000). The State of the California Current, 1999-2000: Forward to a New Regime?. *California Cooperative Oceanic Fisheries Investigations Reports*, 41, 26-52, 0575-3317.

Bograd, S.J. & Lynn, R.J. (2003). Anomalous subarctic influence in the southern California Current during 2002. *Geophysical Research Letters*, 30, 15, 8020, doi:10.1029/2003GL017446.

Cappet, X.J.; Marchesiello, P. & McWilliams, J.C. (2004). Upwelling response to coastal wind profiles. *Geophysical Research Letters*, 13, L13311, doi: 10.1029/2004GL020123.

Casey, K.S. & Adamec, D. (2002). Sea surface temperature and sea surface height variability in the North Pacific Ocean from 1993 to 1999. *Journal of Geophysical Research*, 107, C8, 10.1029/2001JC001060.

Castro, R. & Martínez, J.A. (2010). Spatial and temporal wind field variability. In: *Dynamic of the Pelagic Ecosystem off Baja California, 1997-2007.* Gaxiola-Castro, G. & Durazo, R. (Eds.). INE, CICESE, UABC. Mexico. (In Spanish).

Chelton, D.B.; Schlax, M.G. & Samelson, R.M. (2007). Summertime coupling between sea surface temperature and wind stress in the California Current System. *Journal of Physical Oceanography,* 37, 495–517.

Di Lorenzo, E.; Schneider, N., Cobb, K.M., Chhak, K., Franks, P. J. S., Miller, A. J., McWilliams, J. C., Bograd, S.J., Arango, H., Curchister, E., Powell, T. M. & Rivere, P. (2008). North Pacific Gyre Oscillation links ocean climate and ecosystem change. *Geophysical Research Letters,* 35, L08607, doi:10.1029/2007GL032838.

Durazo, R. & Baumgartner, T. (2002). Evolution of oceanographic conditions off Baja California: 1997-1999. *Progress in Oceanography,* 54, 7-31.

Durazo, R.; Gaxiola-Castro, G., Lavaniegos, B.E., Castro-Valdéz, R., Gómez-Valdés, J. & Mascarenhas Jr., A.S. (2005). Oceanographic conditions west of the Baja California coast, 2002–2003: a weak El Niño and subarctic water enhancement. *Ciencias Marinas,* 31, 3, 537–552, 0185-3880.

Durazo, R. (2009). Climate and upper ocean variability off Baja California, Mexico: 1997-2008. *Progress in Oceanography,* 83, 361-368, doi:10.1016/j.pocean.2009.07.043.

Espinosa-Carreón T.L.; Strub, P.T., Beier, E., Ocampo-Torres, F. & Gaxiola-Castro, G. (2004). Seasonal and interannual variability of satellite derived chlorophyll pigment, surface height, and temperature off Baja California. *Journal of Geophysical Research,* 109, C03039, doi:10.1029/2003JC0020105.

Freeland, H.J.; Gatien, G., Huyer, A. & Smith, R.L. (2003). Cold halocline in the northern California Current: an invasion of subarctic water. *Geophysical Research Letterrs,* 30, 3, 1141. doi:10.1029/2002GL016663.

Gaxiola-Castro, G.; Durazo, R., Lavaniegos, B., De-la-Cruz-Orozco, M.E., Millán-Nuñez, E., Soto-Mardones, L. & Cepeda-Morales, J. (2008). Pelagic ecosystem response to interannual variability off Baja California. *Ciencias Marinas,* 34, 2, 263-270, 0185-3880.

Holm Hansen, O.; Lorenzen, C., Holmes, R. & Strickland, J. (1965). Fluorometric determination of chlorophyll. *J. Cons. Perm. Int. Explor. Mer,* 30, 3-15.

Hoffman, R.N. (1984). SASS wind ambiguity removal by direct minimization. Part II: Use of smoothness and dynamical constraints. *Monthly Weather Review,* 112, 1829–1852.

Huyer, A. (1983). Coastal upwelling in the California Current System. *Progress in Oceanography,* 12, 259-284.

Kahru, M. & Mitchell, B.G. (2000). Influence of the 1997- 98 El Niño on the surface chlorophyll in the California Current. *Geophysical Research Letters* 27, 2937–2940, 0094-8276.

Lavaniegos, B.E. (2009). Influence of a multiyear event of low salinity on the zooplankton from Mexican eco-regions of the California Current. *Progress in Oceanography,* 83, 369-375.

Lavaniegos, B.E.; Ambriz-Arreola, I., Hereu, C.M., Jiménez-Pérez, L.C., Cadena-Ramírez, J.L. & García-García, P. (2010). Seasonal and interannual zooplankton variability. In: *Dynamic of the Pelagic Ecosystem off Baja California, 1997-2007.* Gaxiola-Castro, G. & Durazo, R. (Eds.). INE, CICESE, UABC. Mexico. (In Spanish).

Lavaniegos, B.E. & Ambriz-Arreola, I. (2010). Interannual variability in krill off Baja California in the period 1997-2005. *Progress in Oceanography* (submitted).

Leuliette, E.W. & Wahr, J.M. (1999). Coupled pattern analysis of sea surface temperature and TOPEX/Poseidon sea surface height. *Journal of Physical Oceanography,* 29, 599–611.

Lynn, R.J.; Bumgartner, T., Garcia, J., Collins, C.A., Hayward, T.L., Hyrenbach, K.D., Mantyla, A.W., Murphree, T., Shankle, A., Schwing, F.B., Sakuma, K.M. & Tegner, M.J. (1998). The State of the California Current, 1997-1998: transition to El Niño Conditions. *California Cooperative Ocenaic Fisheries Investigations Reports*, 39, 25-49, 0575-3317.

McClatchie, S.; Goericke, R., Schwing, F.B., Bograd, S.J., Peterson, W.T., Emmett, R., Charter, R., Watson, W., Lo, N., Hill, K., Collins, C., Kahru, M., Mitchell, B.G., Koslow, J.A., Gomez-Valdes, J., Lavaniegos, B.E., Gaxiola-Castro, G., Gottschalck, J., L'Heureux, M., Xue, Y., Manzano-Sarabia, M., Bjorkstedt, E., Ralston, S., Field, J., Rogers-Bennett, L., Munger, L., Campbell, G., Merkens, K., Camacho, D., Havron, A., Douglas, A. & Hildebrand, J. (2009). The State of the California Current, spring 2008-2009: Cold conditions drive regional differences in coastal production. *California Cooperative Fisheries Investigations Reports*, 50, 43-68, 0575-3317.

Perez-Brunius, P.; Lopez M., Pares-Sierra, A. & Pineda, J. (2007). Comparison of upwelling indices off Baja California derived from three different wind data sources. *California Cooperative Oceanic Fisheries Investigations Reports*, 48, 204-214, 0575-3317.

Peterson, P.T. & Schwing, F.B. (2003). A new climate regime in northeast Pacific ecosystems. *Geophysical Research Letters*, 30, 17, doi:10.1029/2003GL017528.

Schwing, F.B.; Bograd, S.J., Collins, C.A., Gaxiola-Castro, G., Garcia, J., Goericke, R., Gomez-Valdes, J., Huyer, A., Hyrenbach, K.D., Kosro, P.M., Lavaniegos, B.E., Lynn, R.J., Mantyla, A.W., Ohman, M.D., Peterson, W.T., Smith, R.L., Sydeman, W.J., Venrick, E. & Wheeler, P.A. (2002). The State of the California Current, 2001-2002: will the California Current System keep cool, or is El Niño looming?. *California Cooperative Oceanic Fisheries Investigations Reports*, 43, 31-68, 0575-3317.

Strub, P.T.; Allen, J.S., Huyer, A., Smith, R.L. & Beardsley, R.C. (1987). Seasonal cycles of currents, temperatures, winds, and sea level over the northeast Pacific continental shelf: 35° N to 48° N. *Journal of Geophysical Research*, 92, C2, 1507-1526.

Strub, P.T. & James, C. (2002). Altimeter-derived surface circulation in the large-scale NE Pacific gyres: Part 1. Seasonal variability. *Progress in Oceanography*, 53, 163–183.

US GLOBEC (1994). A science plan for the California Current. United States GLOBEC Ocean Ecosystem Dynamics, Report 11, Department of Integrative Biology, University of California, Berkeley, California, 124 p.

Venrick, E. & Hayward, T.L. (1984). Determining chlorophyll on the 1984 CalCOFI surveys. *California Cooperative Oceanic Fisheries Investigations Reports*, 25, 74-79, 0575-3317.

Yentsch, C.S. & Menzel, D.W. (1963). A method for the determination of phytoplankton chlorophyll and phaeophytin by fluorescence. *Deep-Sea Research*, 10, 221-231.

Yoder, J. A. & Kennelly, M. A. (2003). Seasonal and ENSO variability in global ocean phytoplankton chlorophyll derived from 4 years of Sea-WiFS measurements. *Global Biogeochemical Cycles* 17, 4, 1112, doi:10.1029/2002GB001942, 0886-6236.

10

Climate change and resilience value of mussel farming for the baltic sea

Ing-Marie Gren
Uppsala, Sweden

1. Introduction

The impact of climatic change on freshwater was investigated already in late 1980s and early 1990s (e.g. Coutant, 1988; Waggoner, 1990). Similar studies on the Baltic Sea were carried out more recently (e.g. Eckersten et al. 2001; Pettersson, 2003; Arrheimer et al., 2005, Blenckner et al. 2007; Lewan and Wallin, 2007; Ulén and Weyhenmeyer, 2007; Zillén, 2008; Klavins et al. 2009; Möllman et al., 2009). In spite of this relatively early concern, the environmental economics literature on climatic change and water quality management is scarce and is mostly applied to water supply management and/or estimation of impacts on agricultural and other water dependent industries (Mendelsohn, 2003; Lacroix, 2005). However, climatic change is likely to affect a variety of ecosystem services related to water quality, such as recreational values and reproduction of fish. On the other hand, certain measures may buffer against large variability in climate and resulting pollutant pressure on a eutrophied sea. The purpose of this paper is to calculate potential values, so called resilience values, of mussel farming from combating eutrophication in the Baltic Sea under climate change conditions.

Potential resilience value of mussel farming emerges from the stochastic nature of pollutant transports in soil and water in drainage basins, which implies risk in reaching predetermined targets in pollutant loads to water recipients. Pollutant emission sources are spatially spread in drainage basins where pollutants transports to the water recipients follow one or several different paths: air, soil, subsurface- and groundwater streams. Therefore, the final impact on the recipient is predicted only under conditions of risk and uncertainty. This uncertainty is one reason for EU water directive's recommendation of expressing water quality targets in precautionary terms where the targets should be obtained with high reliability (EU, 2000). Difficulties of managing stochastic pollution of waters constitute an important cause of aggravation of several types of water quality problems in spite of societies' relatively early perception of the environmental problem. One prominent example is damages from eutrophication caused by nutrient enrichment in several part of the world, such as the Baltic Sea, Black Sea, Missisipi delta, Cheasepeake Bay, and the Mediteranien (see e.g. Turner et al., 1999; NRC, 2000; Bodungen and Turner, 2001). Damages from nutrient enrichment occur from the oxygen depletion that takes place due to biological growth of certain algal species. Huang et al. (1997), Söderqvist (1998), and Markowska and Zylicz (1999) show that people are willing to pay a significant amount of money for reducing damages from nutrient enriched bays.

Holling (1973) is among the first to point to the need of accounting for resilience in ecosystem management. Resilience is then defined as the magnitude of disturbance a system can experience before it shifts into another state with different controls and functions. Although the resilience concept was introduced in early 1970s current considerably large literature in natural science on the role of resilience is still conceptual in nature (see Folke et al. 2004 for a review). Only a few attempts have been made to estimate the value of ecosystems in promoting resilience (Mäler et al., 2007; Walker et al. 2007; Cardona et al., 2008; Sarang et al., 2008; Mäler et al. 2009; Gren 2010).

The analysis and calculations of resilience value of wetlands carried out in this paper are most similar to Gren (2010), where resilience values are calculated for wetlands as nutrient sinks under conditions of stochastic nutrient loads to the Baltic Sea by means of stochastic programming. This study then adds to the scant literature on the value of mussel farming as an abatement option (Hart, 2003; Gren et al. 2009). However, none of the two papers carry out explicit calculations of resilience values of mussel farming, which is the ultimate purpose of this paper. This study thus extends earlier literature in two respects; *i)* on development of methods quantifying resilience values and *ii*) on valuation of nutrient retention of mussel farming.

A few caveats are in order. Due to the focus on the abatement portfolio aspects, the dynamics of water pollution is not included. This neglect is particularly serious for phosphorous pollution due to the long adjustments in the sea to changes in load. As shown in Hart (2003) consideration of dynamics may have significant impact on choice of mitigation or adaptation measures for water pollution. However, the delayed effects of nutrient emission changes in the drainage basins on load to the coastal water is a strong justification for the stochastic framework applied in this paper. Given a relatively short period of time, say five years, it is difficult to determine the load effects from emission changes undertaken in the drainage basin in the beginning of the period. In principle, a long term perspective with a dynamic model would give more precision nutrient transports can be modeled, although the effect in each time period would be difficult to predict.

2. Operational definition of resilience value

This paper applies a risk based approach for the calculations of resilience values, which is characterized by cost effectiveness analysis under stochastic environmental quality (Sarang et al. 2008; Gren 2010). Given environmental targets set by policy makers, resilience value is then estimated as the value of changes in the reliability of reaching the predetermined environmental target(s). More precisely, policy makers are assumed to minimize total costs for achieving a certain environmental quality with a minimum probability at minimum costs. In such a setting, new cleaning technologies may bring about two types of values; replacement and resilience values.

The environmental targets for the Baltic Sea are set by the intergovernmental agreement in autumn 2007 where decreases in nutrient loads to the marine basins of the Sea were determined. Gren et al. (2009) have shown that these targets can be achieved at lower costs if mussel farming is introduced as an abatement option. The estimated value ranged between SEK 0.2 and 2.3 kg^{-1} live mussel depending on assumption of nutrient sequestration by mussels, and option of selling the mussel for food.

In addition to replacement values, mussel farming may generate resilience value. Resilience value is then defined as the decrease in total cost for achieving a certain load reduction target at a minimum probability caused by the introduction of mussel farming as an abatement measure. Resilience value then arises from the existence of uncertainty in reaching the target, which includes stochastic nutrient loads in the drainage basin from the emission sources and also uncertain abatement capacity of mussel farming. It is shown in Gren (2010) that the resilience value is positive only if the introduction of mussel farming reduces overall risk, which is measured as the total variance in nutrient load and abatement by mussel farming. Total risk is reduced only if the covariance between nutrient load to a basin and the abatement capacity of the measure, which is mussel farming in this paper, is positive. Nutrient abatement of mussel farms is then high when load from the drainage basins is high. The associated resilience value is calculated in a similar way as replacement value, which is illustrated in Figure 1.

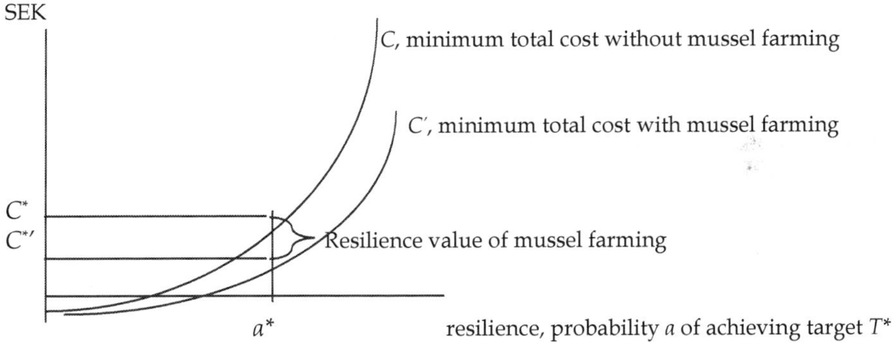

Fig. 1. Illustration of the calculation of resilience value of mussel farming as the reduction in total abatement cost for reaching a certain cleaning target T* with a minimum probability α* from introduction of mussel farming.

The two curves C and C' illustrate resilience provision cost functions which show the minimum cost for achieving the environmental target T^* at different levels of probabilities. The higher the chosen probability, or resilience level, the higher is the cost due to the need for more costly abatement (see Gren 2010 for a further description). The resilience value of mussel farming is then calculated as C-C' for a given predetermined a. As illustrated in Figure 1, the resilience value of mussel farming is C^*-$C^{*\prime}$ for $a= a^*$.
As noticed above, a necessary condition for a positive resilience of mussel farming is a positive co-variance between sequestration by mussels and nutrient load from drainage basins and sediments. However, this is not sufficient since abatement by mussel farming is also uncertain. Resilience value of mussel farming is then positive if the reduction in total risk due to a positive covariance between nutrient loads and abatement by mussel farming is higher then the increase in risk associated with uncertainty in nutrient abatement by mussel farming.

3. Data retrieval

The calculations of resilience values of mussel farming in the Baltic Sea under climate change conditions require data on: nutrient transports to the sea from basins, quantification of climate change, uncertainty quantification of nutrient sequestration by mussel farming, costs of mussel farming and other measures. Data on nutrient transports and costs of measures in the drainage basins, such as reductions in agricultural, household and industrial nutrient loads, are obtained from Gren et al. 2008, and data on costs and effects of mussel farming from Gren et al. 2009. Climate change impacts are quantified as changes in variability in nutrient concentrations in the marine basins of the Baltic Sea.

3.1 Nutrient loads and abatement costs

Although the nine countries with coasts along the Baltic Sea constitute decision units in negotiation processes, the choice of regional division of the Baltic Sea drainage basin is not a self-evident matter. One important reason is the difficulty of matching data on nutrient drainage basin transports with estimates of abatement costs. For that reason, the entire water catchment of approximately 1 745 000 km^2 is divided into 23 drainage basins, which are shown in Figure A1 in the appendix. For each of these regions nutrient emission originates from three types of sources: agriculture, sewage from households and industry, and air deposition. The calculations of nutrient loads to the coastal waters of the Baltic Sea from these emission sources are divided into two steps. First, all emission sources are identified and quantified. Next, these emissions are transformed into loads to the Baltic Sea by means of data on leaching from the root zone into waters stream in the drainage basins, retention of nutrient during the transport to the sea, and air transports of nitrogen oxides and ammonia (see Gren et al. 2008 for a further presentation).

Measurement and data on risk in nutrient loads are obtained from Elofsson (2000), which contains coefficients of variation in nutrient loads for the 23 drainage basins. These are, in turn, based on measurements of nutrient concentration at all river mouths along the Baltic Sea coastal lines. Table 1 presents the coefficients of variation in nutrient loads together with data on nutrient loads to the coastal waters.

Regions	*Nitrogen;*		*Phosphorus:*	
	Kton[1]	*CV*[2]	*Kton*[1]	*CV*[2]
Denmark	44	0.25	1.1	0.27
Finland	49	0.21-0.25	1.7	0.21-0.27
Germany	46	0.2	0.5	0.2
Poland	318	0.18-0.29	22	0.18-0.25
Sweden	74	0.17-0.35	1.6	0.18-0.31
Estonia	56	0.18	1.6	0.18
Latvia	44	0.20	3	0.21
Lithuania	93	0.15	3.5	0.15
Russia	83	0.17-0.39	4	0.39-0.45
Total	*824*	*0.09*	*38.9*	*0.12*

Table 1. Nutrient loads, share of non-point source load, coefficient of variation. Sources: 1. Gren et al. (2007), Table 1 page 13, estimated for year 2005 2. CV (coefficients of variation) Elofsson (2000), Table CV(I, N), page 54

Total calculated annual loads of both nitrogen and phosphorus come relatively close to similar calculations carried out by Helcom (2004). Poland is the country with the largest share of both nitrogen and phosphorus loads, followed by Lithuania and Russia. For most countries, coefficients of variations are larger for nitrogen loads than for phosphorus loads.

Cost estimates of different nutrient abatement measures used in this study are obtained from Gren et al. (2008) and measured in 2007 prices. In addition to mussel farming, the study includes measures for changes in agricultural practices, increased cleaning capacity at sewage treatment plants, and reductions in nitrogen oxide emissions from traffic and industry. More precisely, the measures included in this empirical analysis are: increased nutrient cleaning capacity at sewage treatment plants, sewage treatment at industry and households, phosphate free detergents, catalysts in cars and ships, flue gas cleaning in stationary combustion sources, and reductions in the agricultural deposition of fertilizers and manure. Included land use measures are: change in spreading time of manure from autumn to spring, cultivation of so called catch crops, energy forests, lye grass, and wetland creation. A change of spreading time from autumn to spring implies less leaching since, in spring, there is a growing crop which utilizes the nutrients. Catch crops refer to certain grass crops, which are drilled at the same time as the ordinary spring crop but the growth, and thereby the use of remaining nutrients in the soil, is concentrated to the period subsequent to the ordinary crop harvest. The nutrient abatement cost estimates for fertiliser reductions are based on econometric estimates of panel data. Abatement costs of all other measures are obtained from enterprise budgets.

Costs of nutrient sequestration by mussel farming are obtained from Gren et al. (2009) which, together with Hart (2003), are the only studies estimating the value of mussel farming as an abatement option for a eutrophied sea. The cost of mussel farming depends on type of technology, size and location of the farm, and on the possibility of selling the mussels as human or animal food. Long-line farming is the most common method of mussel farming in Scandinavia. The larvae of the blue mussel settle in early summer on vertical suspenders attached to horizontal long-lines carried up by buoys. The long-lines are typically 200 m long and the suspenders reach close from surface down to about 6 m depth. A varying number of long-lines are collected to a unit which is anchored at both ends. The growth of mussel and, hence, the nutrient sequestration depends to a large extent on the salinity content of the water, which varies in different parts of the Baltic Sea. Mussel production per mussel farm can be twice as large in the southern Baltic Sea as in the northern parts (Gren et al. 2009). The cost of nutrient sequestration also depends on the option of selling mussels for human consumption or for animal food. The calculated constant marginal costs in Gren et al. (2009) varied between SEK 0/kg nutrient cleaning and SEK 635 kg^{-1} and SEK 9000 kg^{-1} for nitrogen and phosphorous cleaning respectively (1 Euro = 9.70 SEK, April 16, 2010). The low marginal cleaning cost occurred for the Kattegat and the Sound marine basins, whereas the largest costs were found in the Northern Baltic Proper basin. In this chapter average marginal costs for each marine basin are used.

A quantification of uncertainty in nutrient abatement by mussel farming is obtained from Gren et al. (2009). A simple estimate is made where the coefficient of variation is calculated as the range in abatement divided by the mean, which gives an estimate of 0.5 for both nitrogen and phosphorus. There is no data on the co-variation between nutrient abatement by mussels and nutrient loads to the coastal waters. Calculations are therefore made with the assumption of a coefficient of variation that equals unity. The estimated resilience values of mussel farming are then the maximum values: A lower correlation coefficient generates lower values, which are zero or negative when the correlation coefficient is zero.

3.2 Climate change effects

Recall from Section 2 that resilience value is calculated as the difference in costs of resilience provision with and without mussel farming. This means that climate change will have effect on the estimated resilience value only if climate change has impacts on variability in nutrient loads. A simplification is made by neglecting effects on costs of abatement measures which can occur through changes in, for example, land and fertilizer prices owing to fluctuations in food demand. A justification for this is the lack of data on risk attitudes, which would be required for incorporating stochastic costs of abatement measures. Another limitation of the study is the neglect of climate-change impacts on the constraints as such. For example, it may turn out that phosphorus concentration and/or water transparency as operational indicators of water quality for different uses, such as for drinking or bathing purposes, need to be changed. This is not, however, accounted for in this study.

There are today a considerable number of studies on climate change effects related to the Baltic Sea (e.g. Eckersten et al. 2001; Pettersson, 2003; Arrheimer et al., 2005, Blenckner et al. 2007; Lewan and Wallin, 2007; Ulén and Weyhenmeyer, 2007; Zillén, 2008; Klavins et al. 2009; Möllman et al., 2009). A common focus of these studies has been to estimate the changes in winter North Atlantic Oscialltions on climate and associated impacts on water temperature, ice conditions, plankton phonology, and nutrient discharges on lakes and water sheds in the Baltic Sea drainage basin and in other parts of Europe.

In spite of the large literature, climatic change impacts, as expressed in terms of changes in variability in nutrient load and water quality, are limited. Studies of small drainage basins reveal that prolonged summer periods may increase phosphorus-recycling from the sediments (Pettersson, 2003), that phosphorus losses from agricultural soils may increase (Ulén and Weyhenmeyer, 2007), and that the nitrogen leaching from arable land and its retention during transport to waters are affected (Lewan and Wallin, 2007). According to Arrheimer et al., (2005) and Eckersten et al. (2001), phosphorus leaching may decrease and nitrogen leaching increase. The estimated range in increase in nitrogen leaching from arable land with current cultivation structure owing to climatic change is 10 and 70 per cent. Wallin (2002) also records increases in leaching of one nutrient, phosphorus, but retention is increased owing to higher biological activity, which implies that the discharges to the coastal waters may decrease or be unaffected. No study has, however, quantified eventual impacts of climate change on variability in loads or water quality.

Owing to the lack of data on climate change effects on variability, minimum cost solutions are calculated for both increases and decreases in nutrient load variability. More precisely, calculations are made for proportional changes in nutrient load variability compared with the reference case presented in Table 1; a decrease in all coefficient of variations by one half and a doubling of the coefficient of variations. It is assumed that the uncertainty in nutrient abatement is subjected to the same proportional changes in variability.

4. Results

Stochastic programming is used for calculating resilience values of mussel farming, where the decision problem includes minimization of costs for pre-specified target(s) of maximum loads under probabilistic constraints (see e.g. Birge and Louveaux, 1997). The algorithm applied for all calculations is GAMS (Brooke et al., 1998).

The target used is the intergovernmental agreement in autumn 2007 on nutrient load reductions to the Baltic Sea, the so-called Helcom Baltic Sea Action Plan (see Helcom, 2007). The targets differ for different marine basins; phosphorus decreases are largest for the Baltic Proper, and the largest nitrogen reductions are needed for Kattegat and the Danish Straits (see Table A1 in appendix for reduction needs and Figure A1 for a map). It is predicted that these reductions will reduce the extension of hypoxic sea bottoms in the Baltic Proper by approximately 1/3, and nitrogen fixation, an indicator of the intensity of cyan bacterial blooms, is expected to decrease by 2/3.

Recall from Section 2 that resilience values of mussel farming are calculated as differences in costs of resilience provision with and without mussel farming. Total abatement costs under different assumptions of climate change impacts and co-variation between nutrient load and nutrient abatement by mussels are shown in Table A2 in the appendix. Costs of resilience provision are calculated as the difference in abatement costs with reliability requirement minus abatement costs without such requirement. Such calculations of resilience provision costs without mussel farming as an abatement option are shown in Figure 2.

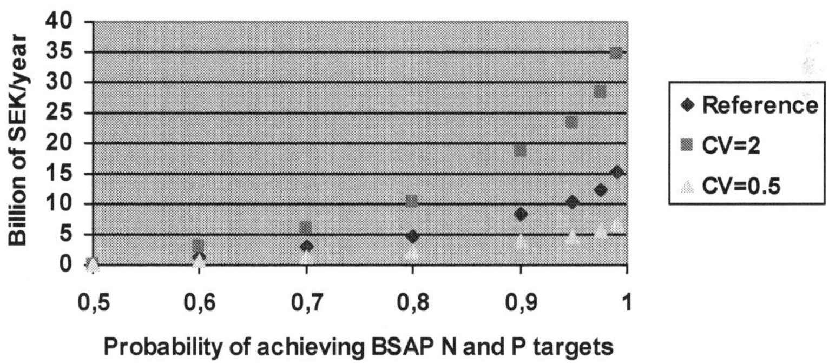

Fig. 2. Costs for resilience provision without mussel farming under alternative impacts of climate change (doubling of total variance, CV=2 and a reduction by one half, CV=0.5), (Source: calculations from data in Table A2 in appendix)

The resilience provision costs in the reference case, when there is no impact on variability in nutrient loads and abatement capacity by mussels from climate change, increase rapidly at probability levels exceeding 0.8; from approximately 5 billion of SEK to 15 billion of SEK. Total costs without reliability concern amount to approximately 25 billion of SEK. Resilience provision cost in the reference case then increases total abatement cost by 20 and 60 per cent compared with the costs without reliability concern (see Table A2 in the appendix). The results presented in Figure 2 also show that the resilience provision cost increases considerably when the total variance in nutrient load is doubled. If instead climate change causes a decline in the variability by one half, resilience provision costs are decreased by almost 2/3 at high resilience levels.

Resilience provision costs when mussel farming is included as an abatement option show a similar pattern as provision costs without mussel farming see Figure 3.

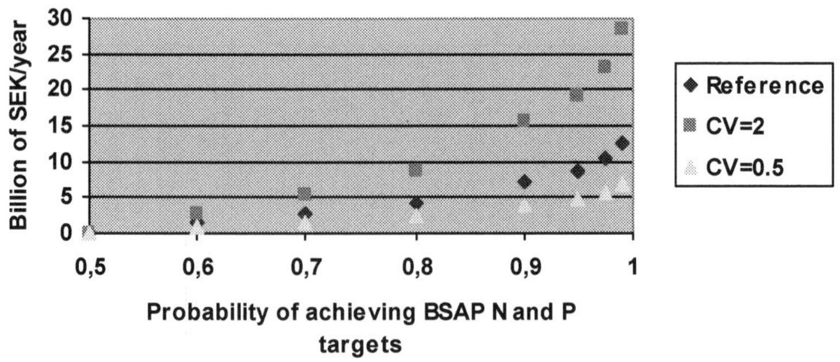

Fig. 3. Costs for resilience provision with mussel farming as an abatement option under alternative impacts of climate change (doubling of total variance, CV=2, and a reduction by one half, CV=0.5). (Source: Calculations based on data in Table A2 in appendix.)

A notable observation is that resilience provision costs with mussel farming are lower than corresponding costs without mussel farming as an abatement option for all three scenarios. Recall that the costs in Figure 3 are calculated with the assumption of a correlation coefficient between nutrient load and nutrient abatement by mussels that equals unity. Cost data in Table A1 show that mussel farming generates no resilience value when the correlation coefficient equals zero.

Calculated resilience values of mussel farming are displayed in Figure 4. These are the maximum possible values since it is assumed that the correlation coefficient between nutrient load and abatement by mussels equals unity.

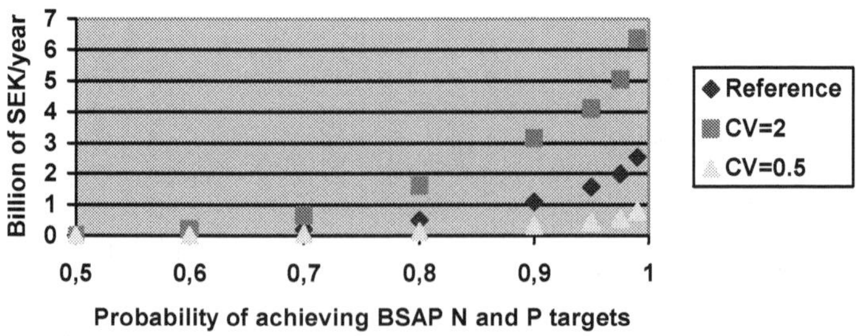

Fig. 4. Maximum resilience values of mussel farming at different resilience levels and impacts of climate change (doubling of total variance, CV=2, and a reduction by one half, CV=0.5). (Source; calculations based on Table A2 in the appendix)

The probability levels along horizontal axis are choices made by decision makers on the minimum probability for achieving the BSAP targets. For a given variability, the resilience value then increases for higher probabilities, since certainty in reaching the targets becomes prioritized by decision makers. This resilience value is then increased by climate change effects if they cause larger total variability, or risk, in nutrient loads. On the other hand, if variability is decreased (which is less likely), the resilience value of mussel farming also decreases.

It is interesting to compare the estimated resilience value of mussel farming with the more conventional replacement values, and also with the sales price of live mussels for human consumption. The conventional replacement value of mussel farming is obtained by subtracting the total abatement cost with mussel farming, which amounts to 24381 mill SEK, from the corresponding cost without mussel farming, which are 25 185 millions of SEK (see Table A1 in the appendix). This gives a total replacement value of 804 mills SEK, or approximately 1 SEK/ kg mussel. The estimated resilience values depend on choice of resilience level and effects of climate change on variability in nutrient loads, see Table 2.

	No resilience[1]	*Reference case*	*CV=2*	*CV=0.5*
Replacement value	1			
Resilience value		0.07 - 3.17	0.24-7.94	0.01-0.94

Table 2. Replacement and resilience values of mussel farming in the Baltic Sea, SEK/kg live mussel under different climatic change impacts 1) Uncertainty in nutrient loads is not of concern

The estimated resilience value ranges between 0.01 and 7.94 SEK kg^{-1} live mussel. It can thus be considerably higher than the replacement value of mussel farming, i.e. the decrease in total abatement costs for achieving the targets caused by the introduction of mussel farming when uncertainty is of no concern for policy makers. It is also interesting to note that the resilience value is twice as large as the market price of live mussel in Sweden 2009, which was approximately 3.50 SEK/kg (Gren et al. 2009).

5. Conclusions

The purpose of this paper has been to estimate the impacts of climate change on resilience values of mussel farming in the Baltic Sea. Resilience value was related to the impact of mussel farming on the exposure to risk in nutrient loadings. Its value is determined by reliability concern; decrease in total risk, and on the cost of mussel farming relative to other abatement measures. Since resilience values in this setting are positive only under conditions of uncertainty, climate change effects were measured in terms of impacts on variability in nutrient loads. Unfortunately, there exist no studies with a systematic assessment of such climate change effects on nutrient loads from all drainage basins. Simplifying assumptions were therefore made; variability is either doubled or reduced by one half due to climate change effects. The results showed that the estimated resilience value ranges between 0.01 and 7.94 SEK kg^{-1} live mussel. This result can be compared with the market retail price of mussel in Sweden in 2009, which amounted to 3.5 SEK kg^{-1}. Thus, when resilience is of concern for policy makers, the value of mussel farming can be considerable.

However, the results must be interpreted with much caution since they rest on several different types of assumptions mainly with respect to quantification of climate change effects, sequestration effects of mussel farming, and uncertainty in nutrient loads. Nevertheless, the results point to the need of considering the role of mussel farming for buffering against high variability in nutrient loads to eutrophied waters under climate change conditions.

6. References

Arheimer, B. Andréasson, J., Fogelberg, S. Johnsson, Pers, C., Persson, K. (2005). Climate change impacts on water quality: Model results from Southern Sweden. *Ambio* 34(7):5-25.

Birge, J., and Louveaux, F. 1997. *Introduction to stochastic programming*. Springer, New York, USA.

Blenckner, T., Adrian, R., Livingston, D., Jennings, E., Weyhenmeyer, G., George, G., Jankowski, T., Järvinen, M., Aonghusa, C., Nöges, T., Straile, D., Teubner, K., (2007). Large-scale climatic signatures in lakes across Europe: a meta analysis. *Global Change Biology* 13: 1314-1326

Cardona OD, Ordaz MG, Marulanda MC, Barbat, AH. (2008) Estimation of probabilistic seismic losses and the public economic resilience - An approach for macroeconomic impact evaluation. *Journal of Earthquake Engineering* 12, supplement 2, 60-70

Coutant, C.C. (1981), Foreseeable effects of CO_2 induced climatic change: Freshwater concerns. *Environmental Conservation* 8, 285-297.

Eckersten, H. Blombäck, K., Kätterer, T., Nyman P. (2001). Modelling C, N, water and heat dynamics in winter wheat under climate change in Southern Sweden. *Agriculture, Ecosystems, and Environment* 86:221-235.

Folke C., Carpenter, S., Walker, B. Sheffer, M., Elmqvist, Gunderson, L.., Holling, CS., (2004). Regimes shifts, resilience, and biodiversity in ecosystem management. *Annual Review of Ecological Evolutionary Systems* 35:557-581.

Elofsson K., (2000). *Cost efficient reductions of stochastic nutrient loads to the Baltic Sea.* Working Paper Series No 2000:6. Department of Economics, Swedish University of Agricultural Sciences, Uppsala, Sweden.

Elofsson K. (2003). Cost effective reductions of stochastic agricultural nitrogen loads to the Baltic Sea. *Ecological Economics,* 47(1): 13-31.

Gren, I-M,, Jonzon, Y., Lindqvist, M., (2008). *Costs of nutrient reductions to the Baltic Sea – Technical report.* Department of Economics, Swedish University of Agricultural Sciences, Working Paper No. 2008:1, Uppsala 2008.

Gren, I-M., Lindahl, O., Lindqvist, M. (2009). The value of mussel cultivation for combating eutrophication in the Baltic Sea. *Ecological Engineering,* 35:935-945.

Gren, I-M. 2010. Resilience value of constructed coastal wetlands for mitigating eutrophication. *Ocean & Coastal Management,* in revision.

Helcom. (2004). The fourth Baltic Sea pollution load compilation (PLC-4). Helsinki Commission. Helsinki, Finland.

Helcom. (2007). An approach to set country-wise nutrient reduction allocations to reach good marine environment of the Baltic Sea. Helcom BSAP Eutro Expo/2007. Helsinki Commission, Helsinki, Finland.

Holling, CS., (1973). Resilience and stability of ecological ecosystems. *Annual Review of Ecological Systems* 4:1-23.

Klavins, M., Briede, A., Rodinov, V., (2009). Long term changes in ice and discharge regime of rivers in the Baltic region in relation to climatic variability. *Climatic Change* 95: 485-498.

Lacroix, A., Beaudoin, N., Makowski, D. (2005). Agricultural water non-point pollution under uncertainty and climatic variability. *Ecological Economics* 52(1), 115-127

Lewan, E.,Wallin, M., (2007). Increased nitrogen leaching in winter seasons (in Swedish). *Miljötrender* 4: 6-7.

Mendelsohn, R., (2003). Efficient adaptation to climate change. *Climate Change* (45): 583-600.

Mäler, KG, LI, C-Z., Destouni, G., (2007). Pricing resilience in a dynamic economy-environmental system: A capital theoretical framework. Beijer Discussion Paper Series No 2008. Beijer International Institute of Ecological Economics, Stockholm.

Mäler, K-G., Anyiar, S., Jansson, Å. (2009). Accounting for ecosystem services. *Environmental and Resource Economics* 42:39-51.

Möllman, C., Diekman, R., Muller-Karulis, B., Korinolovs, G., Plikshs, M., Axe, P. (2009). Reorganization of a large marine ecosystem due to atmospheric pressure: a discountinuous regime shift in the Central Baltic Sea. *Global Change Biology* 15: 1377-1393

Pettersson, K. Grust, K. Weyhenmeyer, G., Blenckner, T. (2003). Seasonality of chlorophyll and nutrient in Lake Erken, effects of weather conditions. *Hydrobiology* 506-509, 75-81.

Sarang, A., Vahedi, A., Shamsai, A., (2008). How to quantify sustainable development: A risk based approach to water quality management. *Environmental Management* 41:200-220.

Ulén, B., Weyhenmeyer, G., (2007). Adapting regional eutrophication targets for surface waters - influence of the EU Water Framework Directive, national policy and climate change. *Environmental Science and Policy* 10: 734-742.

Waggoner, P.E. (1990). *Climate change and U.S. water resources*. John Wiley and Sons, NY, USA.

Walker B, Pearson L, Harris M, Mäler K-G, Li CZ, Briggs R. (2009). *Incorporating resilience in the assessment of inclusive wealth: An example from South East Australia.* Beijer Discussion Papers Series No 209, Royal Swedish Academy of Sciences, Stockholm.

Zillen, L.,. Conley, D., Andrén, T., Endrén, E., Björck, S. (2008). Past occurrences of hypoxia in the Baltic Sea and the role of climate variability, environmental change and human impact. *Earth-Science Reviews* 91:77-92.

Appendix: Tables and Figure

	P	N
Baltic Proper	66	29
Gulf of Finland	29	5
Gulf of Riga	34	
Danish Straits		32
Kattegat		31
Total	*64*	*25*

Table A1. Helcom BSAP basin reduction targets, in %
Source: Helcom (2007) page 2

Probability	Inclusion of mussel farming: Reference CV=2 CV=0.5						Exclusion of mussel farming: Refer. CV=2 CV=0.5		
	ρ= 0	ρ = 1	ρ = 0	ρ=1	ρ = 0	ρ = 1			
0.6	27409	25753	27353	27053	25139	25086	26613	28049	25899
0.7	29309	27102	30609	29797	25909	25779	28106	31255	26640
0.8	32309	28697	35002	33225	26814	26586	29951	35646	27519
0.9	32788	31535	43370	39995	28376	27917	33435	43952	29052
0.95	34788	33058	47991	43609	29219	28620	35427	48529	29863
0.975	36895	34796	53068	47619	30107	29380	37589	53465	30742
0.99	39827	37113	59501	52765	31291	30339	40454	59916	31930
Deterministic cost[1]	24381						25185		

Table A2. Total costs for achieving BSAP targets under different resilience levels (probability of achieving targets), climate change effects on variance (CV), and correlation coefficients between nutrient load and mussel abatement (ρ), in millions of SEK.
1) Abatement cost without any resilience provision.

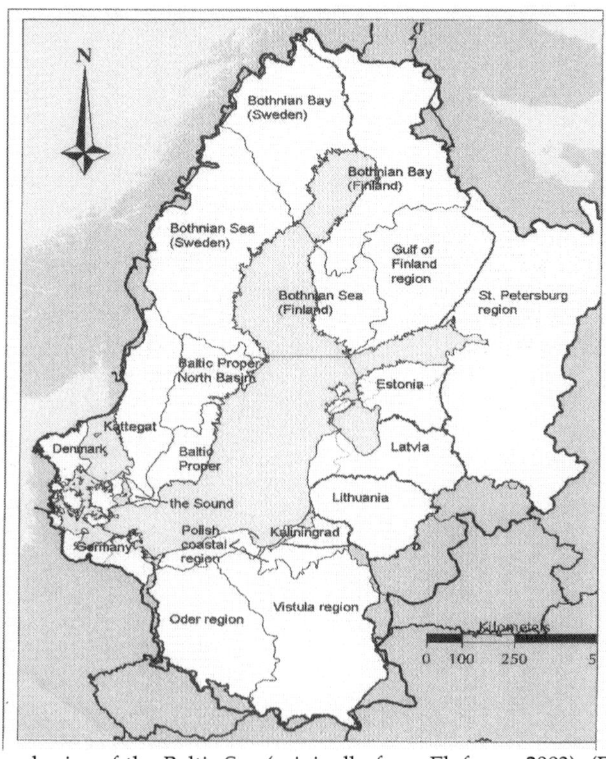

Fig. A1. Drainage basins of the Baltic Sea (originally from Elofsson, 2003). (Drainage basins in Denmark (2), Germany (2), Latvia (2), and Estonia (3) are not provided with names, but are delineated only by fine lines)

11

Temperate forests and climate change in Mexico: from modelling to adaptation strategies

Gómez-Mendoza, Leticia[1] and Galicia, Leopoldo* [2]

[1] *Colegio de Geografía, Facultad de Filosofía y Letras, Circuito Interior s/n, Ciudad Universitaria, C. P. 04510, Coyoacán D. F., México.*

[2] *Departamento de Geografía Física, Instituto de Geografía, Universidad Nacional Autónoma de México. Circuito Exterior s/n, Ciudad Universitaria, C. P. 04510, Coyoacán D. F., México*

**Author for correspondence*

Abstract

At present there is no doubt that climate change is affecting the structure, function and the geographic distribution of temperate ecosystems at regional and global scales; however in Mexico information about impacts of climate change on temperate forest ecosystems are scarce. The present chapter presents two studies of the effect of climate change on distribution of species of pines and oaks, and the change in functional groups composition of temperate forests ecosystems in Mexico; and, finally we resume a experience on identifying adaptation measures in a local forestry sector to climate change with three participatory workshops with key actors in forestry at state level. We realized a projection for the year 2050 that included two scenarios, a severe and a conservative one of how climate change would impact spatial distribution to 17 species of oaks and 17 species of pines at national level. Results of this study showed that the effect of alterations in temperature and precipitation modeled under both climate-change scenarios will reduce the current ranges of geographic distribution of almost all species of oaks and pines. Responses of these species to the different scenarios of climate change are predicted to be species-specific and related to each species affinity. The most affected species under the severe and conservative climate change scenarios will be *P. rudis, P. chihuahuana, P. culminicola, P.oocarpa, Q. acutifolia, Q. crispipilis,* and *Q. peduncularis.* On the other hand we show an analysis of the possible responses of functional groups was based on the construction of an ensemble of eight general circulation models with four scenarios of global emissions, and a Japanese model of regional high resolution (20 x 20 km) for Oaxaca state at South of Mexico. The ensemble of climate change scenarios suggested that by 2050 the temperature of the region will increase between 1.5 and 2.5°C, and rainfall will vary between +5 and -10% on current annual total precipitation. The sensibility analysis pointed that for the climate change scenario in 2050 genera as *Abies* and *Pinus* restricted their distribution area, in contrast, gender or drought-tolerant shrubs are likely to increase their geographic distribution. In another regional study in Tlaxcala (central Mexico) climate change scenarios projected a condition of drier and warmer (lower soil moisture) climate during the spring, thus the risk of forest fires would increase. Besides, losses of climate change of cover forest are another hazard in the state like illegal logging and land use change. It is estimated that by 2080 there will be only about 40% of the actual forest area in Tlaxcala by illegal logging and harvesting practices. Therefore, climate change will accelerate the forests loss in the state and in two or three decades will be very little remaining to preserve. As a result of a series of stakeholders in Tlaxcala we identified adaptation strategies that we grouped under three headings: reforestation, forest conservation and production.

1. Introduction

Some of the key climatic variables that stress structure and functioning and geographic distribution of forest ecosystems are those related to changes in precipitation, temperature, potential evapotranspiration, and increased frequency of fires and storms. Notwithstanding that the structure, composition of the species, and the functionality of the present ecosystems are the result of a series of physical conditions that have made possible its existence (Maslin 2004, Parmesan 2006), it is unknown how will the species adapt to an increase in temperatures between 2 and 5°C in the next 100 years; as well as to all the variations of precipitation between +20 y -20% in average for the planet (Sholze et al. 2006). It is also unknown if the adaptation mechanisms of plants will occur at the same speed than the climate would (IPCC 2007), since future changes will be different in magnitude than the preceding from the last 10,000 years (IPCC 2007). These changes in climate can have present and future impacts on the distribution, diversity of the species, the structure and and function of these ecosystems. Thus, measurement of present and future ecosystems modification, due to the climatic changes, is imperative (Malcolm 1998, Locatelli 2006). The effects on present tendencies on climate in the temperate forests of the world are manifested with greater differential mortality frequency of individuals among species, the presence of plagues and the decrease of productivity in forests (Rebetez & Dobbertin 2004, Mueller et al. 2005).

Climate change can have an influence on the species distribution. The changes on structure, function, species composition and geographic distribution could have profound implications for traditional livelihood, industry, biodiversity, soil and water resources, and hence, agricultural productivity. Moreover, these climate change induced effects would aggravate the existing stresses due to non-climate factors, such as land use changes and the unsustainable exploitation of natural resources.

In Mexico, pine and oak forests occupy 32, 330, 508 ha which represents 17% of the country. These are the richest ecosystems in Mexico with some 7000 plant species, from which about 150 species are pines and 170 are oaks; these represent over 50% of all known pine and oak species. Some species in these temperate forests produce timber of high commercial value in the forestry market. The rate of deforestation (0.5% annually) and illegal logging (21 million m^3 $year^{-1}$) are high. Mexico is the 11th largest emitters of CO_2 from deforestation and contributes 1.6% of global emissions, mainly from its temperate and tropical forests (12.9 and 54.1 Pg C $year^{-1}$, respectively). Temperate ecosystems of Mexico have received a strong influence of anthropogenic activities because they have an important population concentration (51% of the national population), with high growth rate (3%) (Villers & López

1995). The subsequent pressures upon the resources are manifested on the land use change and on the high deforestation rate. Forestal areas have a loss around 500,000 ha yearly (Palacio et al. 2000). Likewise, climate change could influence in the number and distribution of species: for example, likely 1,806 species will be affected in the coniferous forest, 1,309 species in the tropical forests, 1,345 species in the xerophilous scrubland, and 1,060 species in the mesophilous mountainous forests (Arriaga & Gómez 2004). Probably, forestal ecosystems deterioration in Mexico could be greater than the impact of climatic change itself (Gómez et al. 2006). That is why, it is necessary to begin studies on the effect of global warming at a key species scale and its spatial distribution at a national scale, in order to identify its vulnerability to climate change, as it was pointed out by the Fourth National Communication to the Framework Convention on Climate Change (Semarnat 2009).

This chapter is an opportunity to integrate information about the impact of climate change on the species distribution of temperate forests in two spatial and temporal scales; and propose guidelines to help the assisted adaptation process of temperate forests to climate change and decrease the environmental deterioration, both synergistic problematic in temperate forests of Mexico. Therefore, the objectives of this contribution were: i) a description of the simulation results and experiences of species distribution models with coupled models and regionalized country scale species of oaks and pines of wide and narrow geographic distribution; ii) an identification of thresholds identifying functional groups of plants in the mountain range of the Sierra Norte of Oaxaca using regional climate simulations; and iii) finally we present our experience, through government initiatives to identify adaptation measures to climate change in three participatory workshops with key players in forestry in the state of Tlaxcala.

2. Modelling potential distribution of pines and oaks

A diversity of models has been generated to identify the vulnerability of biodiversity to climate change (Stocker 2004; Thuiller et al. 2004; Visser 2004; Araujo et al. 2005; Ohlemuller et al. 2006). Most of these studies use general circulation models (GCM) to generate climate change scenarios through physical functions of energy feedback in climatic change components, as well as greenhouse effects emissions scenarios. However, GCM need to be regionalized or downscaled to get better predict changing values of temperature and rainfall at a higher spatial resolution. Also, models of global and national distribution of species, including climate models and ecological niche are methodological approaches to understand the effect of climate change on vegetation. An example of this is the simulation model GARP (Genetic Algoritm for Rule Set Prediction), based on the ecological requirements of species and logistics regressions to simulate potential distributions of those species (Stockwell & Peters 1999). An exercise was made to simulate the potential effect of climatic change upon the spatial distribution of 17 pine species and 17 oak species in temperate forests of Mexico using the GARP model (Gómez & Arriaga 2007). A projection for the year 2050 was made including two views of how climate would change: a severe one and a conservative one. Through a statistical downscaling process, average temperature change was obtained and the precipitation for the MCG HADCM2 under SRES A1 (severe) scenarios and B2 (conservative).

Results with the conservative emissions scenario suggest that the spatial distribution of pines decreases less than the severe scenario and once again the decrease of the area is variable (0.1 to 50%). Pine species more sensitive to and total precipitacion changes were in its geographic distribution, were *P. oocarpa, P. chihuahuana* and *P. rudis*. On the other hand, moderate sensitive pine species were *P. patula, P. durangensis, P. arizonica, P. teocote, P. ayacahuite,* and *P. culminicola*. It is worth noting that *P. cembroides* is one of the most tolerant species to climatic change, it will only loose 8% of its present distribution (Figure 1). Oaks seem to have less probability of modifying its geographic distribution, because they only decreases between 6 and 27% under the most conservative scenario (Figure 1). Species with high vulnerability to modify its geographic distribution are *Q. peduncularis,* and *Q. acutifolia,* while, the rest of the species will change its distribution between 6.8 and 17.7 %. Significant reductions will be present for *Q. castanea and Q. laeta* (Gómez & Arriaga 2007).

Results of this study showed that long-term vegetation changes can be expected in the temperate forests of Mexico as a consequence of climate change. Alteration in temperature and precipitation modeled under both climate-change scenarios will reduce the current ranges of distribution of almost all species of oaks and pines. Results for the more severe scenarios suggested that the effects will depend upon the species and the reduction of distribution levels have shown variations between 0.2 and 64%. The most sensitive species to change based on its future potential distribution by 2050 were *Pinus rudis, P. chihuahuana, P. oocarpa,* and *P. culminicola*. On the other hand, *P. patula, P. montezumae, P. teocote, P. ayacahuite, P. pseudostrobus, P.leiophylla, P. arizonica* and *P. herrerae,* shown moderate tolerance to future climate change; while *P. cembroides, P. durangensis, P. douglasiana, P. hartwegii,* and *P. strobiformis* are the most tolerant species to climatic change, thus its geographic distribution did not show significant modifications. In contrast, oak species showed a decrease between 11 and 48% of its present distribution for the year 2050; which suggests lower sensitivity than pine species. Oak with more sensitivity to thermical increase and change in rainfall pattern were *Quercus. crispipilis, Q. peduncularis,* and *Q. acutifolia.* On the other hand, *Q. sideroxyla, Q. mexicana, Q. eduardii, Q. castanea, Q. laurina, Q. rugosa, Q. magnoliifolia,* and *Q. crassifolia* resulted to be reasonably tolerant. The most tolerant species were *Q. obtusata, Q. durifolia, Q. segoviensis, Q. elliptica, Q. scytophylla,* and *Q. laeta* (Gómez and Arriaga 2007).

The overall results of this study suggests that species with more geographic distribution range does not have less vulnerability to climatic change, because the geographic distribution change of species seems to be related to climatic similarities of the specie itself. For example, pine species with more vulnerability were the ones found in semi-cold and semi-humid climates; areas or habitats were climate will considerable change with climatic change. Thus, species like *P. rudis, P. chihuahuaza, P. culminicola, Q. peduncularis* and *Q. sideroxyla* that live in these regions will be the ones with greater reductions in its geographic distribution (between 30 and 45%) for the 2050 scenario. Subsequent studies considered that pine species in temperate forests of Mexico, mainly on the regions of central-north, will be more vulnerable to climatic change *P. cembroides* and *P. pseudostrobus* (INE 2009). Together these studies of potential distribution modeling agreed on showing the high level of sensitivity of the species that live in mountainous regions, where temperature changes and reduction of rainfall will affect its development. However, there are still some questions about the environmental tolerance, mainly about climatic envelope that determines the presence of species at a community scale of temperate forests in topographic delimited units enclosed in Mexico.

3. Functional groups and climate change

The term functional group is applied to the group of species that use the same environmental resources class in the same way, this is, those that overlays its ecological niche (Gitay & Noble, 1997; Westoby & Leishman, 1997). In this way, the current climate, being a resource, represents a current climatic tolerance measurement element of species. Such tolerance can be compared with climatic change scenarios to evaluate vulnerability of the functional groups in the future. It is known that under similar climatic parameters in wide geographical levels, the response of the species demonstrate coincide (Retuerto & Carballeira, 2004), because some of the climatic parameters are descriptors of distribution of species (Myklestad & Birks, 1993; Carey *et al.* 1995). However, a more realistic approach requires the application of the regional or local model of the present and future climate so that a suitable policy of conservation for each zone can be applied.

The Sierra Norte of Oaxaca (SNO) has been considered as a priority terrestrial region because of its significance for biodiversity (Dávila et al. 1997; Arriaga et al. 2000). Oaxaca forests take up 8% of its territory (INEGI 2002). This land is considered one of the places with more diversity and endemism for *Pinus* and *Quercus*. Among the more representative species of SNO temperate forests stand out species like *Pinus patula; P. hartwegi, P. ayacahuite* and *P. pseudostrobus,* also *Abies guatemalensis, A. Hickelii* and *A. Oaxacana* (Del Castillo et al. 2004). There are also present *Pinus teocote, P. rudis, P. leiophylla, P. oocarpa, P. oaxacana, P. montezumae, P. douglasiana, P. lawsonii* and *P. pringlei* (Campos et al. 1992; Farjon 1997).

From a SNO inventory of species, with a total of 149,059 records (CONABIO and CIIDIR) connections between the presence of physiognomic dominant species and climate variations (Díaz et al. 1999; Kahmen & Poschlod 2004) were made in order to identify vulnerability to climate change for several types of vegetation: pine forest, *Abies* oak forest, cloud forests, scrubland, evergreen tropical forest tropical forest, dry tropical forest and dry subtropical forest (Table 1) (INEGI 2001). The determination of the possible responses of functional groups was based on the construction of an ensemble of eight general circulation models with four scenarios of global emissions, and a Japanese model (Mizuta et al. 2006), of regional high resolution (20 x 20 km). The ensemble of climate change scenarios suggests that by 2050 the temperature of the region will increase between 1.5 and 2.5°C, and rainfall will vary between +5 and -10% of the current annual precipitation. Finally, functional groups tolerance was identified by type of vegetation to climate change according to its present climatic preference (Gómez et al. 2008).

By means of arithmetic maps techniques, attribute tables of collect sites georeferenced were constructed with map scales of the total annual rainfall with the software ArcView (ESRI Versión. 3.2), the current habitat preference for each set of species grouped by gender was determined. The results indicated that genera like *Quercus, Pinus and Abies* were distributed among the 1,000 and 2,500 mm annual rainfall.

According to the Japanese model of high resolution (Mizuta et al. 2006), by the year 2050 minimal temperatures will increase more during the months of April and November on the SNO, meaning more warm nights. Rainfall will have significant decrease during winter from November through March (could be less than 100 mm per month), and increasing in July up to 150 mm (Figure 2). According to the climatic change scenarios by increasing minimal temperatures up to 3°C on April and December, genera like *Abies, Pinus, Juníperus* and *Quercus* could tolerate this change; because they can live in areas with temperatures up to 14°C. Probably the *Arbutus* in a pine forest and *Abies* and *Amelianchier* in a oak forest could not tolerate this increase on the minimal temperatures, because at the present time they are adapted to -2 a 5°C and from 0 to 6°C, respectively. On the other side, genera of cloud forests, evergreen tropical forest like *Clethra, Dendopanax, Miconia and Percea* have tolerance among minimal temperature of 0 and 14°C. Finally, scrubland genera and dry tropical forest (*Mimosa, Acacia* and *Brahea*) could also tolerate these changes, because they are distributed between -2 and 14°C.

Rainfall change sceneries for the year 2050 show differences among the altitudinal vegetation floors (Figure 2). Rainfall during autumn and winter will decrease in pine forests, while during summer it will be close to the base scenario; in contrast, oak forests will have a rainfall increase during summer. Thus, in the future, SNO pine forests will be dryer and oak forests more humid. This climatic pattern modification suggests that, even though the current temperature has a general increase tendency in the SNO, the differentiation of the anomaly of rainfall could modify the distribution of genera. So, species that require more rainfall levels in pine forests, like *Abies*, could be affected in its geographic distribution.

Regional climatic change scenarios also suggest altitudinal changes on the types of vegetation distribution. The present altitudinal gradient of conifers in the SNO is distributed above the 1,500 m. *Pinus hartwegii* is especially vulnerable to increase in temperature, because it is affected by plagues due to deficiency of low temperatures to eliminate them. *Quercus* is distributed from 150 to 3,500 m in Oaxaca. Species that are distributed at a higher elevation (more than 2700 m) are *Q. crassifolia, Q. laurina,* and *Q. elíptica,* probably these species are the most vulnerable species to climatic change (Gómez et al. 2008).

4. Adaptation capacity building

Once regional scenarios are identified from the assembling of several MCGs, we get close to an identification of future vulnerability of temperate forests of sites geographically enclosed. This way, threats are identified more clearly and adaptation strategies can be generated. However, the real capacity of auto-adaptation in these communities will depend on the no climatic threat magnitude, such as the type of management of forests and land use change. That is why, under the foster of national initiatives a capacity building exercise began with human societies that own, administrate and live in forests on the central region of Mexico. The project Generation Capacity for Adaptation to Climate Change supported by UNDP was to develop case studies to test methodologies, schemes of work disciplines and institutions, and information communication strategies that result in proposals to reduce vulnerability in temperate forest in Tlaxcala, Mexico (Magaña & Neri 2006). The project objective was identifying key actors of the forest sector to understand the condition of vulnerability to climate variability. In this study we work to determine the feasibility of the proposed adaptation strategies, their cost and their effectiveness, so that the methodology could be extended to other regions.

4.1 The forestal sector in the State of Tlaxcala

Tlaxcala State in the central region of Mexico has a surface of 399, 000 ha, from which 16% are forests, 8% are pastures, 74% are cultivable lands, and 1% human settlements. Tlaxcala is one of the states with more erosion index due to high deforestation rates, fire and land use change (Semarnat, 2002). Wood and non-timber products are extracted from Terrenate-Tlaxo municipalities on the North of the State, municipalities like Nanacamilpa and

Calpulalpan, on the West, and the protected natural areas of la Malinche South of the state have problems with clandestine logging (Gobierno del Estado de Tlaxcala 2004). From 1936 to 2000 more than half of the forestal cover has been lost. Under this analysis framework, notwithstanding that silviculture vulnerability points out towards climate, human activities represent the greater threat for the integrity of forests in the area. That is why; non climatic factors have to be considered in an adaptation model in the medium and long time. Climate changes scenarios projected in Tlaxcala drier and warmer condition (lower soil moisture) more frequent in the spring, so the risk of forest fires significantly increases the rate of loss of forest cover. Unless conditions change in the state of Tlaxcala, it is estimated that by 2080 there will be only about 40% of the present area. Therefore, climate change will accelerate forests loss in the state and in two or three decades will be very little remaining to preserve.

4.2 Adaptation Strategies

Through three participative workshops together with key actors and individual surveys measures of adaptation were identified to climatic change through the opinion of local forest producers and managers (Ecology Department, municipalities and SEMARNAT delegations) (Figure 3). Likewise, the feasibility of such measurements in a medium and long term was identified, as its eventual incorporation in the government level strategy. For this study, we applied the Political Framework: APF (UNEP 2004) for the design and execution of the projects to reduce the vulnerability to climate change. Key actors are of extreme importance through the five political adaptation stages marked by APF: 1. Definition and application sphere, 2. Evaluate present vulnerability, 3. Characterize future conditions, 4. Develop adaptation strategies, and 5. Continue with adaptation.

The three participative workshops were held under the monthly session's framework of the Forestal State Council, organization that congregates opinions from agricultural, silvicultural, private and communitarian forests owners, several environmental states institutions and academic representatives related with the study of forestal production and the conservation of state forests (Figure 3). These key factors discussed and prioritized adaptation measurements based on the problematic on climatic change, environmental degradation, and wrong management of forestal resources that they were ready to implement. Measurements of adaptation arrived at by consensus by the different key actors were: conservation, restoration and silvicultural, all of them in a sustainable process framework of forests. Likewise, three application areas at a municipality scale for measurements of adaptation were identified in terms of its benefits, negative impacts, regions, social groups with opportunities, and technical and economic impacts.

a) *Adaptation strategies: Conservation*

Due to the historical deforestation rate in the state, one of the main actions that need to be taken is the conservation of the remaining forest area through different public politic instruments. To this date, there are some federal and state programs that promote environmental services such as the Program for Payment for Hydrological and Environmental Services (PPHES) and the Program for the Environmental Market Services Development of Carbon Capture Derivative of the Biodiversity and the Development of Agroforestry Systems (CONAFOR 2009), that allows conservation of forestal areas in surfaces as in connectivity. Promotion programs state that the owners and land forestal owners are compensated for their services, and environmental services users have to pay them directly or indirectly. Other federal programs are Project of Clean Development Mechanism (CDM) that promotes the National Environmental Secretary (Semarnat). The beneficiaries of these programs are the owners and the owners of forests, academic groups, silvicultural, municipal authorities and communities.

The positive impact of these programs in climatic terms, will be reflected in a greater connectivity between forest surfaces that still are in good conservation in a horizontal and in altitudinal way. This will allow the migration of pine and oak species that will guarantee the permanency of the majority of these species. Likewise, it could help to improve the quality of life (education, health) and the diversity of non-timber products (mushrooms, ecotourism, medicinal plants). All of this promotes the capacity of self-management and it represents an option for creating regional projects sponsored by PPHES, or by international organisms and private businesses independent from federal support.

The feasibility of this measure is high because there are public political instruments that will allow the success of the implementation and monitoring of the conservation strategies. In this case, Development State Plan of Tlaxcala State 2005 – 2010 establishes actions to integrate the regeneration and conservation of forests with the production and planting of young trees. To achieve this success, there is another program for management and fire control for the protected natural areas that allow preserving water and soil. In the same way, there are programs for recovering high erosion areas in the state, conservation, protection and restoration of the forestal mass land of forests and water (Gobierno del Estado de Tlaxcala 2004). These plans and programs have as a final objective, to increase the forestal area for its conservation and management.

b) Adaptation measurement: ecological restoration

An alternative to increase the forestal surface in Tlaxcala is ecological restoration of these ecosystems. The objective of this measurement is to reduce the erosion of the soil, to help the recharge of water and recuperate the biological diversity of arboreal species. Once again, there are public political instruments that guarantee the implementation and monitoring of this adaptation. For example, there is a fiscal stimulus such as the productive reconversion, Temporal Work Program, water capture and reforestation, and the reforestation program in micro basins and the Integral Program for Forestal Resources are just some examples that promote indirectly climatic change adaptation. Mexican Official Rules that can establish mitigation measurements to climatic change, represent an area of opportunity where institutions like INIFAP (Institute of Research on Forest and Agriculture and Livestock) have already started research on genetic optimization processes of species and studies of aptitude for existing varieties under the climatic change scenarios.

c) Adaptation measurements: sustainable forest management

One of the main mitigation and adaptation measurements to climatic change that have been proposed is the sustainable forest management; through the implementation of conservation and carbon capture projects (Cowie et al. 2007). Under the Marrakesh agreements, activities such as afforestation, reforestation, deforestation, forestal management, agricultural management, and grassland management are alternative for mitigating GEI (García-Oliva & Masera, 2004; Cowie et al., 2007). For the implementation of this measurement a State Forestal Program exists for the year 2020, which promotes an increase on the forestal surface under sustainable management. Under this program, the directly beneficiaries are the owners of forests, silvicultural, local authorities and communities. If these measures are established they will be opportunities for the forestal management, the creation of a global state program of natural resources and its link with other productive areas in the State municipalities. Summing up, there are a series of initiatives and programs where a sustainable use of forests can be seen. However, it is still necessary to include the regional climatic change scenarios for Tlaxcala State on the aptitude analysis of the species, stand management use, reforestation programs, erosion decrease practices, and soil recuperation under high erosion, as well as territorial and ecological State level ordination.

5. Conclusions

In Mexico, climatic change is a future threat for the permanency of temperate forests in Mexico, however, the environmental degradation and the inadequate management of forestal resources are the main cause for the loss of these forests in a short term. The increase the increase of temperatures, the variation of rainfall patterns, and change in hydrological balance can have an impact on geographic composition and distribution of species that shelter temperate and temporal forests at different spatial levels. The study at national level suggests that climatic change descriptors will alter the geographic distribution of species; however, the impact was distinctly different between pines and oaks. At a regional scale, a change on the distribution of the species can be detected, on an altitudinal way. The analysis of both spatial resolution scales presented here, suggest that the alteration of climate will change the physiognomic dominant species distribution of temperate forests. However, the factors of local climate, such as geomorphology, orientation, and humid conditions can modify the response of forest communities faced to a climate change. It is important to incorporate the departures of the climate regional models on the behavior studies of the natural species of its own area of importance, for better sustainable use or for the conservation of forest areas in the country.

On the public policy arena at a federal level, it is encouraging to know that there are some attempts to establish adaptation measurements of the forestal sector to climatic change. It is important to point out that measurements proposals have double objective: adaptation to climatic change and environmental degradation reduction, both synergetic problematic in temperate forest of Mexico. This new knowledge, in combination with the ones already obtained from other Mexican scientists, will give bases to generate strategies for sustainable forestal management that will contribute to the reduction of CO_2 carbon emissions and face better the climatic change challenge.

6. References

Araujo, M. B., R. G. Pearson, W. Thuiller, & M. Erhard. 2005. Validation of species-climate impact models under climate change. Global Change Biology. 11:1504–1513.

Arriaga, L. & L. Gómez. 2004. Posibles Efectos del Cambio Climático en algunos Componentes de la Biodiversidad en México. El Cambio Climático: una visión desde México. In Martínez J. A. Fernández & P. Osnaya (compiladores). Instituto Nacional de Ecología-Secretaria de Medio Ambiente y Recursos Naturales. México. 255-266.

Arriaga, L., J. M. Espinosa, C. Aguilar, E. Martínez, L. Gómez & E. Loa. 2000. Regiones Terrestres Prioritarias de México, Comisión Nacional para el Conocimiento y Uso de la Biodiversidad, CONABIO. México.

Campos, A., P. Cortés, P. Dávila, A. García, G. Reyes, G. Toriz, L. Torres & R. Torres. 1992. Plantas y flores de Oaxaca. Cuadernos Núm 18, Instituto de Biología, UNAM, México.

Carey, P. D., C. D. Preston, M. O. Gill, M. B. Usher & S. M. Wright. 1995. An environmentally defined biogeographical zonation of Scotl& designed to reflect species distribution. Journal of Ecology. 88(5) 833-845.

CONAFOR 2009. Comisión Nacional Forestal, 2009. www. conafor.gob.mx

Cowie, A., Schneider, U., & Montanarella, L. 2007. Potential synergies between existing multilateral environments agreements in the implementation of l & use, l & use change & forestry activities. Environmental Science & Policy, 10:353-352.

Dávila, P., L. Torres, R. Torres & O. Herrera.1997. Sierra de Juárez, Oaxaca. In Heywood, V. H. y S. Davis (coords.), Centers of plant diversity. A guide & strategy for their conservation. World Wildlife Fund. 135-138.

Díaz, S., M. Cabido, M. Zak, E. B. Carretero & J. Aranibal. 1999. Plant functional traits, ecosystem structure & l &-use history along a climatic gradient in central-western Argentina. Journal of Vegetation Science. 10: 651-660.

Farjon, A., & B. Styles. 1997. Pinus (Pinaceae). Flora Neotropica.Monograph 75. The New York Botanical Garden, Bronx, New York.

García-Oliva, F., & Masera, O. 2004. Assessment & measurement issues related to soil carbon sequestration in land-use, land-use change, and forestry (LULUCF) projects under the Kyoto protocol. Climate Change, 65:347-364.

Gitay, H. & I. R. Noble. 1997. What are functional types and how should we select them. In Smith, T., H. H. Shugart & F. I. Woodward (eds.), Plant functional types: their relevance to ecosystem properties and global change. International Geosphere-Biosphere Programme Book Series. Cambridge. 3-17.

Gobierno del Estado de Tlaxcala. 2004. Ordenamiento ecológico del estado de Tlaxcala. México.

Gómez Mendoza, L., Aguilar-Santelises, R & Galicia, L. 2008. Sensibilidad de grupos funcionales al cambio climático en la Sierra Norte de Oaxaca, México. Investigaciones Geográficas. 67:76-100

Gómez Mendoza, L. & Arriaga Cabrera, L. 2007. Effects of climate change in Pinus and Quercus distribution in México. Conservation Biology. 21, 6:1545-1555.

Gómez-Mendoza, L. E. Vega-Peña, M. I. Ramírez, J. L. Palacio-Prieto & L. Galicia. 2006. Projecting land-use change processes in the Sierra Norte of Oaxaca, Mexico, Applied Geography. 26:276-290.

INEGI, Instituto Nacional de Estadística Geografía e Informática.2001. Conjunto de datos vectoriales de la carta de Uso de Suelo y Vegetación. Serie II (continuo nacional), escala 1:250 000. México.

INEGI, Instituto Nacional de Estadística Geografía e Informática. 2002. Anuario estadístico del estado de Tlaxcala. México.

IPCC: Intergovernmental Panel of Climate Change, 2007. Climate change 2007. Impacts, adaptation and vulnerability. Working Group II. Contributions to the Intergovernmental Panel of Climate Change. Fourth Assessmente Report. Summary for Policymakers. WMO-UNEP, Geneve.

Kahmen, S. & P. Poschlod. 2004. Plant functional traits responses to grassland succession over 25 years. Journal of Vegetation Science, 15(1). 21-32.

Locatelli, B. 2006. Vulnerabilidad de los bosques y sus servicios ambientales al cambioclimático. Centro Agronómico Tropical de la Investigación y Enseñanza. Grupo de Cambio Climático Global.

Magaña, V. & C. Neri (Comp). Informe de resultados del proyecto Fomento de las capacidades para la etapa II de adaptación al cambio climático en Centroamérica, México y Cuba. UNAM, México.

Malcolm J., A. Diamond, Markham, A. F. Mkanda y A. Starfield. 1998. Biodiversity:species, communities and ecosystems. En United Nation Environmental Programme. Handbook on methods for climate change impact assessment and adoption strategies. Amsterdam. 13-1 - 13-41.

Maslin, M. 2004. Ecological versus climatic thresholds. Science. 306: 2197-2198.

Mizuta, K., H. Yoshimura, K. Katayama, S. Yukimoto, M. Hosaka, S. Kusonoky, H. Kawai and M. Nakagawa. 2006. 20 km mesh global climate simulation using JMA-GSM model. Journal of Meteorological Society of Japan, 84:165-185.

Mueller, R. C., C. M. Scudder, M. E. Porter, R. T. Trotter, C. A. Gehring, & T. G. Whitham. 2005. Differential tree mortality in response to severe drought: evidence for long-term vegetation shifts. Journal of Ecology 93:1085–1093.

Myklestad, Ä. & H. E. J. B. Birks. 1993. A numerical analysis of the distribution of Salix L. species in Europe. Journal of Biogeography. (20)1-32.

Ohlemuller, R., E. S. Gritti, M. T. Sykes, & C. D. Thomas. 2006. Quantifying components of risk for Europeanwoody species under climate change. Global Change Biology. 12:1788–1799.

Palacio-Prieto, J.L; G. Bocco; A. Velásquez, J.F. Mas; F. Takaki-Takaki; A. Victoria; L. Luna-González; G. Gómez- Rodríguez; J. López García: M. Palma; I. Trejo-Vazquez: A. Peralta; J. Prado-Molina; A. Rodríguez: R. Mayorga- Saucedo & F. González. 2000. La Condición Actual de los Recursos Forestales en México: Resultados del Inventario Nacional Forestal 2000. Investigaciones Geográficas 43: 183-203.

Parmesan, C. 2006. Ecological & evolutionary responses to recent climate change. Annual Reviews of Ecology, Evolution, and Systematics 37:637–669.

Rebetez, M., & M. Dobbertin. 2004. Climate change may already threaten Scots pine stands in the Swiss Alps. Theoretical and Applied Climatology 79:1–9.

Retuerto, R & A. Carballeira. 2004. Estimating plant responses to climate by direct gradient analysis and geographic distribution analysis, Plant Ecology, 170(2) 185-202.

Semarnat. 2006. México tercera comunicación nacional ante la Convención Marco de las Naciones Unidas sobre el Cambio Climático, Instituto Nacional de Ecología, México.

Semarnat 2002. Informe de la situación del medio ambiente en México. México.

Semarnat: Secretaria de Medio Ambiente Recursos Naturales y Pesca, 2009. Cuarta Comunicación Nacional ante la Convención Marco de las Naciones Unidas para el Cambio Climático. México.

Sholze, M., W. Knorr., Arnell, N. y Prentice, C. 2006. A climate-change risk analysis for world ecosystems. PNAS. 35: 13116-13120.

Stocker, T. F. 2004. Climate change—models change their tune. Nature 430:737–738.

Stockwell, D., & D. Peters. 1999. The GARP modeling system: problems and solutions to automated spatial prediction. International Journal of Geographical Information Science 13:143–158.

Thuiller, W., L. Brotons, M. B. Araujo, & S. Lavorel, S. 2004. Effects of restricting environmental range of data to project current and future species distributions. Ecography 27:165–172.

UNEP: Programme of United Nations for Development, 2004. Adaptation Policy Frameworks for Climate Change: Developing Strategies, Policies and Measures. Bo Lim y Erika Spanger (Eds) Siegfried. Cambridge University Press.

Villers, L., & I. Trejo. 1998. El impacto del cambio climático en los bosques y áreas naturales protegidas de México. Interciencia 23:10–19.

Visser, H. 2004. Estimation and detection of flexible trends. Atmospheric Environment 38:4135–4145.

Westoby, M. & M. Leishman. 1997. Categorizing plant species into functional types. In Smith (ed.). Plant functional types: their relevance to ecosystem properties and global change. International Geosphere-Biosphere Programme Book Series.

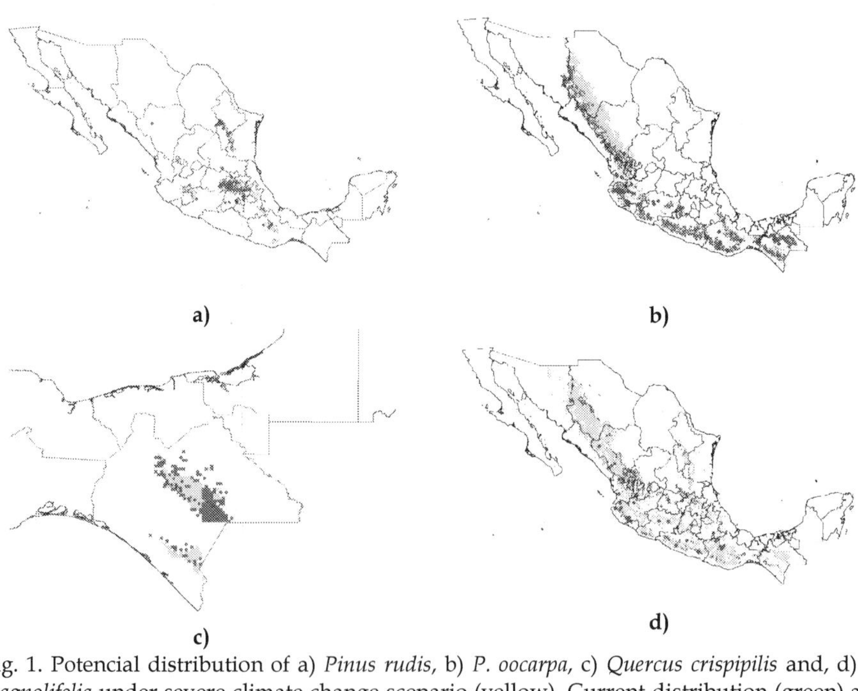

Fig. 1. Potencial distribution of a) *Pinus rudis*, b) *P. oocarpa*, c) *Quercus crispipilis* and, d) *Q. magnolifolia* under severe climate change scenario (yellow). Current distribution (green) and collecting data (red points) are showed.

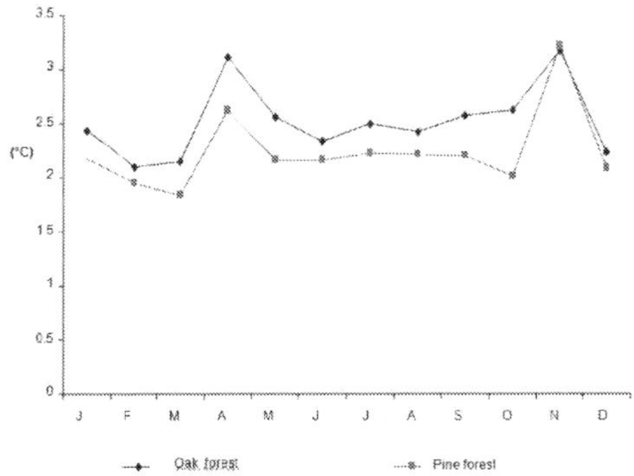

a)

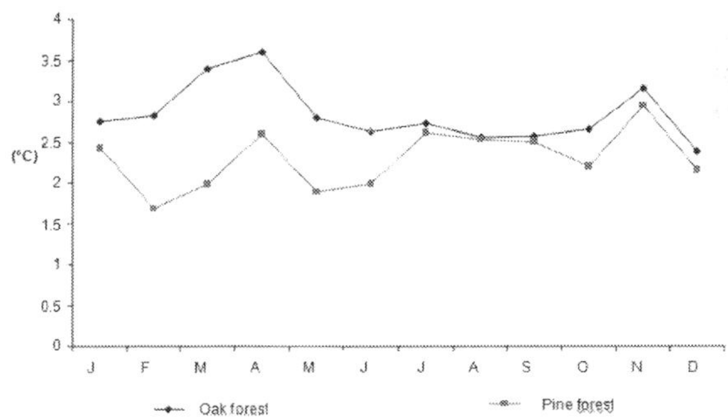

b)

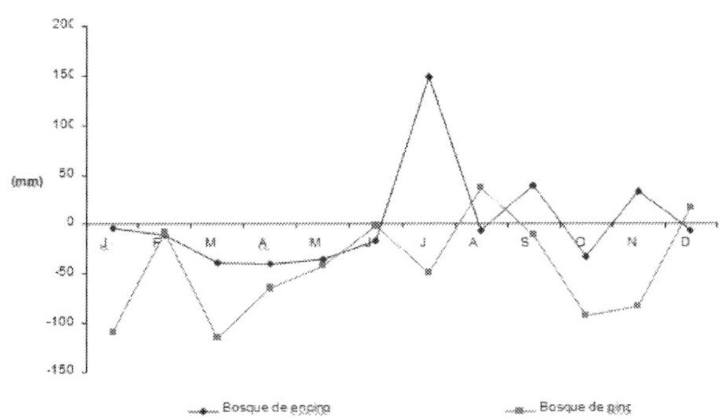

c)

Fig. 2. Climate change scenarios for 2050: a) changes in minimum temperature ; b) changes in maximum temperature and, c) changes in total precipitation in Sierra Norte of Oaxaca, México.

Fig. 3. Participative workshops in Tlaxcala Mexico.

Vegetation type	Physiognomic dominant species	
	Tree species	Brush species
Abies and pinus forest	*Abies hickelii**	*Amelanchier denticulata**
	*Juniperus flaccida**	*Arteostaphylus pungens*
	*Pinus ayacahuite**	*Baccharis heterophylla**
	*Pinus devoniana**	*Bejaria aestuans**
	*Pinus hartwegii**	*Calliandra grandifolia**
	*Pinus oocarpa**	*Gaultheria acuminata**
	*Pinus patula**	*Rhus virens*
	*Pinus teocote**	*Arbutus xalapensis**
	*Quercus crassifolia**	*Comarostaphylis discolor*
	*Quercus elliptica**	*Litsea neesiana**
	*Quercus laeta**	*Roldana sartorio*
Oak forest	*Carpinus caroliniana*	*Comarostaphylis discolor*
	*Quercus elliptica**	*Gaultheria acumina*
	*Quercus laeta**	*Listea glaucescens**
	*Quercus rugosa**	*Lyibua squamulosa*
	*Quercus scytophylla**	*Myrica cerifera*
	*Styrax argenteus**	
Cloud forest	*Clethra sp**	*Calyotranthes schiedeana*
	*Dendropanax populifolius **	*Miconia lonchophylla*
	Ilex discolor	
	Liquidambar styraciflua	
	*Persea americana**	
	*Pinus patula**	
	Podocarpus matudae	
	*Quercus candicans**	
	*Saurauia spp**	
	*Styrax glabrescens**	
	Weinmannia pinnata	

Table 1. Physiognomic dominant species by vegetation types in Sierra Norte of Oaxaca.

12

The influence of climate change on tree species distribution in west part of south-east europe

J. Vukelić[1], S. Vojniković[2], D. Ugarković[1], D. Bakšić[1] and S. Mikac[1]
[1]*Faculty of Forestry University of Zagreb*
Svetosimunska 25,
10000 Zagreb,
Croatia
[2]*Faculty of Forestry University of Sarajevo*
Zagrebačka 20,
71000 Sarajevo,
Bosnia and Herzegovina

1. Introduction

Ecological niche is n-dimensional hipervolume space defined by amount of ecological factors in which population of certain species is able to persist. From ecological point of view there are fundamental (FEN) and realized niche (REN). Fundamental ecological niche is space which certain species occupies in lower competition of other species or in absence of natural antagonists. Opposite to fundamental niche realized niche is space in which certain species accrues and persists in presence of competition with other species (Hutchinson, 1957).

To fully understand ecological niche is rather complicated concept. Ecological niche can be imagined as multidimensional space in which each ecological factor is presented by vector different in amount and direction of action. Amount of ecological factor action is shown in his importance for accruence and development of species, while direction of action can be positive or negative on species prevalence. Entirely understanding influence of each ecological factor on certain species space distribution is fairly difficult since between ecological factors exist certain dependence.

Importance of ecological niche on appearance and accurance of certain species can be observed on few resolution levels. Species during ontogeny development develops different needs toward ecological factors. Seed of all tree species for germination and initial growth needs only temperature and soil moisture, for further development young plant needs higher amount of sun radiation, moisture, nutrients and lower competition of mature trees. If ecological niche is observed from the view of population in certain geographical climate than climatic and geomorphologic (relief) factors have higher importance for spatial distribution and formation of species areal.

Beside air temperature depending on cloudiness and air insulation, precipitation has the highest importance for vegetation development while being the main source of moisture in soil needed for basic physiological processes.
When values of temperature and precipitation are observed should be considered that exactly extreme values (minimum and maximum) of these two ecological factors are limiting for certain species thriving. Extreme values of climate factors determine distribution range of certain species alongside ecological factor gradient.
Climate factors (temperature and precipitation) can be according to their effect divided on local and global that is micro and macroclimatic. Microclimate factors cause anomalies within certain climate area, as for example climate inversion within certain climate area determined by relief.
Climate factor changes cause also change in appearance of vegetation cover certain area. Climate factors effect on dragging certain species in newer areas or higher altitudes while their place is taken by other species that can adapt to changed ecological conditions.
Wills et al. (1999) pointed out significant association between vegetation dynamic and Earth orbital frequencies (Milanković cycles) in amplitudes at about 124000 years. Climate change is normal appearance in nature caused by natural variability. In the last interglaciation period using oxygen isotope $\delta^{18}O$ isolated on Greenland low temperature oscillations were found (Dansgaard et al., 1993). Oscillations in Holocene were ±1,5 °C for average summer and annual temperature (Wick & Tinner, 1997), while precipitation reconstruction showed significant oscillations during Holocene (Magny et al., 2003). Davis (2003) found variations in average temperature from +0,5 to -2,5 °C for the entire Europe during Holocene.
In contest of climate change should be distinguished climate change from climate fluctuation. Climate change is defined as one-way directed change. Contrariwise, climate fluctuation implies rhythmic oscillations around one average value whereat higher or lower amplitudes can appear. When determining climate change there is evident significance of time factor since one-way directed change of certain time period can be, at prolongation of observation sequences, shown as part of climate fluctuation. Therefore is useful to comprise climate change and climate fluctuations with common mark as climate change (Kirigin, 1975).
Nowadays there are many scientific discussions and interpretations of global warming causes. Analyses of ice boreholes are showing high increase of CO_2 concentration in atmosphere and increase of CO_2 during industrial period from 280 ppmv (year 1750.) to 365 ppmv (year 1998.) (Högberg, 2007). Hasselmann (1997) points out that during last century average temperature has increased by 0,5 °C. According to same author theory about anthropogenic cause of global climate changes is still controversial. Loutre (2003) assigns long current interglacial period of almost 5000 years to high CO_2 concentration that prevents development of ice shield and at a same time beginning of new glaciations.
Part of southeast Europe is highly important for overall vegetation of Europe. This geographically small area is abounded with numerous different tree species. Paleonological researches classify it in important refuge during last glacial from where certain tree species have expanded in northern parts of Europe (Willis, 1994).
Willis et al. (1999) indicates on simultaneous existence of subtropical genders as: *Carya, Pterocarya, Liquidambar, Seqouia, Taxoduim, Nissa* etc., alongside with fir (*Abies*) and beech (*Fagus*) on area nowadays Balaton. Second example is specie *Acer monspessulanum* spread over south Europe, and during Ipswichiana period outspread as far as British peninsula

(West, 1980). Same author find out that spruce (*Picea omorika*) in Pastoniana period has grown as far as south of England.

Some tree species appear only in pure forests (composed of only one tree species), while others can be found and in mixed communities (composed of two or more tree species). Some tree species nowadays are widely distributed (*Fagus sylvatica*), while other tree species are only individually incorporated in theirs distribution area (*Acer pseudopaltanus*).

Dominant vegetation types of forests nowadays present in southeast Europe territory are pedunculate oak (*Quercus robur*) forests bounded at larger rivers in Panonian lowland area. In hilly area at 400 m altitude are dominant sessile oak (*Quercus petrea*) forests. In mountain area at 400-800 m of altitude are dominant pure and mixed forests of common beech (*Fagus sylvatica*), and at 700-1100 m of altitude are found mixed forests of common beech, fir and spruce. In pre-mountain region is dominant mountain pine (*Pinus mugo*). In coastal region of Adriatic see dominant are pine species namely Holm oak (*Quercus ilex*) and pubescent oak (*Quercus pubescens).*

Fig. 1. Main vegetation types of forests with dominant tree species in southeast Europe: a) *Pinus mugo*, b) *Picea abies*, c) *Pinus nigra*, d) *Quercus petrea*, e) *Quercus robur*, f) *Pinus sylvestris*, g) *Abies alba*, h) *Quercus ilex*, i) *Fagus sylvatica* and j) *Quercus pubescens*.

Fully understanding of ecological niche it is of crucial importance for getting complete insight in forest ecosystems function. Understanding interaction between climate and vegetation represents one of the most essential segments in researching geographical distribution and development of species.

2. Material and methods

For ecological niche modelling were used available date on tree species distribution obtained from data of terrestrial habitats Republic of Croatia (Oikon d.o.o. for MZOPU, 2004), Flora Croatica Database (http://hirc.botanic.hr), reconstructing vegetation maps (Map of forest vegetation Republic of Bosnia and Hercegovina and Map of forest ecosystems Republic of Croatia) and personal field researches. For ecological niche spatial distribution modelling and their projections under climate changes scenario were chosen climate characteristic tree species on entire are, such as: Holm oak (*Quercus ilex*), pubescent oak (*Quercus pubescens*), pedunculate oak (*Quercus robur*), sessile oak (*Quercus petrea*), common beech (*Fagus sylvatica*), common fir (*Abies alba*), mountain pine (*Pinus mugo*), Austrian pine (*Pinus nigra*) and scots pine (*Pinus sylvestris*).
Climate data used for ecological niche modelling were taken from database WORLDCLIME for period 1950-2000 (http://www.worlclime.org). Spatial distribution of climate data is 30arc-seconds, that equals to the approximately resolution of 1 km^2 (to be exact 0,83 km^2) (Hijmans et al., 2005). Spatial distribution change of ecological niche for main tree species was modelled for year 2080. using climate data obtained from global climate change models CGCM2 (Coupled Global Climate Model-SRES-A2a) and CGCma (Canadian Centre for Climate Modelling and Analysis) (source: http://gisweb.ciat.cgiar.org). This climate model represents second version of first generation global climate model (CGCM1) essentially improved when compared to first version. Parameters of mixing ocean from horizontal vertical scheme were modified to vertiginously mixing (Gent & McWilliams, 1990) and in the second generation model was included sea and ice dynamic (Flato & Hibler, 1992).
For ecological niche modelling was used MAXENT program (Phillips et al., 2004, 2006). Among ecological factors were used extreme values of bioclimatic variables, namely: maximum temperature of the hottest month in year (BIO5), minimum temperature of the coldest month in the year (BIO6), precipitation of the wettest month in the year (BIO13) and precipitation of the driest month in the year (BIO14). Other climate variables were due to high linear correlation coefficient discarded as independent from prediction model of spatial distribution ecological niche. Altitude also was not included in prediction model since it showed extremely high positive correlation with maximum temperature of the hottest month (BIO5). Obtained maps of spatial distribution ecological niche were transformed into grids and visualed in ArcMap 9.2.

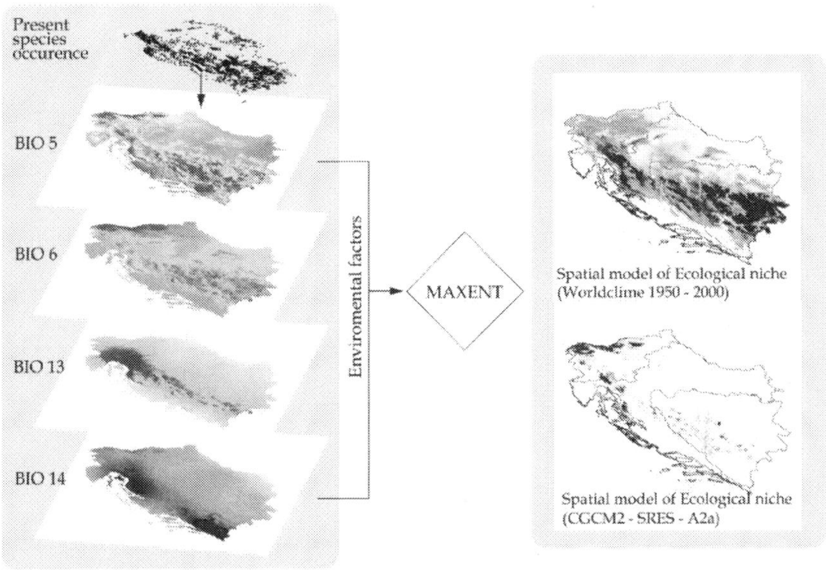

Fig. 2. Process of modelling ecological niche according to chosen climate factors as predictor variables using MAXENT program and projection of the present ecological niche according to the global climate change model CGCM2 for year 2080.

3. Results

The lowest average minimum temperature of the coldest month was determined for mountain pine (*Pinus mugo*) at -5,9 °C and for common fir (*Abies alba*) at -5,3 °C. The highest minimal temperatures of the coldest month were found for holm oak (*Quercus ilex*) at 3,6 °C and pubescent oak (*Quercus pubescens*) at 0,2 °C. The highest values of maximum temperatures of the hottest month were determined for holm oak (*Quercus ilex*) at 28,9 °C, pedunculate oak (*Quercus robur*) at 27,5 °C and for pubescent oak (*Quercus pubescens*) and sessile oak (*Quercus petrea*) at 28,9 °C. The lowest minimum temperature of the hottest month was found for mountain pine (*Pinus mugo*) at 19,1 °C. These temperature values pertain to the period 1950-2008. Global climate model CGCM2 (A2a) showed significant increase of minimum and maximum temperatures for all tree species. Average increase of maximum temperatures the hottest month for all tree species according to used climate change model is 6,9 °C, in range from 6,2 °C for mountain pine (*Pinus mugo*) to 7,6 °C for Austrian pine (*Pinus nigra*). Average increase of minimum temperature the coldest month according to climate change model is 2,0 °C, with the highest change found for holm oak (*Quercus ilex*) at 2,3 °C, and the lowest for sessile oak (*Quercus petrea*) at 1,9 °C.

Variables	WorldClime (1950-2000) (°C)				CGCM2-A2A (2080) (°C)						
	Min. Temp. of Coldest Month		Max. Temp. of Warmest Month		Min. Temp. of Coldest Month		Max. Temp. of Warmest Month				
Species	Means	Std. Dev.	Means	Std. Dev.	Means	Std. Dev.	Means	Std. Dev.	Δmin	Δmax	N
Abies alba	-5,3	1,0	22,3	2,0	-3,2	0,9	29,4	2,1	2,0	7,0	1075
Pinus mugo	-5,9	1,1	19,1	2,0	-3,9	1,1	25,4	1,5	2,0	6,2	56
Pinus nigra	-4,1	1,6	24,6	2,2	-1,9	1,6	32,2	2,1	2,1	7,6	209
Pinus sylvestris	-4,5	1,8	23,8	2,4	-2,4	1,9	31,2	2,4	2,1	7,3	114
Quercus petrea	-3,7	0,7	26,3	1,0	-1,8	0,7	33,2	1,1	1,9	6,9	1557
Quercus pubescens	0,2	2,2	26,3	2,1	2,3	2,1	32,6	2,2	2,1	6,3	1557
Quercus robur	-3,3	0,5	27,5	0,5	-1,4	0,6	34,5	1,0	2,0	7,0	623
Fagus sylvatica	-4,9	1,1	23,1	2,4	-2,9	1,1	30,0	2,4	2,0	6,9	1781
Picea abies	-5,4	1,0	22,0	2,0	-3,4	0,9	29,2	2,2	2,0	7,1	1840
Quercus ilex	3,6	1,4	28,9	1,1	5,8	1,4	35,6	1,4	2,3	6,7	283
All Grps	-3,6	2,7	24,4	2,9	-1,6	2,7	31,3	2,9	2,0	6,9	9095

Table 1. Descriptive date on climate factors minimum temperature of the coldest month and maximum temperature of the hottest month according to the tree species.

The lowest average precipitation of the driest month was found for holm oak (*Quercus ilex*) at 38,6 mm, and the highest for mountain pine (*Pinus mugo*) at 83,9 mm.
Average precipitation quantity in the wettest month for all tree species is 121,6 mm, in range from 99 mm for pedunculate oak (*Quercus robur*) to 142,3 mm for pubescent oak (*Quercus pubescens*).
Area of oncoming mountain pine (*Pinus mugo*) has the highest precipitation quantity in the driest month and also relatively high precipitation quantity in the wettest month (Table 2). Common beech (*Fagus sylvatica*) and fir (*Abies alba*) have equal precipitation quantities in the driest and the wettest month. Holm oak (*Quercus ilex*) accrue in area with extremely low precipitation quantity in the driest month and relatively high precipitation quantity in the wettest month (38,6-137,3 mm).

	WorldClime (1950-2000) (mm)				CGCM2-A2A (2080) (mm)						
	Prec. Of Driest Month		Prec. Of Wettest Month		Prec. Of Driest Month		Prec. Of Wettest Month				
Species	Means	Std. Dev.	Means	Std. Dev.	Means	Std. Dev.	Means	Std. Dev.	Δmin	Δmax	N
Abies alba	72,7	10,8	122,5	18,4	45,4	7,1	118,9	24,2	27,3	3,7	1075
Pinus mugo	83,9	6,1	136,4	14,0	49,4	8,3	134,4	18,8	34,5	2,0	56
Pinus nigra	64,9	7,6	117,8	17,5	37,8	5,1	110,7	17,8	27,1	7,1	209
Pinus sylvestris	64,3	8,7	116,2	16,3	40,5	5,0	107,5	16,7	23,9	8,7	114
Quercus petrea	59,0	9,9	108,4	10,1	42,7	4,3	102,7	14,7	16,3	5,7	1557
Quercus pubescens	59,9	12,8	142,3	15,9	34,2	10,7	135,7	20,2	25,7	6,7	1557
Quercus robur	50,5	5,8	99,0	6,4	40,2	6,1	90,1	10,6	10,3	8,9	623
Fagus sylvatica	69,7	11,1	121,3	17,8	44,2	6,6	116,5	21,9	25,5	4,8	1781
Picea abies	73,2	10,0	120,4	17,2	45,0	7,2	116,4	22,9	28,2	4,0	1840
Quercus ilex	38,6	7,2	137,3	24,6	20,1	5,4	131,5	24,9	18,5	5,8	283
All Grps	64,9	13,5	121,6	20,1	41,4	9,1	116,2	23,9	23,5	5,4	9095

Table 2. Precipitation quantities descriptive data in the driest and the wettest month according to tree species

Precipitation quantities of the driest month according to global climate change model showed significant decrease of average values for 23,5 mm, while average precipitation quantities of the wettest month decreased for 5,4 mm. The highest average precipitation quantities of the driest month were found for mountain pine (*Pinus mugo*) at 34,5 mm or in relative value 41,12 %. The highest decrease in average precipitation quantities of the driest month is found for holm oak (*Quercus ilex*) at 18,5 mm or 48,00 % in relative value. The highest decrease in precipitation of the wettest month was found for pedunculate oak (*Quercus robur*) at 8,9 mm or 8,94 % when compared with data in period 1950-2000.
Statistically significant differences were determined between all values of analysed climate factors apart for pubescent oak (*Quercus pubescens*) and sessile oak (*Quercus petrea*) for maximal temperature of the hottest month (p=0,84494). Statistically significant differences were not found for precipitation quantities of the driest month between spruce (*Picea abies*) and fir (*Abies alba*) (p=0,16283) and between Austrian pine (*Pinus nigra*) and scots pine (*Pinus sylvestris*) (p=0,628145). Significant differences were not found between holm oak (*Quercus ilex*) and mountain pine (*Pinus mugo*) for precipitation quantities of the wettest month (p=0,69797) and for scots pine (*Pinus sylvestris*) and Austrian pine (*Pinus nigra*) (p=0,382217) as well as between spruce (*Picea abies*) and common beech (*Fagus sylvatica*) (p=0,103702).

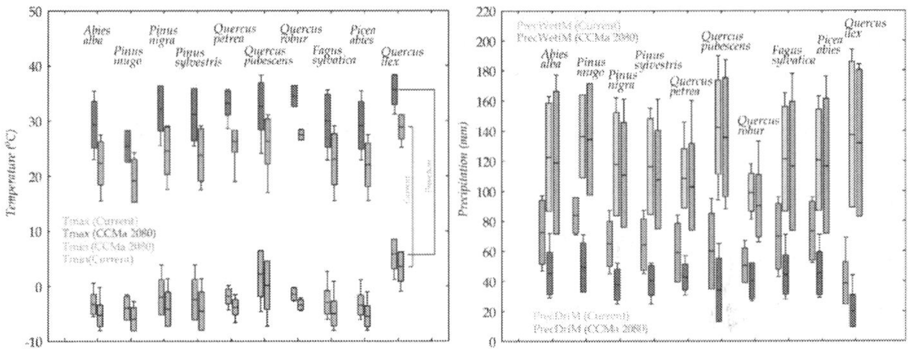

Fig. 3. Box & Whisker plot average values (Mean±1,96 St.Dev) of climate variables used for modelling ecological niche according to tree species for period 1950-2000 and their projected values according to global climate change model CGCM2 with projection for 2080.

Range of tree species population can be monitored depending on every climate factor individual. In Table 3. are presented species range values according to individual climate factor. It can be seen that 95 % of pedunculate oak (*Quercus robur*) population is in very narrow maximum temperature range for the hottest month (T_max M.) at 1,70 °C, while total population has range of 2 °C. In contrary to pedunculate oak, common beech (*Fagus sylvatica*) shows brooder range of total population at 13,6 °C as well as sessile oak (*Quercus petrea*) at 14,10 °C. Common beech is due to wide distribution considered as species of wide ecological valence towards climate factors whereat average range maximum temperatures of the hottest month is at 8,40 °C and 4,30 °C for minimum temperature of the coldest month (T_min M.).

Species	T_max M. (°C)		T_min M. (°C)		Prec. Driest. M. (mm)		Prec. Wettest. M. (mm)	
	95%	Min.-Max.	95%	Min.-Max.	95%	Min.-Max.	95%	Min.-Max.
Abies alba	7,90	12,00	3,80	7,80	38	50	59	68
Pinus mugo	6,90	8,40	3,80	5,10	25	25	40	43
Pinus nigra	8,40	11,50	6,70	8,20	31	42	56	68
Pinus sylvestris	9,00	11,60	7,80	8,50	36	42	56	59
Quercus petrea	4,10	9,10	3,20	5,10	34	44	41	56
Quercus pubescens	8,40	14,10	8,10	11,50	46	59	63	96
Quercus robur	1,70	2,00	1,90	2,30	24	28	21	37
Fagus sylvatica	8,40	13,60	4,30	8,90	44	53	60	73
Picea abies	8,15	12,00	3,65	7,00	36	44	57	68
Quercus ilex	4,70	5,90	5,50	6,90	29	41	92	99

Table 3. Range of values chosen climate factors according to tree species.
Min.-Max. - range between the lowest and the highest value.

Ecological niche analysis considering all variables was done using discriminate analysis (DCA). Results show position of chosen tree species population projected in two-dimensional system where every axis is linear combination all four climate factors (Figure 3).

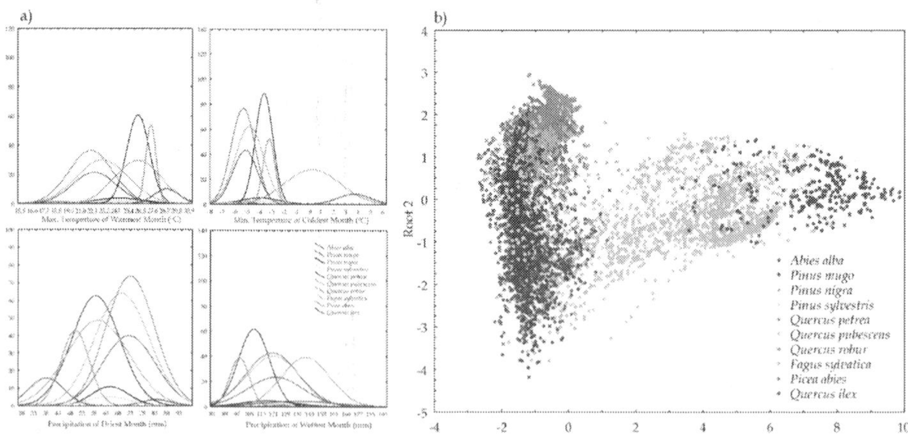

Fig. 3. Frequency distribution according to tree species and chosen climate factors a). Scatter plot canonical values of discriminate analysis b).

Results of discriminate analysis show clear segregation of populations according to tree species in two larger groups. First group includes species holm oak (*Quercus ilex*) and pubescent oak (*Quercus pubescens*) where discriminatory variable is Root 1. Both species clearly segregate from all others and exactly they define geographical appurtenance to Mediterranean region. Second group composed from all other species can be geographically classified in continental region. Discriminatory axis Root 2 shows clear species segregation from pedunculate oak (*Quercus robur*), across sessile oak (*Quercus petrea*) up to mountain pine (*Pinus mugo*), while fir, beech and spruce overlap considering combination of chosen climate factors. In Table 4. are presented values of standard coefficients canonical variables that is effect of each individual climate factor on species segregation.

Variables	Root 1	Root 2	Root 3	Root 4
Max. Temp. Of Warmest Month	-0,94766	1,19337	-1,01995	-0,21000
Min. Temp. of Coldest Month	1,35830	-0,42277	0,38860	0,71913
Prec. Of Driest Month	-0,70913	0,03123	-0,75387	0,97859
Prec. Of Wettest Month	0,22764	-0,17788	-0,60883	-1,08279
Eigenvalue	4,88517	0,92261	0,06825	0,00557
Cum.Prop	0,83059	0,98745	0,99905	1,00000

Table 4. Standard coefficients of canonical variables using discriminatory analysis (DCA).

Spatial distribution of ecological niche according to chosen climate factors for chosen tree species is shown in Figure 4. Proportion of each individual variable in spatial prediction of ecological niche is presented in Table 5. Maximal temperature of the hottest month in the year has the highest effect in fir (*Abies alba*) at 46,5 %, in common beech (*Fagus sylvatica*) 62,4 %, in mountain pine (*Pinus mugo*) 63,3 % and in pedunculate oak (*Quercus robur*) at 50,8 %. Minimal temperature of the coldest month has the highest proportion in spatial prediction of pubescent oak (*Quercus pubescens*) ecological niche at 70,7 % and in holm oak (*Quercus ilex*) at 72,3 %. Precipitations of the driest month in the year have the highest proportion in spatial prediction Austrian pine (*Pinus nigra*) ecological niche at 71,7 % and scots pine (*Pinus sylvestris*) at 44,2 %. Precipitations of the wettest month in year have the highest proportion in prediction pedunculate oak (*Quercus robur*) ecological niche at 40,7 % and in sessile oak (*Quercus petrea*) at 40,0 % (Table 5).

In spatial model of spruce (*Picea abies*) ecological niche equal proportion have maximum temperatures of the hottest month (35,6 %) and minimum temperatures of the coldest month (31,6 %) and precipitations of the driest month in the year (26,6 %).

According to Table 4. can be seen that proportion of climate factors on spatial prediction ecological niche is equal for species: pubescent oak (*Quercus pubescens*) and holm oak (*Quercus ilex*), and for Austrian pine (*Pinus nigra*) and scots pine (*Pinus sylvestris*).

Species	*T_max M. (°C)*	*T_min M. (°C)*	*Prec. Driest. M. (mm)*	*Prec. Wettest. M. (mm)*	AUC
Abies alba	46,2	26,2	21,2	6	0,853
Fagus sylvatica	62,5	20,2	11,8	5,5	0,745
Picea abies	35,6	31,6	26,6	6,3	0,851
Pinus mugo	63,3	34,9	1,2	0,6	0,976
Pinus nigra	3,5	11	71,7	13,8	0,842
Pinus sylvestris	17,5	21,6	44,2	16,6	0,867
Quercus robur	50,8	4,9	3,5	40,7	0,927
Quercus petrea	29,3	23,1	7,6	40	0,800
Quercus ilex	1,1	72,3	23,4	3,2	0,974
Quercus pubescens	4,2	70,7	3,1	22	0,861

Table 5. Relative proportion of each individual climate factor on ecological niche prediction (%). AUC - surface under ROC curve of MAXENT prediction model.

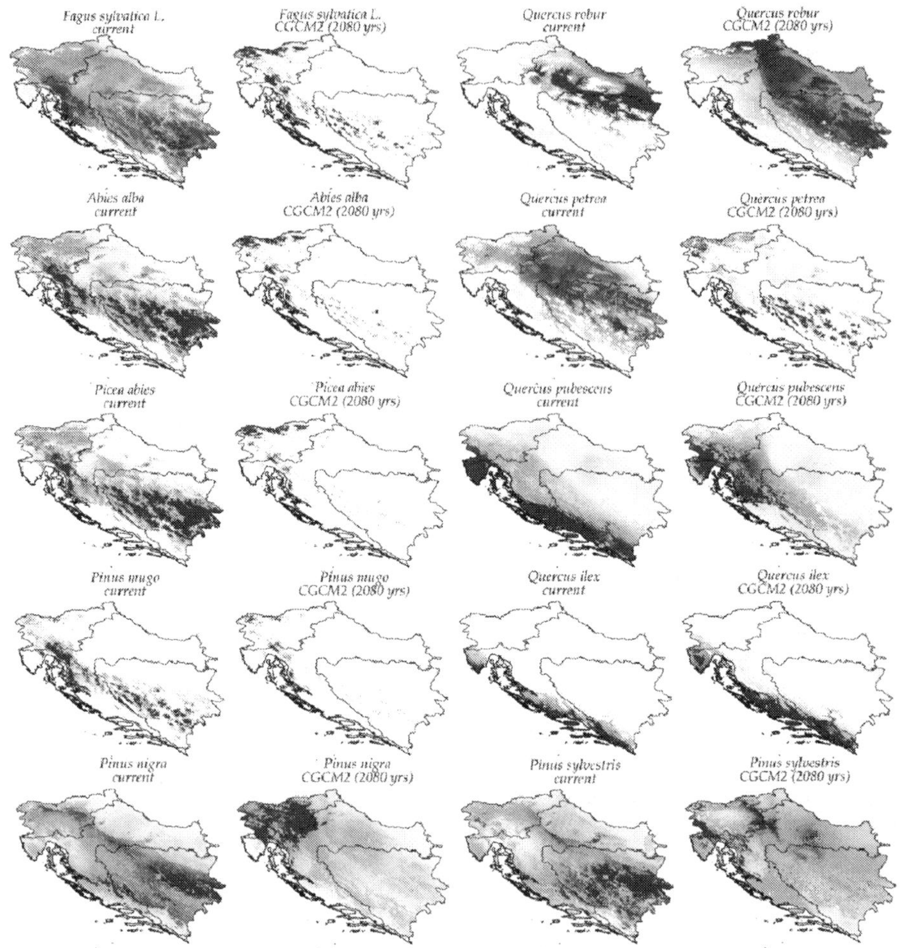

Fig. 4. Spatial distribution of ecological niche according to chosen climate factors with average values for period 1950-2000. and prediction according global climate change model CGCM2 for year 2080. Purple color presents higher (closer 1) prediction probability, while light green presents lower (closer 0), white color presents area out of ecological niche for certain tree species.

Results of spatial distribution ecological niche according to chosen climate factors and projection of the same are shown in Figure 4. Species whose ecological niche overlaps are common beech (*Fagus sylvatica*), fir (*Abies alba*) and spruce (*Picea abies*). According to ecological niche spatial distribution models and their projections according to global climate change model CGCM2 for year 2080. some species show spatial redistribution of ecological niche (pedunculate oak, pubescent oak, holm oak, Austrian pine and Scots pine), while others show vertical stratification of the same (beech, fir, spruce, mountain pine). Sessile oak (*Querus petrea*) ecological niche projection according to climate change model shows horizontal redistribution with vertical ecological niche stratification.

4. Discussion and conclusions

Global climate changes, regardless to the origin (natural condition or anthropogenic effect) are actual appearance in Earth. Models predicting global climate changes contain large entropy because they do not include all those factors and their mutual interaction that have direct or indirect effect on climate.

According to Hays (1976) global climate changes are effected by regular cycles (so called Milanković cycles) in Earth orbit which appear in almost regular amplitudes at 100000 years. Due to it is observed shift of cold and warm periods in Earth history. Warm periods (interglaciations) repeat every 100000 years, and last approximately 10000 years (Berger, 1981). Precisely shift of warm and cold periods in Earth history significantly effects on composition and quantity of vegetation on Earth (Willis et al., 1999).

Due to global changes some tree species show regressive changes (decrease of spatial distribution) of ecological niche, while others show positive direction of changes observed as increase of ecological niche spatial distribution.

Climate factors, such as maximum temperature of the hottest month and minimum temperature of the coldest month, as well as precipitation of the driest and the wettest month represent limiting values of climate factors that effect on accrurence of species. Results of the paleobotanic researches point out significant suppression of one tree species when others are spreading. According to research of Reille (1995), common beech (*Fagus sylvatica*) has during Holocen suppressed common hornbeam (*Carpinus betulus*) dominating species of the temperate zone European forests in Emiana period (130000 years ago) when forest vegetation was distributed up to toadys boreal areas with dominating tundra vegetation. Also, according to West (1980) in Pastonian period (early to middle Pelistocen) species *Picea omorika* was distributed up to today's south of England, and today species population has extremely relict character. When talking about relation between vegetation and climate it should be considered that vegetation indirectly effects and changes climate conditions in certain area, through processes of assimilation and storaging atmosphere CO_2 (Laubhann et al., 2008). Therefore changes of climate that is climate factors cause changes of certain part species ecological niche. Whereat it is difficult to distinguish and completely explain only effect of climate factors on species ecological niche and by that spatial distribution of the same. Example for that is species pedunculate oak (*Quercus robur*) for which accurence especial importance has underground and flood water.

But total pedunculate oak (*Quercus robur*) population from all other species in this research has shown the narrowest range of values according to used climate factors. According to average maximum temperatures of the hottest month pedunculate oak accrue in range at only 2,0 °C that is 2,3 °C for average minimum temperature of the coldest month (climate data in period 1950-2000). In contrary to pedunculate oak, common beech (*Fagus sylvatica*) shows much higher range of average maximum temperatures of the hottest month at 13,6 °C and 8,9 °C for average minimum temperature of the coldest month. Exactly this kind of relations point out wideness of certain tree species ecological niche, so that according to extreme temperature values, common beech (*Fagus sylvatica*) is species of wide ecological range. Alongside common beech (*Fagus sylvatica*) species of wide ecological range according to average maximum temperatures are spruce (*Picea abies*), fir (*Abies alba*), sessile oak (*Quercus petrea*), pubescent oak (*Quercus pubescens*), Austrian pine (*Pinus nigra*) and scots pine (*Pinus sylvestris*). According to the average temperatures of the coldest month value range is narrower than by average temperature of the hottest month. Considering precipitation quantities, pedunculate oak (*Quercus robur*) also shows the narrowest range towards precipitation quantities of the driest and the wettest month from 28 mm to 37 mm. The widest precipitation quantity range in the driest month shows common beech (*Fagus sylvatica*) at 53 mm and pubescent oak (*Quercus pubescens*) at 59 mm. The highest population range considering precipitation quantity in the wettest month shows holm oak (*Quercus ilex*) at 99 mm arriving in the most exothermic conditions in contrary to all other species. Exactly statement about exothermion of species should be observed from the aspect of precipitation distribution all over the year, and not only in average values.

According to global climate change model average maximum temperatures of the hottest month will increase in average for 6,9 °C, and average temperature of the coldest month for 2,0 °C. Increase of the average annual air temperature in territory of Republic Croatia that in 20 century had value from +0,02 °C in 10 years in Gospić to +0,07 °C in 10 years in Zagreb, has continued and strengthened in begging of the 21 century. From 2004 decade trends have been between 0,04 °C to 0,08 °C, and till 2008 between 0,05 °C up to 0,10 °C. Positive trend, present on all territory of Croatia, from beginning of analysed period became especially expressed in last 50 years and even more in last 25 years. Trends of average annual air temperature in 108-year period statistically are significant for all stations besides Osijek, and in last 50 that is 25 years for all analyzed stations. Trend of the annual precipitation quantities shows decrease during 20 century on all territory of Croatia analogous to tendency of dryness in Mediterranean. Decrease is more evident in Adria (Crikvenica:-1,8 % in 10 years, statistically significant and Hvar: -1,2% in 10 year), than in upper land (mountain region-Gospić:-0,8 °C in 10 years, east Slavonija, Osijek:-1,3 % in 10 years, northwest Croatia, Zagreb-Grič:-0,3 % in 10 years). Observed increasing trend of dry days in Croatia raises question about frequency successive dry days. Variation of dry periods are determined analysing data from period 1961-2000 from 25 meteorological stations that evenly cover main climate zones in Croatia (continental, mountain and maritime).

Dominant increase of dry periods in Adria and low expressed trend in continental area contributes that Croatia stays in transitional area between general tendency of precipitation increase in north Europe and decrease in Mediterranean (Fifth national report Republic of Croatia according to Framework convention UN on climate change; UNFCCC, 2009).

Range of temperature values and precipitation quantities for certain species points out importance of climate factors on species arrival. Researches of climate factors effect on certain species ecological niche give essential knowledge's about relation of vegetation, on species level, towards environment especially climate. Former understanding of distinctly high importance of different water forms (underground, flooded and stagnating) on pedunculate oak accruence should be considered in synergy with climate factors, especially temperature.

Forest ecosystems in present time are subjected to changes, whether that they are caused by natural variability or by human activity. Consideration of total synergy all ecological factors on accruence certain tree species represents one of the main segments in researching and understanding how forest vegetation functions. Prognoses of climate change models should be taken with certain precaution and with different possible outcomes. Ecological niche modelling and understanding relations prevail between vegetation and ecological factors, especially climate, should be observed on level of entire species population. However, it should be considered possible spatial parts of the population that during evolution have appated to local climate conditions and their ecological niche differ from other population parts (ecotypes). In that case certainly that future researches relations between climate and vegetation have to include and genetic variability within same species.

Further researches in ecological niche modelling for wood species find application in vegetation and forest ecosystems charting and in analysing interaction between vegetation and ecological factors. More qualitative spatial models in future should be developed and improved with proportion of species in total mixture as depended variable at ecological factors and detail pale botanic researches.

Assumed climate changes can lead to changes in spatial distribution of forest vegetation manifested in abundance of current forest types, possible decay of existing or appearance of new types, changing abundance of certain tree species populations, productivity of forest ecosystems, ecological stability and forest health status as well as in changing total productive and forest function of general benefit.

5. References

Hijmans, R.J.; S.E. Cameron, J.L. Parra, P.G. Jones and A. Jarvis, (2005). Very high resolution interpolated climate surfaces for global land areas. *International Journal of Climatology,* 25: 1965-1978.

Hutchinson, G.E. (1957) The multivariate niche. Cold Spr. Harb. Symp. *Quant. Biol.* 22, 415–421.

Willis, K. J., A. Kleczkowski, S. J. Crowhurst, 1999: 124,000-year periodicity in terrestrial vegetation change during the late Pliocene epoch. *Nature* 397: 685 - 688.

Willis, K. J., 1994: The Vegetational History oh The Balkans. *Quaternary Science Reviews,* 13: 769 - 788.

Dansgaard, W., H. B. Clausen, N. Gundestrup, C. U. Hammer, S. F. Johnsen, P. M. Kristinsdottir, N. Reeh, 1982: A New Greenland Deep Ice Core. *Science,* 218: 1273 - 1277.

Wick, L., W. Tinner, 1997: Vegetation changes and timberline fluctuations in the central airs as indicators of Holocene climatic oscillations. *Arctic and Alpine Research,* 29: 445 - 458.

Magnya, M., C. Bégeot, J. Guiot, O. Peyron, 2003: Contrasting patterns of hydrological changes in Europe in response to Holocene climate cooling phases. *Quaternary Science Reviews,* 22: 1589–1596.

Davis, B. A. S., S. Brewerb, A. C. Stevensona, J. Guiotc, Data Contributors, 2003: The temperature of Europe during the Holocene reconstructed from pollen data. *Quaternary Science Reviews,* 22: 1701 - 1716.

Högberg, P., 2007: Enviromental science: Nitrogen imapcts on forest carbon. *Nature,* 447: 781 - 782.

Hasselmann, K., 1997: Climate-change research after Kyoto. *Nature,* 390, 225 - 226.

Loutre, M. F., 2003: Clues from MIS 11 to predict the future climate: a modeling point of view. Earth and Planetary Science Letters. *Earth Planetary Science Letter* 212: 213 - 224.

West, R. G., 1980: Pleistocene forest history in East Anglia. *New Phytologist* 85: 571 - 622.

Phillips, S. J.,M. Dudík, R. E. Schapire. 2004: A maximum entropy approach to species distribution modeling. In *Proceedings of the Twenty-First International Conference on Machine Learning*, pages 655-662.

Phillips, S. J., R. P. Anderson, R. E. Schapire. 2006: Maximum entropy modeling of species geographic distributions. *Ecological Modelling*, 190: 231-259.

Hays, J., J. Imbrie, N. Shackleton, 1976: Variations in the Earth's Orbit: Pacemaker of the Ice Ages. Science 194: 1121.

Berger, A. L., 1981: The astronomical theory of paleoclimates. Climatic variations: facts and theories (ur. A. L. Berger), str. 501 - 525, Reidel, Dordrecht.

Reille, M., J. L. de Beaulieu, 1995: Pollen analysis of a long upper Pleistocene continental sequence in a Velay maar (Massif Central, France). *Quaternary Research*, 44: 205 - 215.

Laubhann, D., H. Sterba, G. J. Reinds, D. W. Vries, 2008: The impact of atmospheris and climate on growth in Europe monitoring plots: An individual growth model. *Forest Ecology and Management*. In Press.

Gent, P.R. and J.C. McWilliams, 1990: Isopycnal Mixing in Ocean Circulation Models. *J. Phys. Oceanogr.*, 20, 150-155.

Flato, G.M. and Hibler, W.D. III, 1992: Modelling Pack Ice as a Cavitating Fluid. *J. Phys. Oceanogr.*, 22, 626-651.

Kirigin, B., 1975: Kolebanja klimatskih elemenata i sušenje jele na području SR Hrvatske. Radovi, 23: 16-27.

Drugo, treće i četvrto nacionalno izvješće Republike Hrvatske prema Okvirnoj konvenciji Ujedinjenih naroda o promjeni klime (UNFCCC). Ministarstvo zaštite okoliša, prostornog uređenja i graditeljstva, Zagreb. Studeni 2006, str. 96.

13

Climate change impact on vegetation: lessons from an exceptionally hot and dry decade in south-eastern France

Vennetier Michel (1-2)* and Ripert Christian (1)
(1) Cemagref, Aix en Provence, CS 40061, 13182 Aix en Provence Cedex 5 France
(2) ECCOREV, FR 3098, Aix-Marseille Université, BP 80 F-13545 Aix en Provence cedex France

1. Introduction

For the 21st century, all climatic models predict in the Mediterranean basin a faster warming than in most other continental areas of the world, associated with a reduction of rainfall during the growth season (Hesselbjerg-Christiansen & Hewitson, 2007). As warming is likely to be larger in summer, extreme climatic events such the 2003 scorching heat (Ciais et al. 2005; Zaitchik et al. 2006) are prone to be recurrent. Drought being already the main limiting factor for Mediterranean vegetation (Le Houerou, 2005), many species should be at risk with repeated critical water stress during the growth season (Breda et al., 2006).

According to several studies, (Hughes, 2000; Lenoir et al., 2008) the track race between climate change and vegetation is already launched. Many species looking for suitable habitats move towards the poles or upwards in elevation (Walther et al., 2005). However, mean plant dissemination distance is short (Clark et al., 1999). Certain plants may be unable to follow the edges of their potential distribution area, as fast species spread recorded at the end of ice ages (Delacourt & Delacourt, 1987) are slower than the expected limit shift in the 21st century (Thuiller, 2004). Species shift should be checked by biotic interactions (Preston et al., 2008) and competition. Time lags in plant phonology (Menzel & Fabian, 1999) could make them more vulnerable to meteorological extreme events (Morin et al., 2007). Altered architectural development and sexual reproduction (Hedhly et al., 2009 and Thabeet et al., 2009), may also hamper their growth and dissemination. Mediterranean small mountains offer to Alpine or middle-European vegetation fragmented but suitable relict niches mainly near their top (*figure 1*). Inherited from former climate conditions, mixing several biomes in small areas, these niches are biologically very rich (Médail & Quézel, 1999) with a high level of endemism. But future climate warming raises their potential trailing edge over local summits (Trivedi et al., 2008b). In the absence of functional corridors, current reserve networks may be inadequate to ensure the long-term persistence of these species (Araujo et al., 2004). However, on local scale, site conditions including deep soils and steep northern slopes at the highest elevations may create refuges. Such a precise assessment of favourable sites is not easy with existing models. Although many types of models were used to assess the evolution in plant composition with climate change, computing potential distributions, bioclimatic limits or niches (Botkin et al., 2007; Hansen et al., 2001), for individual species (Gaucherel et al., 2008; Heikkinen et al., 2006) or species groups, very few of them tackled local scales (Trivedi et al., 2008b). This is why we recently developed a new bioclimatic model, based on a flora census, taking into account both local and global variables (Table 1) (Vennetier et al., 2008) in order to bridge regional to local scales. One of the possible uses of this model is to assess the potential flora composition turnover with different simulated climate scenarios (Vennetier et al., 2009).

Flora composition is often considered as a good indicator of site conditions, including site and climate parameters (Berges et al., 2006). A hotter and drier climate should lead to a significant flora turnover biased towards heat and drought tolerant plants. However, if the speed of an altitudinal or latitudinal species shift was often assessed in literature, the turnover was rarely documented on local scale. During the decade 1998-2008, south-eastern France experienced an anticipated occurrence of what should be the climate around 2040 according to IPCC B2 or A1B scenarios (Christensen & Christensen, 2007). It was interesting to assess whether these exceptional conditions were reflected in flora composition.

The aims of this study were (i) to measure plant composition evolution in a permanent plot network between 1998 and 2008, (ii) to compare the observed evolution with the potential turnover computed with our model (iii) to disentangle the relationships between observed changes and local site conditions.

2. Material and method

2.1 Study area and sampling

The study area is situated in the French Mediterranean area (*figure 1;* long 4°5' - 6°2' E, lat 43°4', 43°5' N). The climate is characterized by a long summer drought (2-4 months) and mild rainy winters. The mean annual temperature and rainfall range respectively from 15.3°C / 500 mm on the Southwestern coast to 9.5°C / 1000 mm on the highest ridges (around 1100 m), with an average of 13.2°C / 720 mm. *Pinus halepensis* Mill. and *Quercus ilex* L. are the main forest tree species along the coast, at low elevation and on shallow soils, while *Quercus pubescens* Will. is all the more abundant as elevation, continentality and soil depth increase.

The sampling plan was design to be representative of the span of local and global site conditions, crossing the main ecological gradients: soil quality, topography, orientation, climate, continentality (*Table 1*). In order to minimize the role of disturbances in vegetation response, we selected only sites with no registered disturbing activity such as logging, grazing, clearing, or prescribed fire over at least the last 30 years. In most of these sites, dominant trees were more than 70 years old. Initially, 325 forest plots (400m^2 each) were surveyed between 1996 and 1998. A thorough description and measure of site conditions was performed, along with a flora census using Braun-Banquet abundance-dominance scale (Braun-banquet 1932). The flora census was done again in 2008 on a representative sub-sample of 50 plots.

2.2 Model bases

The main output of the model, previously presented by Vennetier et al. (2008), is a bioclimatic index. This index combines two components: the first one is based on variables which can be mapped by GIS from local to regional scale (climate, continentality and orientation); the second one is based on variables which can be precisely observed only on site scale (soil, local topography). In this section, only the bases of the model which are useful for this study are explained. Statistical procedures are described in *annex 1*.

The model was designed in two steps.

The first step was a correspondence analysis (CA) on plant composition with the 325 plots, keeping 192 species present in at least 3 plots. *Figure 2a* shows a synthesis of the main CA plane. When displayed as supplementary variables in this plane, all variables relevant in terms of water balance or temperature, and which could be grouped in four main gradients, were linked to axis 1. All their classes related to unfavourable conditions (low water availability, high temperatures) were found in the left half of the plane and favourable classes in the right half. Considering its dominance (eigenvalue twice the one of second axis), this first axis was retained alone for modelling and considered as a synthetic bioclimatic gradient.

When displayed in the same plane, plant species are sorted along axis 1 in the synthetic gradient. According to their coordinate on this axis, they can be split into five groups of equal number from the left (the most heat and drought tolerant) to the right (water demanding ones) (*figure 2.b*): super xero-thermophilous (sXT), xero-thermophilous (XT), intermediate (Int), slightly mesophilous (Meso) and mesophilous (Meso+). Plots can be displayed in the main CA map. As their position is only determined by their flora composition (plots are at the barycentre of their plants), we considered their coordinate on axis 1 as a Flora index (Fi), sorting them too along the bioclimatic gradient.

The second step consisted in computing a bioclimatic index (Bi) for each plot. Bi is the estimate of Fi with a Partial Least Square regression model using abiotic explanatory variables describing site conditions. *Table 1* presents the height global and six local relevant variables. The model explained 81% of Fi variance. Thanks to the good fit between Fi and Bi, and to the key role played by climate variables in the model, the impact of climate change on plant composition can be assessed as described in paragraph 2.3 below.

<table>
<tr><th></th><th>Coef *</th><th>Variable description</th><th>Gradient in fig 2a</th></tr>
<tr><td rowspan="8">Global variables</td><td>-0.183</td><td>Becker light-climate index (relative received solar energy in % of an horizontal reference plane)</td><td>2</td></tr>
<tr><td>-0.153</td><td>Mean annual temperature</td><td>1</td></tr>
<tr><td>0.131</td><td>Altitude</td><td>1</td></tr>
<tr><td>0.115</td><td>Summer rainfall (cumulated from June to August)</td><td>1</td></tr>
<tr><td>0.082</td><td>Annual rainfall excluding summer, or spring rainfall</td><td>1</td></tr>
<tr><td>0.169</td><td rowspan="2">Maximum altitude between a site and the coastline in two directions: the closest coast line and 247°</td><td rowspan="2">1</td></tr>
<tr><td>0.146</td></tr>
<tr><td>0.106</td><td>Distance to the sea</td><td>1</td></tr>
<tr><td rowspan="2">Local variables</td><td>-0.136</td><td>General topography on landscape and slope scale (5 classes scale)</td><td>4</td></tr>
<tr><td>-0.107</td><td>Topography on local scale (plot size) (5 classes scale)</td><td>4</td></tr>
</table>

	-0.083	Percentage of parent rock outcrops on the plot	4
	0.100	Water holding capacity of earth (mm/cm) based on soil texture	3
	-0.091	Percentage of coarse fragments in the soil	4
	0.119	Total soil depth	4

Table 1. variables describing site conditions and used for the model

* Coef. = Partial regression coefficient in the PLS regression model. All theses coefficients are highly significant (P<0.001). They sort variable by their relative weight in the model.
The initial CA, using the flora census of 1996-98, was computed with the climate of the period 1961-1996.

2.3 Simulation of climate change impact on plant turnover

The simulation used the two steps of the model:
First, we considered that an increasing mean annual temperature or decreasing spring or summer rainfall would favour xero-thermophilous species, and stress mesophilous ones. We simulated this impact (*Figure 3*), changing at plant distribution margins of each plot the composition (*3b*) or the abundance-dominance (*3c*).
Using these modified plots as supplementary observations in the initial CA, we computed the new coordinates of each plot on axis 1 (*Figure 3*), which means new and smaller Fi indices (modified plots shifted to the left in the CA plane).
In the second step, we simulated various climate change scenarios, with lessening summer or spring rainfall (-10 to -30%) and raising temperatures (1 to 2°C). We also used in the model the mean climate observed over respectively the last 10, 20 and 30 years (1978-2007). New and smaller Bi indices for each plot were computed.
The good correlation between Fi and Bi indices enabled linking their respective shift with the flora and climate change simulations, and therefore assessing the plant composition turnover expected for each climate scenario. We did not consider the future of individual or specific plants, but the global turnover as a percentage of plant composition.

2.4 Validation with the new flora census

This census was performed in spring 2008 by the same team using exactly the same protocols. We analysed the flora turnover as the changes in plant presence and abundance-dominance, on one hand with all plants separately, and on the second hand considering the five groups of grading heat and drought tolerance designed in *figure 2a*.

3. Results

3.1 Number of plants per plot

The average number of plants was 25 (7-53, SD= 8) in the 325 initial plots, and 27 (15-48, SD=7.5) in the 50 plots surveyed twice. This average number was unchanged in the last 10 years although its mean absolute variation was 18% (4 - 5 plants), gains in some plots compensating losses in others. The number of plants per plot increased regularly with plot If ($r^2 = 0.22$, $p<0.05$) ranging from 22 for If lowest fourth to 32 for If highest fourth: as a whole, mesophilous sites were richer than drier sites. However, percentage variations in the number of plants per plot did not depend on If.

3.2 Simulation of climate and plant composition changes

Table 2 sorts the If shift for different simulations of climate change and plant composition change. This relation is given as an average for all plots. For a given number of plants changing at the edge of the distribution margins, the richer the plot and the smaller its span, the less sensible it is to these changes. For example, the change of +2 and -2 plants gave an average shift of 0.181 (SD = 0.063) with maximum / minimum values of respectively 0.503 for a plot with 10 species and a wide distribution and 0.088 for a plot with 50 species and a narrow distribution. As the number of plants affected by climate change may be proportional to their total number, we used percentages of flora composition to express changes, making the results independent of plot specific flora.

A reduction of 20% of spring or summer rainfall corresponded to a change of 4 to 5% in flora composition. Each added °C increased by 7 to 8% flora turnover. The mean climate of the last 10 years led to a potential change of one fourth of plant composition.

For a given plot, the mean shift of If observed with the decrease / increase of one point for the Braun-Blanquet coefficient of a plant was 60% of the shift obtained with the suppression of this plant or the addition of the closer plant.

Simulations		% of flora composition	If shift on axis 1 (mean all plots)
Climate change	Flora change		
-10% SpR or SumR			-0.030
	(1) -1 or +1 plant	4%	-0.050
-20% SpR or SumR			-0.052
+1°C *			-0.074
+1°C -10% SpR or SumR			-0.105
	(2) -1 and + 1 plants	8%	-0.106
	Abond-Domin + 1 or -1 for 4 plants		*-0.111*
+2°C *			-0.143
	Abond-Domin +1 or-1 for 6 plants		*-0.154*
	(4) -2 and + 2 plants	16%	-0.181
+2°C -10%SpR and -10%SumR			-0.192
+2°C -20%Spr or SumR			-0.193
+1.4°C -15% SpR and -32% SumR			-0.264
	(6) -3 and + 3 plants	24%	-0.269

Table 2. Fitting flora turnover and climate change impacts on the shift of plot If. The last example of climate change corresponds to the mean climate of the last 10 years (1998-2008) compared to the reference 1961-96.

** Temperature increase is for mean annual temperature.*

SumR = Summer rainfall, SpR = spring rainfall.

3.3 Comparison with the observed shift in plant composition

The observed shift did not occur at random but was biased towards axis 1 bioclimatic gradient. *Figure 4* shows the spatial evolution of flora sorted by plant groups.

In the sXT group, nearly half of the plants were winners, two times more than losers. Conversely in the Meso+ and Meso groups, half of the plants were losers, respectively five times and two times more than winners. In the intermediate and XT groups, the turnover was balanced, losers and winners being as numerous, which means that disappearing species were replaced by plants of the same group. 30% percent of the plants remained spatially stable in the two extreme groups, but only 10 to 15% in other groups. As a whole, mean plant occurrence was unchanged (+0.1), so that disappearing mesophilous plants were replaced by xero-thermophilous ones. But 74 plants lost occurrences for only 55 gaining ground, which means that only a minority of plants took advantage of the new conditions, even within XT groups.

These percentages of spatial turnover may hide very different situations, as a plant was considered as spatially winner or loser whatever the rate of change. This is why the trend on flora composition towards xero-thermophilous species was more precisely assessed with the sums of occurrences and of Braun-Blanquet coefficients within groups (*Figure 5).*

The sXT and XT groups together won 73 occurrences, corresponding to 1.5 plants per plot. But their sum of Braun-Blanquet coefficients won 122 points, nearly 2.5 points per plot. It means that not only new xero-thermophilous plants appeared in plots but also that some plants of these groups, already present in 1996-98, increased in dominance and cover. Meso+ and Meso plants lost together 68 occurrences and a bit more (86) in the sum of Braun-Blanquet coefficients, which means that, for these two groups, losses in cover percentage were mainly due to losses in occurrence. Globally the occurrence turnover between xero-thermophyllous and mesophilous plants reached 3 plants, equivalent to 11.5 % in 10 years. Variations in the sum of Braun-Blanquet coefficients which are not explained by occurrence variations must be taken into account. When their equivalent in occurrences was added (0.6 occurrence for 1 points of coefficients), the total observed turnover excluding the intermediate group was close to 15 %.

This turnover between plant groups was not similar whatever the site. *Figure 6* shows that a biased turnover occurred mainly in mesophilous and intermediate sites, very dry and hot sites remaining more or less stable.

3.4 Mapping Bi index on various scales

Bi index computed with global variables was mapped with a GIS software after being split into nine classes (*figure 7.a*). Each class included 1/9 of Bi total variation interval after exclusion of the 5% extreme values (2.5% at each end). These extreme values were merged with the first and last classes respectively. Range limits of various species match some of these classes. For example the darkest green (class 9) corresponds to the observed niche of Scots pine and related alpine and meso-European relict species, and excludes *Pinus halepensis* because of deep frost in winter. On average soil conditions and with a neutral topography, class 3 (orange) is the extreme limit of *Quercus pubescens*.

4. Discussion

4.1 Validating and quantifying vegetation turnover

A 5 to 10% change in plant composition in the same plot between two following censuses, or with different survey methods, is common without any environmental change (Archaux et al., 2006; 2007). But in such a case this change is not related to a specific gradient, as plants within a given group replace each other. In this study, the most mesophilous and xero-thermophilous plants (Meso+ and sXT groups) were more concerned, respectively loosing and gaining more than intermediate ones. Meso and XT groups also showed a significantly unbalanced turnover.

The real turnover did not concern only the species at the very limit of species distribution for each plot. More distributed variations along the gradient give a lower If shift on axis one, but the same trend as far as species appear and disappear symmetrically. It may explain why the observed turnover was weaker than the simulated one. Although observed patterns of flora turnover were more complex than the simulated ones, they validate the method used for the simulation.

The observed trend is clearly biased towards heat and drought resistant plants. It is opposed to what is expected in aging forest structures. Indeed, in Southern Europe the Mediterranean ecosystems were transformed by several thousand years of disturbances and development (Blondel and Aronson, 1999). The natural evolution of the unmanaged stands of this study, most of them adult but far from senescence, should be a maturation process leading to an increasing dominance of mesophilous and shade tolerant plants and the reduction of light demanding, generally xero-thermophilous species, inherited from past land uses (Tatoni and Roche, 1994). Probably, the adverse climate conditions in the last decade also contributed indirectly to a hotter and drier microclimate in the undergrowth, limiting canopy density though tree mortality (Allen et al., 2009), low branching rates and reduced leaf area (Thabeet et al., 2009).

The potential turnover (25%) simulated on plot scale in this study with the climate expected in the middle of the 21st century hold with previous studies on larger scales: Bakkenes et al (2002) showed that 32% of the European plant species that are present in a grid cell of a few square kilometres in 1990 should disappear from that cell before 2050. High rates of potential extinction among endemic species (average 11%, up to 43%) were forecasted by Malcolm et al. (2006) for the whole Mediterranean basin and other biodiversity hotspots in the world by 2100.

4.2 Flora resistance on landscape and local scales

Plant composition turnover observed in the last decade was significant but not as considerable as simulated by the model. A resistance to climate variations was observed, which may be partly explained by landscape structure. Bi index mapped at any scale is laid out like a patchwork of fragmented bioclimatic classes. When topography and soil are added on local scale, six among the nine classes represented on regional scales with medium site conditions can be found on a single square kilometre of hilly landscape with steep slopes and only one or two hundred meters of difference in elevation. Thanks to that fine grain mosaic, xero-thermophilous plants are scattered everywhere even at high elevation, taking advantage of steep south-facing slopes, shallow and rocky soils. Most of the time, some of them simply remain from degraded ecosystems inherited form former land uses

and fires. They are ready to sprawl from these positions in presently cold and wet areas when mesophilous species become less competitive because of climate change.
Mesophilous species are supposed survive, in dry and hot areas, in scattered niches combining cool expositions, deep soils and favourable topography. Even when killed by extreme climate events, they may come back by seed dispersal from refuges and long lifespan soil seed-banks (Zobel et al., 2007). However, such a hypothesis is not fully supported by our results. The best soil and topographic conditions proved to be no longer sufficient to compensate the climatic water stress in recent years, leading to the extensive dieback of mesophilous species in good sites. As an example among forest trees and representatives of middle-European relict species, Pinus silvestris L., although limited to the highest elevations and north slopes, paid an heavy tall to 2003 scorching heat and following drought (Thabeet et al., 2009). In the future, the increase in water stress should lead this compensation limit to shift left along axis one, and the turnover to start even in the driest and hottest sites of the study area.

4.3 Management issues

Several reserves try to protect Alpine or middle-European flora in Mediterranean mountains. But protected areas are generally fragmented and scattered in developed lands. Corridors are lacking to allow species to move from one to another. This is why it would be necessary to include public and private multi-use lands in conservation practices (Heller & Zavaleta 2009) and to look for the parts of these human-dominated landscapes that should be suitable for forest species. Bioclimatic maps like *figure 7*, if regularly updated, could help designing such strategies.
However, many of the small isolated areas of relict vegetation should be kicked out by the top. According to Trivedi et al. (2008a) 70 to 80% of plant species in similar low mountain conditions should loose most of their potential niche in the next decades, as shown in figure 7.b for Pinus silvestris and associated flora. Moreover, large scale approaches may underestimate this potential climate-based disappearance by far. Extreme weather events may cause larger gaps in already scattered populations (Opdam and Wascher, 2004) and drive them below the critical level of metapopulation persistence. So that *ex-situ* conservation of endemic and rare species or genetic resources should be urgently implemented. For example, *Pinus silvestris* in Provence at low elevation proved to be far more resistant to drought than most other origins of this species. This resource could be useful in the future at higher latitudes, and for genetic improvement programs. But most of the concerned stands were killed between 2003 and 2007 and the last ones should soon be lost.

5. Conclusion

Combined with exceptionally high temperatures, repeated droughts between 1998 and 2008 severely impacted vegetation in south-eastern France. This unexpected experience allowed assessing, three decades sooner, the consequences of the climate forecasted as normal in the future.
We showed that a rapid turnover occurred in ten years on site scale. It was surprisingly faster in the most favourable sites, were mesophilous species could not survive. It may concern all the study area in the future. Accordingly, the simulations with a bioclimatic

model forecasted even more extensive changes with the climate of the last decade. Whatever the quality of models, the long term follow-up of permanent plots is irreplaceable to understand and measure the impact of climate change.

As climate change may accelerate in the future, conservation policies for rare and endangered species, and more generally conservation policies based on fixed reserve networks, should be reconsidered. It is particularly relevant in low mountains were the trailing edge of ecological niches for these species should soon reach the highest ridges.

Landscape structure at a scale fitting with a detailed assessment of topographic and soil variables allows an operational assessment of the change in plant composition and the shift in plant future distribution. Taking into account seed dissemination distances may improve such an assessment, as the scales are of the same order. Many parameters including the real species migration capacity, population dynamics, biotic interactions and community ecology should be included in models (Brooker et al., 2007; Guisan & Thuiller, 2005) to improve the spatial assessment of plant migration in landscapes.

Therefore, such assessments on an operational scale should be multiplied in the main ecosystems and regions, so that large scale approaches could be corrected and better interpreted. Small scales approaches, tailored to specific needs, also enhance local knowledge and encourage dissemination and decision making at operational forest management level. Ecological as well as economical issues are at stake.

Acknowledgements

The initial flora census (1996-98) and model design was funded by the French Ministry for Agriculture and Fisheries. The second census was funded by the French National Research Agency (DROUGHT+ project, N° ANR-06-VULN-003-04). Several students and technicians contributed to field work particularly Jean Stéphane, Estève Roland, Chandioux Olivier and Martin Willy.

6. References

Allen, C.D., Macalady, A.K., Chenchouni, H., Bachelet, D., Mcdowell, N., Vennetier, M., Kitzberger, T., Rigling, A., Breshears, D.D., Hogg, E.H., Gonzalez, P., Fensham, R., Zhang, Z., Castro, J., Demidova, N., Lim, J.-H., Allard, G., Running, S.W., Semerci, A., Cobb, N., 2009. A Global Overview of Drought and Heat-Induced Tree Mortality Reveals Emerging Climate Change Risks for Forests, *Forest Ecology and Management* 259 (4) 660–684

Amato, S., Vinzi, V.E., 2003. Bootstrap-based Q^2kh for the selection of components and variables in PLS regression, *Chemometrics and intelligent Laboratory Systems*, (68): 5-16.

Araujo, M.B., Cabeza, M., Thuiller, W., Hannah, L., Williams, P.H., 2004. Would climate change drive species out of reserves? An assessment of existing reserve-selection methods, *Global Change Biology*, (10) 9: 1618-1626.

Archaux, F., Berges, L., Chevalier, R., 2007. Are plant censuses carried out on small quadrats more reliable than on larger ones?, *Plant Ecology*, 188: 179-190.

Archaux, F., Gosselin, F., Berges, L., Chevalier, R., 2006. Effects of sampling time, species richness and observer on the exhaustiveness of plant censuses, *Journal of Vegetation Science*, (17): 299-306.

Bakkenes M. ; Alkemade J.R.M. ; Ihle F. ; Leemans R. ; Latour J.B. (2002) Assessing effects of forecasted climate change on the diversity and distribution of European higher plants for 2050. *Global Change Biology*, vol. 8, n° 4, p. 390-407.

Berges, L., Gegout, J.C., Franc, A., 2006. Can understory vegetation accurately predict site index? A comparative study using floristic and abiotic indices in sessile oak (Quercus petraea Liebl.) stands in northern France, *Annals of Forest Science*, (63) 1: 31-42.

Blondel, J., Aronson, J., 1999. *Biology and wildlife of the Mediterranean region*, Oxford University Press, Oxford.

Botkin, D.B., Saxe, H., Araujo, M.B., Betts, R., Bradshaw, R.H.W., Cedhagen, T., Chesson, P., Dawson, T.P., Etterson, J.R., Faith, D.P., Ferrier, S., Guisan, A., Hansen, A.S., Hilbert, D.W., Loehle, C., Margules, C., New, M., Sobel, M.J., Stockwell, D.R.B., 2007. Forecasting the effects of global warming on biodiversity, *Biosciences*, (57) 3: 227-236.

Braun-Blanquet J. (1932) *Plant sociology. The study of plant communities*. McGraw-Hill, New-York.

Breda, N., Huc, R., Granier, A., Dreyer, E., 2006. Temperate forest trees and stands under severe drought: a review of ecophysiological responses, adaptation processes and long-term consequences, *Annals of Forest Science*, (63) 6: 625-644.

Brooker, R.W., Travis, J.M.J., Clark, E.J., Dytham, C., 2007. Modelling species' range shifts in a changing climate: The impacts of biotic interactions, dispersal distance and the rate of climate change, *Journal of Theoretical Biology*, (245) 1: 59-65.

Christensen J H and Christensen O H 2007 A summary of the PRUDENCE model projections of changes in European climate by the end of this century, *Climate Change* 81: 7-30

Ciais, P., Reichstein, M., Viovy, N., Granier, A., Ogee, J., Allard, V., Aubinet, M., Buchmann, N., Bernhofer, C., Carrara, A., Chevallier, F., De Noblet, N., Friend, A.D., Friedlingstein, P., Grunwald, T., Heinesch, B., Keronen, P., Knohl, A., Krinner, G., Loustau, D., Manca, G., Matteucci, G., Miglietta, F., Ourcival, J.M., Papale, D., Pilegaard, K., Rambal, S., Seufert, G., Soussana, J.F., Sanz, M.J., Schulze, E.D., Vesala, T., Valentini, R., 2005. Europe-wide reduction in primary productivity caused by the heat and drought in 2003, *Nature*, (437) 7058: 529-533.

Clark J.S. ; Silman M. ; Kern R. ; Macklin E. ; Hillerislambers J. (1999) Seed dispersal near and far: Patterns across temperate and tropical forests. *Ecology* vol. 80, n° 5, p. 1475-1494.

Delacourt, P., Delacourt, H., 1987. *Long-term Forest Dynamics of the Temperate Zone*. Springer, New York.

Emberger, L., 1930. La végétation de la région méditerranéenne : essai d'une classification des groupements végétaux, *Revue Générale de Botanique*, 42: 641-662, 705-721.

Escofier, B., Pages, J., 1994. Multiple Factor analysis, *Computational statistics and data analysis*, 18: 121-140.

Gaucherel, C., Guiot, J., Misson, L., 2008. Changes of the potential distribution area of French Mediterranean forests under global warming, *Biogeoscience*, (5) 6: 1493-1504.

Good, P., 1994. *Permutation tests*, Springer-Verlag, New-York.

Guiot, J., 1991. Methods and programs of statistics for palaeoclimatology and palaeoecology. In: *Quantification des changements climatiques. Méthodes et Programmes, Monographie 1.* Marseille: INSU, Paris, 1991, pp. 258.

Guisan, A., Thuiller, W., 2005. Predicting species distribution: offering more than simple habitat models, *Ecology Letters*, (8) 9: 993-1009.

Hansen, A.J., Neilson, R.R., Dale, V.H., Flather, C.H., Iverson, L.R., Currie, D.J., Shafer, S., Cook, R., Bartlein, P.J., 2001. Global change in forests: Responses of species, communities, and biomes, *Bioscience*, (51) 9: 765-779.

Hedhly, A., Hormaza, J.I., Herrero, M., 2009. Global warming and sexual plant reproduction, *Trends in Plant Science*, (14) 1: 30-36.

Heikkinen, R.K., Luoto, M., Araujo, M.B., Virkkala, R., Thuiller, W., Sykes, M.T., 2006. Methods and uncertainties in bioclimatic envelope modelling under climate change, *Progress in Physical Geography*, (30) 6: 751-777.

Heller, N.E., Zavaleta, E.S., 2009. Biodiversity management in the face of climate change: A review of 22 years of recommendations, *Biological Conservation*, (142) 1: 14-32.

Hesselbjerg-Christiansen, J., Hewitson, B., 2007. Regional Climate Projection. In: *IPCC (2007) Climate Change 2007: The Physical Science Basis. Contribution of Working Group I to the Fourth Assessment Report of the Intergovernmental Panel on Climate Change*. Solomon, S., D. Qin, M. Manning, Z. Chen, M. Marquis, K.B. Averyt, M. Tignor and H.L. Miller (eds)., Cambridge University Press, Cambridge, United Kingdom and New York, NY, USA. chap 11, pp 847-940.

Hughes, L., 2000. Biological consequences of global warming: is the signal already apparent?, *Trends in Ecology and Evolution*, (15) 2: 56-61.

Le Houerou, H.N., 2005. *The Isoclimatic Mediterranean Biomes: Bioclimatology, Diversity and Phytogeography*. Vol. 1 & 2, Copymania Publication, Montpellier.

Lenoir, J., Gegout, J.C., Marquet, P.A., De Ruffray, P., Brisse, H., 2008. A significant upward shift in plant species optimum elevation during the 20th century, *Science*, (320) 5884: 1768-1771.

Malcolm J.R. ; Liu C.R. ; Neilson R.P. ; Hansen L. ; Hannah L. (2006) Global warming and extinctions of endemic species from biodiversity hotspots. *Conservation Biology*, vol. 20, n° 2, p. 538-548.

Médail, F., Quézel, P., 1999. Biodiversity Hotspots in the Mediterranean Basin: Setting Global Conservation Priorities, *Conservation Biology*, (13) (6): 1510 -1513.

Menzel, A., Fabian, P., 1999. Growing season extended in Europe, *Nature*, (397) 6721: 659-659.

Morin, X., Ameglio, T., Ahas, R., Kurz-Besson, C., Lanta, V., Lebourgeois, F., Miglietta, F., Chuine, I., 2007. Variation in cold hardiness and carbohydrate concentration from dormancy induction to bud burst among provenances of three European oak species, *Tree Physiology* (27) 6: 817-25.

Opdam, P., Wascher, D., 2004. Climate change meets habitat fragmentation: linking landscape and biogeographical scale levels in research and conservation, *Biological Conservation*, (117) 3: 285-297.

Preston, K., Rotenberry, J.T., Redak, R.A., Allen, M.F., 2008. Habitat shifts of endangered species under altered climate conditions: importance of biotic interactions, *Global Change Biology*, (14) 11: 2501-2515.

R_Development_Core_Team, 2004. R: A language and environment for statistical computing. R Foundation for Statistical Computing, R Foundation for Statistical Computing.

Tatoni, T., Roche, P., 1994. Comparison of Old-Field and Forest Revegetation Dynamics in Provence, *Journal of Vegetation Science*, (5) 3: 295-302.

Thabeet, A., Vennetier, M., Gadbin-Henry, C., Denelle, N., Roux, M., Caraglio, Y., Vila, B., 2009. Response of Pinus sylvestris L. to recent climate change in the French Mediterranean region, *Trees Structure and Functions*, (28) 4: 843-853.

Thioulouse, J., Chessel, D., Doledec, S., Olivier, J.M., 1997. ADE-4: a multivariate analysis and graphical display software, Stat. Comput., 7: 75-83.

Thuiller W. (2004) Patterns and uncertainties of species' range shifts under climate change. *Global Change Biology*, vol. 10, n° 12, p. 2020-2027.

Trivedi, M.R., Berry, P.M., Morecroft, M.D., Dawson, T.P., 2008a. Spatial scale affects bioclimate model projections of climate change impacts on mountain plants, *Global Change Biology*, (14) 5: 1089-1103.

Trivedi, M.R., Morecroft, M.D., Berry, P.M., Dawson, T.P., 2008b. Potential effects of climate change on plant communities in three montane nature reserves in Scotland, UK, *Biological Conservation*, (141) 6: 1665-1675.

Vennetier, M., Ripert, C., Maillé, E., Blanc, L., Torre, F., Roche, P., Tatoni, T., Brun, J.-J., 2008. A new bioclimatic model calibrated with flora for Mediterranean forested areas, *Annals Forest Science*, (65) 711.

Vennetier M., Christian R. 2009. Forest flora turnover with climate change in the Mediterranean region: A case study in Southeastern France. *Forest Ecology and Management* 258S: 56–63

Walther, G.R., Beissner, S., Burga, C.A., 2005. Trends in the upward shift of alpine plants, *Journal of Vegetation Science*, (16) 5: 541-548.

Zaitchik B.F. ; Macalady A.K. ; Bonneau L.R. ; Smith R.B. (2006) Europe's 2003 heat wave: A satellite view of impacts and land-atmosphere feedbacks. *International Journal of Climatology*, vol. 26, n° 6, p. 743-769.

Zobel, M., Kalamees, R., Püssa, K., Roosaluste, E., Moora, M., 2007. Soil seed bank and vegetation in mixed coniferous forest stands with different disturbance regimes, *Forest Ecology and Management*, (250) 1-2: 71-76.

Annex 1: statistical procedures

We used a classical unweighted CA analysis with Braun-Blanquet coefficients to obtain the Flora index as the coordinates of plots on the first CA axis. We checked the robustness of the CA axes towards potential inaccuracies of floristic censuses due to time or spatial strategies or the observer (Archaux et al., 2007; Archaux et al., 2006) and towards analyses options. This verification was performed with a Multiple Factorial Analysis (MFA) (Escofier and Pages, 1994) testing the stability of CA axes and plots coordinates: on one hand by increasing the number of plots where a plant must be present to be taken into account from 3 to 30, on the other hand comparing presence/absence and Braun-Blanquet coefficients in the analysis.

CA axes 1 to 3 and plots coordinates proved to be particularly stable (r^2 of plot rank on axis 1 > 0.98) whatever the code used (BB coefficients or presence/absence) and up to a limit of 25 occurrences for plant selection.

Some variables seemed to have non-linear relationship with the Flora index (Fi). To optimize the PLS model, we first checked the relation pattern of all relevant variables with Fi using neural networks and transformed some of them (log, sigmoidal or polynomial). Neural networks where used combining variables 6 by 6. Each neural network was optimized with a 200 replications bootstrap, each replication including 10^4 calibration steps. After the optimization of the neural network, the response of each variable was plotted on its whole variation interval, the other variables being maintained at their mean value if they were not correlated with the tested one, or maintained successively to their first, second and third quartile for those which were correlated.

In the last case, the 3 responses where combined in a sliding weighted mean to obtain the global response. We only transformed a variable according to non-linear relation showed by neural networks when this relation was stable throughout these many tests and enhanced as well (i) the weight of the variable in the neural network optimization and (ii) its partial correlation coefficient or the total explained variance in the PLS regression.

For the choice of relevant variables in the model, we used an ascending and descending stepwise PLS regression validated at each step by a permutation test on PLS components and a cross-validation for concerned variables. All variables in the final version of the model were highly significant on the first two PLS components ($p < 0.001$). The number of significant components for PLS regression was chosen with a 10,000 replications permutation test on observations (Good, 1994), keeping components which percentage of explained variance was not passed by more than 5 % of the permutations. With significant components, variables were sorted through a 1,000 resampling cross-validation test (Amato and Vinzi, 2003); only variables which confidence interval (95%) for the partial correlation coefficient excluded 0 were used.

We used ADE4 software (Thioulouse et al., 1997) for CA and related operations (introducing supplementary observations and variables), for MFA and PLS permutation tests, Statgraphics® software for stepwise PLS regression, R software (R_Development_Core_Team, 2004) for the cross-validation of PLS variables, and PPPhalos software (Guiot, 1991) for neural networks.

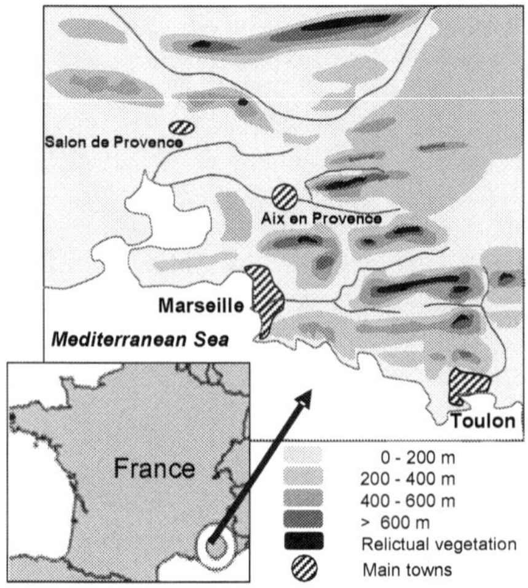

Fig. 1. study area. Alpine and middle-European relict flora remains on steep north slopes near the top of the highest ridges.

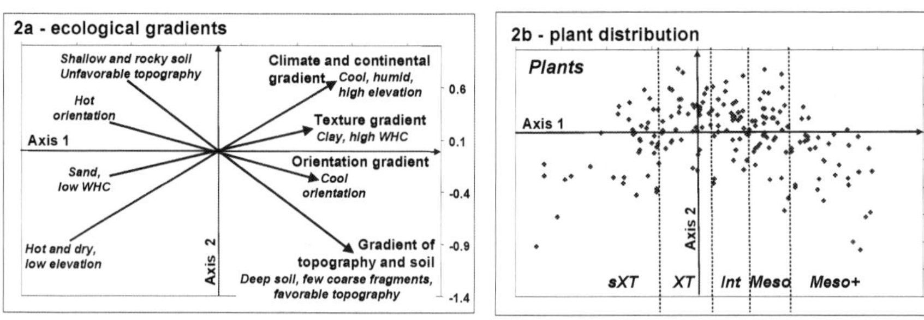

Fig. 2. CA main plane

* fig 2a: synthetic representation of four main ecological gradients related to water balance in the study area, interpreted from the distribution of all variables (see table 1) in CA plane: (1) climate and continentality, (2) orientation, (3) soil texture (water holding capacity), (4) topography and soil quality. These gradients were calculated with a linear regression on the coordinates of the classes of concerned variables in the plane. Axis 1 appeared as the synthesis of these 4 gradients, related to water availability and temperature, integrating local and global scales.

* fig 2b: flora distribution in CA plane. Plant species were split into 5 groups (same number of species in each) according to their coordinate on axis 1.

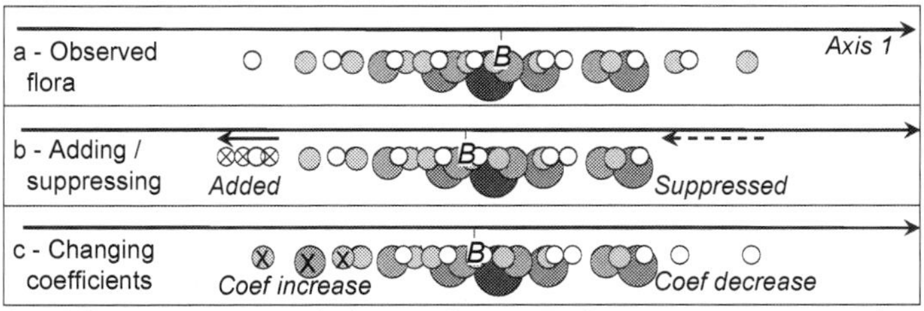

Fig. 3. Simulated flora variations with climate change. Each ring represents one plant species, all the more large and dark as its Braun-Blanquet coefficient is high. B = barycentre of plant distribution on axis one for the plot = plot Fi.

3a - Observed flora.

3b - One to 3 plants are respectively added / suppressed at the lower / upper margin of the distribution. Added plants are those with the closest coordinates to the last observed species.

3c - Braun-blanquet coefficients are respectively raised / reduced for one to 3 plants at the lower / upper margin of the distribution. When a plant initially has the smaller coefficient, reducing it amounts to suppressing this plant.

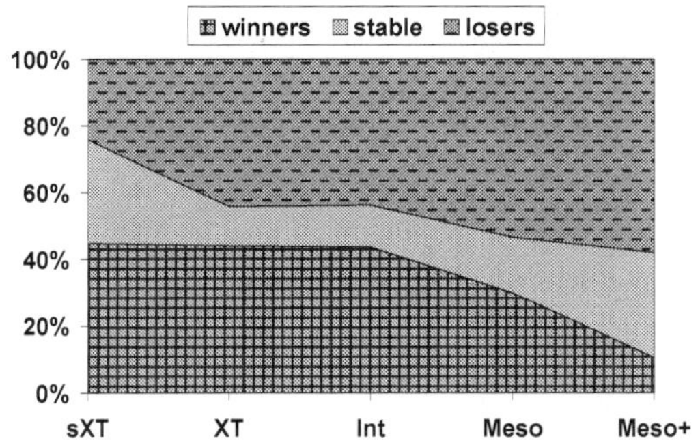

Fig. 4. Variation, between 1996-98 and 2008, of the occurrence of plants according to their tolerance to water stress and heat, as a percentage of the number of plants per group. Winners / losers means plant species respectively present in more / less plots in 2008 than in 1996-98.

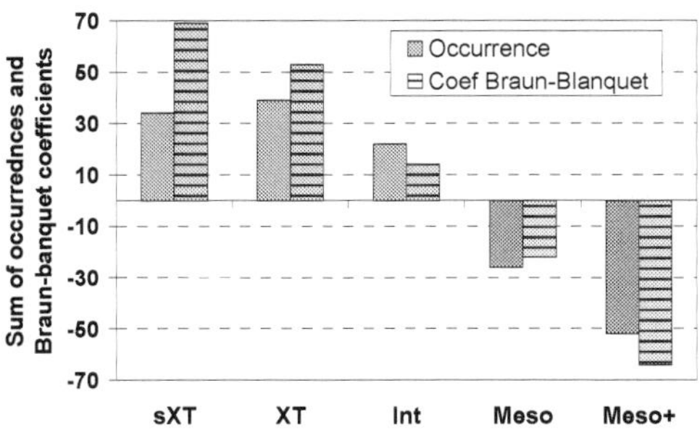

Fig. 1. variations between 1996-98 and 2008 of the sums of occurrences and Braun-Blanquet coefficients per group in the 50 plots revisited in 2008. One occurrence = one species in one plot.

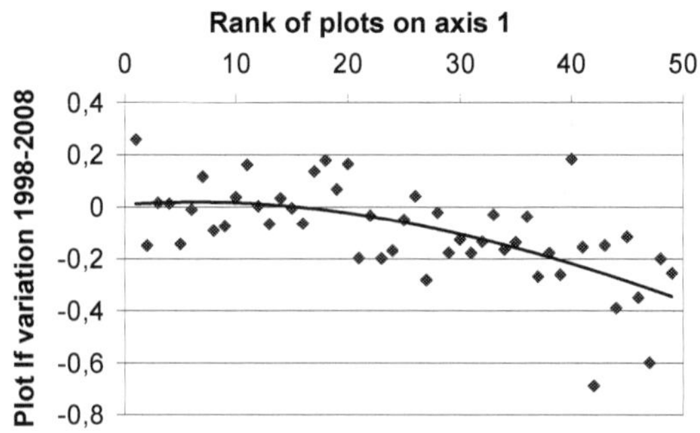

Fig. 6. If variation for the 50 plots resurveyed in 2008 according to their initial position on axis 1.

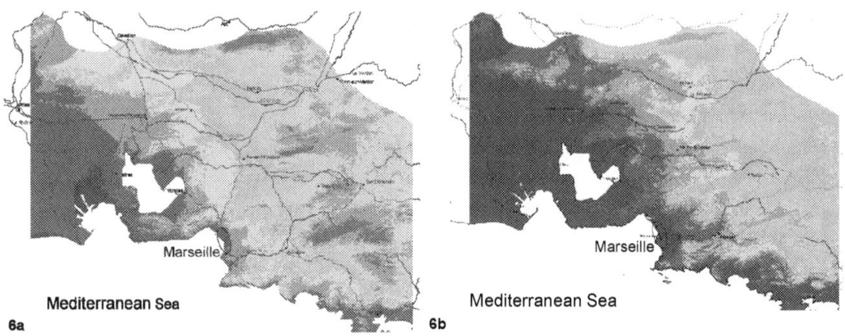

Colour									
Bi class	1	2	3	4	5	6	7	8	9

Fig. 7. Bioclimatic index (Bi) computed with global variables and mapped on regional scale with ArcGIS® software in 9 classes of equal Bi index span. The darkest green (class 9) corresponds to the potential niche of Scots pine and related alpine and meso-European relict species. The three shades of green together draw the area of the supra-Mediterranean bioclimate, the blue to pink intermediate colors the meso-Mediterranean and the red the thermo-Mediterranean bioclimate (Emberger, 1930).
7a: Map with 1961-96 mean climate.
7b: Map with 2050 climate (IPCC, B2 Scenario). Relict vegetation has no potential niche left in the study area. Q. pubescens should move away far from the coast, and the growth of P. halepensis been slower in most of the study area.

Abstract

The ongoing climate change causes a rapid shift of plant distribution at various scales. In South-eastern France from 1998 to 2008, Mediterranean forests experienced an exceptionally hot and dry episode following a regular but more limited warming since the 70's. Flora turnover for this period was both simulated at local scale with a bioclimatic model and measured in permanent plots with two censuses. Model prediction for this turnover with the mean climate of the last 10 years was 25%. A 14% turnover was observed in the permanent plots between the two censuses, fully biased against water demanding species. Changes were all the more fast than sites were favorable (high altitude, cool orientation, deep soils, favorable topography), and were not significant in the driest and hottest sites. This proves that the main changes occurred when the compensation of climatic water deficits by local site favorable conditions was no longer sufficient to allow mesophilous species to survive. Such a threshold should shift towards hottest and driest situations with the future climate. On the landscape scale, various strategies allow a partial plant composition resistance. However, current reserve networks may be inadequate to ensure long-term species persistence. With the measured flora shift, most of the rare species protected in these reserves would potentially disappear from the study area soon in the second half of the 21st century.

Keywords:
Climate change; flora turnover; resurvey; bioclimatic model; ecological niche; reserves

14

Climate change, forest fires and air quality in Portugal in the 21st century

Anabela Carvalho
CESAM & Department of Environment and Planning, University of Aveiro, Aveiro Portugal

1. Introduction

There are regions of the world more vulnerable to climate change and some of those regions are also sensitive to forest fire occurrences. The way climate change interacts with the forests of the world and consequently with forest fire activity is a point of debate among the scientific community.

Several features dominate the forest fires of a given region and these can be described in terms of the main temporal and spatial scales of variation. From a fire event to the definition of the fire regime of a given region several drivers are determinant in the characterization of these dynamics. From the local weather conditions to the large scale weather patterns the forest fires evolve from simple events to the definition of the fire regime of a given region. The short temporal scale and the local characteristics that dominate a fire event are ruled by the local weather conditions that drive the forest fires daily variability. Additionally, the physiographic and the topographic variability that characterize the large temporal and spatial scales variations represent the broader influence of these scales on the fire regime definition.

To correctly assess the inter-connection between all analysed drivers human influence must also be considered. One of the best definitions of this relationship is given by Stephen Pyne (Pyne, 2007) who describes the area burned of a region as "a proxy of climate interacting with people". At a larger temporal and spatial scale human activities may deeply influence climate and subsequently the fire regime of a region. At local scale human activities have a remarkable impact on forest fires mainly through changes in land-use, negligent actions or simply by arson.

Wildland fire is a global phenomenon, and a result of interactions between climate-weather, fuels and people (Flannigan et al., 2009). The influence of the human activities on climate is leading to worldwide forest fire changes and disruptions. The most recent report of the IPCC (IPCC, 2007) discusses the changing of the vegetation structure and composition due to intensified wildfire regimes driven at least partly by the 20th century climate change. Worldwide the wildfire regime is changing. In the United States of America (USA) the number of forest fires is decreasing due to fire prevention strategies but they are becoming larger and consequently more severe. In this sense the fire suppression efforts are escalating (Miller, 2007).

Recently, Europe has experienced a large number of forest fires that have caused enormous losses in terms of human lives, social disturbances, environmental damage and economic disruptions. Most of the fires in Europe take place in the Mediterranean region where over 95% of the forest fire damage occurs (EC, 2003). There are several features that make the landscapes of the European Mediterranean Basin different from those of the rest of Europe. These differences are mainly related to the climate, the long and intense human impact, and the role of fire. The latter is, in turn, influenced by the other two (Pausas and Vallejo, 1999). The interaction between the human activity and the fire is a complex question that drives the majority of fire activity in southern Europe.

Since 1980, the statistics of the annual area burned in Portugal, Spain, France, Italy and Greece, have varied considerably from one year to the next, which can be an indication of how strongly the area burned depends on weather conditions. Fire occurrence increased during the 1990s, but since 2001 the number of fires has remained more or less stable (EC, 2005). This stabilization was possibly due to public information campaigns and improvements in the prevention and fire-fighting abilities of these countries. In Portugal, out of the last 30 years 2003 had the worst fire season, which resulted in the burning of almost 430,000 ha of forested lands and scrublands with global economic losses of 1,200 million Euros (DGRF, 2006). In that year the social costs were most significant with the loss of 20 human lives and the destruction of 117 houses. Due to extreme fire weather conditions (Viegas et al., 2006) the year 2005 also recorded a very high value of area burned, approximately 325,000 ha.

Even the main reason for fire increase is probably changes in land use, climatic factors should be considered as a contributing factor. Fires tend to be concentrated in summer when temperatures are high, and air humidity and fuel moisture are low (Pausas and Vallejo, 1999). Over Portugal and since 1972, there is a general trend towards an increase in the mean annual surface air temperature. Additionally, spring accumulated precipitation has registered a systematic reduction, accompanied by slight increases in the other seasons (Santos et al., 2002). Predictions of climate warming in the Mediterranean basin indicate an increase in air temperature and a reduction in summer rainfall (Christensen and Christensen, 2007)). Although there is uncertainty on the mean and variance of the precipitation changes, all predictions suggest a future increment in water deficit. These changes would lead to an increase in water stress conditions for plants, changes in fuel conditions and increases in fire risk, with the consequent increase in ignition probability and fire propagation (Pausas and Vallejo, 1999).

Since the late 70s biomass burning has been recognized as an important source of atmospheric pollutants (Crutzen et al., 1979). Several works have already discussed the importance of forest fires as a source of air pollutants (Amiro et al., 2001a; Miranda et al., 2005a) and in a changing climatic scenario this contribution can increase dramatically (Amiro et al., 2001b) due to larger area burned. Forest fire emissions, namely particulate matter (PM), ozone (O_3) precursor gases (like nitrogen oxides - NO_x and volatile organic compounds - VOC) and carbon dioxide (CO_2), can significantly impact the ecosystems and the air quality and consequently human health (Riebau and Fox, 2001). Particularly, they can influence plant productivity downwind of fires through enhanced ozone and aerosol concentrations (Sitch et al., 2007). In a changing climate the forest fire emissions can play an important role in all these interactions.

Air quality and its potential impacts namely in the ecosystems, structures and human health is currently one of the main concerns at global, regional and local scales. In Dentener et al. (2006) the troposphere composition change to be expected in the near future (year 2030) is investigated using 26 state-of-the-art global atmospheric chemistry transport models (CTMs) and three different emissions scenarios. Based on the ensemble mean model results, by 2030 global surface ozone is estimated to increase globally by 4.3 ± 2.2 ppb for the IPCC SRES A2 scenario (Nakicenovic et al., 2000). This study shows the importance of enforcing current worldwide air quality legislation and the major benefits of going further. Nonattainment of these air quality policy objectives, such as expressed by the IPCC SRES A2 scenario, would further degrade the global atmospheric environment.
The analysis of the climate change impacts on air quality and its feedback mechanism is nowadays a well recognised approach at the global scale. Nonetheless, studies from the regional to a country scale are not so widespread. The highest number of studies can be found for USA (e.g. Hogrefe et al., 2005). Over Europe some studies have addressed this issue (e.g. Szopa et al., 2006) pointing that by 2030 estimated ozone levels, in July, may increase up to 5 ppb. In Europe and at country level these studies are still reduced and only applied for episodic situations (e.g. Borrego et al., 2000). Additionally, the interaction between climate change, forest fire emissions and air quality is still poorly discussed. Spracklen et al. (2009) firstly investigated the potential impacts of future area burned on aerosol concentrations over the United States. In this scope, the main objective of this chapter is to investigate the role of climate change in forest fire activity and its impacts on air quality patterns over Portugal through the projection of future area burned and pollutants emissions under the IPCC SRES A2 climatic scenario.

2. Fire activity data in Portugal

Forested lands in Portugal occupy 5.4 millions of hectares and represent two-thirds of Portugal's surface area (DGRF, 2006). Eleven percent of the Portuguese territory is occupied by Maritime Pine stands or lands (*Pinus pinaster*), followed by Eucalypt (*Eucalyptus globulus*) (8 %) and Cork Oak (*Quercus suber*) (8 %). The Holm Oak (*Quercus rotundifolia*) represents 5 %, and the oak tree (*Quercus faginea*) and Stone Pine (*Pinus pinea*) exhibit 1 % each. Figure 1 shows the Portuguese districts identification and the dominant forest types. Maritime Pine is mostly common in the Castelo Branco, Coimbra, Leiria and Viseu districts. Castelo Branco, Aveiro and Santarém districts have higher forest lands of Eucalypt. On the other hand, the southern districts of Évora, Portalegre, Santarém and Setúbal have the majority of the Cork Oak in Portugal. The oak tree is most common in the northern districts of Vila Real, Bragança, and Guarda.
The Portuguese population is mainly concentrated in the urban and sub-urban areas of the coastal regions. The north region contains 35 % of the population, the Lisbon area 26 % and the central part 23 %. The remaining Portuguese regions have occupation levels below 8 % (INE, 2003). This represents a considerable population asymmetry that certainly influences forest fire ignitions and spreading.
Some aspects of the property regime in the north and centre of Portugal, namely the high number of land owners (most of them unknown) and the absence of adequate property records, have important negative consequences concerning forest management. An increase of population within the forested lands greatly enhances the forest fire risk and,

consequently, the destruction of goods and human lives and creates difficulties for the fire fighting operations. Land abandonment, due mainly to the aging of the land owners, also creates difficulties in the management of forested properties, leading to an increase in the fuel load and consequently in forest fire risk. In the southern part of the country the districts of Beja, Évora and Portalegre have a different demographic pattern. The populations are more concentrated and not spread among the forested areas, additionally the dominant forest types are resistant to forest fires. These are the regions of Portugal which reach the highest temperatures during the summer period and have lower precipitation rates throughout the year.

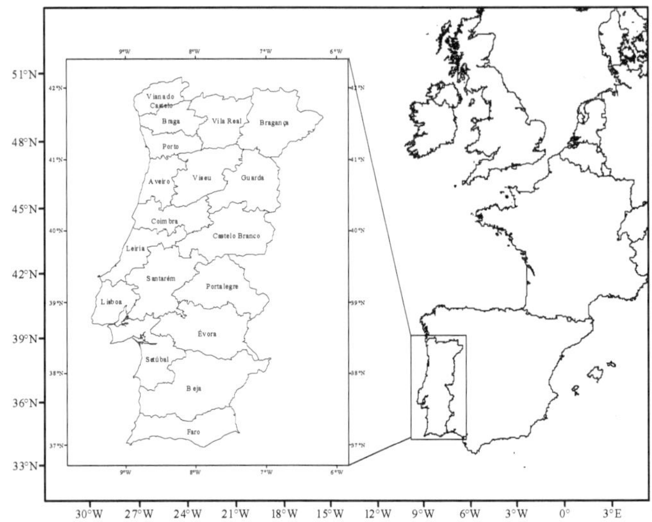

District name	Dominant forest type (% district area)
Bragança	Maritime Pine (5%)
Vila Real	Maritime Pine (17%)
Viana do Castelo	Maritime Pine (17%)
Braga	Eucalyptus (16%)
Porto	Eucalyptus (18%)
Aveiro	Eucalyptus (25%)
Viseu	Maritime Pine (23%)
Guarda	Maritime Pine (12%)
Coimbra	Maritime Pine (30%)
Castelo Branco	Maritime Pine (24%)
Leiria	Maritime Pine (30%)
Santarém	Cork Oak (18%)
Portalegre	Cork Oak (23%)
Lisboa	Eucalyptus (10%)
Setúbal	Cork Oak (25%)
Évora	Holm Oak (20%)
Beja	Holm Oak (16%)
Faro	Cork Oak (8%)

Fig. 1. – Location of Portugal in the Iberian Peninsula, Portuguese districts identification and dominant forest types as a percentage of district area.

The forest fire database for Portugal is provided by the Autoridade Florestal Nacional-AFN. This database constitutes the national component of the European Forest Fire Information System (EFFIS) created by the European Commission in 1994. The Commission Regulation (EC) 804/94 (now expired) established a Community system of information on forest fires for which a systematic collection of a minimum set of data on each fire occurring, the so-called "Common Core", had to be carried out by the Member States participating in the system. According to the currently in force Forest Focus, Regulation (EC) 2152/2003, concerning monitoring of forests and environment interactions in the community, the forest fire common core data should continue to be recorded and notified in order to collect comparable information on forest fires at the Community level.

At national level, the recorded information includes daily area burned and daily number of fires per district, among other variables. From 1980 to 2009, the dataset record illustrates a total of 3.1×10^6 ha of area burned, approximately 30 % of Portugal's total area, and 487,000 of forest fire occurrences. The AFN database is based on *in situ* information provided by the

Ministry of Agriculture and the National Civil Protection Service. Since 1990, the annual area burned is mapped based on satellite information.

Simple statistics for forest fire activity in Portugal were performed in order to better understand its main characteristics in terms of spatial and temporal distribution. Figure 2 represents the annual area burned and number of fires between 1980 and 2009. The maximum number of annual forest fires occurred in 1995, 1998, 2000 and 2005, reaching 35,000 occurrences. In terms of area burned the year of 2003 reached the highest value ever registered - 430,000 ha, followed by 2005 with 337,000 ha. It is interesting to observe that between 1980 and 2000 the annual number of forest fires registered an increase from year to year except in some specific years. In addition to this and according to the Portuguese Meteorological Institute since 1974 there is a clear increase in the average temperature values in Portugal. The years of 1995, 1997, 1998 and from 2000 to 2006 present higher average temperatures than the normal. From 1995 to 2000 the number of forest fires registered a clear increase, although since 2001 they tend to remain almost constant except in 2005 and 2009.

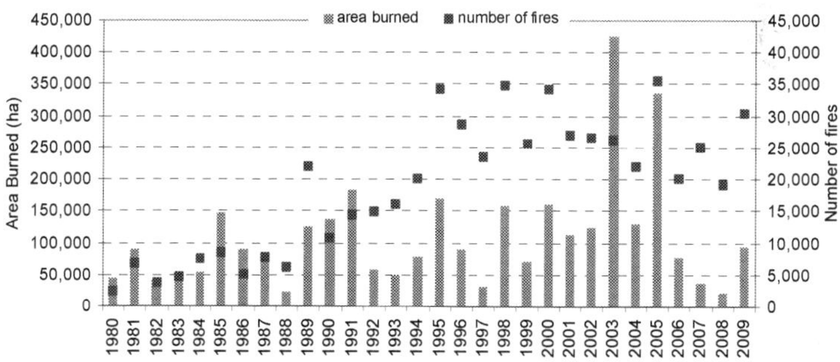

Fig. 2. Annual area burned (bars) and number of fires (square points) in Portugal between 1980 and 2009.

The number of fire occurrences in the months of June (8 %), July (22 %), August (32 %), and September (20 %) represent 82 % of the yearly total (not shown). The area burned peak is observed in August accounting for 45 % of the yearly total (not shown) with the districts of Guarda, Castelo Branco, Viseu, and Coimbra presenting the highest area burned values in Portugal (not shown). According to the National Plan for Forest Fires Prevention, the ignitions peak occurs along the weekend, and especially during the afternoon, denoting an important human influence on fire starts (APIF, 2005).

The annual number of forest fires is higher in the most urban and sub-urban districts (Aveiro, Braga, Lisboa, Porto, Viana do Castelo, and Setúbal) (not shown). In terms of forest fire occurrences, the Porto district (urban/sub-urban region) represents the highest percentage of fire occurrences reaching almost 22 % of the total. Additionally, the Guarda district (rural region) accounts for almost 18 % of the area burned in Portugal, followed by Castelo Branco with 11 % (not shown).

An analysis performed for the period between 1993 and 2003 revealed that 97 % of the forest fire ignitions were due to human influence with 37 % to arson, 28 % to negligence, and 32 % to unknown causes (APIF, 2005). Arson is mainly related to fraud, hunting conflicts, and building construction interests, and is most notorious in the northern part of the country especially in the coastal regions. Negligence is the most important cause in the south mainly due to clearance activities. In the southern districts of Beja, Évora, and Portalegre the principal cause of negligence is related to agricultural machinery use. Specific regional characteristics are also responsible for forest fires starts such as fireworks activity in the northern districts of the country (APIF, 2005). Portugal, like the majority of the southern European countries, has fewer forest fires due to natural causes because phenomena such as lightning have a low frequency of occurrence during the summer period.

3. Observed fire weather risk in Portugal

The Canadian Forest Fire Weather Index (FWI) System is a system that monitors forest fire risk and supplies information to support fire management. The components of the FWI System can be used to predict fire behaviour and can be used as a guide to policy-makers in developing actions to protect life, property and the environment.

The FWI system was developed for Canadian forests but has also been applied in other countries and environments such as Mexico, Southeast Asia, Florida and Argentina (Moriondo et al., 2006). For the Mediterranean basin, several studies showed that the FWI system and its components were well suited to the estimation of fire risk for the region (Viegas et al., 1999). Moreover, the FWI is currently the fire risk index used by the Joint Research Centre (JRC) to map fire risk at the European Union level (JRC, 2006). The success of this system is due both to the simplicity of its calculation procedures and to the simulation of the moisture of a generalized fuel type, which has been successfully applied to model fire potential in a broad range of fuel types (Van Wagner, 1987).

The FWI System (Figure 3) is a weather-based system that models fuel moisture using a dynamic bookkeeping system that tracks the drying and wetting of distinct fuel layers in the forest floor. There are three moisture codes that represent the moisture content of fine fuels (fine fuel moisture content, FFMC), loosely compacted organic material (duff moisture code, DMC), and a deep layer of compact organic material (drought code, DC). The drying time lags for these three fuel layers are 2/3 of a day, 15 days, and 52 days respectively for the FFMC, DMC, and DC under normal conditions (temperature 21.1°C, relative humidity 45%). These moisture indexes are combined to create a generalized index of the availability of fuel for consumption (build up index, BUI). The FFMC is combined with wind speed to estimate the potential spread rate of a fire (initial spread index, ISI). The BUI and ISI are combined to create the FWI which is an estimate of the potential intensity of a spreading fire. The daily severity rating (DSR) is a simple exponential function of the FWI intended to increase the weight of higher values of FWI in order to compensate for the exponential increase in area burned with fire diameter (Van Wagner 1970).

For the purposes of this study, the FWI System components were computed using daily mean values of temperature, relative humidity, wind, and daily total precipitation. The Statistical Analysis System (SAS) version 9.1.3 (SAS, 2004) was used for the FWI System components estimation. The FWI System components have been estimated for the period between 1980 and 2005 were meteorological data was available at 12 synoptic sites across Portugal.

Figure 4 presents the daily mean fire weather index (FWI) from 1980 to 2005. Porto district exhibits the lowest values. The southern region formed by Portalegre, Évora, and Beja districts presents the highest interquartilic interval (interquartile range from 25th percentile to 75th percentile) of FWI values. In terms of yearly distribution, 2005 (Figure 4b) presents the highest interquartilic interval of values but the maximum FWI index was attained in 2004. According to the monthly distribution (Figure 4c), July presents the maximum FWI value but the interquartilic interval remains almost the same between July and August. As expected, the period between May and October presents the highest FWI values.

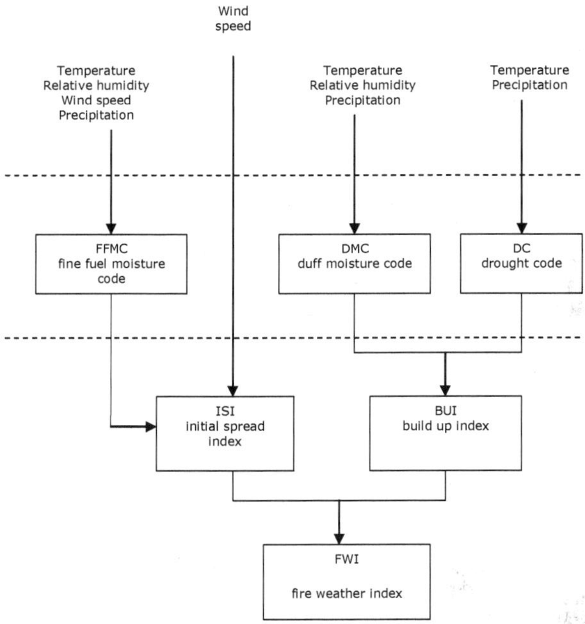

Fig. 3. Canadian Fire Weather Index (FWI) System components (adapted from Van Wagner, 1987).

The 12 stations fire weather data was used to estimate the FWI spatial pattern for Portugal. A Geographic Information System (GIS) software was used to compute a spline interpolation (Schumaker, 1981) over the monthly means of the FWI daily values estimated between 1980 and 2005. Figure 5 presents the FWI spatial distribution over Portugal, from May to October, between 1980 and 2005. According to Figure 5 July and August register the highest values and the southern regions are also the main affected. The monthly mean FWI values range from 1 to 32 depending on the region giving an indication on the associated fire danger. According to Viegas et al. (2004) the highest FWI values registered in the southern region are associated to low fire danger classes. On the contrary, in July and August the districts of the centre and north interior present FWI values ranging from 13 to 29, which are related to moderate to high level of, fire danger. The coastal regions in the north and Centre show the lowest FWI values ranging from 1 to 12. In these regions the obtained FWI values are related to a low to moderate level of fire danger.

The FWI index spatial distribution has a markedly NW-SE gradient. The NW region of Portugal exhibits the lowest FWI values and the SE the highest. This gradient is in agreement with the temperature patterns and the mean sea level pressure field of the summer climatology observed over the Iberian Peninsula (Hoinka et al., 2009).

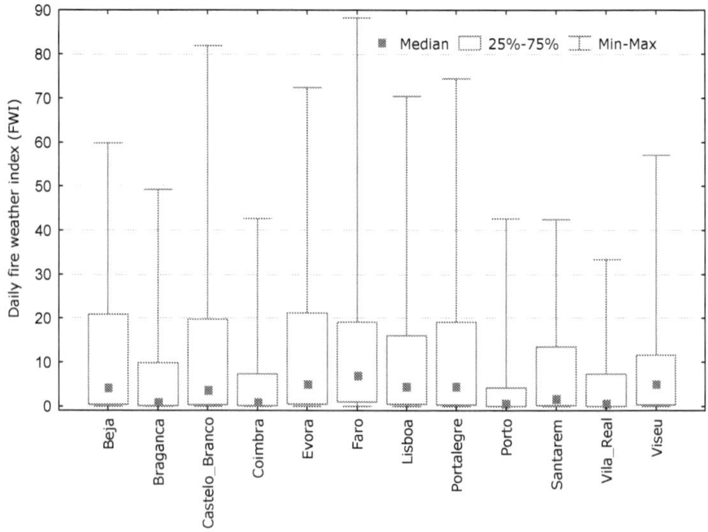

a)

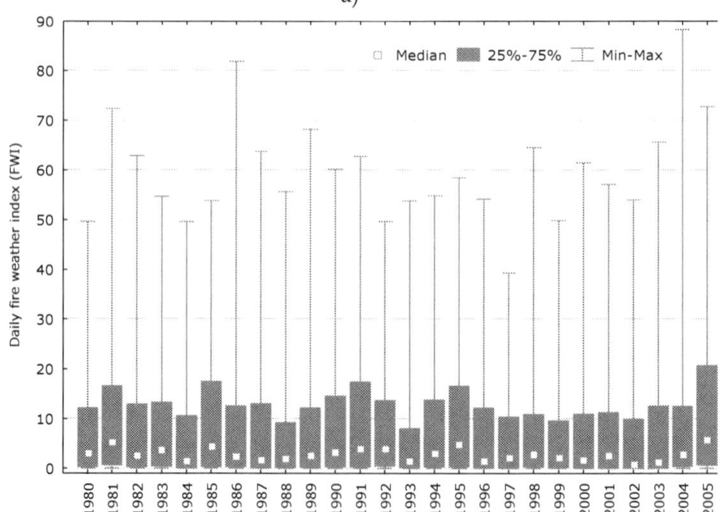

b)

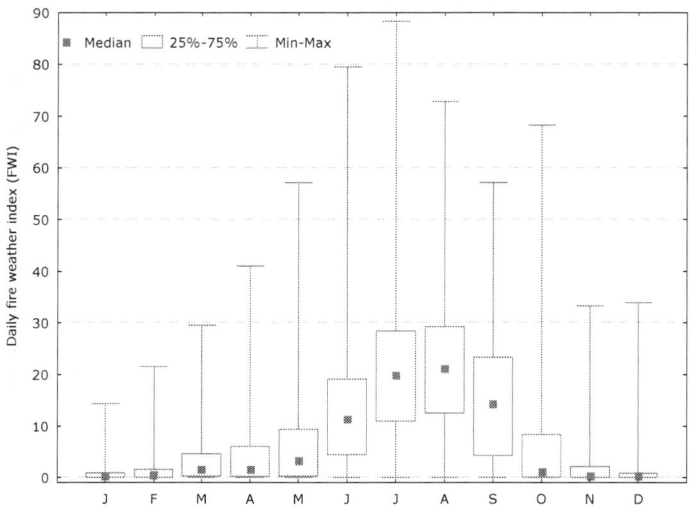

c)

Fig. 4. Daily fire weather index (FWI) between 1980 and 2005 by a) district, b) year and c) month.

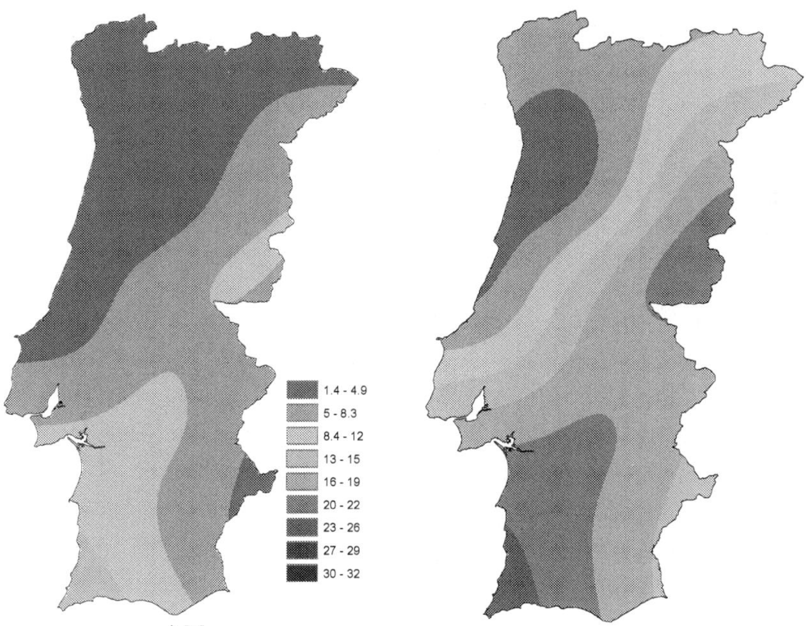

a) May

b) June

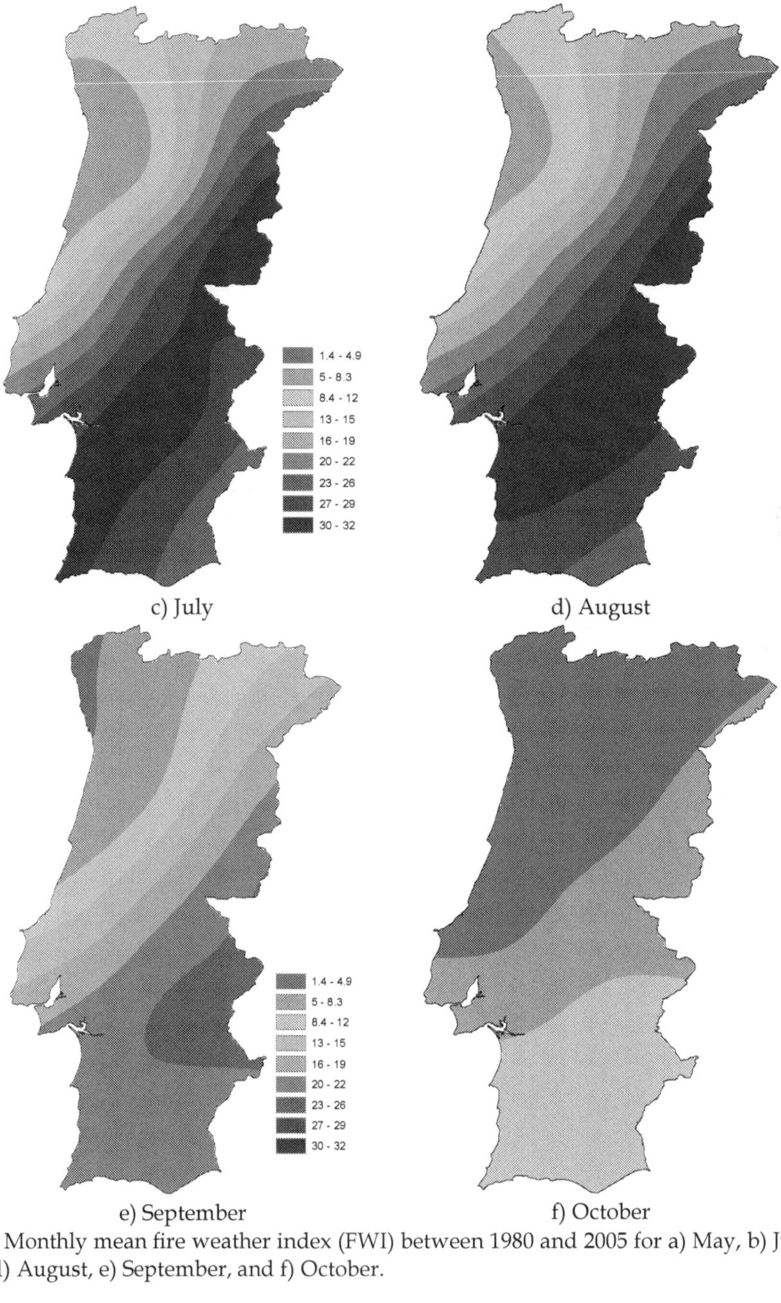

Fig. 5. Monthly mean fire weather index (FWI) between 1980 and 2005 for a) May, b) June, c) July, d) August, e) September, and f) October.

4. Climate change impacts on fire weather risk

To estimate the impacts of climate change on the fire weather risk, daily climatic data were collected from the regional climate model HIRHAM (Christensen et al., 1996), at 12 km spatial resolution from the Prediction of Regional scenarios and Uncertainties for Defining EuropeaN Climate change risks and Effects - PRUDENCE - project (Christensen and Christensen, 2007) considering the IPCC Special Report on Emissions Scenarios (SRES) A2 scenario. The IPCC SRES A2 scenario is characterized by a very heterogeneous world with a continuously increasing global population. The economic development is primarily regionally oriented and per capita economic growth and technological change is more fragmented and slower than in other IPCC scenarios. In this sense, the A2 is considered a high emission scenario. For the analysed time slices the IPCC SRES A2 is consistent to a 2 x CO_2 climatic scenario.

A detailed validation of the HIRHAM outputs was performed for 12 sites across Portugal between 1980 and 1990 (11 year period for which observed data were available). Using SAS version 9.1.3 monthly mean values of the simulated daily mean temperature, daily maximum temperature, daily mean wind speed, total precipitation, and daily mean relative humidity and fire weather risk variables were evaluated. According to the validation procedure the relative humidity presented significant differences against the observed data (Carvalho et al., 2010a). Concerning relative humidity the HIRHAM model presents drier values than the observed at the weather stations. In order to correct the relative humidity field the dew point temperature was evaluated and statistical significant differences were found. The HIRHAM model presents a cold bias in the dew point temperature fields and especially in the south of Portugal and in autumn. A correction factor was applied based on a monthly discrimination (Carvalho et al., 2010a). The correction factor was applied to the reference and to the future climate simulations. This approach has already been used in other works and is considered an adequate calibration procedure (e.g., Flannigan et al., 2005). The corrected climatic fields were used to estimate the FWI System components for reference (1961-1990) and future (2071-2100) scenarios.

The HIRHAM projections over Portugal point to an increase of the mean temperature in all seasons especially in summer, reaching almost 6 °C in the inner districts of the country ($p<0.0001$) (Figure 6a). The daily precipitation decreases in all seasons especially in spring. The north and central part of Portugal will register the highest reductions in rainfall amounts (Figure 6b). These projections will deeply influence the fuel moisture conditions in future climate.

Concerning fire severity, all seasons experience an increase in the FWI component by the end of 21st century. The summer months of June, July and August show the highest FWI values. May registers the highest relative increases, October and November also exhibit high increases in the FWI index. This could lead to a clear anticipation of the fire season starting and an increase in its length. There is also a clear FWI increasing trend from north to south (Carvalho et al., 2010a).

Figure 7 exhibits the FWI cumulative frequency distribution for each scenario by district. To help on the discussion the districts are organized by north, centre and south of Portugal.

The obtained cumulative distribution functions clearly show the FWI shift to attain higher values in a future climatic scenario. The districts of the north formed by Viana do Castelo, Bragança, Vila Real, Braga, and Porto show an increase of the maximum FWI range of values from 26-53 to 45-76. The 50th percentile also shows an increase but not so pronounced. The districts in the Centre like Viseu, Guarda, Aveiro, Coimbra, Castelo Branco and Leiria, also present an increase in the FWI maximum values ranging from 39-55 to 55-71 from the reference to the future climatic scenario. The southern districts of Santarém, Portalegre, Lisboa, Évora, Setúbal, Beja and Faro present the highest FWI maximum values in the reference scenario and the same is verified in the future climate. The FWI values between the 25th percentile and the maximum show a clear increase in all southern districts. In this part of the country the FWI ranges from 50-71 and in future climate these values increase to 57-76.

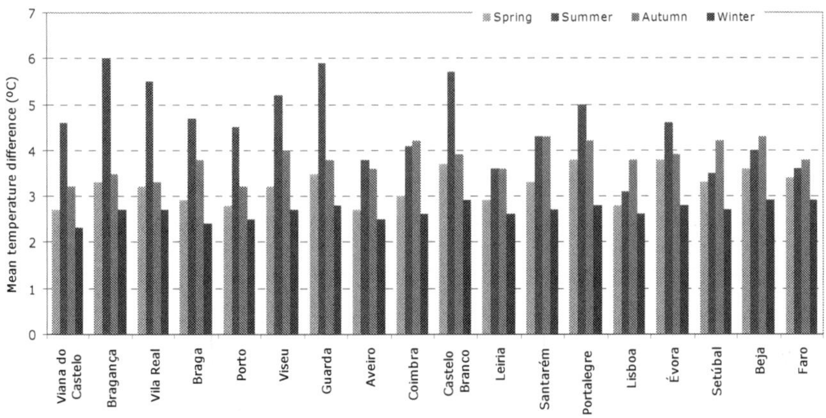

a)

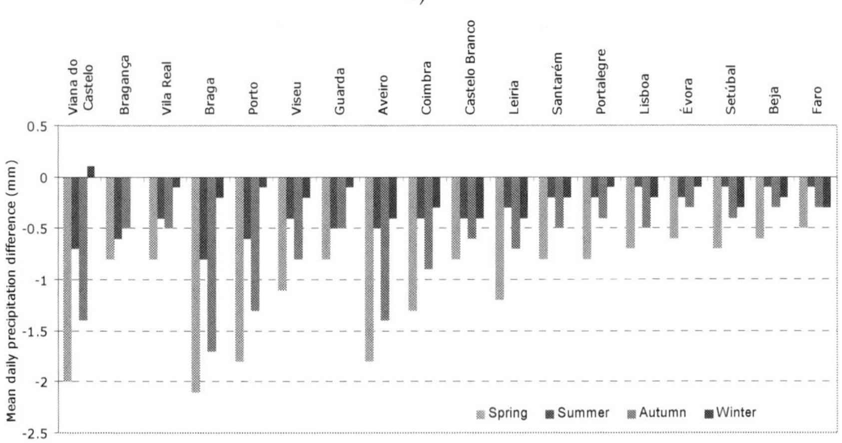

b)

Fig. 6. Differences between future (2071-2100) and reference (1961-1990) climatic scenarios for a) daily mean temperature and b) daily precipitation, by season and for each Portuguese district.

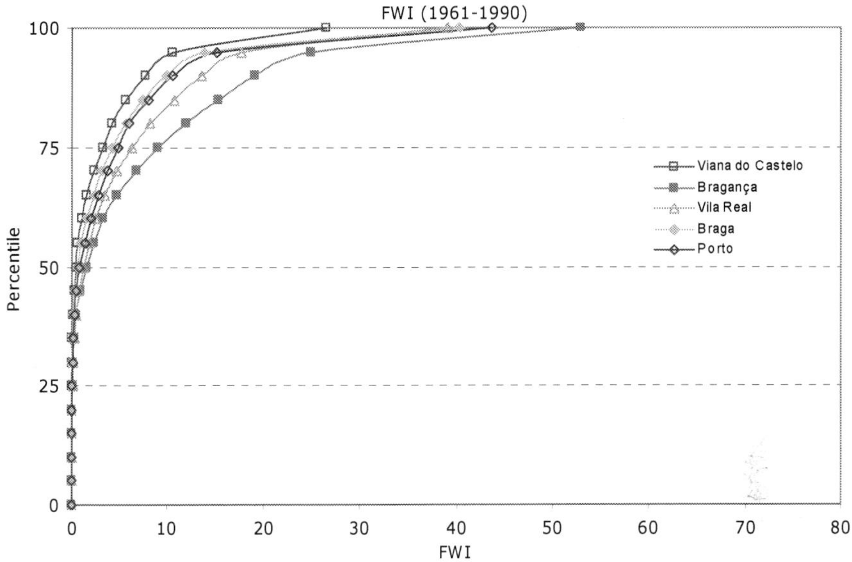
FWI (1961-1990)
Percentile
FWI
Viana do Castelo
Bragança
Vila Real
Braga
Porto
0
25
50
75
100
0
10
20
30
40
50
60
70
80

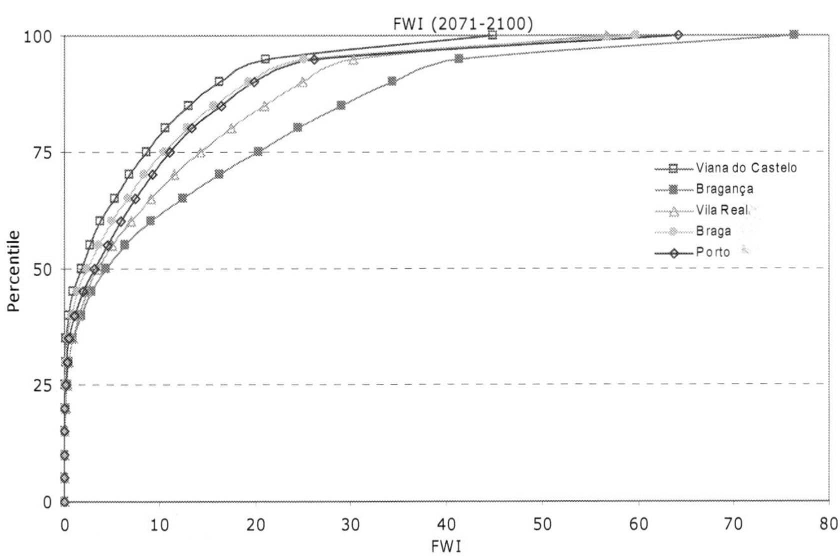
FWI (2071-2100)
Percentile
FWI
Viana do Castelo
Bragança
Vila Real
Braga
Porto
0
25
50
75
100
0
10
20
30
40
50
60
70
80

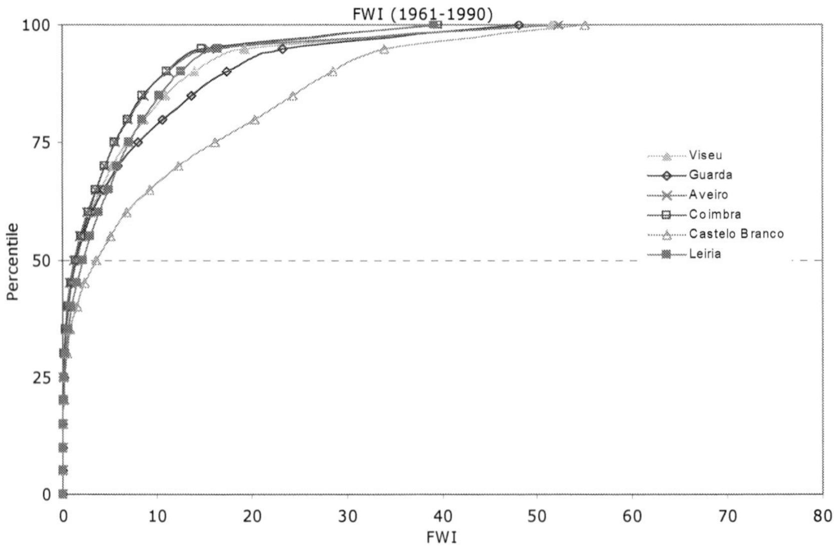
FWI (1961-1990)
100
75
50
25
0
Percentile
0
10
20
30
40
50
60
70
80
FWI
Viseu
Guarda
Aveiro
Coimbra
Castelo Branco
Leiria

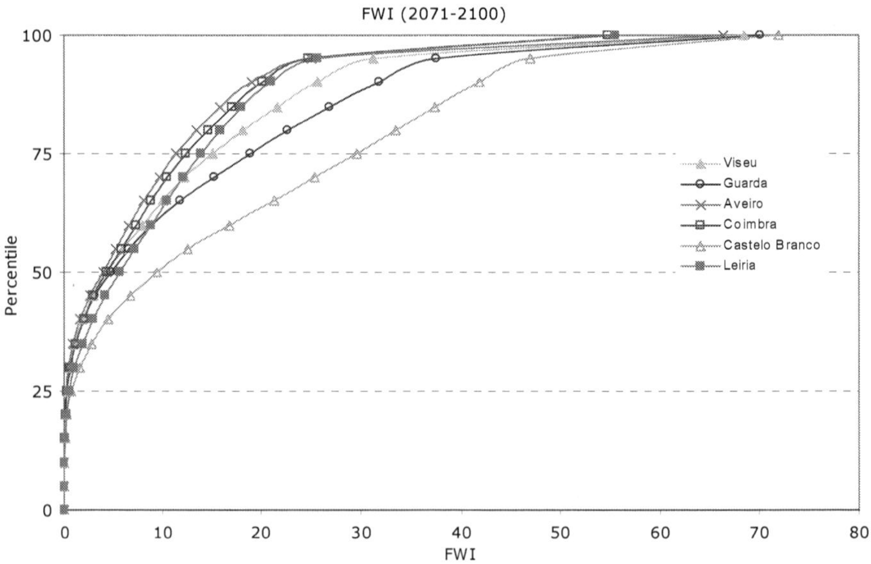
FWI (2071-2100)
100
75
50
25
0
Percentile
0
10
20
30
40
50
60
70
80
FWI
Viseu
Guarda
Aveiro
Coimbra
Castelo Branco
Leiria

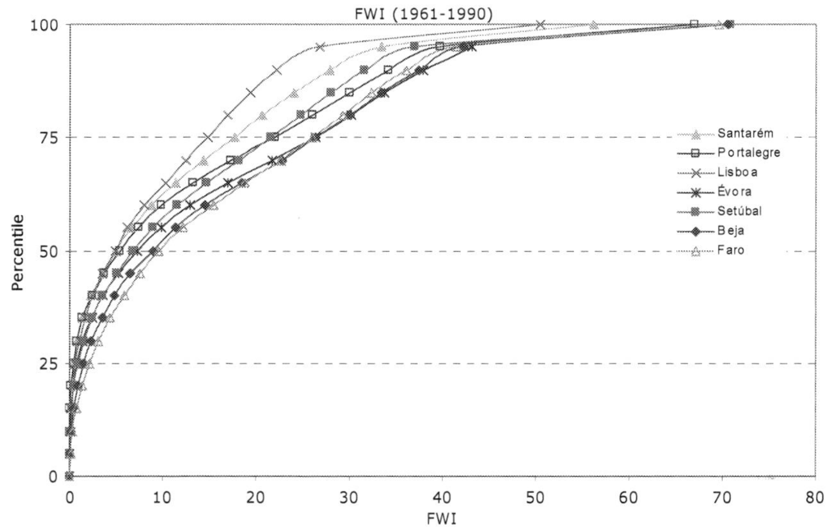

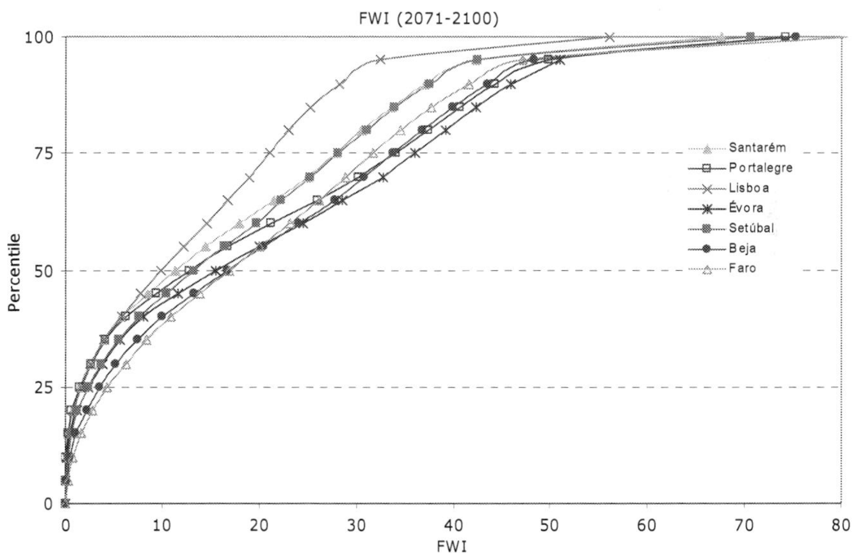

Fig. 7. Cumulative frequency distribution of the daily FWI component by district and for each climatic scenario (reference and 2 x CO_2). The districts are organized by north, centre and south of Portugal.

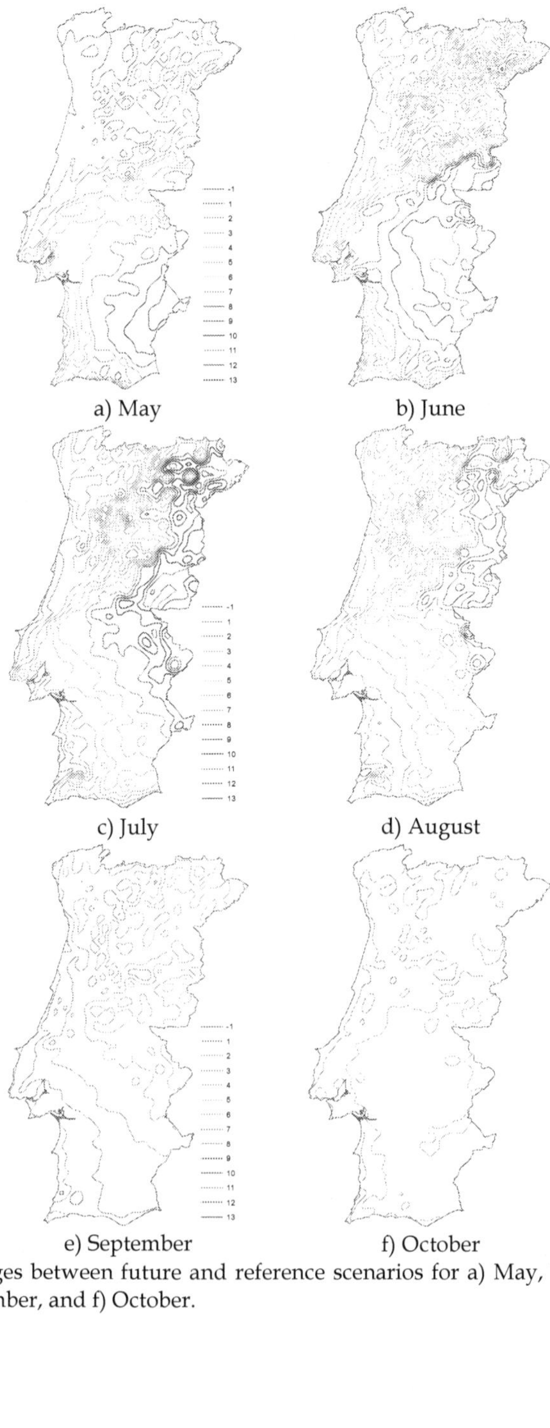

a) May b) June

c) July d) August

e) September f) October

Fig. 8. FWI changes between future and reference scenarios for a) May, b) June, c) July, d) August, e) September, and f) October.

Figure 8 presents the differences on the FWI patterns for May, June, July, August, September, and October between future and reference climatic scenarios. As previously noted, all the districts across Portugal experience an increase on the FWI component but this is more pronounced in the inner regions. July and August show the highest increases namely in the districts of Bragança, Guarda, and Castelo Branco. These districts form a regional elongated pattern that goes from north to the centre just close to the Spanish border. September and October exhibit a very homogeneous pattern of increase from north to south.

The area burned and the number of fires in Portugal is strongly dependent on the weather conditions. So, it is expected that fire activity will increase with a changing climate. In Carvalho et al. (2008) statistically significant relationships were established between the area burned, the number of fires and the weather for different Portuguese districts. The authors have concluded that the weather explains the majority of the variance of the area burned and of the number of fires in Portugal. The obtained statistical models were used to estimate the area burned and the number of fires for future climate. The results point to a substantial increase in the area burned and on the number of fires ranging from 238 % to 643 % and 111 % to 483 %, respectively, depending on the Portuguese district (Carvalho et al., 2010a). The regions in the central part of the country will be the most affected. The monthly distribution of the area burned and number of fires indicates that an earlier fire season starting may be expected under future climate.

5. Forest fires and climate change impacts in particulate matter and ozone levels

The impact of climate change and future area burned on forest fire emissions and consequently on regional air quality was assessed through the application of the MM5/CHIMERE numerical modelling system. This numerical system has been widely tested and successfully used over Portugal (Monteiro et al., 2005; Monteiro et al., 2007; Monteiro, 2007). The HadAM3P (Jones et al., 2005) simulations for the reference and the IPCC SRES A2 climatic scenario were used to driven the MM5/CHIMERE modelling system. The forest fire emissions for both scenarios were estimated and considered in the air quality simulations over Portugal.

The air quality modelling application was performed using the chemistry-transport model CHIMERE (Schmidt et al., 2001; Bessagnet et al., 2004), forced by the musicale model MM5 (Grell et al., 1994). The MM5 model has been used worldwide in several regional climate studies (e.g., Boo et al., 2004; Van Dijck et al., 2005). The MM5/CHIMERE modelling system has already been used in several studies that investigated the impacts of climate change on air pollutants levels over Europe (Szopa et al., 2006) and specifically over Portugal (Carvalho et al., 2010b).

The Fifth-Generation Penn State University/National Center for Atmospheric Research (PNU/NCAR) Mesoscale Model, known as the MM5, is a no hydrostatic, vertical sigma-coordinate model designed to simulate musicale atmospheric circulations. MM5 has multiple nesting capabilities, availability of four-dimensional data assimilation (FDDA), and a large variety of physics options. The selected MM5 physical options were based on the already performed validation and sensitivity studies over Portugal (Ferreira et al., 2004; Aquilina et al., 2005; Carvalho et al., 2006) and over the Iberian Peninsula (Fernandez et al.,

2007). A detailed description of the selected simulation characteristics is presented in Carvalho et al. (2010b). The MM5 model generated the several meteorological fields required by CHIMERE model, such as wind, temperature, water vapour mixing ratio, cloud liquid water content, 2 m temperature, surface heat and moisture fluxes and precipitation.

CHIMERE is a tri-dimensional chemistry-transport model, based on the integration of the continuity equation for the concentrations of several chemical species in each cell of a given grid. It was developed for simulating gas-phase chemistry (Schmidt et al., 2001), aerosol formation, transport, and deposition (Bessagnet et al., 2004; Vautard et al., 2005) at European and urban scales. The meteorological input variables driven by the MM5 model are linearly interpolated to the CHIMERE grid. In addition to the meteorological input, the CHIMERE model needs boundary and initial conditions, emission data, and the land use and topography characterization. The non-methane volatile organic compounds (NMVOCs) are disaggregated into 227 individual VOCs according to the speciation suggested by Passant (2002) for each activity sector. The methodology for biogenic emissions of isoprene and terrenes is described in Schmidt et al. (2001). The land use database comes from the Global Land Cover Facility (Hansen et al., 2000), providing the grid cell coverage of coniferous and broadleaf forests. The Stohl et al. (1996) methodology is used for biogenic emissions of nitrogen monoxide (NO) from fertilized soils. The gas-phase chemistry scheme, derived from the original complete chemical mechanism MELCHIOR (Lattuati, 1997), has been extended to include sulphur aqueous chemistry, secondary organic chemistry and heterogeneous chemistry of HONO and nitrate (Hodzic et al., 2005). The model simulates the concentration of 44 gaseous species and 6 aerosol chemical compounds.

The CHIMERE model requires hourly spatially resolved emissions for the main anthropogenic gas and aerosol species. For the simulation over Europe, the anthropogenic emissions for NO_x, CO, SO_2, NMVOC and NH_3 gas-phase species, and for PM2.5 and PM10 are provided by the EMEP (Co-operative Programme for Monitoring and Evaluation of the Long-range Transmission of Air Pollutants in Europe) (Vestreng, 2003) with a spatial resolution of 50 km. The national inventory INERPA was used over the Portugal domain (Monteiro et al., 2005). This inventory takes into account annual emissions from line sources (streets and highways), area sources (industrial and residential combustion, solvents and others) and large point sources (with available monitoring data at each industrial plant). Time disaggregation was calculated by the application of monthly, weekly, and hourly profiles obtained in the scope of the GENEMIS Project (GENEMIS, 1994).

In the present analysis, the CHIMERE model was applied first at the European scale (with 50 x 50 km^2 resolution) and then over Portugal using the same physics and a simple one-way nesting technique, with 10 x 10 km^2 horizontal resolution. The vertical resolution consists of eight vertical layers of various thicknesses extending from ground to 500 hPa, with the first layer at 50 m. Lateral and top boundaries for the large-scale run were obtained from the LMDz-INCA (gas species) (Hauglustaine et al., 2004) and GOCART (aerosols) (Ginoux et al., 2001) global chemistry transport models. Transport of Saharan dust from the GOCART boundary conditions, as well as within-domain erosion, are considered using the formulation of Vautard et al. (2005). For the Portugal domain, boundary conditions are provided by the European scale simulation.

In order to simulate the impact of climate change on air quality the MM5/CHIMERE modelling system was forced by the Hadley Centre global atmospheric circulation model HadAM3P (Jones et al., 2005). Reference (1990) and the IPCC SRES-A2 climatic scenario

(2100) over Europe and over Portugal were simulated by dynamical downscalling using the outputs of HadAM3P, as initial and boundary conditions to the MM5 model. The MM5 model requires initial and time evolving boundary conditions for wind components, temperature, geopotential height, relative humidity and surface pressure. MM5 also requires the specification of SSTs Carvalho et al. (2010b).

The integration between the HadAM3P outputs and the MM5 model was set through a programming stage that was implemented in order to convert, interpolate and generate the pressure levels and the data formats requirements needed for the MM5 simulations. The downscaling of the global model outputs to the MM5 model has already been carried out for the regional climate change simulations over South America (Solman et al., 2007).

To better evaluate the influence of the future fire activity on air quality, the anthropogenic emissions were kept constant in the simulations for the 2100 scenario. The emissions were not scaled in accordance to the IPCC SRES A2 scenario. The air quality simulations assumed no changes in regional anthropogenic emissions of the chemical species primarily involved in the chemical reactions of ozone formation and destruction, but only accounted for changes in the climate. This idealized regional model simulation provides insights into the contribution of possible future climate changes on ozone and particulate matter concentrations. The forest fire emissions were just included in the simulation over Portugal. The European domain simulation did not take into account these emissions.

Based on future area burned projections (Carvalho et al., 2010a) it was possible to estimate forest fire emissions. Forest fire emissions depend on multiple and interdependent factors like forest fuels characteristics, burning efficiency, burning phase, fire type, meteorology, and geographical location. The forest fire emissions concerning CO_2, CO, methane (CH_4), non methane hydrocarbons (NMHC), PM2.5, PM10 and NO_x were estimated based on the methodology investigated in previous studies over Portugal (Miranda et al., 2004; Miranda et al., 2005b). The annual forest fire emissions were estimated for 1990 and for 2100 climates and then included in the CHIMERE model application over Portugal (Carvalho, 2008). Monthly and hourly profiles of forest fire activity were considered in the emissions temporal disaggregation for both climates. All Portuguese districts suffer a substantial increase in emissions due to the projected increases on the area burned (Carvalho, 2008). This increase on the forest fire emissions is proportional to the area burned projections. The annual CO_2 equivalent emissions derived from forest fires account for 1.27 Mton for the 1980-1990 period and 7.44 Mton in a 2 x CO_2 scenario. This represents an overall increase of approximately 500 %.

The MM5/CHIMERE simulations were conducted from May 1st to October 30th for 1990 and 2100. Over Portugal the simulation design comprised three approaches:

- Control simulation (C1) - 1990 climate and 1990 forest fire emissions;
- Scenario 1 (S1) - 2100 climate and 1990 forest fire emissions;
- Scenario 2 (S2) - 2100 climate and 2100 forest fire emissions.

In this sense and in order to assess the impact on air quality it is possible to analyse the changes only due to climate change and the impact of both climate change and future forest fire emissions.

Figure 9 exhibits the monthly mean O_3 changes over Portugal due to climate change alone and to climate change and future forest fire emissions for July, August and September. The O_3 levels are exhibited as differences between future and reference scenarios (S1-C1 and S2-C1).

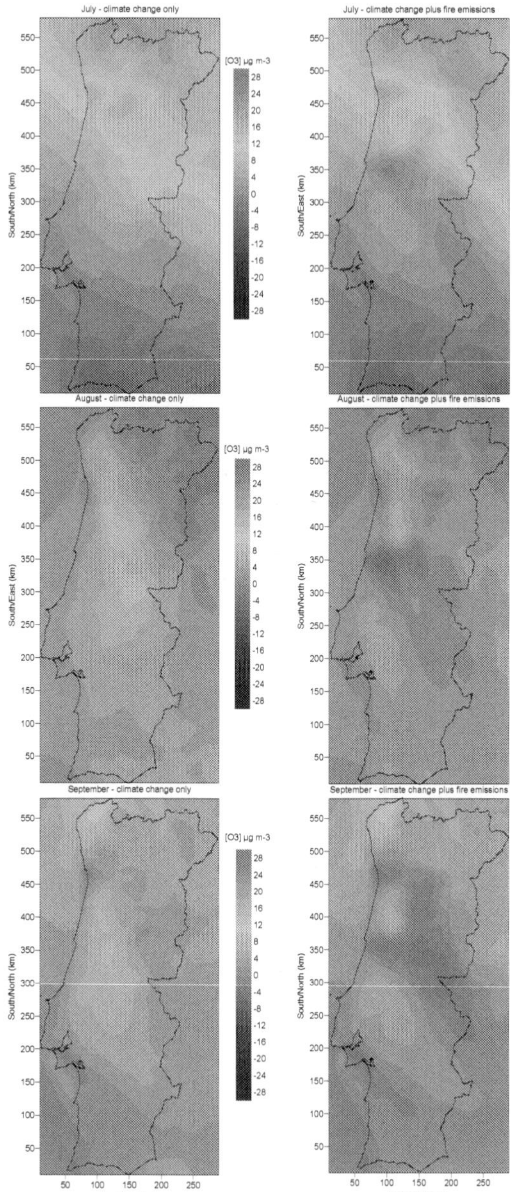

Fig. 9. Monthly mean surface O_3 changes simulated over Portugal considering only climate change (S1 - C1) and climate change and future fire emissions (S2 - C1) for July, August, and September.

The highest increase in O_3 concentrations are detected in July but this is clearly influenced by the air quality boundary conditions settled in the north of Portugal by the European domain (Carvalho et al. (2010b)). In July there is an increase of approximately 20 μg.m-3 in the O_3 levels in the north and central region of Portugal only due to climate change Carvalho et al. (2010b). If future forest fire emissions are considered the regions in the centre of Portugal especially over Coimbra and Porto in the north experience a smaller increase or even a reduction in the O_3 concentrations (-1.2 μg.m-3 in July, -4.9 μg.m-3 in August and -3.8 μg.m-3 in September). This feature is probably due to the O_3 consumption promoted by the O_3 precursor's emissions (like NO_x, CO and VOC) released by the forest fires in these regions.

The area burned projections (Carvalho et al., 2010a) highlight the district of Coimbra as the main affected. Consequently the future forest fire emissions are highest in this region. The O_3 precursor's emissions may also lead to its depletion (e.g. through NO titration) and the overall balance may conduct to the diminishing of the O_3 levels in the atmosphere. It is also expectable an increase of the O_3 concentrations downwind of the fire due to the dispersion of the emitted pollutants and their chemical transformation (Stich et al., 2007). In Figure 9 it is possible to see the depletion of the O_3 levels in July, August, and September, although the monthly average analysis does not allow verifying a clear increase of the O_3 concentrations downwind of the fire locations.

The ozone levels in the atmosphere present a markedly daily profile closely connected to the photochemical activity that reaches its maximum during the afternoon. In this sense and in order to make a more detailed discussion on the O_3 concentrations change along the diurnal cycle the O_3 average values at 12, 15 and 18 UTC are discussed for August (Figure 10).

At noon, the highest O_3 level is observed for the S2 scenario reaching 108 μg.m-3 in the inner part of the country (not shown). Considering the climate change impacts only, it is possible to see an increase of the O_3 concentrations and its plume extension in the districts of Porto and Coimbra.

The highest levels of ozone in the atmosphere are observed at 15 UTC, reaching almost 130 μg.m-3 in the S2 scenario (not shown). It can also be detected the increase of the pollutant plume with higher concentrations and its spreading towards the centre and the southern part of the country. At this time Coimbra district registers a decrease on the ozone levels (S2-C1) (-10 μg.m-3) that may be related with the higher amounts of forest fire emissions released in this region. It is also possible to see a clear increase on the O_3 plume concentrations northern and southern of Coimbra. The increase on the O_3 levels reaches approximately 27 μg.m-3 and 28 μg.m-3 for the S1 and S2 scenario, respectively.

By the end of the afternoon, at 18 UTC, the ozone plume differences (scenario S2-C1) with higher concentrations diminish its extension and a decrease pattern can be clearly identified in the centre of Portugal (districts of Viseu, Coimbra, and Castelo Branco). The ozone precursor's emissions due to forest fire activity are consuming the ozone that was previously produced. This decrease can reach up to -30 μg.m-3. It can also be observed the decrease of the ozone concentrations over a larger extension in the surroundings of the main Portuguese cities like Porto and Lisbon.

The monthly and the hourly analysis of the average ozone patterns over Portugal allow verifying that climate change alone may significantly impact the pollutant levels in the atmosphere especially in July and August. For instance, the projected increases on temperature in summer may deeply influence the kinetic rates of the atmospheric chemical cycles. The projected impacts of climate change on the boundary height, wind speed and relative humidity may also influence the obtained ozone concentration patterns (Carvalho et al., 2010b).

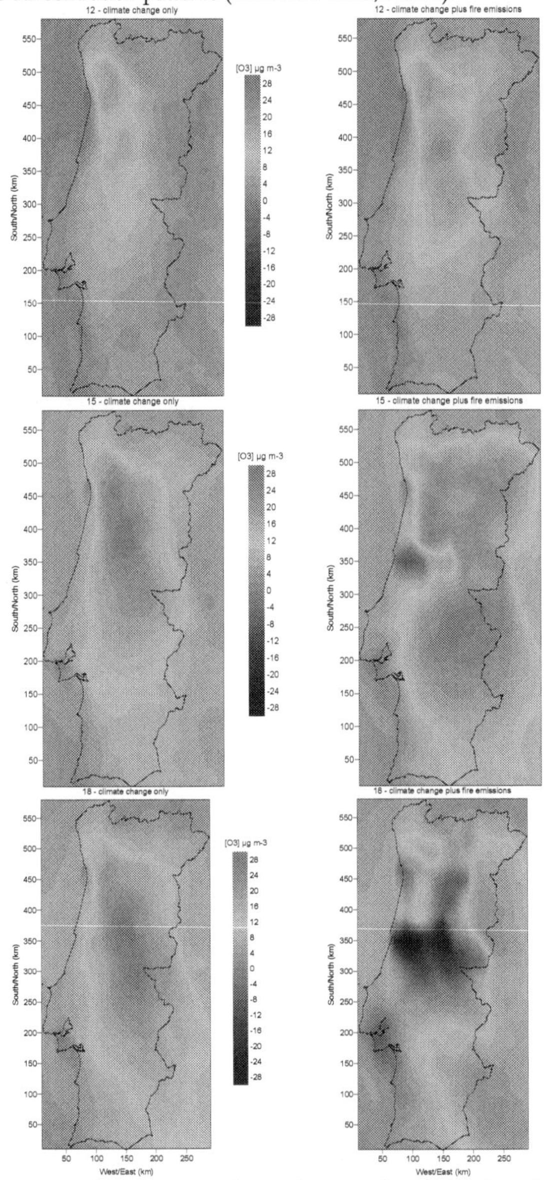

Fig. 10. Hourly average O_3 concentrations changes for August at 12, 15 and 18 UTC considering climate change only (S1-C1) and climate change and future forest fire emissions (S2-C1).

The interaction between the emitted pollutants and the overall chemical reactions in the atmosphere under a changing climate may lead to increases and decreases of ozone values depending on the region. The hourly average of the ozone daily profile gives important information regarding the pollutant patterns distribution in the vicinity of the fires and distant from their main locations. It is clearly that there is a decrease of the ozone concentrations just close to the forest fires and an increase in the surrounding areas. The diurnal evolution of the obtained ozone differences is also closely connected to the forest fire emissions hourly profiles considered in the numerical modeling that allocates the highest percentage of the released emissions from noon to 18 UTC (Carvalho, 2008). After 18 UTC the forest fire pollutants emitted to the atmosphere are leading to the O_3 consumption.

Particulate matter is an important pollutant emitted from forest fires that can lead to severe air pollution episodes and visibility impairment (Valente et al., 2007). Figure 11 depicts the monthly average of the PM10 changes for July, August, and September considering only climate change and climate change and future forest fire emissions.

Only due to climate change impact the PM10 concentrations diminish almost 36 $\mu g.m^{-3}$ over Porto region in May. In the rest of the country the values register a maximum increase of 4 $\mu g.m^{-3}$. In June the range of variation of the PM10 concentrations goes from -10 $\mu g.m^{-3}$ over Porto region to +10 $\mu g.m^{-3}$ along the coastal regions (Carvalho et al., 2010b).

In July it is possible to see the different plume patterns between simulations considering or not the future forest fire emissions. The PM10 levels increase 20 $\mu g.m^{-3}$ over Porto region due to climate change and future forest fire emissions. Only due to climate change the PM10 levels in July may raise up to 18 $\mu g.m^{-3}$. The influence of future forest fire emissions is visible in the PM10 plume extension presenting higher concentrations over Porto, Coimbra, and Viseu districts. It is clearly visible that the registered increases are located in the north and centre part of Portugal. The PM10 values diminish over the Atlantic Ocean and over the coast in Sines region.

In August the PM10 plume shows its highest concentrations again over the centre of Portugal and in Bragança district for the simulation considering climate change and future forest fire emissions. The maximum increase in PM10 values is 15 $\mu g.m^{-3}$ considering only climate change and 16 $\mu g\ m^{-3}$ under climate change and future forest fire emissions. The PM10 dispersion plume clearly shows the influence of the forest fires emissions on the atmospheric concentrations of this pollutant.

September and October register the highest increases on the PM10 values reaching 30 $\mu g.m^{-3}$ and 26 $\mu g.m^{-3}$, respectively, just due to climate change (Carvalho et al., 2010b). The maximum increases are always observed over Porto region. In September, the increase on the PM10 values due to climate change and future forest fire emissions are visible in the pollutants dispersion plume with higher values over Coimbra, Viseu, and Castelo Branco districts with a maximum increase of 32 $\mu g.m^{-3}$.

In summary, the monthly PM10 average values revealed that climate change may deeply impact its levels in the atmosphere. The Porto region is the most affected one in terms of PM10 increases (Carvalho et al., 2010b). Nowadays Porto region faces specific air quality problems closely related to the high levels of PM10 that are registered at the monitoring network (Borrego et al., 2010).

The months of July, August, and September present a clear increase in the PM10 levels due to climate change and the inclusion of future forest fire emissions reaches concentration increases of 32 $\mu g.m^{-3}$. Climate change alone may increase the PM10 average levels in

30 μg.m-3 (Carvalho et al., 2010b). At some extent this should be related to the different dispersion characteristics that may prevail in future climate namely related to boundary layer height, relative humidity and wind speed (Carvalho, 2008). Using a global chemical transport model Spracklen et al. (2009) have estimated that climate change will increase summertime organic carbon (OC) aerosol concentrations over the western United States by 40% and elemental carbon (EC) concentrations by 20% from 2000 to 2050. Most of this increase (75% for OC and 95% for EC) is caused by larger wildfire emissions with the rest caused by changes in meteorology and for OC by increased monoterpene emissions in a warmer climate.

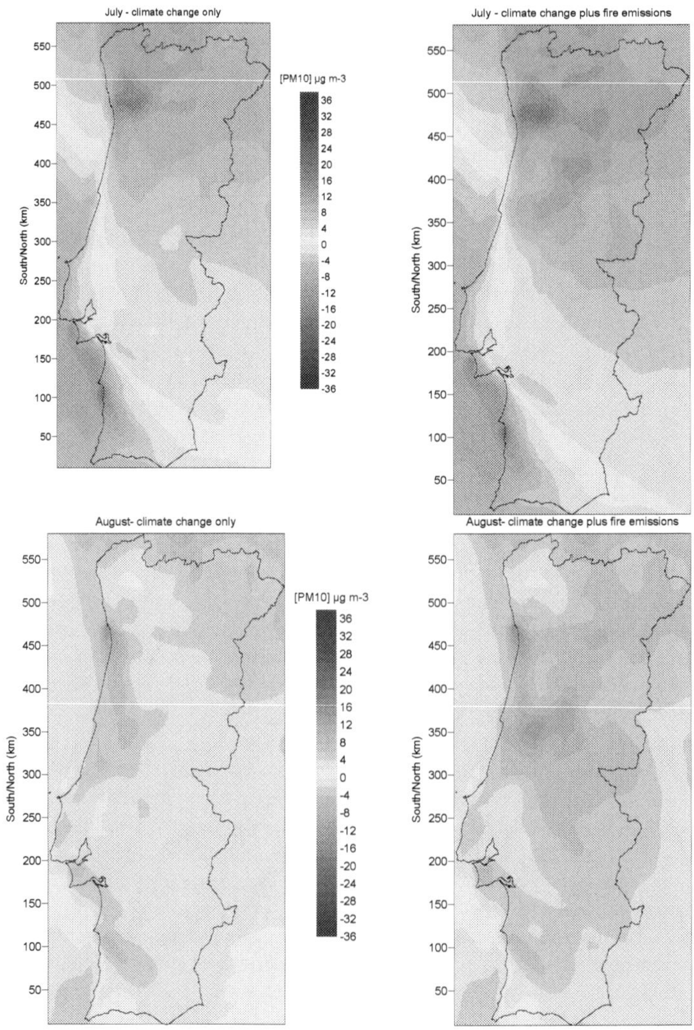

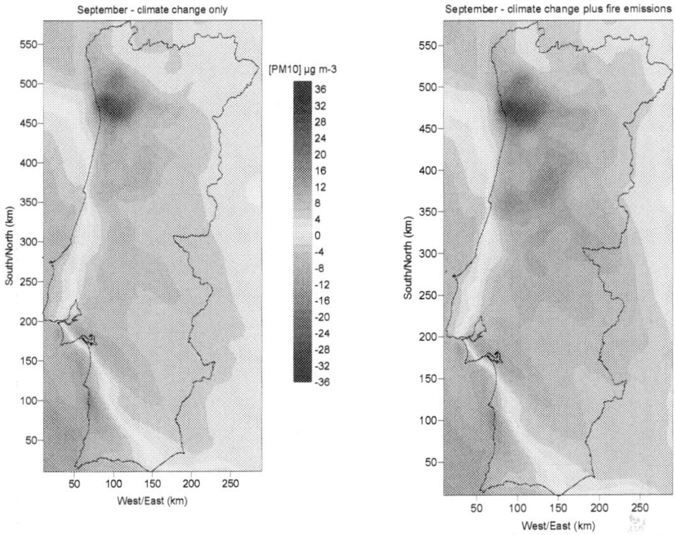

Fig. 11. Monthly average PM10 concentrations over Portugal considering climate change only (S1 - C1) and climate change and future fire emissions (S2 - C1) for July, August and September.

6. Final remarks

The main aim of this study was to evaluate the impacts of the IPCC SRES A2 climatic scenario on forest fire activity and on air quality over Portugal.

The analysis indicates an increase in the average and extremes values of the FWI component in all Portuguese districts. The fire weather severity attains higher values in a future climatic scenario. The districts of the north and centre of Portugal show the highest raises in the FWI component. Almost all the Portuguese districts will face at least 100 % increase on the fire weather risk during spring.

The percentile estimations of the FWI system components were evaluated for both climates and for each Portuguese district. The obtained cumulative frequency functions clearly show the fire weather severity shifts to attain higher values in a future climatic scenario. The districts of the north and centre of Portugal show the highest increases in the FWI cumulative frequency distribution. This fact may be closely related to the occurrence of a higher number of extreme events under the SRES A2 scenario. The occurrence of these extreme weather conditions will dramatically influence forest fire activity over Portugal.

The estimated impact on future fire weather risk will dramatically increase the future area burned and number of fires. The implication of the projected area burned on future forest fire emissions and its impacts on air quality was assessed. The MM5/CHIMERE simulations pointed out that there may be significant increases of O_3 and PM10 levels in the atmosphere under climate change conditions but decreases over specific regions may also be registered.

This analysis revealed that an increase on future forest fire emissions does not directly mean that ozone levels in the atmosphere will increase. The interplay between pollutants concentrations (like NO_x and VOC), surface emissions, and meteorology leads to strong nonlinearities for the atmospheric ozone chemistry. The interaction between ozone precursor's emissions and ozone formation and depletion may be deeply impacted under future climatic scenarios. The knowledge of these relationships constitutes an important tool to correctly evaluate the role of forest fires on air quality under a changing climate.

The projected impacts of forest fire emissions on O_3 and PM10 levels in the atmosphere raise the concern regarding the application of prescribed burning as a management tool. It is recognized that forest fires release high amounts of pollutants to the atmosphere that, in the short term, may lead to acute air pollution episodes with important human health injuries. An adequate prescribed burning planning should also consider the potential impacts of forest fire emissions on the air quality of a region. The obligation for the fulfilment of the European and national air quality standards is an important issue to be taken into account during these initiatives.

The achieved results point to dramatic consequences of climate change on future forest fire activity and on air quality over Portugal. Future developments should consider other variables that could better represent the relationship between climate change, forestry dynamics, land-use change and future human activities. The use of dynamic vegetation models and/or landscape models could better represent the interaction between weather, vegetation changes, forest fires and human activities. The application of today's developed statistical models implies that the relationships between forest fires and weather would remain the same under future climatic scenario and this may not correspond to the truth. A dynamic analysis of these interactions could lead to a better representation of the weather, fire and climate relationships.

The human influence on forest fire activity is another variable that should be addressed. Due to lack of information it was not possible to effectively assess the influence of human activities and human behaviour on forest fire numbers. This variable may change dramatically in future and thus influencing the forest fire statistics and their related impacts.

The application of more than one climatic scenario gives the opportunity to better characterize the range of possible changes that can be detected in future. An ensemble of the several possible scenarios for future climate may give important information regarding uncertainty analysis and promote a better characterization of the future forest fire activity and air quality over Portugal. The use of an ensemble approach will be particularly important to provide uncertainty information and bracket the response. This would represent an important added value to the already projected changes. The analysis of the impacts of climate change and designed pollutant emissions reduction policies would constitute an important step forward to effectively assess the impact of the implemented measures on the air quality of the next 20 to 30 years.

This work represents an important attempt to relate climate change, forest fires and air quality over Portugal. The achieved results and main outcomes constitute an adequate scientific tool to support the implementation of measures and plans in the forest fire management and in the air quality fields.

7. References

Amiro, B.D.; Todd, J.B., Wotton; B.M., Logan, K.A.; Flannigan, M.D.; Stocks, B.J.; Mason, J.A.; Martell, D.L. & Hirsch, K.G. (2001a). Direct carbon emissions from Canadian forest fires, 1959 to 1999. *Canadian Journal of Forestry Research* 31, 512-525.

Amiro, B.D.; Stocks, B.J.; Alexander, M.E.; Flannigan, M.D. & Wotton, B.M. (2001b). Fire, climate change, carbon and fuel management in the Canadian boreal forest. *International Journal of Wildland Fire* **10**, 405–413.

APIF - Agência para a Prevenção de Incêndios Florestais (Agency for the Forests Fires Prevention) (2005). Proposta Técnica de Plano Nacional de Defesa da Floresta contra Incêndios - Plano de Acção. Vol. II, (Lisboa, Portugal).

Aquilina, N.; Dudek, A.V.; Carvalho, A.; Borrego, C. & Nordeng, T.E. (2005). MM5 high resolution simulations over Lisbon. Geophysical Research Abstracts. Vol. 7, 08685, SRef-ID: 1607-7962/gra/EGU05-A-08685. European Geosciences Union 2005.

Bessagnet, B.; Hodzic, A.; Vautard, R.; Beekmann, M.; Cheinet, S.; Honore; C.; Liousse, C. & Rouil, L. (2004). Aerosol modeling with CHIMERE— Preliminary evaluation at the continental scale. *Atmospheric Environment* 38, 2803- 2817.

Boo, K.; Kwon, W.; Oh, J. & Baek, H. (2004). Response of global warming on regional climate change over Korea: An experiment with the MM5 model. *Geophysical Research Letters* 31, L21206, doi: 10.1029/2004GL021171

Borrego, C.; Miranda, A.I.; Carvalho, A.C. & Fernandez, C. (2000). Climate change impact on the air quality: the Portuguese case. *Global Nest - the International Journal* 2(2), 199-208.

Borrego, C.; Valente, J.; Carvalho, A.; Sá, E.; Lopes, M. & Miranda, A.I. (2010). Contribution of residential wood combustion to the PM10 levels in the atmosphere. *Atmospheric Environment* 44, 642-651, DOI:10.1016/j.atmosenv.2009.11.020

Brown, T.J.; Hall, B.L. & Westerling, A.L. (2004). The impacts of twenty-first century climate change on wildland fire danger in the western United States: an application perspective. *Climatic Change* **62**, 365-388.

Carvalho, A. (2008). Forest fires and air quality under a climate change scenario. PhD Thesis. Department of Environment and Planning. University of Aveiro. Aveiro.

Carvalho, A.; Flannigan, M.; Logan, K.; Gowman, L.; Miranda, A.I. & Borrego C. (2010a). The impact of spatial resolution on area burned and fire occurrence projections in Portugal under climate change. *Climatic Change* **98**, 177–197 DOI: 10.1007/s10584-009-9667-2

Carvalho, A.; Flannigan, M.; Logan, K.; Miranda, A.I. & Borrego, C., 2008: Fire activity in Portugal and its relationship to weather and the Canadian Fire Weather Index System. *International Journal of Wildland Fire* **17**, 328-338.

Carvalho, A., Monteiro, A., Solman, S., Miranda, A.I., Borrego, C., 2010b. Climate-driven changes in air quality over Europe by the end of the 21st century, with special reference to Portugal. Environment Science & Policy, DOI: 10.1016/ j.envsci.2010.05.001.

Carvalho, A.C.; Carvalho, A.; Gelpi, I.; Barreiro, M.; Borrego, C.; Miranda, A.I. & Perez-Munuzuri, V. (2006). Influence of topography and land use on pollutants dispersion in the Atlantic coast of Iberian Peninsula. *Atmospheric Environment* 40 (21), 3969-3982.

Christensen, J.H. & Christensen, O.B. (2007). A summary of the PRUDENCE model projections of changes in European climate by the end of this century. *Climatic Change*, doi: 10.1007/s10584-006-9210-7.

Christensen, J.H.; Christensen, O.B.; Lopez, P.; van Meijgaard, E. & Botzet, M. (1996). The HIRHAM4 Regional Atmospheric Climate Model. Scientific Report 96-4, DMI, Copenhagen.

Crutzen, P.; Heidt, L.; Krasnec, J.; Pollock, W. & Seiler, W. (1979). Biomass burning as a source of atmospheric gases CO, H2, N2O, NO, CH3Cl and COS. *Nature* 282(5736), 253-256.

Dentener, F.; Stevenson, D.; Ellingsen, K.; van Noije, T.; Schultz, M.; Amann, M.; Atherton, C.; Bell, N.; Bergmann, D.; Bey, I.; Bouwman, L.; Butler, T.; Cofala, J.; Collins, B.; Drevet, J.; Doherty, R.; Eickhout, B.; Eskes, H.; Fiore, A.; Gauss, M.; Hauglustaine, D.; Horowitz, L.; Isaksen, I.; Josse, B.; Lawrence, M.; Krol, M.; Lamarque, J. F.; Montanaro, V.; Müller, J. F.; Peuch, V. H.; Pitari, G.; Pyle, J.; Rast, S.; Rodriguez, J.; Sanderson, M.; Savage, N.; Shindell, D.; Strahan, S.; Szopa, S.; Sudo, K.; Wild, O. & Zeng, G. (2006). The Global Atmospheric Environment for the Next Generation. *Environment Science and Technology* **40**, 3586-3594.

DGRF - Direcção Geral dos Recursos Florestais (Forestry Resources General Directorate), 2006: Inventário Florestal Nacional de 1995-1998 (3ª Revisão). Divisão de Planeamento e Estatística, Direcção Geral dos Recursos Florestais. (Lisboa, Portugal)

EC - European Commission (2003). Forest Fires in Europe: 2002 fire campaign. Directorate-General Joint Research Centre, Directorate-General Environment, S.P.I.03.83 EN. (Ispra, Italy)

EC - European Commission (2005). Forest Fires in Europe 2004. Directorate-General Joint Research Centre, Directorate-General Environment, S.P.I.05.147 EN. (Ispra, Italy)

Fernández, J.; Montávez, J.P; Sáenz, J.; González-Rouco, J.F. & Zorita, E. (2007). Sensitivity of the MM5 mesoscale model to physical parameterizations for regional climate studies: Annual cycle. *Journal of Geophysical Research* 112, D04101, doi:10.1029/2005JD006649.

Ferreira, J.; Salmim, L.; Monteiro, A.; Miranda, A. I. & Borrego, C. (2004). Avaliação de episódios de ozono em Portugal através da modelação fotoquímica. In: Actas da 8ª Conferência Nacional de Ambiente, 27-29 Outubro, Lisboa, Portugal, 383-384. (Proceedings in CD Rom).

Flannigan, M.D.; Logan, K.A.; Amiro, B.D.; Skinner, W.R. & Stocks, B.J. (2005). Future area burned in Canada. *Climatic Change* **72**, 1-16.

Flannigan, M.D.; Krawchuk, M.A.; de Groot, W.J.; Wotton, B.M. & Gowman, L.M. (2009). Implications of changing climate for global wildland fire. *International Journal of Wildland Fire* 18, 483–507.

GENEMIS - Generation of European Emission Data for Episodes Project (1994). EUROTRAC Annual Report, 1993, Part 5. EUROTRAC International Scientific Secretariat, Garmisch-Partenkirchen.

Ginoux, P.; Chin, M.; Tegen, I.; Prospero, J.M.; Holben, B.; Dubovik, O. & Lin, S.J. (2001). Sources and distributions of dust aerosols simulated with the GOCART model. *Journal of Geophysical Research* 106, 20255 - 20273.

Grell, G.A.; Dudhia, J. & Stauffer, D.R. (1994). A description of the fifth-generation Penn State/NCAR Mesoscale Model (MM5), Tech. Rep. NCAR/TN-398+STR, Natl. Cent. for Atmos. Res., Boulder, Colorado.

Hansen, M.C.; DeFries, R.S.; Townsend, J.R. & Sohlberg, R. (2000). Global land cover classification at 1 km spatial resolution using a classification tree approach. *International Journal of Remote Sensing* 21(6,7), 1331-1364.

Hauglustaine, D.A.; Hourdin, F.; Jourdain, L.; Filiberti, M.-A.; Walters, S.; Lamarque, J.-F. & Holland, E.A. (2004). Interactive chemistry in the Laboratoire de Météorologie Dynamique general circulation model: description and background tropospheric chemistry evaluation. *Journal of Geophysical Research* 109, D04314, doi:10.1029/2003JD003957.

Hodzic, A.; Vautard, R.; Bessagnet, B.; Lattuati, M. & Moreto, F. (2005). Long-term urban aerosol simulation versus routine particulate matter. *Atmospheric Environment* 39, 5851-5864.

Hogrefe, C.; Leung, L.R.; Mickley, L.J.; Hunt, S.W. & Winner, D.A. (2005). Considering climate change in U.S. air quality management. *Environmental Manager*, 19-23.

Hoinka, K.; Carvalho, A. & Miranda, A.I. (2009). Regional-scale weather patterns and wildland fires in Central Portugal. *International Journal of Wildland Fire* 18, 36–49 DOI: 10.1071/WF07045

INE - Instituto Nacional de Estatística (2003). XIV Recenseamento Geral da População, resultados definitivos. INE, Estimativas Provisórias de População Residente para 31.12.2002, aferidas dos resultados definitivos dos Censos 2001, ajustados com as taxas de cobertura. Instituto Geográfico Português (IGP), Carta Adminisrativa Oficial de Portugal. (Lisboa, Portugal)

IPCC - Intergovernmental Panel on Climate Change (2007). Climate Change 2007: The Physical Science Basis. Contribution of Working Group I to the Fourth Assessment Report of the Intergovernmental Panel on Climate Change. S. Solomon, D. Qin, M. Manning, Z. Chen, M. Marquis, K.B. Averyt, M. Tignor & H.L. Miller (Eds.), Cambridge University Press, Cambridge, 996 pp.

Jones, R.G.; Murphy, J.M.; Hassel, D.C. & Woodage, M.J. (2005). A high resolution atmospheric GCM for the generation of regional climate scenarios. Hadley Center Technical Note 63, Met Office, Exeter, UK.

Jones, R.G.; Murphy, J.M.; Hassel, D.C. & Woodage, M.J. (2005). A high resolution atmospheric GCM for the generation of regional climate scenarios. Hadley Center Technical Note 63, Met Office, Exeter, UK.

Lattuati, M. (1997). Contribution à l'étude du bilan de l'ozone troposphérique à l'interface de l'Europe et de l'Atlantique Nord: modélisation lagrangienne et mesures en altitude. Thèse de sciences, Université Paris 6, France.

Miller, M. (2007). The San Diego Declaration on Climate Change and Fire Management. 4th International Wildland Fire Conference, 13-17 May, Seville, Spain. (Proceedings in CD Rom)

Miranda, A.I.; Borrego, C.; Santos, P.; Sousa, M. & Valente, J. (2004). Database of Forest Fire Emission Factors. Departamento de Ambiente e Ordenamento, Universidade de Aveiro: 2004, AMB-QA-08/2004. Deliverable D251 of SPREAD Project [EVG1-CT-2001-00043].

Miranda, A.I.; Ferreira, J.; Valente, J.; Santos; P.; Amorim, J.H. & Borrego, C. (2005a). Smoke measurements during Gestosa 2002 experimental field fires. *International Journal of Wildland Fire* **14**, 107–116.

Miranda, A.I.; Borrego, C.; Sousa, M.; Valente, J.; Barbosa, P. & Carvalho, A. (2005b). Model of Forest Fire Emissions to the Atmosphere. Deliverable D252 of SPREAD Project (EVG1-CT-2001-00043). Department of Environment and Planning, University of Aveiro AMB-QA-07/2005, Aveiro, Portugal, 48 pp.

Monteiro, A.; Miranda, A.I.; Borrego, C.; Vautard, R., Ferreira, J. & Perez, A.T. (2007). Long-term assessment of particulate matter using CHIMERE model. *Atmospheric Environment*, doi:10.1016/j.atmosenv.2007.06.008

Monteiro, A.; Vautard, R.; Borrego, C. & Miranda, A.I. (2005). Long-term simulations of photo oxidant pollution over Portugal using the CHIMERE model. *Atmospheric Environment* **39**, 3089-3101.

Moriondo, M.; Good, P.; Durão, R.; Bindi, M.; Giannakopoulos, C. & Corte-Real, J. (2006). Potential impact of climate change on fire risk in the Mediterranean area. *Climate Research* **31**, 85-95. doi: 10.3354/cr031085

Nakicenovic, N.; Alcamo, J.; Davis, G.; de Vries, B.; Fenhann, J.; Gaffin, S.; Gregory, K.; Grübler, A.; Jung, T. Y.; Kram, T.; La Rovere, E. L.; Michaelis, L.; Mori, S.; Morita, T.; Pepper, W.; Pitcher, H.; Price, L.; Raihi, K.; Roehrl, A.; Rogner, H-H.; Sankovski, A.; Schlesinger, M.; Shukla, P.; Smith, S.; Swart, R.; van Rooijen, S.; Victor, N. & Dadi, Z. (2000). IPCC Special Report on Emissions Scenarios, Cambridge University Press, Cambridge, United Kingdom and New York, NY, USA, 599 pp.

Passant, N.R. (2002). Speciation of U.K. emissions of non-methane VOC, AEAT/ENV/0545.

Pausas, J.G. & Vallejo, V.R. (1999). The role of fire in European Mediterranean Ecosystems. In: Chuvieco E. (ed.) Remote sensing of large wildfires in the European Mediterranean basin, Springer-Verlag, 3-16.

Pyne, S. (2007). Megaburning: The Meaning of Megafires and the Means of the Management. 4th International Wildland Fire Conference, 13-17 May, Seville, Spain (http://www.wildfire07.es/doc/cd/INTRODUCTORIAS_ST/Pyne_ST1.pdf)

Riebau, A. & Fox, D. (2001). The new smoke management. *International Journal of Wildland Fire* 10, 415–427.

Santos, F.D.; Forbes, K. & Moita, R. (2002). Climate Change in Portugal. Scenarios, Impacts and Adaptation Measures - SIAM Project, Gradiva, Lisboa, Portugal, 454 pp.

SAS Institude Inc. (2004). SAS OnlineDoc®, Version 9.1.3, SAS Institude Inc., Cary, NC.

Schmidt, H.; Derognat, C.; Vautard, R. & Beekmann, M. (2001). A comparison of simulated and observed ozone mixing ratios for the summer of 1998 in Western Europe. *Atmospheric Environment* 35(36), 6277- 6297.

Schumaker, L.L. (1981). Spline functions, basic theory. Wiley-Interscience, 553 pp.

Sitch, S.; Cox, P.; Collins, W. & Huntingford, C. (2007). Indirect radiative forcing of climate change through ozone effects on the land-carbon sink. Nature 448(7155), 791-794.

Solman, S.; Nuñez, M. & Cabré, M.F. (2007). Regional climate change experiments over southern South America. I: present climate. *Climate Dynamics*, doi: 10.1007/s00382-007-0304-3

Spracklen, D.V.; Mickley, L.J.; Logan, J.A.; Hudman, R.C.; Yevich, R.; Flannigan, M.D. & Westerling, A.L. (2009). Impacts of climate change from 2000 to 2050 on wildfire activity and carbonaceous aerosol concentrations in the western United States. *Journal of Geophysical Research* **114**, D20301, doi:10.1029/2008JD010966.

Stohl, A.; Williams, E.; Wotawa, G. & Kromp-Kolb, H. (1996). A European inventory for soil nitric oxide emissions and the effect of these emissions on the photochemical formation of ozone. *Atmospheric Environment* 30, 374-3755.

Szopa, S.; Hauglustaine, D.A.; Vautard, R. & Menut, L. (2006). Future global tropospheric ozone changes and impact on European air quality. *Geophysical Research Letters* 33, L14805, doi:10.1029/2006GL025860.

Valente, J.; Miranda, A.I.; Lopes, A.G.; Borrego, C.; Viegas, D.X. & Lopes, M. (2007). A local-scale modelling system to simulate smoke dispersion. *International Journal of Wildland Fire* 16, 196-203.

Van Dijck, S.; Laouina, A.; Carvalho, A.; Loos, S.; Schipper, A.; Kwast, H.; Nafaa R.; Antari, M.; Rocha, A.; Borrego, C. & Ritsema, C. (2005). Desertification in Northern Morocco due to effects of climate change on groundwater recharge. Desertification in the Mediterranean Region. A Security Issue. Eds. Kepner, W., Rubio, J., Mouat, D., Pedrazzini, F., Springer New York, 614 p., ISBN:1-4020-3758-9.

Van Wagner, C.E. (1987). Development and Structure of the Canadian Forest Fire Weather Index System. Canadian Forest Service, Forestry Technical Report 35, Ottowa, Canada

Vautard, R.; Bessagnet, B.; Chin, M. & Menut, L. (2005). On the contribution of natural Aeolian sources to particulate matter concentrations in Europe: testing hypotheses with a modelling approach. *Atmospheric Environment* 39 (18), 3291-3303

Vestreng, V. (2003). Review and revision of emission data reported to CLRTAP, EMEP Status Report, July 2003.

Viegas, D.X.; Reis, R.M.; Cruz, M.G. & Viegas, M.T. (2004). Calibração do Sistema Canadiano de Perigo de Incêndio para Aplicação em Portugal (Calibration of the Canadian Fire Weather Index System for application over Portugal). Silva Lusitana 12(1), 77-93.

Viegas, D.X.; Sol. B.; Bovio, G.; Nosenzo, A. & Ferreira, A.D. (1999). Comparative study of various methods of fire danger. *International Journal of Wildland Fire* **9**(4), 235-246.

Viegas, D.X.; Abrantes, T.; Palheiro, P.; Santo, F.E.; Viegas, M.T.; Silva, J. & Pessanha, L. (2006). Fire weather during the 2003, 2004 and 2005 fire seasons in Portugal. In V International Conference on Forest Fire Research. Ed D.X. Viegas, Figueira-da-Foz, 2006. Proceedings in CD.

15

The role of mycorrhizas in forest soil stability with climate change

Simard[1], Suzanne W., and Austin[2], Mary E.
[1]*University of British Columbia, Vancouver, Canada, and* [2]*Corvallis, USA*

1. Introduction

Global change and the related loss of biodiversity as a result of explosive human population growth and consumption are the most important issues of our time. Global change, including climate change, nitrogen deposition, land-use change and species invasions, are altering the function, structure and stability of the Earth's ecosystems (Vitousek, 1994; Lovelock, 2009). Climate change specifically has been marked by an 80% increase in atmospheric CO_2 levels and a 0.74 °C increase in average global near-surface temperature over the period 1906–2005, with average temperature projected to increase by an additional 1 to 6°C by 2100 (IPCC, 2007). Warming is expected to continue for centuries, even if greenhouse gas emission are stabilized, owing to time lags associated with climate processes and feedbacks (IPCC, 2007). Precipitation patterns have changed along with temperature, with average annual increases up to 20% in high-latitude regions but decreases up to 20% in mid- and low-latitudinal regions. The changes in temperature and precipitation patterns have resulted in higher sea levels, decreases in the extent of snow and ice, earlier timing of species spring events, upward and poleward shifts in species ranges, increases and earlier spring run-offs, and increases in forest disturbances by fires, insects and diseases. Of critical importance are the effects of global change on soils. Soils store one-third of the Earth's carbon and, therefore, small shifts in soil biogeochemistry could affect the global carbon balance (Schlesinger & Andrews, 2004). The effects of global change on soils are complex, however, with multiple feedbacks across broad spatiotemporal scales that have the potential to further amplify climate change effects on the ecology of the Earth. Changes in soils are already occuring as a result of climate change, and include increased soil temperatures, increased nutrient availability, melting of permafrost, increased ground instability in mountainous regions, and increased erosion from floods (IPCC, 2007).

Forests are especially important in the carbon balance of the Earth. Even though forests comprise only 30% of the terrestrial ecosystems, they store 86% of the above-ground carbon and 73% of the world's soil carbon (Sedjo, 1993). On average, forests store two-thirds of their carbon in soils, where much of it is protected against turnover in soil aggregates or in chemical complexes (FAO, 2006). Forest soils not only absorb and store large quantities of carbon, they also release greenhouse gases such as CO_2, CH_4 and N_2O. The carbon sink and source strengths of soils have been considered relatively stable globally, with the strong sink strength of northern-mid latitudes roughly balanced by the strong source strength of the tropics (Houghton et al., 2000). However, climate change can upset the soil carbon balance, or its functional stability, by reducing carbon storage and causing a large positive feedback to atmospheric CO_2 levels. Indeed, the amount of CO_2 emissions being sequestered by terrestrial ecosystems is declining and they may become a source by the middle of the 21st century (Cox et al., 2000; Kurz et al., 2008a). When this happens, the atmospheric carbon trajectory will become less dependent on human activities and more so on the much larger carbon pools in terrestrial ecosystems and oceans (Cox et al., 2000). To underscore the gravity of this shift, the magnitude of total belowground respiration is already approximately 10 times greater than fossil fuel emissions annually (Lal, 2004). The effect of climate change on soil functional stability is particularly concerning in high latitude ecosystems (boreal forests, taiga, tundra and polar regions) because these systems store 30% of the Earth's carbon, and are currently warming at the fastest rates globally (IPCC, 2007; Schuur et al., 2009). The tundra-polar regions recently became a net source of atmospheric CO_2 (Apps et al., 2005). The functional stability of soils or ecosystems is defined in this paper as the maintenance of soil or ecological complexity within certain bounds so that key processes (e.g., carbon cycling, productivity) are protected and maintained (Levin, 2005).

Although the climate change forecasts by the IPCC (2007) have the illusion of predictable and steady change over the next century, the real changes in climate will likely be sudden and unexpected (Lovelock, 2009). Indeed, non-linearity, unpredictability and disequilibrium characterize the Earth and its ecosystems as complex systems (Levin, 2005). Congruently, the IPCC (2007) is predicting an increase in the frequency of climatic extremes, such as heavy rains, heat waves and hot days/nights. These will affect disturbances caused by fire, drought, hurricanes, windstorms, icestorms, insect and disease outbreaks, and invasion by exotic species, and these are projected to increase in frequency, extent, severity and intensity as climate changes (Dale et al., 2001). Changes in natural disturbance regimes have the potential to increase the uncertainty in climate change projections because of their large effects on terrestrial carbon pools (Houghton et al., 2000; Kurz et al., 2008b). Disturbances could greatly overshadow the direct incremental effects of climate change on forest soil and ecosystem stabililty, or the effect of mitigation efforts (Kurz et al., 2008b). Large increases in forest fire and insect disturbances in Canada since 1980 have already reduced ecosystem carbon storage (Kurz & Apps, 1999). Disturbances not only kill plants and affect soil carbon storage, but they also accelerate nutrient cycling, alter mycorrhizal communities, and change soil foodweb dynamics.

Carbon storage in soils involves complex feedbacks between plants and soil organisms. Carbon storage depends on the balance between carbon inputs through photosynthesis and outputs through autotrophic (root and mycorrhiza) and heterotrophic (soil microbial) respiration (Bardgett et al., 2008). Both photosynthesis and respiration are directly affected by climate change factors; including atmpospheric CO_2 level, soil nutrient availability, and temperature and precipitation patterns. They are also clearly affected by tree mortality. The direct effects of these climate change factors on plants then feed back to indirectly affect the structure and activity of soil microbial communities, which drive nutrient cycling, soil carbon storage, and soil stability (Bever et al., 2002a). The intimate cascading interaction between plants and soil microbes in their response to climate change factors is likely of critical importance in predicting the consequences of climate change to ecosystem stability and the carbon balance. Although the feedbacks are complex and poorly understood, we are already measuring climate change effects on soil carbon in high latitude ecosystems (Apps et al., 2005; Schuur et al., 2009) as well as on the composition and activity of soil communities involved in soil nutrient cycling in northern forests (Treseder, 2008).

Of the soil microbes, mycorrhizal fungi are likely the most intimately involved and responsive to carbon fluxes between plants, soils and the atmosphere, and hence are important to consider in climate change impacts on terrestrial ecosystems. This is because of their pivotal position at the root-soil interface, where they link the aboveground and belowground components of biogeochemical cycles. Mycorrhizal fungi are obligate symbionts with all forest tree species, where they scavenge soil nutrients and water from the soil in exchange for photosynthate from the tree. Without their fungal symbionts, most trees cannot acquire enough soil resources to grow or reproduce; without the trees, the fungi have insufficient energy to carry out their life cycle. Because of this obligatory exchange, mycorrhizal fungi are considered the primary vectors for plant carbon to soils (Talbot et al., 2008) and, conversely, the primary vectors of soil nutrients to plants (Hobbie & Hobbie, 2006). The fungal partner plays a role in other essential services as well, such as increasing soil structure, protecting soil carbon against mineralization, and protecting tree roots against disease or drought. A single mycorrhizal fungus can also link different plants together, thus forming mycorrhizal networks. These networks have been shown to facilitate regeneration of new seedlings, alter species interactions, and change the dynamics of plant communities (Selosse et al., 2006). As such, mycorrhizas are considered key players in the organization and stability of terrestrial ecosystems (Smith & Read, 1997; Simard, 2009).

The objective of this synthesis paper is to review the role of mycorrhizas and mycorrhizal networks in the stability of forest ecosystems and forest soils as climate changes. We start by reviewing the role mycorrhizal fungi play in soil carbon flux dynamics. We then review some of the direct effects of climate change factors (specifically increased CO_2, nutrient availability, temperature and drought) on plants and mycorrhizal fungi. Next, we briefly review the current and potential effects of climate change on forests in North America. The crux of our review, however, is on the role of mycorrhizas and mycorrhizal networks in helping to mitigate the effects of climate change through their stabilizing effects on forest ecosystems. We use our own research in the interior Douglas-fir forests of western North America to illustrate these stabilizing effects, including the role of mycorrhizal networks in forest recovery following disturbance and in soil carbon flux dynamics. We then discuss the potential roles that management can play in helping maintain forest stability as climate changes. The body of studies suggests that mycorrhizal fungi, and their capacity to stabilize forests, will have a significant impact on the terrestrial portion of the global carbon budget.

2. The role of mycorrhizal fungi in soil carbon fluxes

The soil carbon pool is 3.3 times larger than the atmospheric carbon pool and 4.5 times larger than the biological carbon pool (Lal, 2004). As a result, the global carbon balance is strongly influenced by soil carbon flux dynamics. The global soil carbon pool is 2500 Gt, and is comprised of 1500 Gt organic carbon (70%) and 950 Gt inorganic carbon (30%) (Schlesinger & Andrews, 2004; Lal, 2004). The organic portion of the soil pool is comprised of plant roots, fungal biomass, microbial biomass, and decaying residues. It includes fast-cycling sugars, amino acids and proteins, and slow-cycling cellulose, hemicellulose and lignin. The soil organic pool is highly dynamic, variable, and greatly influenced by land use practices (Rice et al., 2004).

There are three functional groups of fungi in soils: mycorrhizal (i.e., mutalists), saprotrophic (i.e., decomposers) and pathogenic (i.e., parasitic) fungi. Of these, the mycorrhizal (plant-fungal) symbiosis is ancient, having evolved over 4.5 million years into a tight mutualism (generally speaking, but there is a continuum in the symbiosis between mutualism and parasitism; Jones & Smith, 2004). The mycorrhizal symbiosis involves thousands of fungal species world-wide (Molina et al., 1992). Mycorrhizas are universally present in all terrestrial biomes, including native forests, woodlands, savannas, grasslands and tundra (Smith & Read, 1997). The three dominant groups are ectomycorrhizas (ECM, primary on trees and shrubs in boreal and temperate ecosystems), ericoid mycorrhizas (ERM, primarily on Ericaceae species in high latitude and high altitude ecosystems), and arbuscular mycorrhizas (AM, primarily on grasses, herbs and tropical tree species).

Many mycorrhizal taxa associate with a broad range of plant species, and thus are considered host generalists. Most fungi in the AM group are considered host generalists, whereas fungi in the ECM and ERM groups include both host generalists and host specialists (i.e., that associate with a narrow group of plant species) (Molina et al., 1992). The low host specificity of many mycorrhizal taxa allows a single mycorrhizal fungal mycelium to link the roots of two or more plants of one or more species in a mycorrhizal network. Increasingly, mycorrhizal networks are recognized as ubiquitous in terrestrial ecosystems, including tropical, temperate and boreal forests (van der Heijden & Horton, 2009). Mycorrhizal networks can function in the mycorrhizal colonization of new seedlings, spread of fungal mycelia, or transfer of carbon, nutrients or water between plants (Simard et al., 2002), thus affecting plant and fungal community dynamics. The architecture of mycorrhizal networks can follow regular, random or scale-free models. In both regular and random networks, links (e.g., fungi) tend to be distributed equally among nodes (e.g., trees). In scale-free models, however, some nodes (e.g., hub trees) are highly linked (Bray, 2003). The architecture of the network reflects its resilience against disturbance (e.g., removal of trees).

All mycorrhizas take up nutrients and water from the soil in exchange for photosynthate carbon from host trees. Photosynthate carbon has been shown to transfer from host plants to mycorrhizal hyphae within hours (Johnson et al., 2002) and this drives half of the belowground microbial activity, with the rest fueled by heterotrophic metabolism of dead organic matter (Högberg & Högberg, 2002). Plants invest photosynthate carbon in mycorrhizas (instead of building their own roots) because the small and profuse hyphae have 60 times more absorptive area than fine roots (Simard et al., 2002). Generally, as nutrient and water limitations increase, plants allocate more photosynthate to mycorrhizal hyphae to increase soil resource uptake. This explains their increasing dominance (relative to bacteria) in high latitude, high altitude or upslope ecosystems (Hobbie, 2006; Högberg et al., 2007). In turn, colonization by mycorrhizal fungi has been shown to up-regulate photosynthesis (Rygiewicz & Anderson, 1994; Miller et al., 2002).

A large portion of photosynthate carbon is allocated belowground and metabolized by roots, mycorrhizal fungi and heterotrophic organisms. The proportion of carbon that is allocated belowground to roots and mycorrhizas has been shown to range from 27-68% of net primary productivity (NPP) in ECM culture studies (Hobbie, 2006). The proportion of carbon allocated directly to mycorrhizal fungi ranges from 1-21% of total NPP (Hobbie, 2006). The amount allocated to root exudation represents 1-10% of NPP, and is important in fueling soil foodwebs and soil organic matter formation (Cardon & Gage, 2006).

Mycorrhizal fungi have a diversity of functions in carbon metabolism. They not only directly access mineral nutrients and water in soil in exchange for photosynthate, they can also decompose soil carbon for energy and nutrient uptake. For example, mycorrhizal fungi have been shown to assimilate simple organic compounds (e.g., amino acids) from the soil solution while in symbiosis (Näsholm et al., 1998). Recently, ECM fungi, ERM fungi and, to a lesser extent AM fungi, have also been found to act as decomposers of larger organic molecules (e.g., proteins, chitin, pectin, hemicellulose, cellulose, polyphenols) by producing extracellular enzymes (e.g., proteases, polyphenol oxidases) (Read & Perez-Moreno, 2003; Tu et al., 2006; Talbot et al., 2008). Talbot et al. (2008) proposed three conditions under which mycorrhizas act as decomposers: (1) when plant photosynthate is low (e.g., in shade, winter-early spring, or when plants are declining) and mycorrhizas require an alternative energy source, (2) when soils are highly organic (e.g., at high latitude or high altitude) and mycorrhizas are required to mine organic nutrients, or (3) when plant productivity is high (especially in crop plants), and mycorrhizal decomposition is primed by large belowground photosynthate carbon fluxes. The model of Talbot et al. (2008) differs from the traditional decomposition model where saprotrophic fungi were considered exclusively responsible for all soil organic matter decomposition. In addition to saproptrophs, however, different taxa of mycorrhiza fungi are now recognized as targeting different carbon sources, implying niche partitioning. This niche partitioning can help explain why such a dazzling diversity of fungi are involved in carbon and nutrient metabolism in soils (Hansen et al., 2008). It also points to the importance of understanding the diverse roles of plants and fungi in global carbon flux dynamics.

Arbuscular mycorrhizal fungi generally do not break down soil organic matter, but they do play important roles in promoting soil aggregation and soil carbon storage. Soil aggregation occurs when hyphae pervade soil pores and entwine soil particles. Mycorrhizal hyphal growth in soils is extensive, with mycelial lengths reaching 111 m cm^{-3} (0.5 mg g^{-1}, or up to 900 kg ha^{-1}) in a prairie soil (Miller et al., 1995). Though AM hyphae turn over quickly (in days to a few months), they also deposit significant quantities of relatively recalcitrant carbon compounds such as chitin and glomalin. Glomalin is a carbon-, nitrogen- and iron-rich glycoprotein produced in fungal cell walls (Treseder & Turner, 2007). When it is deposited during decomposition, glomalin joins hyphae in binding small soil particles, thus promoting aggregation and soil stability. Although it constitutes only 0.4-6% of hyphal biomass, glomalin accumulates in soil macro-aggregates at much higher masses (e.g., >100 mg g^{-1}) than does hyphae. In soil aggregates, glomalin carbon is protected from decomposition by chemicals and soil organisms, allowing it to remain in soils for decades and accumulate over time (Rillig et al., 2001; Zhu & Miller, 2004). Carbon in bulk soil, by contrast, is more vulnerable to decomposition. Hence, AM glomalin represents a large pathway for storage of stable carbon in soils. Glomalin content of soils generally increases with the abundance of AM plants and carbon allocation to AM hyphae, and has been shown to represent 3-8% of soil carbon in undisturbed AM grassland and chaparral communities (Rillig et al., 2001).

The composition of mycorrhizal communities shifts with changes in the balance of carbon and nutrients in soils because of fungal species variation in demands for carbon, nitrogen and phosphorus. For example, increases in carbon allocated belowground with CO_2 enrichment or warming may shift the mycorrhizal community toward dominance by high biomass fungi with proteolytic or long-distance exploration capabilities that enable them to compete for scarce nutrients or contribute to soil carbon storage (Treseder, 2005; Hobbie & Hobbie, 2006). These fungi are also considered important in forming mycorrhizal networks with high transfer capacity (Simard & Durall, 2004). In the next section we discuss how climate change can trigger such shifts in the mycorrhizal fungal community.

3. Effects of climate change factors on mycorrhizal fungi

Climate change is resulting in increasing atmospheric CO_2 concentrations, increasing soil nutrient availability, regional warming and regional drying as a result of fossil fuel burning, land use change, and nutrient pollution. These changes are having multi-faceted effects on plants, mycorrhizal fungi and ecosystems. In this section, we review the key climate factors individually and their potential effects on mycorrhizal fungi.

3.1 Atmospheric CO_2 enrichment

Carbon as atmospheric CO_2 has increased from a pre-industrial level of 280 ppm to 392 ppm in 2010 (Keeling, 1998; IPCC, 2007; http://co2now.org/). The most important effects of atmospheric CO_2 enrichment on mycorrhizal fungi are expected to be indirect through their impacts on plants (Staddon & Fitter, 1998). Plants generally respond to CO_2 enrichment with increased photosynthesis, decreased stomatal conductance, and increased net primary productivity (Poorter, 1993). They also distribute greater amounts of carbon belowground to roots, mycorrhizas, soil foodwebs and exudates (Pritchard et al., 2008; Drigo et al., 2008), due either to greater productivity or shifts in allocation patterns (Zak et al., 2000). Increased availability of carbon to mycorrhizas belowground is considered an important strategy for plants to meet their increasing needs for nutrients and water (Bazazz, 1990; Rogers et al., 1994). In addition to these predicted shifts, mycorrhizal function may also change with increasing CO_2, resulting in lower net carbon costs or increased nutrient-uptake benefits for host plants (Johnson et al., 2005). In addition to acquiring carbon, mycorrhizal fungi also mediate the return of CO_2 to the environment through metabolism and decomposition. The degree to which the increased carbon allocated belowground is rapidly released as CO_2 or allocated to a more recalcitrant soil pool is not well understood.

In keeping with the above predictions, Treseder (2004) found in a meta-analysis of field studies that mycorrhizal abundance increased on average by 47% (84% for AM fungi; 19% for ECM fungi) with increased atmospheric CO_2 concentration; these increases occurred irrespective of biome, level of CO_2 enrichment or measurement method. Meta-analyses are powerful tools that can be used to detect general responses in ecosystems that are often difficult to isolate in individual studies. Individual studies are still critical, however, in uncovering sources of variation and response mechanisms. In long-term CO_2 enrichment experiments, for example, Allen et al. (2005) and Treseder et al. (2003) were able to determine that AM fungal abundance response increased with CO_2 enrichment and peaked at 550-650 ppm. (Some caution is needed in interpreting such experiments because abrupt rises in CO_2 enrichment can over-estimate mycorrhizal responses (Klironomos et al., 2005)). Treseder et al. (2003) also determined that net ecosystem exchange to the atmosphere declined with increasing CO_2, where the extra carbon was added to bulk soil and, to a greater degree, soil macro-aggregates through increased AM hyphal growth and glomalin production. Staddon et al. (1999) also showed that AM fungi stimulated carbon flow belowground with elevated CO_2, but they estimated that most of this belowground carbon was respired. Allen et al. (2005) found that the standing crop of fungi, bacteria and soil organisms did not increase with elevated CO_2 in arid chaparral ecosystems, but they speculated that microbial turnover increased in response to increased carbon allocation belowground. A few other studies have found no effect or even reduced AM fungal colonization with increased CO_2 levels (Staddon & Fitter, 1998). Though the meta-analysis of Treseder (2004) showed strong trends, clearly there are multiple environmental and species influences on mycorrhizal responses to elevated CO_2 that remain to be explored.

Enrichment of CO_2 is expected to cause shifts in mycorrhizal community composition. These shifts will depend on the relative abilities of different fungal taxa to exploit carbon, nitrogen and phosphorus pools, or to acclimatize to the changing environment. Where elevated CO_2 increases belowground carbon allocation and stimulates nutrient deficiencies, "late stage" or medium or long distance "exploration types" of mycorrhizal fungi may be favoured because of their specific exploration strategies for accessing immobile or distant nutrient patches (Agerer, 2001; Hobbie & Agerer, 2010). Where phosphorus is limiting in particular, fungi that invest more carbon into hyphal branching should be favoured because of the relative immobility of this nutrient (Treseder, 2005). In environments where nitrogen is more limiting, however, fungal groups that invest in rhizomorphs that forage over long distances to nitrogen-rich patches should be favoured. Other mycorrhizal fungal taxa may also be favoured in these low nutrient environments because of their adaptations for producing extracellular enzymes to decompose soil organic complexes (see above), or for cultivating associative N-fixing bacteria in their hyphosheres in exchange for nitrogen (Agerer, 2001; Treseder, 2005). Studies examining ECM communities under the low nutrient conditions expected under CO_2 enrichment have found shifts toward morphotypes dominated by extraradical hyphae, rhizomorphs and thin fungal sheaths, and to communities dominated by *Cortinarius, Suillus, Tricholoma* or *Cenococcum* (Lilleskov et al., 2001 and 2002; Treseder, 2005). These exploration fungal types are also considered important in the formation of mycorrhizal networks and transfer of nutrients between plants, suggesting that elevated CO_2 may favour the development of more extensive networks that link plants over long distances.

In addition to its effect on the composition of mycorrhizal communities, elevated CO_2 has also been shown to alter the composition of the broader soil microbial community (Allen et al., 2005). Increased carbon allocation to roots and mycorrhizas stimulates soil foodweb activity, but variation in the amount and quality of carbon can favour specific members of the foodweb. For example, saprotrophic fungi have been shown to increase in abundance with rising CO_2 because of greater inputs of root and leaf litter to the soil (Parrent & Vilgalys, 2007). Modifications in litter chemistry, including increases in lignin concentrations with increasing CO_2 levels (Norby et al., 2001), should also have consequences for soil microbial communities (Bradley et al., 2007).

3.2 Soil nutrient enrichment

Nutrient availability is generally increasing in two ways with global change: through localized anthropogenic nutrient deposition via fertilization and pollution and, to a lesser extent, through increased microbial decomposition with soil warming. Although global change is having the strongest impact on the nitrogen cycle, soil warming has the potential to affect the availability of all soil nutrients. Nitrogen deposition specifically has increased by 3-5 times through industrial fixation and fossil fuel burning, and now exceeds levels of natural nitrogen fixation world-wide (Vitousek, 1994). Because plant productivity is nitrogen-limited globally, NPP has increased and plant distributions have shifted in response to nitrogen enrichment (Vitousek, 1994; Treseder et al., 2005). Currently most nitrogen deposition is in the temperate regions of the USA and Europe, where nitrogen is considered most limiting, but future nitrogen deposition is expected to increasingly occur in the tropics (Dentener et al., 2006). Nutrient enrichment through soil warming can result in increased NPP, but this may ultimately be limited by depletion of the soil nutrient capital (e.g., phosphorus).

In global change studies, scientists have investigated nitrogen enrichment effects on plants and mycorrhizas along nitrogen deposition gradients, in fertilization experiments, and in experiments that have artificially increased soil temperature. These studies have generally shown positive effects of nitrogen enrichment on aboveground plant productivity but negative effects on the belowground foodweb (Treseder et al., 2004). As soil nutrient availability increases, plants have less need for investing carbon into roots, mycorrhizas and microbial activity for nutrient uptake, and therefore they allocate more carbon to aboveground biomass. A recent meta-analysis has indeed shown that industrial nitrogen deposition not only stimulated aboveground forest growth (Thomas et al., 2010), but also reduced soil microbial activity, diversity and soil organic matter decomposition, thus stimulating carbon sequestration in temperate forests (Janssens et al., 2010). Congruently, in a meta-analysis of field fertilization studies, (Treseder, 2004) found that mycorrhizal biomass declined on average by 15% with soil nitrogen enrichment (25% decline in AM biomass versus 5% decline in EM biomass) and by 32% with soil phosphorus enrichment. Similar declines (15%) in total microbial biomass (fungi plus bacteria) with nitrogen additions were observed in a separate meta-analysis of 82 field studies, with greater declines where fertilizer was added over longer periods and at higher amounts (Treseder et al., 2008). Janssens et al. (2010) caution, however, that saturating levels of nitrogen deposition could lead to declines in forest productivity, both above- and belowground, because of soil acidification, leaching of ions and nitrogen, and increasing phosphorus deficiencies. These negative effects may overwhelm any positive effects of nitrogen deposition world-wide, particularly in tropical forests where phosphorus is the primary limitation to tree growth.

Fertilization studies suggest that smaller changes tend to occur in ECM than AM fungal communities and in deciduous than coniferous forests (Peter et al., 2001; Aber et al., 2003; Treseder et al., 2007; Vitousek et al., 2008). Correspondingly, ECM fungal diversity and richness declined in coniferous forests along a nitrogen deposition gradient (Lilleskov et al., 2002), but appeared to decline to a lesser degree in deciduous forests (Arnolds, 1991). The differences in these responses is likely related to the degree to which plant species are nitrogen or phosphorus limited, the diversity of associated fungal species, and the availability of soil mineral and organic nitrogen (Talbot et al., 2008). For example, many deciduous tree species are more nutrient-rich than coniferous species (Simard et al., 1997a; Jerabkova et al., 2006), and should therefore be less sensitive to nutrient additions. In addition, AM plants generally occur in more nutrient rich environments (Smith & Read, 1997), but the wider diversity of fungi that ECM plants host for accessing nutrients may provide a degree of functional similarity that buffers the community against increases in nutrient availability (Jones et al., 2010).

Where nutrients are elevated, "early stage", contact or short distance "exploration types" of mycorrhizal fungi may be favoured because of their ability to rapidly colonize new seedlings and exploit nutrient-rich environments (Deacon & Donaldson, 1983; Hobbie & Agerer, 2010). When plants are initially establishing on disturbed or enriched sites, carbon can be briefly limiting to mycorrhizal growth. Under these conditions, mycorrhizal taxa that allocate more biomass to exchanges sites, such as arbuscules in AM fungi, or the Hartig net in ECM fungi, or those taxa that can acquire carbon from alternate sources, may also be favoured (Treseder, 2005). The decline in ECM fungal diversity observed by Lilleskov et al., (2001, 2002) along a nitrogen deposition gradient corresponded with a shift toward early successional fungi such as *Laccaria, Paxillus* and *Lactarius* that posess these characteristics. Early successional fungi have been shown to form mycorrhizal networks in forests and facilitate carbon and nitrogen transfer over short distances (Simard et al., 1997a), but to a smaller degree than later successional fungi (Teste et al., 2009a). Reductions in mycorrhizal richness, whether involving early or later successional fungi, reduces the complexity of mycorrhizal networks, which has corresponded with lower rates of nutrient transfer and survival of establishing seedlings in temperate forests (Teste et al., 2009a).

Nitrogen deposition can not only reduce mycorrhizal activity and diversity, but it can also favour specific saprotrophic communities (Janssens et al., 2010). After 19 years of annual fertilization at Toolik Lake, Alaska, for example, Deslippe et al. (2010) found an increase in the abundance of saprotrophs and small changes in the ECM fungal community. The increasing group of saprotrophs (as discussed by Janssens et al. (2010)) can be superior at producing cellulose-decomposing and phosphate-acquiring enzymes, but not be very efficient at producing lignin-degrading enzymes. Ironically, the saprotrophs can therefore leave more recalcitrant organic matter, ultimately leading to greater accumulation of soil carbon and reducing respiration. Studies show that a large fraction of this soil organic matter is chemically or physically protected from further microbial decay, particularly where it is associated with clay particles. It is important to note that the more decay resistant carbon is the result of saprotrophic biochemical transformations rather than increased soil aggregation; this is because mycorrhizal abundance and rhizodeposition generally decline with increasing nitrogen availability. The long-term stability of these changes in soil carbon is therefore uncertain.

3.3 Soil warming

Plant growth generally increases with soil temperature, but it can also decline where nutrient deficiencies are induced or soil water availability is reduced through increased rates of evapo-transpiration (Pendall et al., 2004). Where plant productivity increases with soil temperature, mycorrhizal and microbial activity are also predicted to increase to help meet increasing nutrient and water demands (Pendall et al., 2004). In keeping with these predictions, mycorrhizal fungal abundance has been shown to increase with soil warming. However, they have also declined initially where limiting thresholds of nutrient or water availability were exceeded (Rustad et al., 2001). Thus, temperature effects on mycorrhizal activity can be mediated through nutrient and water cycles.

Plants and mycorrhizas are not necessarily limited by the same resources at the same time, and feedbacks between climate change factors will mediate plant, mycorrhizal and soil responses to warming (Hobbie, 2000; Pendall et al., 2004). Moreover, plants and mycorrhizal fungi may acclimate to soil temperature changes (Allison et al., 2010). This suggests we should expect variable effects of soil warming on mycorrhizal fungi depending on the type of plant community, the length of time since warming, and feedbacks among different climate processes. For example, mycorrhizal fungal growth increased following 14 years of warming in the Arctic tundra at Toolik Lake, Alaska (Clemmenson et al., 2006) but declined in mature black spruce forests of Alaska (Allison and Treseder, 2008). After 19-years in the warming treatment at Toolik Lake, Deslippe et al. (2010) found that ECM colonization of the dominant tundra shrub, *Betula nana*, had returned to control levels, suggesting the ECM community had acclimatized to the new conditions. However, they also found an increase in high biomass mycorrhizal fungi with proteolytic capacity, especially *Cortinarius*, and a reduction in fungi with high affinities for nitrogen, especially *Russula*, supposedly reflecting *Betula nana*'s increased demand for nutrients bound in soil organic matter with warming. In the black spruce forest, the decline in fungal biomass likely coincided with reductions in soil moisture and increases in nitrogen availability. Exceeding minimum thresholds in soil moisture due to evapotranspiration appears to constrain the predicted mycorrhizal increases with warming, and this ought to occur more commonly in dry than moist forests or than moist tundra underlain by permafrost.

The positive effects of soil warming on the abundance of mycorrhizal fungal taxa with high biomass and long distance exploration types found by Deslippe et al. (2010) suggests that soil warming should promote development of larger mycorrhizal networks. At Toolik Lake, Deslippe & Simard (2010) found that mycorrhizal networks transferred fixed carbon between *Betula nana* shrubs, but not to other plant species, and the amount was potentially sufficient to affect the performance of *Betula nana*. Development of larger mycorrhizal networks with warming should therefore favour community dominance by *Betula nana* and may help explain its current expansion on the Arctic tundra with warming. The increase in carbon uptake throught expansion of *Betula nana* in the tundra, however, will likely be exceeded by carbon release resulting from permafrost thawing effects (Schuur et al., 2009).

In addition to general increases in mycorrhizal biomass with soil temperature, increases in microbial activity should lead to increases in soil organic matter decomposition. This is because microbes produce extracellular enzymes that catalyze the conversion of soil organic matter to dissolved organic carbon, which is the rate-limiting step in decomposition (Allison et al., 2010). The increase in decomposition with soil warming is speculated to offset the increases in carbon allocation belowground with increased atmospheric CO_2. In keeping with this expectation, soil CO_2 and CH_4 emissions have been found to initially increase in soil warming experiments. However, emission rates have been found to then decline back to control levels within a few years once microbes acclimate to the elevated temperature and allocate less carbon toward biomass growth (i.e., they reduce their carbon-use efficiency) (Allison et al., 2010). The short-term nature of respiration increases have also resulted partly from rapid depletion of labile carbon pools (Bradford et al., 2008).

3.4 Reduced soil water availability

Reduced precipitation predicted for mid-latitude ecosystems will likely result in increasing water limitations to plant growth (IPCC, 2007). When soil water availability declines, plants should allocate more carbon to mycorrhizal fungi so that they can access scarce soil water (Augé, 2001). Conversely, in areas where precipitation increases, less plant carbon should be allocated to mycorrhizal growth. Studies show that drier conditions have tended to increase arbuscular mycorrhizas as predicted but have had variable effects on ectomycorrhizas (Allison & Treseder, 2008). While water limitations are generally expected to increase with climate change, increases in water use efficiency may buffer some of the negative effects of drought. For example, colonization with mycorrhizal fungi can increase plant water use efficiency due to improved phosphorus nutrition (Augé, 2001). Water-use efficiency of plants has also been shown to increase with atmospheric CO_2 (Bazazz, 1990).

3.4 Overall climate change effects

The inter-related effects of climate change factors on forest ecosystems, plants and mycorrhizal fungi are complex and difficult to predict. The results of field studies generally suggest that increased CO_2, soil warming and soil drying should increase plant carbon allocation to mycorrhizas and shift the fungal community to species characterized by high biomass, long distance exporation strategies, and proteolytic capabilities for meeting nutrient demands. These mycorrhizal types should favour development of mycorrhizal networks, which would promote tree species establishment and survival, but they should also have greater decomposition capabilities. Conversely, nutrient enrichment should reduce mycorrhizal fungal abundance and favour early successional mycorrhizal species that are nitrophillic and with more limited networking capacity. This may limit growth of older trees or promote invasion of weedy plants that are less reliant on mycorrhizal fungi for meeting their resource needs. On balance, the present state of knowledge regarding climate change trajectories suggests that forest health will decline in the future and forest soils will become a net source of atmospheric CO_2 (Jones et al., 2004; Pendall et al., 2008; Kliejunas et al., 2009; Kurz et al., 2008a). Changes at high latitudes, including thawing and warming of Arctic and boreal soils are especially at risk of strong positive CO_2 and CH_4 feedbacks to the atmosphere (Schuur et al., 2009), as is evidenced by the recent shift in Arctic soils from being a net carbon sink to a net carbon source (Apps et al., 2005). This has the potential to greatly amplify climate change in the near future (Schuur et al., 2009).

Climate change factors could also alter the functional roles of mycorrhizal species in soil carbon dynamics (e.g., as vectors, scavengers or decomposers) (Talbot et al., 2008). Similarly, these changes could shift the compatibility and cooperation between hosts and fungi along the mutualism-parasitism continuum, and the relative fitness of various mycorrhizal fungi and other microbes that currently protect roots or suppress root disease (Kiers & van der Heijden, 2006; Hoeksema & Forde, 2008; Kliejunas et al., 2009). Specific changes in plant growth and physiology, population genetics, and interactions with changes in the mycorrhizal community, will also affect interplant interactions, plant community composition, and mycorrhizal fungal community composition. Therefore, the direct and indirect effects of climate change on both plants and mycorrhizas should have direct consequences for the global carbon balance.

4. Effects of climate change on forests and their mycorrhizal communities

The effects of climate change on forests are expected to be profound (Aber et al., 2001; Dale et al., 2001). Climate change is expected to affect tree species and forest distributions, forest dynamics and succession, the interactions and co-evolution between trees, mycorrhizal fungi and other mutualists, and ecosystem function (Malcolm et al., 2006; Hamann & Wang, 2006; Whitham et al., 2006). Forest productivity can change slowly in response to the relatively slow and directional changes in mean CO_2 levels, temperature and precipitation, but it can also change rapidly in response to extreme events (e.g., drought, fire, insect outbreaks), which are occurring with greater frequency and severity world-wide (IPCC,

2007; Liu et al., 2010). There are currently 120 documented cases of forest dieback worldwide that are directly attributed to climate change (Allen et al., 2010). Forest decline resulting from climate change, whether due to slow increases in stress or sudden diebacks, has the potential to transform soil microbial communities and cause massive CO_2 feedbacks to the atmosphere. The dieback of 12 million hectares of lodgepole pine due to the mountain pine beetle epidemic in British Columbia, for example, has changed these forests from a net sink to a net source of carbon to the atmosphere (Kurz et al., 2008a).

In the Northern Hemisphere, the distribution of plant communities is expected to change dramatically and idiosyncratically over the next century. The IPCC (2007) generally predicts a northward migration of the boreal and temperate forests, an expansion of prairie and shrub-lands, and a dramatic reduction in the taiga and Arctic tundra. In British Columbia, climate models predict that the sub-boreal, montane and subalpine forests will almost disappear by 2100, while the grasslands and temperate forests will greatly expand to the north (Hamann & Wang, 2006; Spittlehouse, 2008). For each 1°C increase in temperature, forest zones will have to move 160 km (Petit et al., 2004; Hamann & Wang, 2006); for an increase in 4°C over the next century, species in the Northern Hemisphere may have to move northward by 500 km (or 500 m higher in altitude), or a few kilometers per year to find a suitable habitat (IPCC, 2007). This far outpaces the historical tree migration rate of 100-200 m per year estimated from pollen records and chloroplast DNA analyses (MacLachlan & Clark, 2004). If tree species are unable to migrate as predicted or adapt rapidly, they will face extirpation (Aitken et al., 2008). Gene flow of pre-adapted alleles from warmer climates will help tree species migrations at the leading edges of their ranges; however, populations at the rear edges will have greater chance of dieback due to lags in adaptation and migration ability (Aitken et al., 2008). Diebacks and declines are already evident in North American species such as paper birch, trembling aspen, ponderosa pine, pinyon pine and lodgepole pine (Hogg et al., 2002; Mueller et al., 2005; Bouchard et al., 2008; Heineman et al., 2010). Conversely, northward or upward migration is evident in lodgepole pine, white spruce and green alder (Johnstone & Chapin, 2003; Danby & Hik, 2007).

Mycorrhizal fungi, through their obligate role in tree establishment, survival and growth, will play a key role in the conservation of core native forests, minimizing diebacks of forests at the trailing edges of tree species ranges, and facilitating migration of tree species at the leading edges of their ranges. Mycorrhizal networks, and their role in mycorrhization and mediation of resource distribution among trees according to need, will likely play a key role in maintaining both the integrity and reorganization of old and new forests. In the next section, we describe how mycorrhizal networks play an important role in the self-organization and stability of forests.

5. Role of mycorrhizal networks in forest stability with climate change

In this section, we argue that the most important role of mycorrhizas with climate change may be in their stabilizing effects on forests that are under increasing environmental stress. The functional significance of mycorrhizal fungi at these higher levels of ecosystem organization is increasingly recognized, including the role of mycorrhizal networks in forest regeneration, succession and resistance against exotic invasions (Nara & Hogetsu, 2004; Simard & Durall, 2004; Selosse et al., 2006; McGuire et al., 2007; Simard, 2009). Mycorrhizal networks may thus provide a community-based model for feedback pathways that promote forest stability with climate change. We explore this concept with our research on mycorrhizal networks in the interior Douglas-fir forests of British Columbia.

Interior Douglas-fir forests vary widely in composition and structure, from predominantly single-species, uneven-aged forests in the arid and cool climatic regions, to multi-species, even-aged forests in the moist, warm climatic regions of British Columbia. Regardless of this variation in forest composition and structure, Douglas-fir is a dominant tree species. The composition of the fungal community changes with succession, where a few pioneering taxa such as *Wilcoxina rehmii* and *Mycelia atrovirins radicans* dominate the roots of Douglas-fir germinants in the first few years following wildfire or harvesting disturbances (Jones et al., 1997; Teste et al., 2009a; Barker et al., 2010). This is followed by rapid succession to a more diverse, late-stage ECM fungal community increasingly dominated by the *Rhizopogon vinicolor/R. visiculosus* complex (Twieg et al., 2007). The *Rhizopogon* complex joins up to 63 other ectomycorrhizal species in a complex fungal community colonizing interior Douglas-fir (Twieg et al., 2007). Even with shifts in ECM fungal species composition with disturbance and succession, there is enough functional similarity among taxa that total enzyme production by the community remains unchanged (Twieg et al., 2009; Jones et al., 2010). Congruently, seedling nutrient uptake and growth remain stable over a wide range of disturbance severities (Barker & Simard, 2010).

Early and late successional ECM fungi form mycorrhizal networks linking together Douglas-fir trees of many ages in the arid temperate forests (Teste et al., 2009b, Beiler et al., 2010). Douglas-fir can also form linkages with several other tree and shrub species in these forests (Simard et al., 1997b; Hagerman et al., 2001; Twieg et al., 2007). In a dry, uneven-aged interior Douglas-fir forest, we used multi-locus, microsatellite DNA markers to determine that all Douglas-fir trees were interconnected and that the young trees had regenerated within the extensive *Rhizopogon* network of old veteran Douglas-fir trees (Beiler et al., 2010). Most of the young trees were linked to large, old hub (i.e., highly connected) trees, suggesting the network had scale-free properties; thus, the hub trees were important in self-regeneration of the old-growth forests. In similar forests nearby, we examined this experimentally and showed that seedling establishment success increased by four times where they had full access to the mycorrhizal network of older Douglas-fir trees (Teste & Simard, 2008; Teste et al., 2009b). Access to the network not only improved seedling survival and physiology, but seedlings were colonized by a more complex fungal community and received carbon, nitrogen and water transferred from the older trees (Schoonmaker et al., 2008; Teste et al., 2009a,b). The finding that the network had scale-free properties suggests they were robust against random removal or death of individual trees, which would have little effect on the connectivity of the network (Bray, 2003). By contrast, targeted removal of hub trees, such as through high-grade logging, or insects that selectively attact large trees (e.g., bark beetles), would have negative effects on the regeneration system. In fact, network models have shown that removal of highly connected nodes can cause the scale-free network to stop functioning (Bray, 2003). Random networks, where links are distributed equally among nodes, such as those found in the widely spaced *Quercus garryana* forests of California (Southworth et al., 2005) are conversely more likely to unravel from random tree removal.

The mixed Douglas-fir - paper birch stands in the moist, warm Interior Cedar-Hemlock forests are more productive and regenerate more readily after disturbance than do the pure Douglas-fir stands of the dry forests (Simard et al., 2005), but mycorrhizal networks also play a role in Douglas-fir regeneration. In the understory of century-old paper birch and Douglas-fir mixtures, establishment success of Douglas-fir has increased where seedlings were linked into the mycorrhizal network of older trees (Simard et al., 1997b). There, greater regeneration success was associated with seedling colonization by a more complex mycorrhizal network associated with the mature trees (Simard et al., 1997b). In nearby clearcuts, Douglas-fir seedlings have also benefited from simple mycorrhizal networks by receiving carbon from neighbouring paper birch, particularly where Douglas-fir was shaded (Simard et al., 1997a). Net carbon transfer followed a source-sink photosynthate gradient, from carbon- and nutrient-rich paper birch source seedlings to increasingly light-stressed Douglas-fir sink seedlings. Traditional models of forest dynamics predict that regeneration patterns are controlled by competitive interactions with neighbours (Oliver & Larson, 1997), but this study showed that facilitation by networks increased regeneration performance and affected interspecific interactions between paper birch and Douglas-fir, encouraging a more diverse tree community. These tree-species-rich forests are also more resilient to insect attack and disease than pure Douglas-fir forests, as shown when deciduous species are removed by weeding or thinning (Morrison et al., 1988; Baleshta et al., 2004; Simard et al., 2005).

Forest ecosystems are dynamic, and this is illustrated by dynamic patterns and processes in mycorrhizal networks. Not only do the complexity and composition of mycorrhizal networks change over time (Twieg et al., 2007), but belowground fluxes of nutrients change over the growing season with shifts in source-sink gradients among networked plants (Lerat et al., 2002). Using dual $^{13}C/^{14}C$ labelling in the field, Philip (2006) found that the direction of net carbon transfer reversed twice over the growing season: (1) from shooting Douglas-fir to bud-bursting birch in spring; (2) then reversing, from nutrient and photosynthate-enriched paper birch to stressed understory Douglas-fir in summer; and (3) reversing again, from still-photosynthesizing Douglas-fir to senescent paper birch in the fall. The carbon moved back-and-forth between birch and fir through multiple belowground pathways, including mycorrhizal networks, soils, and a non-networked mycorrhizal-soil pathway (Philip et al., 2010). Here, there appears to be a dynamic interplay between birch, fir and the interconnecting fungi, with carbon and nutrients moving in the direction of greater need over the growing season, resulting in an integrated, dynamic system.

Where severe disturbances remove forest floor and trees in dry climates, mycorrhizal networks are disrupted, resulting in greater reliance of new germinants on mycorrhizal colonization by spores or mycorrhizal fragments in the soil (Teste et al., 2009a; Barker et al., 2010). Although severe disturbances are part of the historic mixed fire regime in the interior Douglas-fir forests, they have usually been infrequent and restricted to small patches (Klenner et al., 2008). As the climate of these forests becomes warmer and drier (Hamann & Wang, 2006), severe disturbances are expected to increase in extent and frequency (Dale et al., 2001), raising concerns about mycorrhizal spore production and dispersal into the disturbed areas. Production of fruiting bodies declines in dry summers (Durall et al., 2006), and belowground dispersal of spores by truffle-forming species such as *Rhizopogon* may be limited over extensive openings. Douglas-fir seed rain and regeneration have also been sporadic in these forests (Vyse et al., 2006), and the resulting regeneration lags in dry summers could cause local extinction of *Rhizopogon* and other network-forming ECM fungal taxa. Nevertheless, early successional ECM fungi are host-generalists, and they will continue to play a critical role in seedling establishment following disturbance.

In spite of the greater risk of severe fires disrupting networks in arid climates, the stress-gradient hypothesis suggests that biotic facilitation of Douglas-fir regeneration by mycorrhizal networks should be even greater in stressed environments. We tested this hypothesis along an environmental stress gradient caused by soil disturbance in the dry interior Douglas-fir forests. We found that naturally regenerated Douglas-fir seedlings received more transferred carbon through mycorrhizal networks from their neighbours where soils were disturbed by forest floor removal and compaction than where soils were undisturbed, but only where the seedlings were initially well colonized by EM fungi (Teste et al., 2010). Here, disturbance created a sufficient source-sink gradient between seedlings for carbon transfer to occur, but receiving seedlings also had to be healthy and colonized well enough to generate adequate sink strength. We are also testing network facilitation along a regional precipitation gradient across the interior Douglas-fir forests, from the very dry climate of the Interior Douglas-fir zone to the moist climate of the Interior Cedar-Hemock zone (M. Bingham, unpublished data); this regional climate gradient is serving as a proxy for climate change. Early results suggest that, as expected, older Douglas-fir trees transferred more carbon through *Rhizopogon*-dominated mycorrhiza networks, and thus facilitated tree regeneration more strongly, in dry than in wet climates. A decade of drought combined with western spruce budworm and Douglas-fir bark beetle attack, however, has resulted in extensive dieback of these older trees (Campbell et al., 2003; Maclauchlan et al., 2007). Extensive hub tree mortality in some stands may be exceed thresholds where *Rhizopogon* networks are no longer sufficiently intact to facilitate regeneration (M. Bingham, unpublished data). Simpler networks comprised of early successional fungi, however, ought to continue to play a critical role in regeneration after disturbance (Barker et al., 2010).

6. Facilitating the stabilizing effects of mycorrhizas through forest management

6.1 Historical management practices

Forest management practices that sequester carbon include conservation of native forest, siliviculture practices that emulate natural processes, reforestation of crop-lands, manipulations of tree chemistry to favour lignin, and changes to the soil microbial community. Conversely, those forest management practices that result in net losses of soil organic matter include deforestation or conversion of native forests to plantations (Giller et al., 1997; Lal, 2004; Guo & Gifford, 2002). In an analysis of forest harvesting studies, Nave et al. (2009) showed that harvesting of native forests reduced soil carbon storage by an average of 8%, with considerably more lost from the forest floor (30%) than the mineral soil (no consistent change). In general, forest practices that result in carbon sequestration favour fungi over bacteria (e.g., since fungi have half the respiration rate of bacteria (30-40% versus 60%)). They should also favour fungal taxa that produce prodigious mycorrhizal networks for increased soil aggregation and connectivity, or that produce decay-resistant compounds.
In the interior Douglas-fir forests of British Columbia, forest management practices have generally ignored the importance of ECM fungi or mycorrhizal networks in the natural forest regenerative capacity following disturbance. Over the past century, clearcut or high-grade harvesting along with severe insect attacks and severe fire have taken over mixed fire and insect attacks as the primary disturbance agents in these forests (Campbell et al., 2003; Maclauchlan et al., 2007; Klenner et al., 2008). The standard harvesting practice has been to

remove the tallest, straightest, largest diameter stems (i.e., the hub trees) for their economic value, and leave patches of smaller residual trees and advance regeneration to grow and disperse seed into the harvested gaps (Vyse et al., 2006). These management practices are characterized by high mortality of establishing seedlings and patchy regeneration (40% survival of planted seedlings in the very dry forests; Simard, 2009). High-grading not only compromises mycorrhizal networks and regenerative capacity but probably also affects genotypic diversity of the trees comprising the forests. Indeed, the high-grading management approach, combined with summer drought (Hamann & Wang, 2006), episodic seed dispersal (Vyse et al., 2006) and gap-phase disturbance regime characteristic of interior Douglas-fir forests (Klenner et al., 2008), has lead to variable natural regeneration success across the dry climatic zones of interior Douglas-fir (Vyse et al., 2006; Stark et al., 2006).

In the moist, warm forests, or at the upper elevations of the dry forests, clearcutting followed by planting has become the most common practice. The interior Douglas-fir nursery stock that is planted on to these sites is grown under high watering and nutrient regimes and thus seedlings are non-mycorrhizal when lifted for planting (Kazantseva et al., 2009). Historically, regeneration of interior Douglas-fir under these conditions has been more or less successful provided site preparation is suitable, frost pockets recognized and weather conditions are favourable. However, the recent increase in extended summer droughts and more variable weather conditions in the wetter forest types, combined with the cumulative high-grading effects and increased severity of natural disturbances (Flannigan et al., 2005; Maclauchlan et al., 2007; Klenner et al., 2008; Kurz et al., 2008a), have changed the structure of these forests and lead to greater uncertainty in regeneration outcomes. Tree mortality is expected to increase even further at species trailing edges as summer drought and disturbance severity increase (Dale et al., 2001; Campbell et al., 2003; Parmesan, 2006). Mycorrhizal colonization and networks should become increasingly important to the recovery of these forests from disturbance as they become increasingly drought-stressed during the summer months as predicted by climate models (Hamann & Wang, 2006).

Forest practices, therefore, can play an important role in carbon sequestration and forest stability under climate stress. Conservation of whole intact forests should be a global priority given the alarming trends in climate change and loss of biodiversity. Where harvesting is necessary, however, retention of hub trees and their mycorrhizal networks should help maintain the strong carbon storage capacity of forests that is critical to the global carbon balance. Just as conserving living trees plays a critical role in conserving mycorrhizal diversity and function, mycorrhizas in turn play a critical role in the self organization and productivity of forests. By contrast, large-scale clearcutting not only increases greenhouse gas emissions (Kurz et al., 2008b), it also removes critical hub trees, threatens biodiversity (Jones et al., 2003; Martin et al., 2004) and could promote decline of nearby forests.

6.2 Future management practices

Forests may shift to new stability domains with climate change (Suding et al., 2004). As discussed in Section 4, climate models predict a dramatic shift in tree species ranges in North America (Parmesan, 2006; Hamann & Wang, 2006), typically with northward or eastward migration at the leading edges and extensive mortality at the trailing edges. To help mitigate lags in forest re-assembly and minimize the potential for large carbon pulses to the atmosphere, humans can play an important role in mitigating the decline of existing forests and in assisting tree migrations (Rehfeldt et al., 2001). At the trailing edges of tree species ranges, conservation of hub trees forming complex mycorrhizal networks should increase ecosystem stability by facilitating natural regeneration. Conserving forests in these warmer ecosystems will be very important for the source of pre-adapted alleles they provide for currently colder climates (Aitken et al., 2008), as well as for their strong carbon storage capacity. At the leading edges, an important potential barrier in tree migration may be in the colonization of non-local tree genotypes by weakly compatible local mycorrhizal fungi. Although most temperate trees are colonized by both host-specific and host-generalist ECM fungi (Molina et al., 1992), thus providing insurance against negative fungal community composition shifts, the symbiosis of specific plant and fungal pairings can range from mutualistic to parasitic (Bever, 2002a; Klironomos, 2003; Jones & Smith, 2004). Moreover, recent research shows that plants and fungi benefit more frequently with locally adapted associates (Johnson et al., 2010). Indeed, strong feedbacks between compatible symbionts have historically contributed to species coexistence and stability in plant and fungal communities (Bever, 2002b; Klironomos, 2003), whereas weak or antagonistic feedbacks have resulted in forest plantation failures. This research suggests trees that are migrated to new environments may suffer from poor matchings with local mycorrhizal fungi as well as intense competition with existing plants, leading to uncertain performance within local, existing mycorrhizal networks. Such poor marriages could limit the success of assisted tree migrations and contribute to the loss of forest stability (Suding et al., 2004). The loss could be magnified where management practices fail to conserve a diversity of tree and fungal genotypes (Levin, 2005). By contrast, conserving a genetically diverse community of mycorrhizal fungi at the leading edge of tree species ranges may reduce the risk of deleterious matchings and facilitate regeneration of genetically diverse forests with high adaptive capacity (*sensu* Whitham et al., 2006).

Even with good management and assisted migrations, mature and juvenile tree mortality is expected to increase from disturbances associated with climate change (IPCC, 2007). Mortality can be managed in a manner that eases the transition from one forest type to another, however, by conserving the structural and functional legacies of the original forest and establishing the new forest before the old trees are completely dead. It is well established that healthy plants can transfer nutrients to other healthy plants directly through mycorrhizal networks (Simard & Durall, 2004; Selosse et al., 2006), but even larger amounts may transfer from dying trees to healthy roots (Simard et al., 2002; Pietikäinen & Kytöviita, 2007). Where the dying native forest is protected (i.e., not salvage logged) until the new generation or community of trees is established, the new seedlings may be poised to capture nutrients released from the mycorrhizal network of the dying trees before they are acquired by soil microbes. Where new seedlings are not established during the dying process, the organic compounds exuded from senescing roots may be rapidly immobilized by the rhizosphere microbial community and, through turnover, the CO_2 respired back to the atmosphere and the inorganic nutrients released to the soil solution for microbial uptake, other plant uptake, or leaching. If germinants of native plants can avoid competition with soil microbes by acquiring carbon and nutrients directly from dying trees through a mycorrhizal network, they may establish more rapidly, thus increasing competiveness with non-networking invasive plants and reducing CO_2 feedback to the atmosphere. Mycorrhizal networks connecting new generations with old in forests under climate stress may thus be important in conserving existing forests, facilitating native plant establishment and migration, providing barriers to weed invasion, and mitigating large carbon losses.

7. Conclusions

Forests have been diminishing world-wide because of land-use changes and are experiencing additional stress from climate change. While CO_2 enrichment, warming and nutrient pollution are increasing forest productivity and belowground carbon sequestration in North America and Europe, increases in drought, extreme weather events and deforestation practices are also pushing disturbance regimes outside of their natural range. Wide-spread forest diebacks or decline are already occuring in response to increasing drought, wildfire, and insect and disease attacks with climate change. These have the potential to outweigh any positive effects of climate change factors on increased belowground carbon allocation to mycorrhizas, soil microbes or roots. Recovery of these forests is uncertain given the changing dynamics between climate, trees and their mutualists, as well the changing severity and extent of disturbances. The potential for positive feedbacks from dying forest respiration to atmospheric CO_2 levels is high. Humans can play an important role in mitigating forest mortality and assisting migration of species, thus dampening the impacts of climate change.

Mycorrhizas play an important role in the recovery and organization of forests, and it therefore follows that conservation of mycorrhizal fungal communities should help stabilize forests and soils with climate change. Mycorrhizal networks form rapidly following disturbance in the interior Douglas-fir forests of British Columbia, providing critical water and nutrients to establishing seedlings. In mature Douglas-fir forests, most trees, even those of different species, ages and sizes, are connected by a mycorrhizal network. The extensive networks of large hub trees facilitate regeneration of younger trees in the understory, helping them tolerate the stressful environmental conditions. Mycorrhizal networks and hub trees are foundational to the organization of forests because they create favorable local conditions for tree establishment and growth. Therefore, conserving hub trees and mycorrhizal networks appears important to the conservation, regeneration and restoration of forests. Conserving forests, mitigating or managing forest diebacks or declines, and assisting migration of tree species are all important strategies for adapting to the effects of climate change.

8. References

Aber, J.D., Neilson, R.F., McNulty, S., Lenihan, J.M., Bachelet, D. & Drapek, R.J. 2001. Forest processes and global environmental change: predicting the effects of individual and multiple stressors. BioScience 51: 735–751.

Aber, J.D., Goodale, C.L., Ollinger, S.V., Smith, M.-L., Magill, A.H., Martin, M.E., Hallett, R.A. & Stoddard, J.L. 2003. Is nitrogen deposition altering the nitrogen status of north-eastern forests? BioScience 53: 375-389.

Agerer, R. 2001. Exploration types of ectomycorrhizae: A proposal to classify ectomycorrhizal mycelial systems according to their patterns of differentiation and putative ecological importance. Mycorrhiza 11: 107–114.

Aitken, S.N., Yeaman, S., Holliday, J., Wang, T. & Curtis-McLane, S. 2008. Adaptation, migration or extirpation: climate change outcomes for tree populations. Evolutionary Applications 1: 95-111.

Allen, C.D., Macalady, A.K., Chenchouni, H., Bachelet, D., McDowell, N., Vennetier, M., Kitzberger, T., Rigling, A., Breshears, D.D., Hogg, E.H., Gonzalez, P., Fensham, R., Zhang, Z., Castro, J., Demidova, N., Lim, J.-H., Allard, G., Running, S.W., Semerci, A., & Cobb., N. 2010. A global overview of drought and heat-induced tree mortality reveals emerging climate change risks for forests. Forest Ecology & Management 259: 660-684.

Allen, M.F., Klironomos, J.N., Treseder, K.K. & Oechel, W.C. 2005. Responses of soil biota to elevated CO_2 in a chaparral ecosystem. Ecological Applications 15: 1701-1711.

Allison, S.D. & Treseder, K.K. 2008. Warming and drying suppress microbial activity and carbon cycling in boreal forest soils. Global Change Biology 14: 2898-2909.

Allison, S.D., Hanson, C.A. & Treseder, K.K. 2007. Nitrogen fertilization reduces diversity and alters community structure of active fungi in boreal ecosystems. Soil Biology & Biochemistry 39: 1878-1887.

Allison, S.D., Czimczik, C.I. & Treseder, K.K. 2008. Microbial activity and soil respiration under nitrogen addition in Alaskan boreal forest. Global Change Biology 14: 1156-1168.

Allison, S.D., Wallenstein, M.D. & Bradford, M.A. 2010. Soil-carbon response to warming dependent on microbial physiology. Nature Geoscience 3: 336-340.

Apps, M. J., Kurz, W.A., Luxmoore, R.J., Nilsson, L.O., Sedjo, R.A., Schmidt, R., Simpson, L.G. & Vinson, T.S. 2005. Boreal forests and tundra. Water, Air, & Soil Pollution 70: 39-53.

Arnolds, E. 1991. Decline of ectomycorrhizal fungi in Europe. Agr., Ecos. & Env. 35: 209-244.

Augé, R.M. 2001. Water relations, drought and vesicular-arbuscular mycorrhizal symbiosis. Mycorrhiza 11: 3–42.

Baleshta K., Simard, S.W., Guy, R.D. & Chanway, C. 2005. Reducing paper birch density increases Douglas-fir growth and Armillaria root disease incidence. Forest Ecology & Management 208: 1-13.

Bardgett, R.D., Freeman, C. & Ostle, N.J. 2008. Microbial contributions to climate change through carbon cycle feedbacks. ISME Journal 2: 805-814.

Barker, J.S. & Simard, S.W. 2010. Natural regeneration potential of Douglas-fir along wildfire and clearcut severity gradients. In preparation.

Barker, J.S., Simard, S.W., Jones, M.D. & Durall, D.M. 2010. The influence of wildfire and clear-cutting on ectomycorrhizas of naturally regenerating interior Douglas-fir. In preparation.

Bazazz, F.A. 1990. The response of natural ecosystems to rising global CO_2 levels. Annual Review of Ecology and Systematics 21: 167-196.

Beiler, K.J., Durall, D.M., Simard, S.W., Maxwell, S.A. & Kretzer, A.M. 2010. Mapping the wood-wide web: mycorrhizal networks link multiple Douglas-fir cohorts. New Phytologist 185: 543-553.

Bever, J.D. 2002a. Host-specificity of AM fungal population growth rates can generate feedback on plant growth. Plant & Soil 244: 281–290.

Bever, J.D. 2002b. Negative feedback within a mutualism: Host-specific growth of mycorrhizal fungi reduces plant benefit. Proc. Royal Soc. London B 269: 2595–2601.

Bouchard, M., Kneeshaw, D. & Bergeron, Y. 2008. Ecosystem management based on large-scale disturbance pulses: A case study from sub-boreal forests of western Quebec (Canada). Forest Ecology & Management 256: 1734-1742.

Bradford, M.A., Davies, C.A., Frey, S.D., Maddox, T.R., Melillo, J.M., Mohan, J.E., Reynolds, J.F., Treseder, K.K. & Wallenstein, M.D. 2008. Thermal adaptation of soil microbial respiration to elevated temperature. Ecology Letters 11: 1316-1327.

Bradley, K.L., Hancock, J.E., Giardina, C.P., Pregitzer, K.S. 2007. Soil microbial community responses to altered lignin biosynthesis in *Populus tremuloides* vary among three distinct soils. Plant & Soil 294: 185-201.

Bray, D. 2003. Molecular networks: the top-down view. Science 301: 1864–1865.

Campbell, R., Smith, D.J. & Arsenault, A. 2003. Multicentury history of western spruce budworm outbreaks in interior Douglas-fir forests. Can. J. For. Res. 36: 1758–1769.

Cardon, Z.G. & Gage, D.J. 2006. Resource exchange in the rhizosphere: molecular tools and the microbial perspective. Ann. Rev. Ecol., Evol. Syst. 37: 459-488.

Clemmensen, K.E., Michelsen, A., Jonasson, S. & Shaver, G.R. 2006. Increased ectomycorrhizal fungal abundance after long-term fertilization and warming of two Arctic tundra ecosystems. New Phytologist 171: 391–404.

Cox P.M, Betts R.A, Jones C.D, Spall S.A, Totterdell I.J. 2000 Acceleration of global warming due to carbon-cycle feedbacks in a coupled climate model. Nature 408: 184–187.

Dale, V.H., Joyce, L.A., McNulty, S., Neilson, R.P., Ayres, M.P., Flannigan, M.D., Hanson, P.J., Irland, L.C., Lugo, A.E., Peterson, C.J., Simberloff, D., Swanson, F.J., Stocks, B.J & Wotton, B.W. 2001. Climate change can affect forests by altering the frequency, intensity, duration, and timing of fire, drought, introduced species, insect and pathogen outbreaks, hurricanes, windstorms, ice storms, or landslides. BioScience 51: 723-734.

Danby, R.K. & Hik, D.S. 2007. Variability, contingency and rapid change in recent subarctic alpine tree line dynamics. Journal of Ecology 95: 352-363.

Deacon, .JW. & Donaldson, S.J. 1983. Sequences and interactions of mycorrhizal fungi on birch. Plant & Soil 71: 257–262.

Dentener, F., Drevet, J., Lamarque, J.F., Bey, I., Eickhout, B., Fiore, A.M., Hauglustaine, D., Horowitz, L.W., Krol, M., Kulshrestha, U.C., Lawrence, M., Galy-Lacaux, C., Rast, S., Shindell, D., Stevenson, D., Van Noije, T., Atherton, C., Bell, N., Bergman, D., Butler, T., Cofala, J., Collins, B., Doherty, R., Ellingsen, K., Galloway, J., Gauss, M., Montanaro, V., Müller, J.F., Pitari, G., Rodriguez, J., Sanderson, M., Solmon, F., Strahan, S., Schultz, M., Sudo, K., Szopa, S. & Wild, O. 2006. Nitrogen and sulfur deposition on regional and global scales: A multimodel evaluation. Global Biogeochemical Cycles 20:

Deslippe, J.R. & Simard, S.W. 2010. Carbon transfer through mycorrhizal networks may facilitate shrub expansion in Low-Arctic tundra. Ecology Letters, submitted.

Deslippe, J.R., Haartman, M., Mohn, W.W. & Simard, S.W. 2010. Long-term experimental manipulation of climate alters the ectomycorrhizal community of *Betula nana* in Arctic tundra. Global Change Biology, in press.

Drigo, B., Kowalchuk, G.A. & van Veen, J.A. 2008. Climate change goes underground: effects of elevated atmospheric CO_2 on microbial community structure and activities in the rhizosphere Biology & Fertility of Soils 44: 667-679.

Durall, D.M., Gamiet, S., Simard, S.W., Kudrna, L. & Sakakibara, S.M. 2006. Effects of clearcutting and tree species composition on the diversity and community composition of epigeous fruit bodies formed by ectomycorrhizal fungi. Can. J. Bot. 84: 966-980.

FAO. 2006. Global forest resources assessment 2005. FAO Forestry Paper 147, Rome.

Flannigan, M.D., Logan, K.A., Amiro, B.D., Skinner, W.R. & Stocks, B.J. 2005. Future area burned in Canada. Climate Change 72: 1–16.

Giller, K.E., Beare, M.H., Lavelle, P., Izac, -M.N. & Swift, M.J. 1997. Agricultural intensification, soil biodiversity and agroecosystem function. Applied Soil Ecology 6: 3-16.

Guo, L.B. & Gifford, R.M. 2002. Soil carbon stocks and land use change: a meta analysis. Global Change Biology 8: 345-360.

Hagerman, S.M., Sakakibara, S.M. & Durall, D.M. 2001. The potential for woody understory plants to provide refuge for ectomycorrhizal inoculum at an interior Douglas-fir forest after clear-cut logging. Can. J. For. Res. 31: 711-721.

Hamann, A. & Wang, T. 2006. Potential effects of climate change on ecosystem and tree species distribution in British Columbia. Ecology 87: 2773-2786.

Hansen, C.A., Allison, S.D., Bradford, M.A., Wallenstein, M.D. & Treseder, K.K. 2008. Fungal taxa target different carbon sources in forest soil. Ecosystems 11: 1157-1167.

Heineman, J.E., Sachs, D.L., Mather, W.J. & Simard, S.W. 2010. Investigating the influence of climate, site, location and treatment factors on damage to young lodgepole pine in British Columbia. Can. J. For.Res. 40: 1109-1127.

Hobbie, E.A. 2006. Carbon allocation to ectomycorrhizal fungi correlates with belowground allocation in culture studies. Ecology 87: 563-569.

Hobbie, E.A. & Agerer, R. 2010. Nitrogen isotopes in ectomycorrhizal sporocarps correspond to belowground exploration types. Plant & Soil 327: 71-83.

Hobbie, J.E. & Hobbie, E.A. 2006 ^{15}N in symbiotic fungi and plants estimates nitrogen and carbon flux rates in Arctic. Ecology 87: 816-822.

Hobbie, S.E. 2000.Interactions between litter lignin and soil nitrogen availability during leaf litter decomposition in a Hawaiian montane forest. Ecosystems 3: 484-494.

Hoeksema, J.D. & Forde, S. E. 2008. A meta-analysis of factors affecting local adaptation between interacting species. Americal Naturalist 171: 275–290.

Högberg, P. & Högberg, M.N. 2002. Extramatrical ectomycorrhizal mycelium contributes one-third of microbial biomass and produces, together with associated roots, half of the dissolved organic carbon in a forest soil. New Phytologist 154: 791-795.

Högberg, P., Högberg, M.N., Göttlicher, S.G., Berson, N.R., Keel, S.G., Metcalfe, D.B., Campbell, C., Schindlbacher, A., Hurry, V., Lundmark, T., Linder, S. & Näsholm, T. 2007. High temporal resolution tracing of photosynthate carbon from tree canopy to forest soil microorganisms. New Phytologist 177: 220-228.

Hogg, E.H., Brandt, J.P. & Krochtubajda, R. 2002. Growth and dieback of aspen forests in northwestern Alberta in relation to climate and insects. Can. J. For. Res. **32**: 823-832.

Houghton, R.A., Skole, D.L., Nobre, C.A., Hackler, J.L., Lawrence, K.T. & Chomentowski, W.H. 2000. Annual fluxes of carbon from deforestation and regrowth in the Brazilian Amazon. Nature 403: 301-304.

IPCC. 2007. Climate Change 2007: Synthesis Report. Contribution of Working Groups I, II and III to the Fourth Assessment Report of the Intergovernmental Panel on Climate Change [Core Writing Team, Pachauri, R.K. & Reisinger, A. (eds.)]. IPCC, Geneva, Switzerland, 104 pp.

Janssens, I.A., Dieleman, W., Luyssaert, S., subke, J.A., Reichstein, M., Ceulemans, R., Ciais, P., Dolman, A., Grace, J., Matteucci, G., Papale, D., Piao, S.I., schulze, E.D., Tang, J., & Law, B.E. 2010. Reduction of forest soil respiration in response to nitrogen deposition. Nature GeoScience 3: 315-322.

Jerabkova, L., Prescott, C.E. & Kishchuk, B.E. 2006. Nitrogen availability in soil and forest floor of contrasting types of boreal mixedwood forests. Can. J. For. Res. 36: 112–122.

Johnson, D., Leake, J.R., Ostle, N., Ineson, P. & Read, D.J. 2002. In situ $^{13}CO_2$ pulse-labelling of upland grassland demonstrates a rapid pathway of carbon flux from arbuscular mycorrhizal mycelia to the soil. New Phytologist 153: 327-334.

Johnson, N.C., Wolf, J., Reyes, M.A., Panter, A., Koch, G.W. & Redman, A. 2005. Species of plants and associated arbuscular mycorrhizal fungi mediate mycorrhizal responses to CO_2 enrichment. Global Change Biology 11: 1156–1166.

Johnson, N.C., Wilson, G.W.T., Bowker, M.A., Wilson, J.A. & Miller, R.M. 2010. Resource limitation is a driver of local adaptation in mycorrhizal symbioses. PNAS 107: 2093–2098.

Johnstone, J.F. & Chapin, F.S. 2003. Non-equilibrium succession dynamics indicate continued northern migration of lodgepole pine. Global Change Biology 9: 1401-1409.

Jones, C., McConnell, C., Coleman, K., Cox, P., Falloon, P., Jenkinson, D. & Powlson, D. 2004. Global climate change and soil carbon stocks; predictions from two contrasting models for the turnover of organic carbon in soil. Global Change Biology 11: 154-166.

Jones, M.D. & Smith, S.E. 2004. Exploring functional definitions of mycorrhizas: Are mycorrhizas always mutualisms? Botany 82: 1089–1109.

Jones, M.D., Durall, D.M., Harniman, S.M.K., Classen, D.C. & Simard, S.W. 1997. Ectomycorrhizal diversity on *Betula papyrifera* and *Pseudotsuga menziesii* seedlings grown in the greenhouse or outplanted in single-species and mixed plots in southern British Columbia. Can. J. For. Res. 27: 1872-1889.

Jones, M.D., Durall, D.M. & Cairney, J.W.G. 2003. Ectomycorrhizal fungal communities in young forest stands regenerating after clearcut logging. New Phytologist 157: 399-422.

Jones MD, Ward V, Twieg BD, Durall DM, & Simard SW. 2010. Functional diversity and redundancy for extracellular enzyme activity of Douglas-fir ectomycorrhizas after wildfire or clearcut logging. Functional Ecology, in press.

Kazantseva, O., Bingham, M.A., Simard, S.W. & Berch, S.M. 2009. Effects of growth medium, nutrients, water and aeration on mycorrhization and biomass allocation of greenhouse-grown interior Douglas-fir seedlings. Mycorrhiza 20: 51-66.

Keeling, C.D. 1998. Rewards and penalties of monitoring the Earth. Annual Review of Energy and the Environment 23: 25-82.

Kiers, E.T. & van der Heijden, M.A. 2006. Mutualistic stability in the arbuscular mycorrhizal symbiosis: exploring hypotheses of evolutionary cooperation. Ecology 87: 1627-1636.

Klenner, W., Walton, R., Arsenault, A. & Kramseter, L. 2008. Dry forests in the southern interior of British Columbia: historic disturbances and implications for restoration and management. Forest Ecology & Management 206: 1711-1722.

Kliejunas, J.T., Geils, B.W., Glaeser, J.M., Goheen, E.M., Hennon, P., Kim, M.-S., Kope, H., Stone, J., Sturrock, R. & Frankel, S.J. 2009. Review of literature on climate change and forest diseases of western North America. Gen. Tech. Rep., PSW-GTR-225. Albany, CA: U.S. Department of Agriculture, Forest Service, Pacific Southwest Research Station. 54 p.

Klironomos, J.N. 2003. Variation in plant response to native and exotic mycorrhizal fungi. Ecology 84: 2292-2301.

Klironomos, J.N., Allen, M.F., Rillig, M.C., Piotrowski, J., Makvandi-Nejad, S., Wolfe, B.E. & Powell, J.R. 2005. Abrupt rise in atmospheric CO_2 overestimates community response in a model plant-soil system. Nature 433: 621-624.

Kurz, W.A. & Apps, M.J. 1999. A 70-year retrospective analysis of carbon fluxes in the Canadian forest sector. Ecological Applications 9: 526-547.

Kurz, W.A., Stinson, G. & Rampley, G.J. 2008a. Could increased boreal forest ecosystem productivity offset carbon losses from increased disturbances? Phil. Trans. R. Soc. B 363: 2259-2268.

Kurz, W.A., Dymond, C.C., Stinson, G., Rampley, G.J., Neilson, E.T., Carroll, A.L., Ebata, T. & Safranyik, L. 2008b. Mountain pine beetle and forest carbon feedback to climate change. Nature 452: 987-990.

Kurz, W.A., Stinson, G., Rampley, G.J., Dymond, C.C. & Neilson, E.T. 2008c. Risk of natural disturbances makes future contribution of Canada's forests to the global carbon cycle highly uncertain. PNAS 105: 1551-1555.

Lal, R. 2004. Soil carbon sequestration impacts on global climate change and food security. Science 304, 1623-1627.

Lerat, S., Gauci, R., Catford, J.G., Vierheilig, H., Piché, Y. & Lapointe, L. 2002. ^{14}C transfer between the spring ephemeral *Erythronium americanum* and sugar maple saplings via arbuscular mycorrhizal fungi in natural stands. Oecologia 132: 181–187.

Levin, S.A. 2005. Self-organization and the emergence of complexity in ecological systems. BioScience 55: 1075-1079.

Lilleskov, E.A., Fahey, T.J. & Lovett, G.M. 2001. Ectomycorrhizal fungal aboveground community change over a nitrogen deposition gradient. Ecological Applications 11: 397-410.

Lilleskov, E.A., Fahey, T.J., Horton, T.R. & Lovett, G.M. 2002. Belowground ectomycorrhizal fungal community change over a nitrogen deposition gradient in Alaska. Ecology 83: 104-115

Liu, Y., Stanturf, J. & Goodrick, S. 2010. Trends in global wildfire potential in a changing climate. Forest Ecology & Management 259: 685-697.

Lovelock, J. 2009. The vanishing face of Gaia. Basic Books, New York.

Maclauchlan, L., Cleary, M., Rankin, L., Stock, A. & Buxton, K. 2007. 2007 Overview of Forest Health in the Southern Interior Forest Region. BC Min. For., Kamloops, Canada.

McLachlan, J.S. & Clark, J.S. 2004. Reconstructing historical ranges with fossil data at continental scales. Forest Ecology & Management 197: 139-147.

Malcolm, J.R., Liu, C., Neilson, R.P., Hansen, L. & Hannah, L. 2006. Global warming and extinctions of endemic species from biodiversity hotspots. Conservation Biology, 20: 538–548.

Martin, K., Aitken, K.E.H. & Wiebe, K.L. 2004. Nest sites and nest webs for cavity-nesting communities in interior British Columbia, Canada: nest characteristics and niche partitioning. The Condor 106: 5-19.

McGuire, K.L. 2007. Common ectomycorrhizal networks may maintain monodominance in a tropical rain forest. Ecology 88: 567–574.

Miller, R.M., Jastrow, J.D. & Reinhardt, D.R. 1995. External hyphal production of vesicular-arbuscular mycorrhizal fungi in pasture and tallgrass prairie. Oecologia 103: 17-23.

Miller, R.M., Miller, S.P., Jastrow, J.D. & Rivetta, C.B. 2002. Mycorrhizal mediated feedbacks influence net carbon gain & nutrient uptake in *Andropogon gerardii*. New Phytologist 155:149-162.

Molina, R., Massicotte, H. & Trappe, J.M. 1992. Specificity phenomenon in mycorrhizal symbiosis: community-ecological consequences and practical implications. Page 357-423 *in* Allen, M.F., ed. Mycorrhizal Functioning: An Integrative Plant-Fungal Process. Chapman Hall, New York, NY, USA.

Morrison, D.J., Wallis, G.W. & Weir, L.C. 1988. Control of *Armillaria* and *Phellinus* root diseases: 20-year results from the Skimikin stump removal experiment. Can. For. Serv., Pac. For. Cen., Victoria, B.C. Inf. Rep. BC-X-302.

Mueller, R.C., Scudder, C.M., Porter, M.E., Trotter III, R.T., Gehring, C.A. & Whitham, T.G. 2005. Differential tree mortality in response to severe drought: evidence for long-term vegetation shifts. Journal of Ecology 93: 1085-1093.

Näsholm, T., Ekblad, A., Nordin, A., Giesler, R, Högberg, M. & Högberg, P.1998. Boreal forest plants take up organic nitrogen. Nature 392: 914-916.

Nara, K. and Hogetsu, T. 2004. Ectomycorrhizal fungi on established shrubs facilitate subsequent seedling establishment of successional plant species. Ecology 85: 1700–1707.

Nave, L.E., Vance, E.D., Swanston, C.W., & Curtis, P.S. 2010. Harvest impacts on soil carbon storage in temperate forests. Forest Ecology & Management 259: 857-866.

Norby, R.J., Cotrufo, M.F., Ineson, P., O'Neill, E.G., Canadell, J.G. 2001.Elevated CO2, litter chemistry, and decomposition: a synthesis. Oecologia 127:153–165.

Oliver, C. & Larson, B.C. 1997. Forest stand dynamics: updated edition. Wiley, New York.

Parmesan, C. 2006. Ecological and evolutionary responses to recent climate change. Annual Rev. Ecol. Evol. Syst. 37: 637–669.

Parrent, J.L. & Vilgalys, R. 2007. Biomass and compositional responses of ectomycorrhizal fungal hyphae to elevated CO_2 and nitrogen fertilization. New Phytologist 176: 164 - 174.

Pendall, E., Bridgham, S., Hanson, P.J., Hungate, B., Klicklighter, D.W., Johnson, D.W., Law, B.E., Luo, Y., Megonigal, J.P., Olsrud, M., Ryan, M.G. & Wan, S. 2004. Belowground process responses to elevated CO_2 and temperature: A discussion of observations, measurement methods, and models. New Phytologist 162: 311-322.

Pendall, E., Rustad, L. & Schimel, J. 2008. Toward a predictive understanding of below-ground process responses to climate change: have we moved any closer? Functional Ecology 22: 937-940.

Peter, M., Ayer F. & Egli, S. 2001. Nitrogen addition in a Norway spruce stand altered macromycete sporocarp production and below-ground ectomycorrhizal species composition. New Phytologist 149: 311-325.

Petit, R.J., Aguinagalde, I., de Beaulieu, L., Bittkau, C., Brewer, S., Cheddadi, R., Ennos, R., Fineschi, S., Grivet, D., Lascoux, M., Mohanty, A., Müller-Starck, G., Demesure-Musch, B., Palmé, A., Martin, J.P., Rendell, S. & Vendramin, G.G. 2003. Glacial refugia: hotspots but not melting pots of genetic diversity. Science 300: 1563-1565.

Pietikäinen, A. & Kytöviita, M.-M. 2007. Defoliation changes mycorrhizal benefit and competitive interactions between seedlings and adult plants. Journal of Ecology 95: 639-647.

Philip, L.J. 2006. Carbon transfer between ectomycorrhizal paper birch (*Betula papyrifera*) and Douglas-fir (*Pseudotsuga menziesii*). PhD thesis, UBC, Vancouver, Canada.

Philip, L.J., Simard, S.W. & Jones, M.D. 2010. Pathways for belowground carbon transfer between paper birch and Douglas-fir seedlings. Plant Ecology & Diversity, in press.

Poorter, H. 1993. Interspecific variation in the growth response of plants to an elevated ambient CO2 concentration. Vegetatio 104/105: 77-97.

Pritchard, E. T. G., Strand, A.E., McCormack, M.A., Davis, M.A., Finzi, A.C., Jackson, R.B., Roser, M., Rogers, H.H. & Oren, R. 2008. Fine root dynamics in a loblolly pine forest are influenced by free-air-CO2- enrichment: a six-year-minirhizotron study. Global Change Biology 14: 1–15.

Read D. J. & Perez-Moreno, J. 2003. Mycorrhizas and nutrient cycling in ecosystems - a journey towards relevance? New Phytologist 157: 475–492.

Rehfeldt, G.E., Wykoff, W.R. & Ying, C.C. 2001. Physiological plasticity, evolution and impacts of a changing climate on *Pinus contorta*. Climatic Change **50**: 355-37.

Rice, C.W., White, P.M., Fabrizzi, K.P. & Wilson, G.W.T. 2004. Managing the microbial community for soil carbon management. Supersoil 2004: 3rd Australian New Zealand Soils Conference 5-9 December 2004, University of Sydney, Australia.

Rillig, M.C., Wright, S.F., Nichols, K.A., Schmidt, W.F. & Torn, M.S. 2001. Large contribution of arbuscular mycorrhizas to soil C pools in tropical forest soils. Plant & Soil 233: 167-177.

Rogers, H.H., Runion, G.B., Krupa, S.V. 1994. Plant responses to atmospheric CO_2 enrichment with emphasis on roots and the rhizosphere. Environmental Pollution 83: 155-189.

Rustad, L.E., Campbell, J.L., Marion, G.M., Norby, R.J., Mitchell, M.J., Hartley, A.E., Cornelissen, J.H.C. & Gurevitch, J. 2001. A meta-analysis of the response of soil respiration, net nitrogen mineralization, and aboveground plant growth to experimental ecosystem warming. Oecologia 126: 542–562.

Rygiewicz, P.T. & Anderson, C.P. 1994. Mycorrhizae alter quality and quantity of carbon allocated below ground. Nature 369: 58-60.

Schlesinger, W.H. & Andrews, J.A. 2004. Soil respiration and the global carbon cycle. Biogeochemistry 48: 7-20.

Schoonmaker, A.L., Teste, F.P., Simard, S.W. & Guy, R.D. 2007. Tree proximity, soil pathways and common mycorrhizal networks: their influence on utilization of redistributed water by understory seedlings. Oecologia 154: 455-466.

Schuur, E.A.G., Vogel, J.G., Crummer, K.G., Lee, H., Sickman, J.O. & Osterkamp, T.E. 2009. The effect of permafrost thaw on old carbon release and net carbon exchange from tundra. Nature 459: 556-559.

Sedjo, R. 1993. The carbon cycle & global forest ecosystem. Water, Air, Soil Poll. 70: 295-307

Selosse, M.-A., Richard, F., He, X. & Simard, S.W. 2006. Mycorrhizal networks: les liaisons dangeureuses? Trends in Ecology & Evolution 21: 621-628.

Simard, S.W. 2009. The foundational role of mycorrhizal networks in the self-organization if interior Douglas-fir forests. Forest Ecology & Management 258S: S95–S107.

Simard, S.W. & Durall, D.M. 2004. Mycorrhizal networks: a review of their extent, function, and importance. Botany 82: 1140–1165.

Simard, S.W., Perry, D.A., Jones, M.D., Myrold, D.D., Durall, D.M. & Molina, R. 1997a. Net transfer of carbon between ectomycorrhizal tree species in the field. Nature 388: 579-582.

Simard, S.W., Perry, D.A., Smith, J.W. & Molina, R. 1997b. Effects of soil trenching on occurrence of ectomycorrhizae on *Psuedostuga menziesii* seedlings grown in mature forests of *Betula papyrifera* and *Psuedotsuga menziesii*. New Phytologist 136: 327-340.

Simard, S.W., Jones, M.D. & Durall DM. 2002. Carbon and nutrient fluxes within and between mycorrhizal plants. Pages 33-61 *in* M. van der Heijden and I. Sanders, eds. Mycorrhizal Ecology. Springer-Verlag, Heidelberg. Ecological Studies, Vol. 157.

Simard, S.W., Hagerman, S.M., Sachs, D.L., Heineman, J.L. & Mather, W.J. 2005. Conifer growth, *Armillaria ostoyae* root disease and plant diversity responses to broadleaf competition reduction in temperate mixed forests of BC. Can. J. For. Res. 35: 843-859.

Smith, S.E. & Read, D.J. 1997. Mycorrhizal symbiosis, 2nd ed. Academic Press, London, UK.

Southworth, D., He, X.-H., Swenson, W. & Bledsoe, C.S. 2005. Application of network theory to potential mycorrhizal networks. Mycorrhiza 15: 589–595.

Spittlehouse, D.L. 2008. Climate change, impacts, and adaptation secenarios: climate change and forest and range management in British Columbia. B.C. Min. For. Range, Res. Br., Victoria, BC. Tech. Rep. 045.

Staddon, P.L., & Fitter, A.H. 1998. Does elevated atmospheric carbon dioxide affect arbuscular mycorrhizas? Trends in Ecology & Evolution 13: 455-458.

Staddon, P.L., Fitter, A.H. & Robinson, D. 1999. Effects of mycorrhizal colonization and elevated atmospheric carbon dioxide on carbon fixation and below-ground carbon partitioning in *Plantago lanceolata*. Journal of Experimental Botany 335: 853-860.

Stark, K.E., Arsenault, A. & Bradfield, G.E. 2006. Soil seed banks and plant community assembly following disturbance by fire and logging in the interior Douglas-fir forests of south-central British Columbia. Botany 84: 1548-1560.

Suding, K.N., Gross, K.L. & Houseman, G.R. 2004. Alternative states and positive feedbacks in restoration ecology. Trends in Ecology & Evolution 19: 46-53.

Talbot, J.M., Allison, S.D. & Treseder, K.K. 2008. Decomposers in disguise: mycorrhizal fungi as regulators of soil carbon dynamics in ecosystems under global change. Functional Ecology 22: 955-963.

Teste, F.P. & Simard, S.W. 2008. Mycorrhizal networks and distance from mature trees alter patterns of competition and facilitation in dry Douglas-fir forests. Oecologia **158**: 193-203.

Teste, F.P., Simard, S.W. & Durall, D.M. 2009a. Role of mycorrhizal networks and tree proximity in ectomycorrhizal colonization of planted seedlings. Fungal Ecology **2**: 21-30.

Teste, F.P., Simard, S.W., Durall, D.M., Guy, R.D., Jones, M.D. & Schoonmaker, A.L. 2009b. Access to mycorrhizal networks and tree roots: importance for seedling survival and resource transfer. Ecology 90: 2808–2822.

Teste FP, Simard SW, Durall DM, Guy RD, & Berch SM. 2010. Net carbon transfer under soil disturbance between *Pseudostuga menziesii* seedlings in the field. Journal of Ecology 98:429-439.

Thomas, C.D., Cameron, A., Green, R.E. Bakkenes, M., Beaumont, L.J., Collingham, Y.C., Erasmus, B.F.N., de Siqueira, M.F., Grainger, A., Hannah, L., Hughes, L., Huntley, B., van Jaarsveld, A.S., Midgley, G.F., Miles, L., Ortega-Huerta, M.A., Peterson, A.T., Philips, O.L. & Williams, S.E. 2004. Extinction risk from climate change. Nature 427: 145-148.

Thomas, R.Q., Canham, C.D., Weathers, K.C. & Goodale, C.L. 2010. Increased tree carbon storage in response to nitrogen deposition in the US. Nature Geoscience 3: 13-17.

Treseder, K.K. 2004. A meta-analysis of mycorrhizal responses to nitrogen, phosphorus, and atmospheric CO_2 in field studies. New Phytologist 164: 347-355.

Treseder, K.K. 2005. Nutrient acquisition strategies of fungi and their relation to elevated atmospheric CO_2. In: The Fungal Community: Its Organization and Role in the Ecosystem, Third Edition. Edited by: J. Dighton, J.F. White & P. Oudemans. Taylor & Francis Group, LLC. ISBN: 978-1-4200-2789-1. Chapter 26, pp 713-731.

Treseder, K.K & Turner, K. 2007. Glomalin in ecosystems. Soil Sci. Soc. Am. J. 71: 1257-1266.

Treseder, K.K. 2008. Nitrogen additions and microbial biomass: a meta-analysis of ecosystem studies. Ecology Letters 11: 1111-1120.

Treseder, K.K., Egerton-Warburton, L.M., Allen, M.F., Cheng, Y. & Oechel, W.C. 2003. Alteration of soil carbon pools and communities of mycorrhizal fungi in chaparral exposed to elevated carbon dioxide. Ecosystems 6: 786-796.

Treseder, K.K., Czimczik, C.I., Trumbore, S.E. & Allison, S.D. 2008. Uptake of an amino acid by ectomycorrhizal fungi in a boreal forest. Soil Biology & Biochemistry 40: 1964-1966.

Tu, C., Booker, F.L., Watson, D.M., Chen, X., Rufty, T.W., Shi, W. & Hu, S.J. 2006. Mycorrhizal mediation of plant N acquisition and residue decomposition: impact of mineral N inputs. Global Change Biology 12, 793–803.

Twieg, B., Durall, D.M. & Simard, S.W. 2007. Ectomycorrhizal fungal succession in mixed temperate forests. New Phytologist 176: 437-447.

Twieg, B, Durall DM, Simard SW, Jones MD. 2009. Influence of stand age and soil properties on ectomycorrhizal communities in mixed temperate forests. Mycorrhiza 19: 305-316.

Van der Heijden, M.G.A. & Horton, T.R. 2009. Socialism in soil? The importance of mycorrhizal fungal networks for facilitation in natural ecosystems. Journal of Ecology 97, 1139–1150.

Vitousek, P.M. 1994. Beyond global warming: ecology and global change. Ecology 75: 1861-1876.

Vitousek, P.M., Aber, J.D., Howarth, R.W., Likens, G.E., Matson, P.A., Schindler, D.W., Schlesinger, W.H. & Tilman, D.G. 1997. Human alteration of the global nitrogen cycle: sources and consequences. Ecological Applications 7: 737-750.

Vyse, A., Ferguson. C., Simard, S.W., Kano, T. & Puttonen, P. 2006. Growth of Douglas-fir, lodgepole pine, and ponderosa pine underplanted in a partially-cut, dry Douglas-fir stand in south central British Columbia. Forestry Chronicle 82: 723-732.

Whitham, T.G., Bailey, J.K., Schweitzer, J.A., Shuster, S.M., Bangert, R.K., LeRoy, C.J., Lonsdorf, E.V., Allan, G.J., DiFazio, S.P., Potts, P.M., Fischer, D.G., Gehring, K.A., Lindroth, R.L., Marks, J.C., Hart, S.C., Wimp, G.M. & Wooley, S.C. 2006. A framework for community and ecosystem genetics: from genes to ecosystems. Nature Review Genetics 7: 510-523.

Zak, D.R., Pregitzer, K.S., Curtis, P.S., Vogel, C.S., Holmes, W.E. & Lussenhop, J. 2000. Atmospheric CO_2, soil-N availability, and allocation of biomass and nitrogen by *Populus tremuloides*. Ecological Applications 10: 34-46.

Zhu, Y.-G. & Miller, R.M. 2004. Carbon cycling by arbuscular mycorrhizal fungi in soil-plant systems. Trends in Plant Science 8: 407-409.

16

Impact of temperature increase and precipitation alteration at climate change on forest productivity and soil carbon in boreal forest ecosystems in Canada and Russia: simulation approach with the EFIMOD model

Oleg Chertov[1], Jagtar S. Bhatti[2] and Alexander Komarov[3]

[1] *Bingen University of Applied Sciences, Berlin Str. 109, 55411, Bingen am Rhein Germany*

[2] *Canadian Forest Service, Northern Forestry Centre, 5320 122 Street, Edmonton, Alberta T6H 3S5, Canada*

[3] *Institute of Physicochemical and Biological Problems in Soil Science, Russian Academy of Sciences, Institutskaya 2, 142290, Pushchino Russia*

Abstracts

The results of long-term EFIMOD simulations for black spruce (*Picea mariana* [Miller]) and Jack pine (*Pinus banksiana* Lamb.) forests in Central Canada show that climate warming, fire, harvesting and insects significantly influence net primary productivity (NPP), soil respiration (Rh), net ecosystem production (NEP) and pools of tree biomass and soil organic matter (SOM). The effects of six climate change scenarios demonstrated similar increasing trends of NPP and stand productivity. The disturbances led to a strong decrease in NPP, stand productivity, soil organic matter (SOM) and nitrogen (N) pools with an increase in CO_2 emission to the atmosphere. However the accumulated NEP for 150 years under harvest and fire fluctuated around zero. Net ecosystem productivity becames negative only at a more frequent disturbance regime with four forest fires during the period of simulation. Climate change with temperature and precipitation rise leads to the increasing of forest productivity but it reduces SOM pool that can be an indication of ecosystem resilience. The results from this study show that changes in climate and disturbance regimes might substantially change the NPP as well as the C and N balance, resulting in major changes in the C pools of the vegetation and soil under black spruce forests. Soil conditions, especially the potential productivity, as determining by the N pool, modify the effect of climate change and disturbances: poor soils contrasting relative effect of climate change and damages, contrariwise more rich soil mitigates the effect of damages and climate change. The same results were obtained for some West European and Russian boreal forest. Moreover, the EFIMOD runs show that atmospheric nitrogen deposition and especially various silvicultural regimes strongly modify the impact of climate changes on boreal forests. Nitrogen deposition can mitigate the negative impact of temperature rise on forest soils, while overexploitation has the same effect as forest disturbances in Canada.

Keywords: Boreal forests, climate change, productivity, carbon budget, silvicultural regimes, disturbances, atmospheric nitrogen deposition

1. Introduction

Bounded between the northern tundra and the southern grassland or broad-leaved forests, the boreal or "northern" forest is a very large biome in Northern hemisphere and it occupies about 35% of the total Canadian and Russian land area and almost 70% of total forest lands in both countries. Forest biomass, coarse woody debris and soil are the three major pools of carbon in forest ecosystems. Changes in forest ecosystem C pools are mainly driven by the dynamics of the living biomass. Accumulations of organic C in litter and soil change significantly in respect to forest development stages (forest succession) and stands disturbances, such as fire, insects and harvesting. Forest primary succession (from pioneer tree species to old-growth uneven-aged forest) and secondary succession (forest restoration after disturbances and cutting) leads to consistent increase of soil C (Chertov, Razumovsky, 1980; Bobrovsky, 2004). Disturbances transfer biomass C to detritus and soil C pools where it decomposes at various rates over the years following the disturbance. In Canadian boreal forests, the disturbance frequency has increased over the past three decades - a trend that appears to be consistent with that expected from climatic warming - and this has caused significant changes in the net carbon balance at the national scale (Kurz and Apps, 1999). Numerous investigators have also examined the effects of disturbances on carbon balance, with particular focus as to whether they represent a significant carbon source to the atmosphere (Amiro et al., 2001; Wei et al., 2003; Hatten et al., 2005).

Projected climate change scenarios for the boreal forest generally predict warmer and somewhat drier conditions, and are expected to change the disturbance pattern. Fire and insect predation regimes, for example, are historically sensitive to climate and are expected to change considerably under global warming (Wotton et al., 1998). Altered boreal forest disturbance regimes - especially increases in their frequency, size and severity - may release soil C at higher rates. Will the net effect of such changes result in positive feedback to climate change and thereby accelerate global warming?

Two aspects of the impact of climate change on forest ecosystems can be distinguished: (a) direct impact of temperature growth and precipitation alteration on the ecosystem processes (tree growth and soil dynamics); (b) catastrophic impact of increased frequency of the ecosystem disturbances (increased fire hazard at draught and forest breakdown at extreme atmospheric events, e.g. storms, hurricanes, landslides etc.). Harvesting regimes are also linking up the second aspect. The first aspect relates mostly to forest stand level, the second one – to the landscape and regional level.

The primary objective of this work was a long term simulation by EFIMOD model to specify direct effects of temperature and precipitation changes at climate change on tree growth and net primary productivity (NPP), soil C dynamics and soil heterotrophic respiration (Rh) and total carbon budget (net ecosystem productivity, NEP) at climate change in boreal forests of Central Canada and European Russia. The effects of disturbances (forest fires) and various silvicultural regimes were also taken into account.

2. EFIMOD model

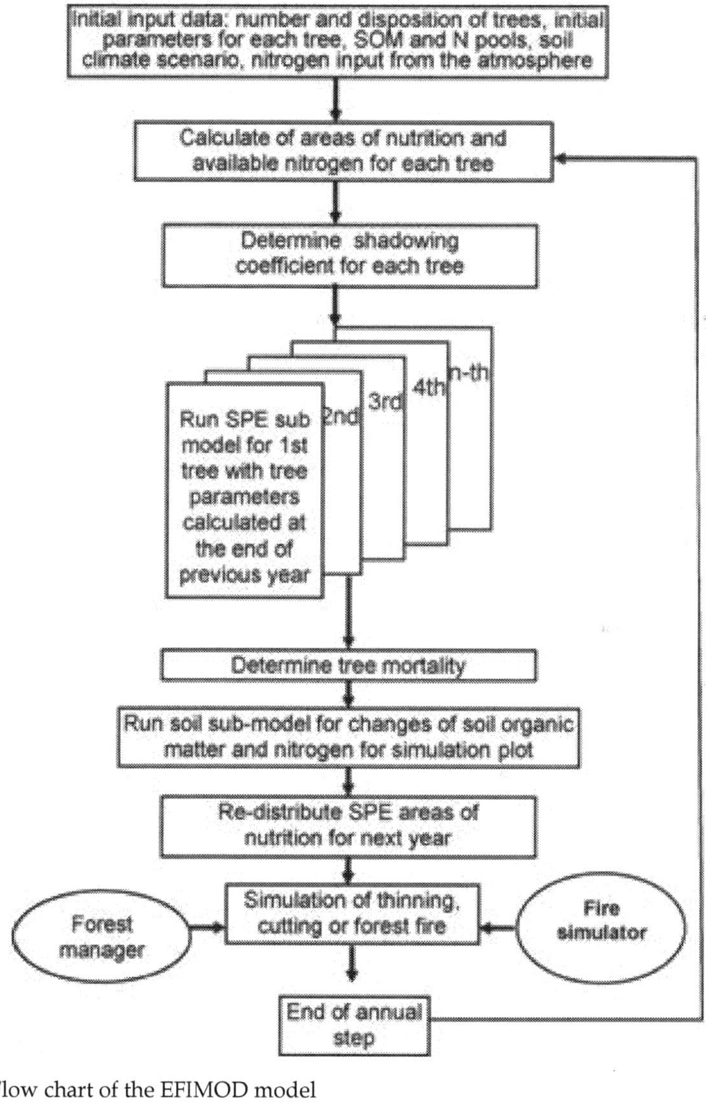

Fig. 1. Flow chart of the EFIMOD model

The model of forest growth and elements cycling in forest ecosystems EFIMOD (Chertov et al., 1999; Komarov et al., 2003, 2007) is an individual-based spatially explicit simulator of tree-soil system that calculates parameters of carbon (C) balance and standard forest inventory characteristics: NPP, Rh, soil available nitrogen (N), tree and stand biomass by tree compartments, soil organic matter (SOM) and N pools, stand density, height, DBH, growing stock and some other parameters. It includes soil model ROMUL as an important component (Chertov et al., 2001) that is driven by soil water, temperature and SOM parameters. The EFIMOD allows for a calculation the effect of silvicultural operations (Fig. 1, "forest manager") and forest fires ("fire simulator").

There is a positive promising experience for the implementation of models ROMUL and EFIMOD at a wide range from East Europe till North America in combination with regional forest databases for the estimation of the components of carbon balance (Chertov et al., 2002, 2005; Nadporozhskaya et al., 2006; Shaw et al., 2006; Komarov et al., 2007; Bobrovsky et al., 2010; Yurova et al., 2010). They were also implemented for Germany in a frame of the EU Project RECOGNITION (Komarov et al., 2007; Kahle et al., 2008). The special version of the EFIMOD model (IMPACT, Chertov et al., 2003) is now implementing in Finland for ecological certification of forest products to calculate C, N and energy losses from forest ecosystems due to forest exploitation. The EFIMOD was also implemented for evaluation of the different forestry regimes in terms of their impact on carbon budget and forest productivity (Mikhailov et al., 2004; Komarov et al., 2007) and for modelling carbon balance in a frame of the Program of Russian Academy of Sciences "Change of Environment and Climate".

Both the SOM model ROMUL and the ecosystem model EFIMOD were previously comprehensively calibrated and validated for European boreal and temperate forests in a frame of the European Forest Institute (EFI) projects, EU Project RECOGNITION (Kahle et al., 2008) and later for Canadian boreal forests (Shaw et al., 2006; Chertov et al., 2009; Bhatti et al, 2009).

3. Objects and methods of simulation

The objects of EFIMOD simulation for determination of climate change effects on boreal forests were Central Canadian boreal forests at the Boreal Forest Transect Case study (BFTCS) of Canadian Forest Service, some permanent sample plots in West Europe and a part of forest enterprise in European Russia. Due to a strong difference of natural and economic conditions in North America and Europe the simulation scenarios for climate change in Canada and Europe are slightly different: the scenarios for Canada accentuate an importance of forest fires and insect attacks with constant cutting regime; the scenarios for Europe and Central European Russia emphasize the various cutting regimes and N deposition from the atmosphere (without fire and insect damage).

Canadian sites. The EFIMOD model was parameterized and calibrated for jack pine (*Pinus banksiana* Lamb.) and black spruce (*Picea mariana* [Miller]) forests along BFTCS (Shaw et al., 2006; Bhatti et al., 2009; Chertov et al., 2009). The BFTCS was established with the primary goal of understanding the response of boreal forest ecosystems to climate change and how this is affected by natural and anthropogenic disturbances (Price and Apps, 1995). The 1000-km transect has a set of permanent sites.

Jack pine is a typical post-fire pioneer tree species that forms pure stands of low productivity on dry sites. The jack pine sites along the BFTCS have a sandy to sandy-loam soil with rapidly drained conditions with a thin raw humus layer and low soil C concentration in mineral topsoil. Black spruce is widespread in the Canadian boreal ecoregions where it forms late-succession forests (Gower et al., 1997). In the Canadian boreal forest, black spruce occupies both upland and lowland sites. Commonly it grows in pure stands on organic soils and in mixed stands on mineral soils. In the absence of fire, the accumulation of organic matter forms a thick forest floor layer dominated by feather moss and sphagnum (Oechel, Van Cleve, 1986).

Initial forest stand parameters for all the simulations were identical. Stand density were 2500 trees ha^{-1} for jack pine and 10000 trees ha^{-1} for black spruce, age of seedlings was 5 years, their height 0.3 (s.d. 0.1) m with initially random pattern of the seedlings on the simulated plot. The same characteristics were used for simulation of forest regeneration after harvesting, insect and fire disturbances.

The model validation was performed using the stand and soil parameters of 14 sample plots at BFTCS sites as an experimental dataset. For atmospheric N deposition, we used values reported by Shaw et al. (2006) as 2.04 kg [N] ha^{-1} $year^{-1}$. Additionally, the published data on NPP, Rh, and NEP estimated by Nakane et al. (1997), Bond-Lamberty et al. (2004), Howard et al., (2004), Wang et al. (2002, 2003) and Zha et al., (2009) were used as well.

To study the effects of climate change and disturbances, the simulations were carried out with initial soil C and N data from Candle Lake BFTCS site situated approximately in the centre of this transect.

For the climate change simulations, we used the 150-year scenarios compiled by Price et al. (2004) with three General Circulation Models (GCMs): the Canadian Climate Centre for Modelling and Analysis, CGCM2; the UK Hadley Centre, HadCM3; and the Australian CSIRO Mark 2 GCM. For each GCM scenario, we used two IPCC SRES carbon dioxide emissions scenarios (A2 and B2) for the period 1901-2100. In all scenarios, the data for 1961-1990 are identical, and were extracted for the BFTCS sites from the complied climatic database described at http://www.glfc.cfs.nrcan.gc.ca/landscape/climate_models_e.html. The climate change scenarios with altered values of temperature and precipitation begin only in 1991 only. Additionally, a constant climate scenario (i.e. before the period of major, human-induced climate change) was compiled from the data for the period 1901-1975 that was repeated twice to reach a 150-year time series. It should be pointed out that all three GCMs showed increasing trends of monthly air temperature and precipitation, although the UK model had the lowest rate of increase in these parameters and the CSIRO model had the highest. The data from these GCMs, on minimal and maximal monthly air temperature and precipitation, for 150-year period starting from 1961 were processed by the statistical climate generator SCLISS (Bykhovets and Komarov, 2002) to compile soil climate time series (soil temperature and moisture for organic and mineral soil horizons) which was required for EFIMOD runs. Finally, a set of seven soil climate scenarios was obtained: constant climate; CGCM A2; CGCM B2; SCIRO A2; SCIRO B2; HADCM A2; HADCM B2.

Model simulations were carried out for stand replacing disturbances; namely harvesting, fire and insect disturbances as defined by Kurz et al. (1992). Harvesting represents one thinning at the age of 40 (30% of stand biomass cutting), and clear cutting at age 70. All residues from the mid-rotation thinning remained on the site for decomposition. At harvest, the 90% of stem wood and 10% for branches and leaves were removed from the forest. Two rotations were simulated.

The intensity of crown (canopy) forest fire was the following (as percentage combustion during fire): foliage 100; twigs 60; wood 5; fine roots 30; forest floor (L horizon) 100; forest floor (F+H horizons) 25.

In the simulation of insect-induced disturbance, 90% of the foliage was transferred to the forest floor as insect excrement and 10% was transferred into insect biomass plus their expenses for respiration.

Trees killed by fire and insect attack were not removed from the forest. After harvest, fire, or insect attack, we simulated successful forest regeneration five years following the disturbances. Simulations were conducted under a total of seven different disturbance regimes resulting in a matrix of 49 simulation scenarios. The parameters of C balance used to analyses of the simulation results which included: net primary productivity, soil and deadwood respiration, and loss of C from disturbances (harvested wood, burned trees and forest floor). The C balance was calculated as net ecosystem productivity:

NEP = NPP - [Rh+DIST], where NEP, NPP and, Rh defined above, and DIST is C loss with disturbances. We did not calculate standard deviation because NPP and NEP values are strongly variable due to disturbance events, and the C losses due to disturbance have a pulsating character.

Simulations were conducted under a total of 7 different disturbance regimes (No disturbance for 150 years, Two harvests at 70 and 145 years, Two fires at 70 and 145 years, Four fires at 32, 70, 107 and 145 years, One harvesting at 70 and one fire at 145 years, One fire at 70 and one harvesting at 145 years, One insect attack at 70 and one harvesting at 145 years), each in combination with 7 climate scenarios (Constant climate, CGCM2 A2 and B2, CSIRO A2 and B2, HADCM A2 and B2).

European sites. The Russian, Finnish and West European experimental data (from the EU RECOGNITION Project) were used for the validation and calibration of EFIMOD model (Chertov et al., 2003; Komarov et al., 2003; Van Oijen et al., 2008). Then EFIMOD was implemented for the analysis of impact of climate warming in combination with atmospheric N deposition in a frame of RECOGNITION Project. The Project was devoted to growth trends in European forests to specify factors affecting consistent increasing of forest productivity in the second half of 20th century in Europe (Kahle et al., 2008; Komarov et al., 2007).

Seven sites with long-term experimental data on tree growth were selected (4 Scots pine, *Pinus sylvestris* L. and 3 Norway spruce, *Picea abies* L. [Karst.]): 2 from Finland, 2 from Sweden, 2 from Germany and 1 from Scotland) to represent North Scandinavian and Central West European forests. The forests were represented by pure stands of these coniferous trees on well drained soils with rather high soil C both in forest floor and mineral topsoil. We analyzed a set of scenarios for 80 years simulation for scenarios of natural development (no thinning) and managed forest with 5 thinning and final clear cutting.

The climatic scenarios for the simulation were as follows: actual climate and nitrogen deposition for 20th and 21st centuries - measured and predicted by climatic models (*actual climate, actual N*), stable initial climate and stable low N deposition as at the beginning of 20th century (*low climate, low N*), stable initial climate and actual N deposition (*low climate, actual N*), actual initial climate and stable low N deposition (*actual climate, low N*). For 21st century, we used time courses of weather variables from simulations run by the Hadley Centre in the

UK using the HadCM3 GCM. (Mitchell et.al., 2004; van Oijen et al., 2008). Three start times were used: 1920, 1960 and 2000 to cover periods with different combination of climates and nitrogen deposition for two centuries.

The initial tree parameters were as follows: age 3 years, height 0.3 (s.d. 0.04) m, initial tree density was 10000 tree per ha for German sites, 5000 for Swedish and Scottish sites, and 3000 for Finnish sites. The initial site specific soils parameters were the same for the runs with different start time.

At the analysis of the results, we postulated that the difference between scenarios starting in 1960 and in 1920 reflects the effects of increasing nitrogen deposition, because there are no still strong temperature changes in the scenarios. The comparison of the parameters between scenarios starting in 2000 and in 1960 demonstrates the effects of temperature increasing because both scenarios have rather high nitrogen deposition (but not absolutely the same). The comparison of the ecosystem parameters between scenarios starting in 2000 and 1920 shows the cumulative effects of nitrogen deposition and temperature increasing. The results were aggregated in two clusters: North Europe (Finland and Sweden, "North" on Fig. 4-6) and the rest of Europe (Germany and Scotland, "South" on these figures).

The Russian site for forest management regimes and climate change study at landscape level (Mikhailov et al., 2004, 2007; Komarov et al., 2007) represents a part of forest enterprise that located 100 km south of Moscow on the East European Plain. It possesses a continental climate and contains both coniferous and mixed forests (Mikhailov et al., 2004, 2007). The State Forest "Russky Les" occupies the left bank of the Oka River with sandy and loamy well drained soils (Alfisols). These forests were intensively exploited since the 17th century, and overexploited in the 20th century. Secondary forests are now widespread in the "Russky Les". Silver birch (*Betula pendula* L.), Scots pine (*Pinus sylvestris* L.) and Norway spruce (*Picea abies* L. Karst.) mixed stands dominate the forests. Young stands (<40 years of age) occupy 12% of the enterprise area, mean-aged stands (40–60 years) occupy 53%, and pre-mature stands (60–80 years) cover 35%. Generally, the forests have high density and productivity.

Four management blocks in the "Russky Les" state forest were selected for the case study. They contain 104 forest compartments (stands) comprising 300 ha. The selected forest is typical among forest enterprises with regard to stand composition, forest age, and soils. Current inventory data were used as initial input parameters for the simulations.

Four simulation scenarios were compiled: 1. *Natural development* (NAT). This scenario prevents cutting in all forest compartments. 2. *Russian legal system* (LRU). This scenario permits managed forests with four thinning (at 5, 10, 25, and 50 years), a final clear cutting (90-year age for conifer and oak, 60-year age for birch and lime), and natural regeneration by the target species with a mixture of deciduous species. In these forests, clear cutting must be followed by obligatory forest regeneration, either natural undergrowth or forest planting. 3. *Selective cutting system* (SCU). This scenario creates a managed forest with two thinning in young and mean-aged stands, and then selective cuttings after the stand reaches the age of 80 years (each 30 years in uneven-aged stands, intensity is 30% of basal area from above). 4. *Illegal practice* (ILL). This represents heavy upper thinning and removing of the best trees, and clear cutting without careful natural regeneration, often dominated by deciduous stands. All residues after the final harvest (leaves and branches) in LRU and ILL scenarios are removed (burning on clear-cut area). This treatment follows the Russian legislation, but causes a loss of carbon and nitrogen from the forest ecosystem. These scenarios reflect existing and theoretically possible silvicultural regimes in the simulated forest. A 200-year

period was selected because it is a period when so-called 'managed' even-aged forests will be transformed into 'close-to-nature' uneven-aged forests in the NAT scenario.
The model's runs were performed with current climate and with climate change for 150 years using British GCM HADCM3 A1 Fi (Mitchell et.al., 2004) that predicts in this region about 4°C increase of mean annual temperature for 21st century.

4. Results

EFIMOD validation for Canadian sites. The early EFIMOD validations (including the RECOGNITION project) demonstrated a good correlation between simulated and measured dendrometric parameters, soil C and N pools (Chertov et al., 2003; Komarov et al., 2003), The results of validation at Canadian BFTCS also showed correspondence of measured and calculated dendrometric parameters (Shaw et al., 2006). The validation of functional ecosystems characteristics represented at Fig. 2. The deviation of simulated and measured values is a result of variation of stand parameters and SOM pools and unknown stand history at the used experimental dataset
General impact of climate change on boreal forest without disturbances and cutting in Canada and Russia. The results of 100-year simulation of tree growth and soil dynamics at constant and changing climate are represented in Tables 1 and 2 and Fig. 3.

Tree species	Constant climate		Changed climate		%% of changes at changed climate
	mean	s.d.	mean	s.d.	
Black spruce	278.5	87.1	314.7	96.4	12.9
Norway spruce	360.5	113.9	406.1	118.2	12.6
Jack pine	134.4	43.4	156.1	42.15	16.4
Scots pine	403.3	68.2	301.8	37.6	-25.2

Table 1. Simulated growing stock, m^3 ha^{-1}, at 100 years without disturbances; results of Monte-Carlo simulations

The data of simulation clearly show that expected climate change will mostly lead to the increase of forest stand productivity both for Biomass (not shown) and especially for growing stock (Table 1). Proportional increase in growing stock is similar for both spruce species in spite of significant differences in their total productivity. Both spruces grow on wet sites that can minimise the effects of precipitation change and possible water deficit on tree growth. However, growth of American and European pine species is strongly different. There is a strong increase of growing stock in unproductive jack pine stands growing in cold continental climate of Central Canada. From the other hand, rather productive Scots pine forest on dry sandy site in European Russia near the south border of boreal forests is predicted to lose just a quarter of growing stock at climate change.

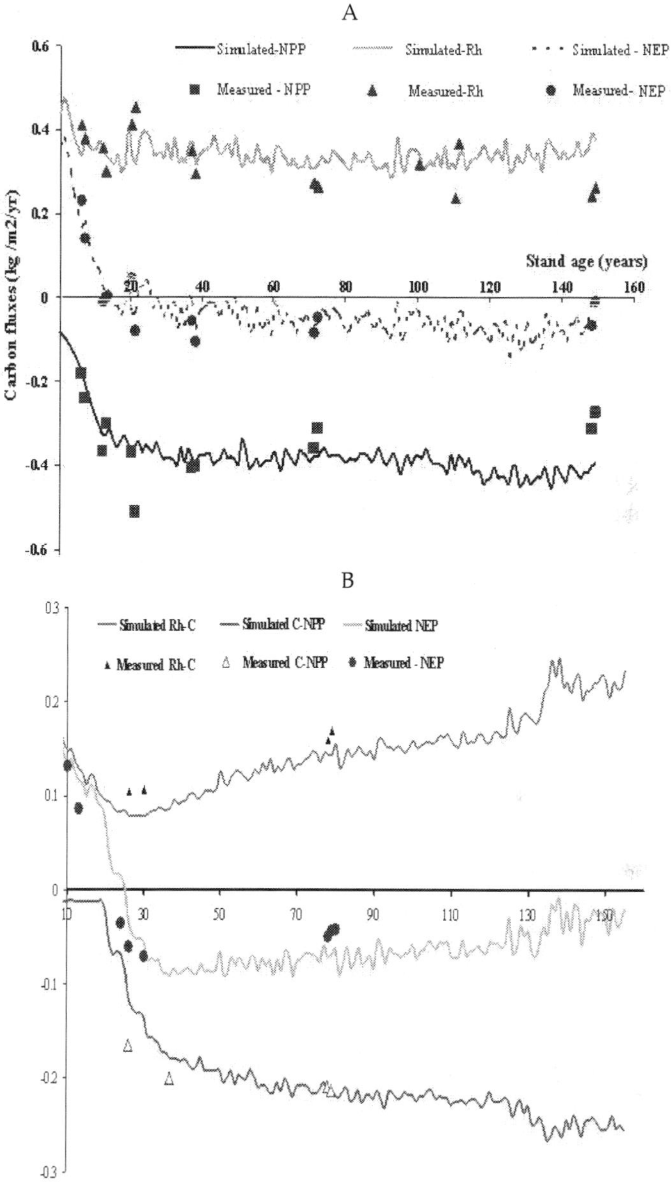

Fig. 2. EFIMOD validation against experimental data at Boreal Forest Transect Case Study (BFTCS) sites, Central Canada; components of carbon budget: A, black spruce (Chertov et al., 2009); B, jack pine (Bhatti et al., 2009). NPP, net primary productivity; Rh, soil respiration; NEP, net ecosystem productivity. Rh includes dead wood respiration.

Soil C pools respond similarly in Canada and Russia, with all soils losing organic matter at climate change. In Canada, the scale of soil C reduction is comparable for black spruce and jack pine. In Russia, the C loss in Norway spruce forest on wet site is relatively low, while it is very high in the Scots pine forest. It happens simultaneously with a strong decrease of stand productivity determining the input of litter fall in soil and the trends of soil C dynamics.

Tree species	Constant climate		Changed climate		%% of changes at changed climate
	mean	s.d.	mean	s.d.	
Black spruce	8.89	2.42	7.72	2.08	-13.2
Norway spruce	5.61	1.53	5.36	1.28	-4.5
Jack pine	3.14	0.81	2.71	0.83	-13.7
Scots pine	8.93	1.01	6.07	0.59	-32.2

Table 2. Simulated soil C pools, kg m-2, at 100 years without disturbances; results of Monte-Carlo simulations;

The temporal dynamics of tree and soil C under constant climate and climate change scenarios can be compared in Fig. 3. The curves of tree biomass growth are of monotonous type while soil dynamics demonstrates soil C decrease in young stands due to low litter fall production at this age. Clear positive trends of forest floor and mineral soil C increase are typical for post-fire forest ecosystems. These figures also show that the effect of climate change becomes clearly visible mostly in mean-aged and old forests.

A

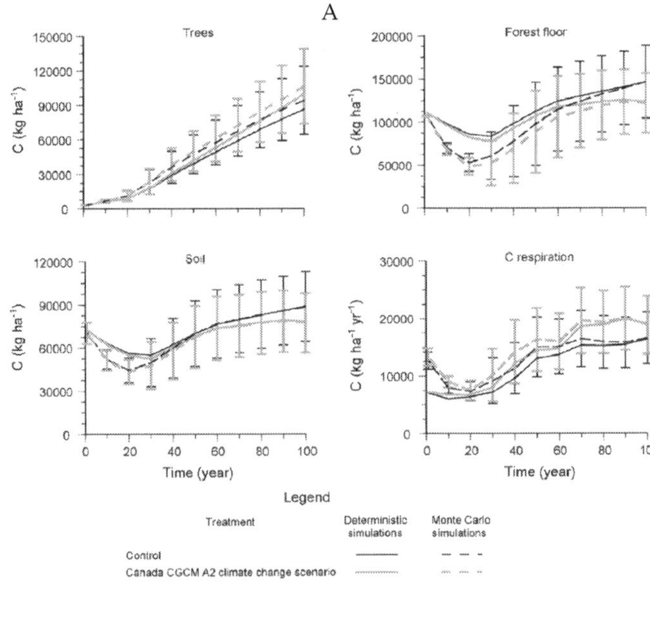

B

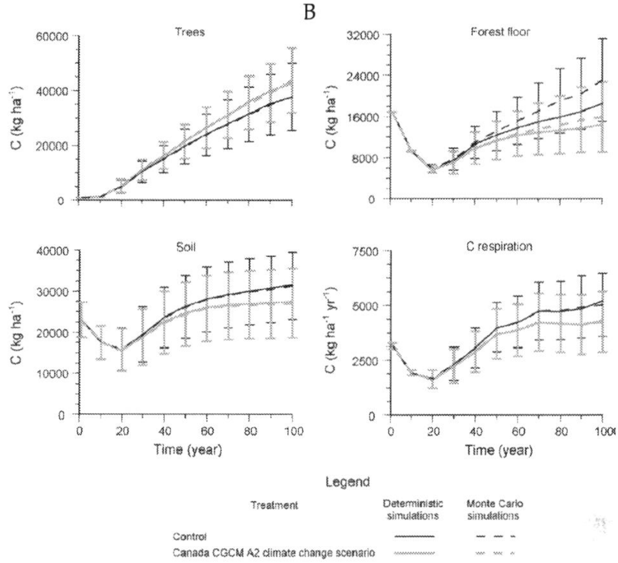

C

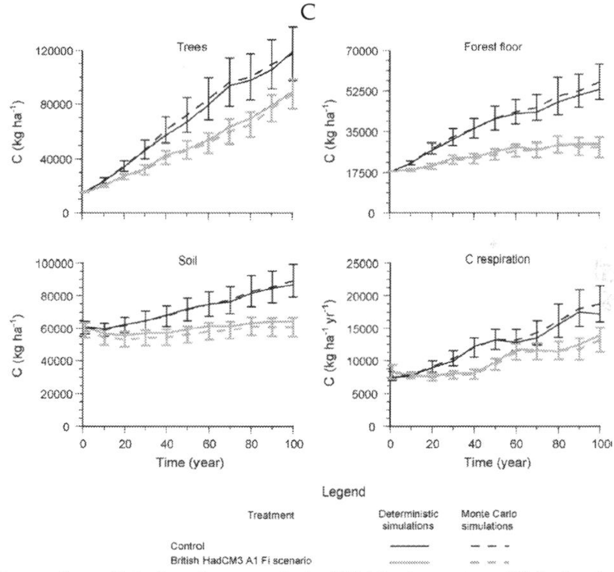

Fig. 3. Comparison of predicted carbon pools in (A) black spruce, (B) jack pine and (C) Scots pine forest ecosystem located in Canada and Russia using EFIMOD, following prescribed and Monte Carlo simulations. Error bars represent the standard deviations computed from the Monte Carlo predictions (Larocque et al., 2008). The dynamic pattern of Norway spruce in Russia is the same as black spruce in Canada; therefore we don't put Norway spruce on the graph.

In conclusion we should point out that impact of climate warming is also site-specific. It is positive on mesic and wet forest sites with productive soils. However, on dry and poor soils it can lead to a strong decrease of forest productivity and soil C pools.

Climate scenarios	Jack pine		Black spruce	
	Growing stock, m^3 ha^{-1}	% changes	Growing stock, m^3 ha^{-1}	% changes
Constant	242	0	366	0
CGCM A2	280	16	439	20
CGCM B2	273	13	429	17
CSIRO A2	282	17	455	24
CSIRO B2	278	15	450	23
HADCM A2	258	7	397	8
HADCM B2	256	6	394	8

Table 3. Growing stock, m^3 ha^{-1}, at harvesting without disturbances and six climate change scenarios, sum of two rotations over 150 years

Tree species	Climate scenario	Harvesting	2 fires	4 fires	Harvest-fire	Fire-harvest	Insect
Jack pine	Constant	0.126	0.114	0.043	0.128	0.113	0.130
	CGCM	0.143	0.133	0.055	0.149	0.131	0.151
	CSIRO	0.148	0.133	0.054	0.149	0.132	0.152
	HADCM	0.136	0.123	0.049	0.137	0.122	0.140
Black spruce	Constant	0.323	0.288	0.122	0.326	0.285	0.323
	CGCM	0.388	0.354	0.154	0.393	0.350	0.389
	CSIRO	0.402	0.367	0.161	0.407	0.363	0.404
	HADCM	0.351	0.322	0.139	0.355	0.318	0.351

Table 4. Mean net primary productivity (NPP) over 150 years of simulation (kg m^{-2} $year^{-1}$).

Impact of climate change on jack pine and black spruce forests with cutting, fire and insect attacks in Central Canada. The results of these simulations show that the effects of six different climate change scenarios demonstrated the similar trends of stand productivity increase in 21st century (Table 3). The highest increase of forest productivity was found with Australian CSIRO GCM, the lowest one was with British HADCM.

The same dynamic trends exhibit NPP data (Table 4). Moreover they show a strong effect of disturbance regimes on forest NPP: it is significantly lower at all fire scenarios and about 3-fold lower at the 4-fires scenario. However, the ecological effect of "harvest-fire" and "insect attack" is the same as harvesting because all burned and killed tree biomass did not remove from the forest in these scenarios. The important aspect is that the absolute values of disturbance effects are significantly higher of the effects of climate change on NPP.

Tree species	Climate scenario	Harvesting	2 fires	4 fires	Harvest-fire	Fire-harvest	Insect
Jack pine	Constant	0.092/ 0.122	0.089/ 0.106	0.037/ 0.051	0.092/ 0.117	0.089/ 0.111	0.107/ 0.120
	CGCM	0.109/ 0.143	0.106/ 0.124	0.046/ 0.060	0.110/ 0.136	0.104/ 0.129	0.125/ 0.147
	CSIRO	0.110/ 0.145	0.107/ 0.125	0.045/ 0.059	0.112/ 0.138	0.106/ 0.130	0.126/ 0.148
	HADCM	0.099/ 0.131	0.097/ 0.115	0.042/ 0.055	0.100/ 0.126	0.097/ 0.119	0.115/ 0.135
Black spruce	Constant	0.278/ 0.320	0.250/ 0.293	0.125/ 0.154	0.277/ 0.320	0.251/ 0.286	0.294/ 0.318
	CGCM	0.343/ 0.392	0.311/ 0.357	0.152/ 0.183	0.341/ 0.388	0.312/ 0.357	0.362/ 0.393
	CSIRO	0.356/ 0.408	0.324/ 0.370	0.158/ 0.189	0.355/ 0.403	0.325/ 0.372	0.376/ 0.409
	HADCM	0.308/ 0.353	0.280/ 0.325	0.139/ 0.169	0.306/ 0.350	0.282/ 0.324	0.324/ 0.350

Table 5. Mean soil CO_2-C emission/Total C loss with soil emission and disturbances over 150 year simulation, kg m^{-2} year^{-1}

Tree species	Climate scenario	Harvesting	2 fires	4 fires	Harvest-fire	Fire-harvest	Insect
Jack pine	Constant	0.005	0.007	-0.007	0.011	0.003	0.006
	CGCM	0.004	0.009	-0.005	0.012	0.002	0.004
	CSIRO	0.003	0.008	-0.005	0.011	0.002	0.004
	HADCM	0.005	0.008	-0.006	0.011	0.003	0.005
Black spruce	Constant	0.003	-0.005	-0.032	0.006	-0.001	0.005
	CGCM	-0.004	-0.003	-0.029	0.005	-0.005	-0.004
	CSIRO	-0.006	-0.003	-0.028	0.004	-0.009	-0.005
	HADCM	-0.002	-0.003	-0.030	0.005	-0.006	0.001

Table 6. Mean net ecosystem productivity (NEP) over 150 year simulation, kg m^{-2} year^{-1};

Climate warming has the same pattern in relation to soil heterotrophic respiration (CO_2 emission) as a main feedback from ecosystem to the atmosphere (Table 5): it becomes about 10-20% higher in a case of climate change. Again, a significant impact of disturbances on soil C emission was simulated here.

The values of total C loss from forest ecosystems (sum of Rh and C loss with fires, insect and harvested wood) repeat the differences of Rh data. They are the highest in scenarios with wood harvest but lowest in fire scenarios where burned wood remains in the forest for decomposition. There is one seeming contradiction in the data on total C loss: the scenario with four fires has twice lower values of C loss in comparison with other ones though, logically thinking, we could expect an opposite picture. It happened because all four fires were simulated in 37-year old stands with a low tree biomass.

The data on net ecosystem productivity (NEP, Table 6) integrate all C flows and represent a general C budget of the ecosystem. At a glimpse, all values are fluctuating around zero. However, there is irregular difference between climate change scenarios, disturbances (four fires regime leads to maximal C loss) and tree species (negative values in black spruce are dominating).

Additional simulations for Canadian jack pine and black spruce sites with variation of soil C and N pools showed that soil conditions, especially its productive potential determining by the N pool, modify the effect of climate change and disturbances: poor soils contrasting relative effect of climate change and damages, contrariwise more rich soil mitigates the effect of damages and climate change.

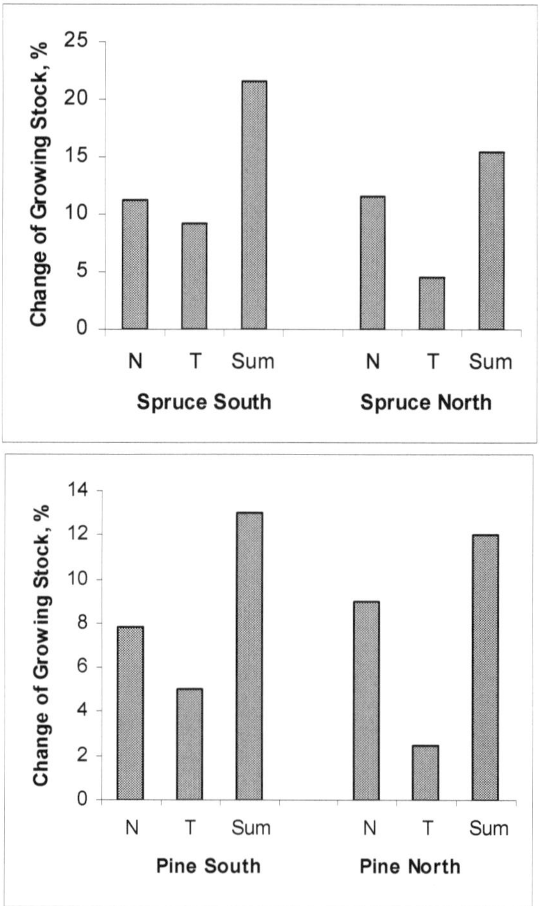

Fig. 4. Differences in growing stock as affected by climate change and atmospheric nitrogen. Symbols on the figure (here and below): N - nitrogen deposition increasing, T - temperature increasing, Sum - cumulative effect of nitrogen deposition and temperature increasing.

The simulated results show significant increases in NPP and Rh under changing climates as compared to a constant climate, suggesting a strong increase in biological C cycle capacity under climate change. The data on the effects of climate change shows differences among climatic scenarios. The greatest increases in ecosystem processes took place under the Australian (CSIRO) scenarios, while the UK scenarios (HADCM) showed slower ecosystem changes. Therefore, the projected climate change led to increases in wood production, reaching 17-24% in the Australian (CSIRO A2) scenarios. The simulated results show that increased concentration of atmospheric CO_2 (high at A2 and twice higher at B2) significantly smoothed the influence of difference of climatic scenarios themselves in relation to the forest productivity.

The effects of climate change and nitrogen deposition on European forests. The simulation data at seven forest sample plots across West Europe show rather interesting results on the comparative effects of climate change and atmospheric N deposition mostly due to industrial pollution (Chertov et al., 2006; Chertov, Komarov, 2007). Initially, we should call attention to the absence of principal difference in ecosystem responses for naturally developed and managed forests with regular thinning.

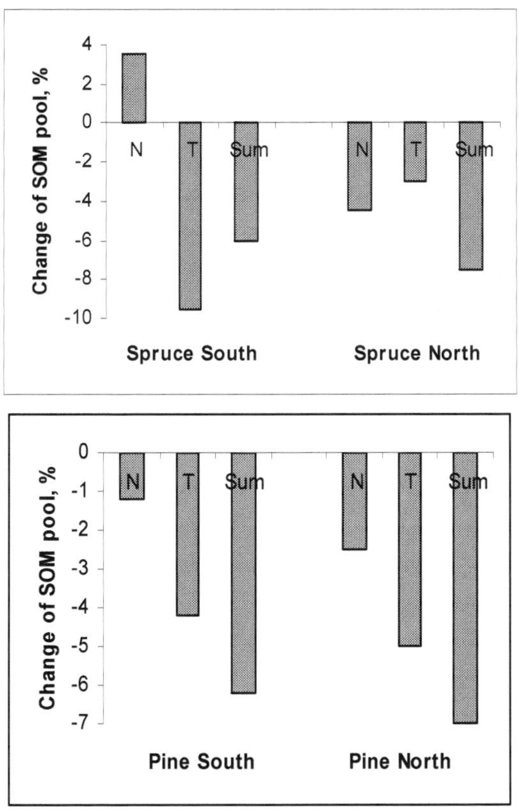

Fig. 5. Differences in soil organic matter (SOM) pools as affected by diverse ecological factors. Symbols corresponds to Fig. 4

First of all we should point out that the differences in stand height and basal area are totally positive for all comparisons reflecting effect of different factors. These positive changes vary from 0.5 to 10%. Both species in south sites demonstrate a little bit higher height changes, although basal area has no so clear changes. Unexpectedly, the differences of growing stock for both species are significantly higher of height or basal area changes (Fig. 4) reaching in some cases 22%. The effect is stronger for south sites and in Norway spruce stands. The nitrogen response was found to be sufficiently higher of temperature response.

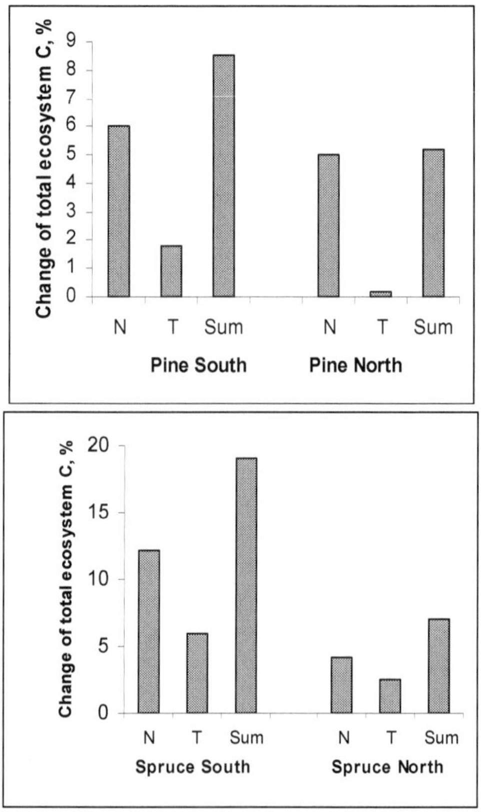

Fig. 6. Differences in total ecosystem carbon pools (sum of trees, dead wood and soil carbon) as affected by diverse ecological factors. Symbols corresponds to Fig. 4

The differences of SOM pools (Fig. 5) have the opposite trends in comparison with stand height, basal area and growing stock. The higher is the positive impact of temperature growth and N deposition the stronger is a loss of organic matter in the soil due to more intensive soil C mineralization. In terms of SOM, the effect of temperature increasing is stronger of the effect of N deposition.

The responses of total ecosystem carbon pools to climate warming and N deposition are very expressive (Fig. 6). In spite of the decrease of soil organic matter, all ecosystems accumulate carbon because significantly higher rise of stand productivity. The differences are larger in managed forest. South forests and Norway spruce stands demonstrate significantly higher response to the climate warming.

Simulation of climate change and forest management regimes at landscape level in European Russia. The generalised data of 200-year EFIMOD runs for 300-ha forest area with 108 forest compartments representing various coniferous and mixed stands are reflected on Fig. 7 and 8. The Fig. 7 shows that the more intensive is forest harvest regime the less is difference of C pools in tree biomass and soil in comparison with their values at constant climate. However, the rise of tree biomass C is higher the loss of soil C at all forestry regimes. The increase of atmospheric N deposition slightly mitigates the negative impact of forest overexploitation on soil C pools. The data on C balance (NEP, Fig. 8) clearly exhibit a positive effect of climate change at different silvicultural regimes. Selective cutting and Russian legislative scenarios have just a zero C balance in current climate with low and high N deposition. Though, they become C sequestering regimes at climate change. The scenario of legislation breach with forest overexploitation remains a C source even at climate change with increase of forest productivity.

The results of this simulation also show that other environmental factors (N deposition) and human-generated disturbances (forestry regimes) strongly modify the impact of climate change on forest territory.

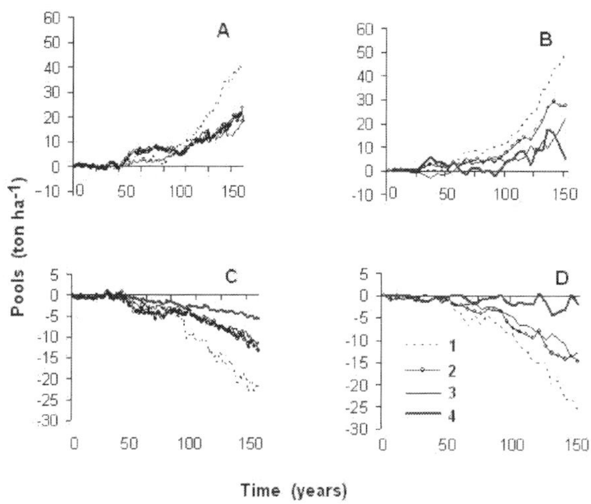

Fig. 7. Difference of C accumulation between scenarios with climate change and constant climate in forest landscape/unit in European Russia at two levels of nitrogen deposition [6 (A, C) and 12 kg ha^{-1} yr^{-1} (B, D)] for trees (A, B) and soil (C, D). Scenarios of forest management: 1, natural development, 2, selective cutting, 3, Russian legislative forest management, 4, legislation breach (Mikhailov et al., 2007)

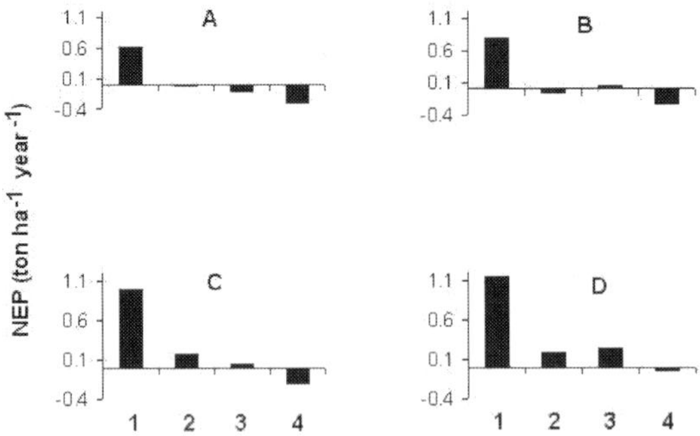

Fig. 8. Carbon balance (net ecosystem productivity, NEP) of forest landscape at constant climate (A, B) and climate change (C, D) in forest landscape/unit in European Russia at two levels of atmospheric N deposition [6 (A, C) and 12 kg ha^{-1} yr^{-1} (B, D)]. Scenarios of forest management: 1, natural development; 2, selective cutting; 3, Russian legislative forest management; 4, legislation breach (Mikhailov et al., 2007)

5. Discussion and conclusion

The results of EFIMOD model simulations to specify a possible effect of forthcoming climate warming allowed for preliminary quantification of the effects of this environmental change on boreal forests in North America and Europe. In Central Canada, the black spruce and jack pine forests respond to climate warming, fire, harvesting and insects by significant modification of net primary productivity (NPP), soil respiration (Rs), net ecosystem production (NEP) and pools of tree biomass and soil organic matter (SOM). The simulated effects of six climate change scenarios demonstrated the similar increasing trends of NPP and stand productivity. The disturbances led to a strong decrease in NPP, stand productivity, soil organic matter (SOM) and nitrogen (N) pools with an increase in CO_2 emission to the atmosphere. However the accumulated NEP for 150 years under harvest and fire fluctuated around zero. In jack pine forest, it becames negative only at a more frequent disturbance regime with four forest fires. In black spruce stands, it is slightly negative also in a case of four fire scenario as well as for harvest and two fires during the period of simulation. The results from this study show that climate warming and disturbance regimes might substantially change the NPP as well as the C and N balance, resulting in major changes in the C pools of the vegetation and soil under black spruce forests. Soil conditions, especially its productive potential determining by the N pool, modify the effect of climate change and disturbances: poor soils contrasting relative effect of climate change and damages, contrariwise more rich soil mitigates the effect of damages and climate change.

In Russia, the effect of climate change in Norway spruce forests has the same dynamic patters as in Canadian black spruce forests. However, Scots pine forests on dry sandy sites lose both soil C and forest productivity. This reaction reflects a difference of environmental condition in Canadian jack pine stands in central boreal forest with harsh continental climate and Russian Scots pine stands in rather warm climate near the border of boreal and temperate broad-leaved forests.

The results of selective analysis of EFIMOD simulations using experimental data of long-term ecological researches (Recognition project, Kahle et al., 2008) show the uniformity of forest ecosystem reactions to climate warming and N deposition changes. We can conclude that this reaction in the Central Europe is more expressive than in the North Europe, especially for growing stock in Scots pine and Norway spruce. Low response of the forest ecosystems to temperature increasing in these simulations can be explained by the lower nitrogen deposition during most part of 20th century in comparison with prognosis for 21st century. This circumstance can smooth the effect of temperature growth.

The landscape level simulation in Central European Russia shows that silvicultural regimes and atmospheric N deposition strongly modify the impact of climate warming on forest ecosystems. The higher is harvesting intensity (maximal at legislation breach scenario) the less is relative effect of climate change. Therefore the regime of heavy disturbances or forest overexploitation strongly reduces absolute amplitude of the impact of climate change (warming with slight changes of precipitation) on boreal forests. Actually the negative impact of increased disturbances (together with overexploitation) can outbalance the positive effect of climate change on forest productivity. It can lead to the decrease of carbon sequestration by forests with a positive feedback on carbon dioxide concentration in the atmosphere and climate change.

We should point out that the decrease of soil C pools at climate change was also widely reported and simulated both for forest and agricultural ecosystems (Bondeau et al., 2007; Schulze, Freibauer, 2005; Smith et al., 2005, 2006; Yurova et al., 2010). The loss of soil C can be a sign of decrease of forest stability because soil system plays a role of ecosystem stabiliser in a case of any disturbances.

The EFIMOD has no response function to simulate CO_2 fertilization effect on NPP. Its influence was evaluated only by the change in temperature and precipitations trends affecting the rate of soil processes and consumption of soil N. The increase in boreal NPP due to CO_2 fertilization remains a subject of debate (Kurz et al., 2007). Recently Bond-Lamberty et al. (2007) observed that CO_2 fertilization have no significant effect on net biome productivity of central Canadian boreal forest expect its interaction with climate (temperature and precipitation) which significantly influence the fire disturbance regime. Percy et al. (2002), from their Aspen FACE experiment which simulate open-air environment for tree growth, also concluded that positive effect of CO_2 was being overestimated insofar as the productivity of these northern boreal forests is concern.

Some ecophysiological models (Van Oijen et al., 2008) take into account the effect of CO_2 concentration on photosynthesis. Peng and Apps (1999) and Kang et al. (2006) used a simple empirical function to simulate NPP under CO_2 fertilizer effect without explicit calculation of N budget in plant and soil in boreal forests. Tree growth in these models is calculating by consistent multiplication by growth multipliers including CO_2 response. In the EFIMOD model tree growth is calculating by Liebig's bottle neck principle. So, if we take into account CO_2 influence on potential NPP, but the ecosystem has nitrogen deficit, then the tree growth will be determined only by the pool of available nitrogen. In this case, effect of NPP increase due to carbon dioxide fertilization will be equal to zero. Therefore, the including of any dependence of NPP on CO_2 concentration will also not influence tree growth in the EFIMOD.

We suppose that increasing of CO_2 concentration in the atmosphere and corresponding increasing of photosynthetic rate in forest stands on nitrogen-deficit soils can slightly increase wood increment (by formation mainly a spring lignin-poor part of tree ring). But the main mechanism of CO_2 influence supposes to be an indirect effect by the acceleration of soil processes. There is a well known "priming effect" (Blagodatskaya, Kuziakov, 2008) as the impact of easy decomposable organic compounds on the increase of slow decomposable soil C mineralization. It can take place if excessive primary photosynthetic assimilates that cannot be transformed into tree biomass (because nitrogen and other elements deficit) go to soil as root exudates. It will lead to the intensification of soil N mineralization with the increase of available for tree growth N and correspondingly to the NPP rise.

Many simulation models have been used to understand the influence of climate change and disturbances on C dynamics in boreal forests. However, EFIMOD carried out these simulations with significant feedback from the soil dynamic processes, specifically the N feedback mechanism and soil microclimatic conditions over stand age since EFIMOD uses soil climatic scenarios for soil temperature and moisture with a monthly time step allowing to include the climate change data in the simulation. The performance of the EFIMOD for boreal forests in Canada and Russia exemplifies the satisfactory results of this model implementation for these conditions. The impacts of disturbances and harvesting regimes on ecosystem C dynamics were substantial as compared to climate change alone. The feedback mechanism between aboveground NPP and changes in soil N availability under different climate change and disturbance scenarios had pronounced effects on C cycle and therefore these processes need to be more fully represented in models of forest ecosystem responses to climate change.

The data presented here describes generally a picture of forest productivity growth with reduction of soil C at climate warming with small changes of precipitation. In reality, the boreal forests along the climatic gradients have the same pattern of forest and soil changes. In Canada, the comparison of old jack pine forest at climatic gradient with identical soils along BFTCS (between Thompson and Candle Lake) with a slope of mean annual temperature 3.2°C (from -3.4 to -0.2°C) and the same summer precipitation shows a significant increase of forest productivity at warmer site (Shaw et al., 2006). The stand height increases from 10.1 to 14.7 m, DBH from 10.3 to 13.9 cm, stem C from 1.7 to 4.0 kg m^{-2}. Though this gradient is about twice higher of predicted changes by GCMs this comparison gives a clear picture of a significant rise of forest productivity at climate warming in boreal forest.

The same is clearly seen at East European Plane in Russia (Molchanov, 1961; Kurnaev, 1973; Usoltsev, 2001): there is an increase of forest productivity from North to Central taiga by one site index (Bonität) and from Central to South taiga also by one site index but with change of forest vegetation and especially soil. In ecological optimum (mesic relatively productive sites), there are raw humus gley-podzols with poor plant diversity in North taiga and modern and mull podzolic soils with rich plant diversity in South taiga. It is a main functional difference of natural climatic gradients with projected ecosystem changes reported here. It seems that fast warming does not still lead to qualitative changes of forest vegetation and soil in contrast with forest ecosystems along the natural climate gradients. Perhaps we can tell on some kind of "climatic succession" in a case of fast climate change with slow invasion of flora and fauna of warmer ecosystems and rather fast rise of net ecosystem exchange and carbon sequestration.

The very last conclusion that is obvious on the base of this simulation experiment is as follows: the NEP increase in boreal forest ecosystems in optimal edaphic conditions at climate change is an important negative feedback of the biosphere on environmental change.

Acknowledgements

This work was supported in part by the Federal Panel on Energy Research and Development (PERD) Canada, Canadian Forest Service, EU INTAS Projects Silvics and Podzol, EU Projects Recognition, OMRISK and eLUP, the Russian Fund of Basic Research Grant 09-04-01209 and by the President's of Russian Federation Program, Governmental

Contract 02.740.11.5197. Great thanks to the Sciyo Editorial Board and Ms. Ruth Errington, CFS, for the comprehensive editing of the manuscript.

6. References

Amiro, B.D., Stocks, B.J., Alexander, M.E., Flannigan, M.D., Wotton, B.M., 2001. Fire,climate change, carbon and fuel management in the Canadian boreal forest.International Journal of Wildland Fire 10, 405–413.

Bhatti, J. Chertov, O. Komarov, A. 2009. Influence of climate change, fire, insect and harvest on C dynamics for jack pine in central Canada: simulation approach with the EFIMOD model. Int. J. Climate Change: Impacts and Responses 1, 43-61.

Blagodatskaya, E.V; Kuzyakov, Y. 2008. Mechanisms of real and apparent priming effects and their dependence on soil microbial biomass and community structure: critical review, Biology and Fertility of Soils 45, 115-131.

Bobrovsky, M.V. 2004. Forest soils: biotic and antropogenic factors of their formation. In: Smirnova O.V., Ed. East European Forests. Book 1. Nauka, Moscow, pp. 381-427. In Russian.

Bobrovsky, M., Komarov, A., Mikhailov, A., Khanina, L. 2010. Modelling dynamics of soil organic matter under historical land-use management in European Russia. Ecological Modelling 221, 953-959. doi:10.1016/j.ecolmodel.2009.12.013

Bondeau A., P.C. Smith, S. Zaehle, S. Schaphoff, W. Lucht, W. Cramer, D. Gerten, H. Lotze-Campen, C. Müller, M. Reichstein, and B. Smith. 2007. Modelling the role of agriculture for the 20th century global terrestrial carbon balance. Global Change Biology 13, 679–706.

Bond-Lamberty, B., Wang, C., Gower, S.T., 2004. Net primary production and net ecosystem production of a boreal black spruce wildfire chronosequence. Global Change Biology 10, 473–487.

Bond-Lamberty, B., Peckham, S.D., Ahl, D.E., Gower, S.T., 2007. The dominance of fire in determining carbon balance of the central Canadian boreal forest. Nature 450, 89–92.

Bykhovets, S.S., Komarov, A.S. 2002. A simple statistical simulator of soil climate. Eurasian Soil Sci. 35, 443-452.

Chertov, O.G., Komarov, A.S. 2007. Model analysis of the reasons for increase of the forest growth in Europe at second half of 20th century. Project Recognition. In Kudeyarov, V.N. Responsible ed. Modelling Dynamics of Organic Matter in Forest Ecosystems. Nauka, Moscow, pp. 305-314. In Russian with English contents.

Chertov, O.G., Razumovsky, S.M. 1980. On the ecological trends of soil forming processes. J. Obshchei Biol., Moscow (J. General Biol.). 41, 386-396. In Russian with English summary.

Chertov, O., Bhatti, J., Komarov, A., Mikhailov, A., Bykhovets, S. 2009. Influence of climate change, fire and harvest on the carbon dynamics of black spruce in Central Canada. Forest Ecology Management 257, 941-950. DOI:10.1016/j.foreco.2008.10.038

Chertov, O.G. Komarov, A.S., Tsiplianovsky, A.V. 1999. A combined simulation model of Scots pine, Norway spruce and Silver birch ecosystems in European boreal zone. Forest Ecology and Management 116, 189-206.

Chertov, O.G., Komarov, A.S., Nadporozhskaya, M.A., Bykhovets, S.A. and Zudin, S.L. 2001. ROMUL - a model of forest soil organic matter dynamics as a substantial tool for forest ecosystem modelling. Ecological Modelling, 138, 289-308.

Chertov, O.G., Komarov, A.S., Bykhovets, S.S. and Kobak, K.I. 2002. Simulated soil organic matter dynamics in forests of the Leningrad administrative area, northwestern Russia. Forest Ecology and Management 169 (1-2), 29-44. DOI:10.1016/j.ecolmodel. 2008.07.024

Chertov, O., Komarov, A., Kolström, M., Pitkänen, S., Strandman, H., Zudin, S., Kellomäki, S. 2003. Modelling the long-term dynamics of populations and communities of trees in boreal forests based on competition for light and nitrogen. Forest Ecol. Management 176, 355-369.

Chertov, O., Komarov, A., Mikhailov, A., Andrienko, G., Andrienko, N., Gatalsky. P. 2005. Geovisualisation of forest simulation modelling results: a case study of carbon sequestration and biodiversity. Computers and Electronics in Agriculture 49, 175–191.

Chertov O., Komarov A., Loukianov A., Mikhailov A., Nadporozhskaya M., Zubkova E. 2006. The use of forest ecosystem model EFIMOD for research and practical implementation at forest stand, local and regional levels. Ecological Modelling 194, 227-232.

Gower, S.T., Vogel, J., Stow, T.K., Norman, J.M., Steele, S.J., Kucharik, C.J. 1997. Carbon distribution and above-ground net primary production of upland and lowland boreal forest in Saskatchewan and Manitoba. Journal of Geophysical Research 104, 29029–29041.

Hatten, J., Zabowski, D., Scherer, G. and E. Dolan. 2005. A comparison of soil properties after contemporary wildfire and fire suppression. Forest Ecology and Management 220(1-3), 227-241.

Howard, E.A., Gower, S.T., Foley, J.A., and Kucharik, C.J. 2004. Effects of logging on carbon dynamics of a jack pine forest in Saskatchewan, Canada. Global Change Biology 10, 1-18.

Kahle, H.-P., Karjalainen, T., Schuck, A. , Ågren, G., Kellomäki, S., Mellert, K., Prietzel, J., Rehfuess, K.E., Spiecker, H. Eds. Eds. Causes and Consequences of Forest Growth Trends in Europe - Results of the RECOGNITION Project. EFI Res. Rep. 21. Brill, Leiden, Boston.

Kang, S., Kimball, J.S., Running, S.W., 2006. Simulating effects of fire disturbance and climate change on boreal forest productivity and evapotranspiration. Science of The Total Environment 362, 85–102.

Komarov, A., Chertov, O., Zudin, S., Nadporozhskaya, M., Mikhailov, A., Bykhovets, S., Zudina, E., Zoubkova. 2003. EFIMOD 2 - - A model of growth and elements cycling in boreal forest ecosystems. Ecological Modelling 170, 373-392.

Komarov, A.S., Chertov, O.G., Mikhailov, and Autors' Collective (14 names). 2007. Modelling Dynamics of Organic Matter in Forest Ecosystems [Responsible editor V.N. Kudeyarov]. Nauka, Moscow. 380 p. In Russian with English contents. ISBN 5-02-034053-7.

Kurnaev, S.F. 1973. Forest Zoning of the USSR. Nauka, Moscow. 220 p. In Russian.

Kurz, W.A., Apps, M.J., 1999. A 70-year retrospective analysis of carbon fluxes in the Canadian forest sector. Ecological Applications 9, 526–547.

Kurz, W. A., Apps, M. J., Webb, T. M. & McNamee, P. J. 1992. The Carbon Budget of the Canadian Forest Sector: Phase I. Information Report NOR-X-326 (Forestry Canada, Northwest Region, Northern Forestry Centre, Edmonton.

Kurz, W.A., Stinson, G., Rampley, G., 2007. Could increased boreal forest ecosystem productivity offset carbon losses from increased disturbances? Philosophical Transactions of the Royal Society of London Series B, doi:10.1098/rstb.2007.2198.

Larocque, G.R., Bhatti, J. S., Boutin, R., Chertov, O. 2008. Uncertainty analysis in carbon cycle models of forest ecosystems: Research needs and development of a theoretical framework to estimate error propagation. Ecological Modelling 219, 400–412.

Mikhailov, A., Komarov, A., Chertov, O. 2004. Simulation modelling of forest ecosystem development under the different forest management scenarios. Proc. of the 4th European Conference on Ecological Modelling, ECEM 2004. Bled, Slovenia, 95-97.

Mikhailov A.V., Lukianov, A.M., Bykhovets, S.S., Priputia, I.V. 2007. Impact of atmospheric nitrogen deposition and climate change on different regimes of forest management. In Modelling Dynamics of Organic Matter in Forest Ecosystems [Responsible editor V.N. Kudeyarov]. Nauka, Moscow, pp. 297-305. In Russian with English contents.

Mitchell, T.D., Carter, T.R., Jones, P.D., Hulme, M., New, M. 2004. A comprehensive set of high-resolution grids of monthly climate for Europe and the globe: the observed record (1901–2000) and 16 scenarios (2001–2100). Tyndall Centre for Climate Change Research. Working Paper No. 55. 25 p.

Molchanov, A.A. 1961. Forest and Climate. Nauka, Moscow. 278 p. In Russian.

Nadporozhskaya, M.A. Mohren, G.M.J., Chertov, O.G., Komarov, A.S., Mikhailov, A.V. 2006. Soil organic matter dynamics at primary and secondary forest succession on sandy soils in The Netherlands: an application of soil organic matter model ROMUL. Ecological Modelling Vol. 190, 399-418. DOI:10.1016/j.ecolmodel. 2005.03.025

Nakane, K., Kohno, T., Horikoshi, T., Nakatsubo, T., 1997. Soil carbon cycling at a black spruce (Picea mariana) forest stand in Saskatchewan. Canadian Journal of Geophysical Research 102, 28785–28793.

Oechel,W.C., Van Cleve, K., 1986. The role of bryophytes in nutrient cycling in the taiga. Forest Ecosystems in the Alaskan. In: Taiga K., Van Cleve, Chapin, III, F.S., Flanagan, P.W., Viereck, L.A., Dyrness, C.T. (Eds.), Ecological Studies 57. Springer-Verlag, New York, pp. 121–137.

Peng, C., Apps, M.J., 1999. Modelling the response of net primary productivity (NPP) of boreal forest ecosystems to changes in climate and fire disturbance regimes. Ecological Modelling 122, 175–193.

Percy, K.E., Awmack, C.S., Lindroth, R.L., et al., 2002. Altered performance of forest pests under atmospheres enriched by CO_2 and O_3. Nature 420, 403–407.

Price, D.T., Apps, M.J. 1995. The boreal forest transect case study: global change effects on ecosystem processes and carbon dynamics in boreal Canada. Water. Air. Soil Pollution, 82, 203-214.

Price, D.T., McKenney, D.W., Papadopol, P., Logan, T., Hutchinson, M.F., 2004. High resolution future scenario climate data for North America. In: Proceedings of the American Meteorological Society. 26th Conference on Agricultural and Forest Meteorology. Vancouver, B.C., August 23–26 13 pp. (CD-ROM).

Shaw C., Chertov O., Komarov A., Bhatti J., Nadporozskaya M., Apps M., Bykhovets S., Mikhailov A. 2006. Application of the forest ecosystem model EFIMOD 2 to jack pine along the Boreal Forest Transect Case Study. Canadian J. Soil Sci. 86, 171-185.

Schulze, E.-D., Freibauer, A. 2005. Carbon unlocked from soils. Nature 437, 205-206.

Smith, J.U., Smith, P., Wattenbach, M., Zaehle, S., Hiederer, R., Jones, R.J.A., Montanarella, L., Rounsevell, M.D.A., Reginster, I. , Ewert, F. 2005. Projected changes in mineral soil carbon of European croplands and grasslands, 1990-2080. Global Change Biology, 11, 2141-2152

Smith, P., Smith, J.U., Wattenbach, M., Meyer, J., Lindner, M., Zaehle, S., Hiederer, R., Jones, R., Montanarella, L., Rounsevell, M., Reginster, I., Kankaanpää, S. 2006. Projected changes in mineral soil carbon of European forests, 1990-2100. Canadian Journal of Soil Science, 86(2 Special Issue SI), 159-169

Usoltsev, V.A. 2001. Phytomass of the North Eurasian Forests: Database and Geography. Ural Centre of Russian Academy of Sci., Ekaterinburg. 708 p. In Russian.

Van Oijen, M., Ågren, G., Chertov, O., Kellomäki, S., Komarov, A., Mobbs, D., Murray. M. 2008. Methodology for the application of process-based models to analyse changes in European forest growth. In: Kahle, H.-P., Karjalainen, T., Schuck, A. , Ågren, G., Kellomäki, S., Mellert, K., Prietzel, J., Rehfuess, K.E., Spiecker, H. Eds. Eds. Causes and Consequences of Forest Growth Trends in Europe - Results of the RECOGNITION Project. Brill, Leiden, Boston, Köln. Chapter 3.2, pp. 67-80.

Wang, C., Bond-Lamberty, B., Gower, S.T., 2002. Soil surface CO2 flux in a boreal black spruce fire chronosequence. Journal of Geophysical Research 108 WFX5-1–WFX5-8.

Wang, C., Bond-Lamberty, B., Gower, S.T., 2003. Carbon distribution of a well- and poorly drained black spruce fire chronosequence. Global Change Biology 9, 1–14.

Wei, X., Kimmins, J.P. and G. Zhou. 2003. Disturbances and the sustainability of long-term site productivity in lodgepole pine forests in the central interior of British Columbia — an ecosystem modeling approach. Ecological Modelling 164, 239–256.

Wotton, B.M., Stocks, B.J., Flannigan, M.D., Laprise, R., Blanchet, J.P., 1998. Estimating current and future fire climates in the boreal forest of Canada using a Regional Climate Model. In: Proceedings of the 3rd International Conference on Forest Fire Research and 14th Conference on Fire and Forest Meteorology, Luso, Portugal, November 16–20, pp. 1207–1221.

Yurova, A.Yu., Volodin, E.M., Ågren, G.I., Chertov, O.G., Komarov, A.S. 2010. Effects of variations in simulated changes in soil carbon contents and dynamics on future climate projections. Global Change Biology16, 823-835. DOI: 10.1111/j.1365-2486.2009.01992.x

Zha, T., Barr, A. G., Black, T. A., McCaughey, J. H., Bhatti, J., Hawthorne, I. Krishnan, P., Kidston, J., Saigusa, N., Shashkov, A., Nesic, Z. 2009. Carbon sequestration in boreal jack pine stands following harvesting. Global Change Biology. 15, 1475-1487. doi: 10.1111/j.1365-2486.2008.01817.x

17

Simulating peatland methane dynamics coupled to a mechanistic model of biogeochemistry, hydrology, and energy: Implications to climate change

Takeshi Ise[1,5], Allison L. Dunn[2,6],
Steven C. Wofsy[3,7] and Paul R. Moorcroft[4,8]

[1] *Japan Agency for Marine-Earth Science and Technology, 3173-25 Showa-machi, Kanazawa-ku, Yokohama, Kanagawa, 236-0001, Japan*

[2] *Department of Physical and Earth Sciences, Worcester State College, 486 Chandler Street, Worcester, MA 01602, U.S.A.*

[3] *Department of Earth and Planetary Sciences, Harvard University, 20 Oxford Street, Cambridge, MA 02138, U.S.A.*

[4] *Department of Organismic and Evolutionary Biology, Harvard University, 22 Divinity Avenue, Cambridge, MA 02138 U.S.A.*

[5] *Correspondence author. ise@jamstec.go.jp, phone: +81-45-778-5595, fax: +81-45-778-5706, Mailing address: 3173-25 Showa-machi, Kanazawa-ku, Yokohama, Kanagawa, 236-0001, Japan*

Abstract

Northern peat lands have a strong potential to modify climate through changes in soil organic carbon (SOC) and methane (CH_4). A dynamic interaction among climate, soil physical properties (e.g., temperature and moisture), and biogeochemistry (e.g., quantity and quality of SOC) determines the peat land system. Due to this interaction, CH_4 production, oxidation, and transport dynamics changes dramatically under climate change. To appropriately study the future CH_4 in a predictive manner, a simulation model must be able to reproduce the inherent dynamic interaction in the peat land system.

Here, these complex interactions were simulated simultaneously in a biogeochemical peat land model coupled with mechanistic soil hydrology and thermal dynamics (ED2.0-peat). The model successfully reproduced soil physical profiles and the resultant SOC and CH_4 observed in a poor fen of northern Manitoba. With an experimental simulation of 4°C warming, a significant long-term decline in CH_4 emission was found, caused by a loss in substrate and prevalence of aerobic conditions. However, there was a transient increase in CH_4 emission shortly after warming because of time lag between the temperature dependence of microbial activity (a fast response to climate change) and the loss in peat depth (a slow response).

1. Introduction

High latitudinal regions are predicted to undergo particularly intensive warming (Holland and Bitz, 2003), with SOC in northern peat lands likely to be strongly affected (Ise et al., 2008). Peat decomposition has a potential to induce a strong biogeochemical feedback to the global climate system by adding excess CO_2 to the atmosphere (Friedlingstein et al., 2006). In addition, northern peat lands are a large source of CH_4, (~10%; Schlesinger, 1997; Frolking et al., 2006), another major greenhouse gas. Developing a process-based understanding of peat land dynamics therefore is critical to predicting peat land responses to climate and feedbacks onto global climate.

Soil thermal and moisture conditions strongly controls wetland CH_4 dynamics, in addition to SOC accumulation and decomposition (Pelletier et al., 2007). The position of water table is the essential determinant of peat land CH_4 dynamics because CH_4 production by methanogens is a strictly anaerobic process, and CH_4 oxidation by methanotrophs consumes CH_4 in aerobic layers (Fig. 1). In addition, those biological processes are dependent on temperature of each layer. To accurately simulate CH_4 dynamics of a peat land, appropriate reproduction of short-term (sub-daily), depth-structured soil physics (i.e., hydrology and energy) is needed.

Moreover, in long-term, changes in peat quantity and quality that alter substrate conditions should be appropriately represented. Biogeochemical and physical dynamics in organic soils are markedly different from those in mineral soils. In particular, the amount of SOC stored in peat affects its physical properties, with peat surfaces rising and falling as SOC changes. During periods of SOC accumulation, the area becomes moister as humified peat generally has a higher water holding capacity and a lower hydraulic conductivity than mineral soils (Clymo, 1984). This causes the water table to rise, and the resultant anaerobic condition slows SOC decomposition, leading to further accumulation (positive feedback). The rate of water table rise does not keep up with the peat growth rate indefinitely, however, and thus this eventually acts as a negative feedback on further SOC accumulation (the proportion of SOC below water table becomes smaller through time, and the resultant higher decomposition rate restricts peat growth; Ise et al., 2008). Water table depth, which is a critical factor regulating decomposition rates, should be simulated from hydrological changes in the physical characteristics of peat, thereby capturing those feedbacks on SOC accumulation and decomposition. To our knowledge, only ED2.0-peat (Ise et al., 2008) has examined those processes simultaneously and mechanistically. Thus the mechanistic calculation of soil energy and moisture by ED2.0-peat is advantageous to study the future trend of CH_4 in a predictive manner, under various environmental conditions.

In this paper, we introduce CH_4 production, oxidation, and transportation sub models into ED2.0-peat—a peat carbon model with dynamic hydrological and thermal processes—to simulate peat land CH_4 dynamics in a predictive manner, together with dynamically-changing soil conditions. The peat biogeochemical model is coupled with a soil physics model, allowing peat depth to change with SOC, and simulate the resultant changes in soil properties. As a continuation from Ise et al. (2008), we parameterize the model with field observations from a poor fen in Manitoba, Canada and evaluate the model with the observed SOC, CH_4 emission rate, soil temperature, and water table depth. Using an experimental warming, responses of peat SOC to a warming climate are then studied. It should be noted that the explicit reproduction of the dynamic peat land system (i.e., vertical profiles of temperature, moisture, and SOC) is the prerequisite for the predictive study of CH_4 dynamics, and we describe the model structure and outputs of ED2.0-peat accordingly.

2. Methods

ED2.0-peat was applied to the Northern Study Area (NSA) Fen site of the Boreal Ecosystem and Atmosphere Study (BOREAS; Sellers et al., 1995; Trumbore and Harden, 1997) in northern Manitoba, Canada. This study area is close to the northern limit of boreal forest, and in the discontinuous permafrost zone (Veldhuis et al., 2002). The area is periodically covered by standing water especially in spring.

The site-specific parameters were obtained and summarized from the BOREAS dataset. The average bulk density of fibrous peat is 0.092 gSoil/cm^3. The average bulk density of humic peat (0.38 gSoil/cm^3) is taken from the modal estimate for Oh horizons in BOREAS Northern Study Area, among the range of 0.13 gSoil/cm^3. The carbon fraction of fibrous peat is 0.4 gC/gSoil, and that of humic peat is 0.3 gC/gSoil (Apps and Halliwell, 1999). Litter input rate to fibrous layer was estimated from the net primary productivity (NPP) reported by Bond-Lamberty et al. (2004; 0.26 kgC m^{-2} y^{-1}) and partitioned into labile (0.17 kgC m^{-2} y^{-1}) and recalcitrant (0.09 kgC m^{-2} y^{-1}) fractions according to a relationship between nitrogen density and lignin fraction of plant biomass described by Parton et al. (1987), using a generalized lignin fraction of boreal forest species including bryophytes calculated by Ise and Moorcroft (2006). Meteorological data (1994-2005) from the eddy covariance tower at the Old Black Spruce site (OBS; 7 km southwest of the Fen site; Dunn et al., 2007) were used to drive the simulation with a 30-minute time step. Missing meteorological data (<10%) were filled with data from the closest weather station at Thompson Airport, Manitoba (39 km east of the site).

ED2.0-peat model structure and data acquisition are described in detail by Ise et al. (2008). To evaluate the model performance, ED2.0-peat was first driven by the meteorological data of 1994, the year of the BOREAS field campaign for soil thermal and hydrological observation at the Fen site. Mean water table depth, soil temperature, and CH_4 emission data of 4 observation subsides at Fen site for year 1994 were calculated to evaluate the model. 12 years of meteorological data (1994-2005) from the OBS flux tower were then used repeatedly to drive long-term simulations (see Supplementary Information in Ise et al. (2008) for the 12-year variability in simulation). To obtain insights about possible effects of future climate change, transient responses of the peat land system under the climate change and SOC sensitivities under a new climate regime were studied. We simulated CH_4 production, oxidation, transport, and emission to the atmosphere at NSA Fen site based on Walter and Heimann (2000) and parameterization was identical to their application to a poor fen in Michigan unless otherwise noted below.

The dynamic peat depth is characteristic to ED2.0-peat, and the CH_4 dynamics simulation is applied to each peat layer separately (Fig. 1). The position of water table determines the ratio between CH_4 production and oxidation. CH_4 production rates are different among substrates (fibrous and humid peat). The CH_4 production rate of the fibrous layer $R_{prod f}$ [μmol m^{-2} h^{-1}] is:

$$R_{prod \cdot f} = R_0 \cdot f_{labile \cdot f} \cdot f_{in} \cdot TD_f \cdot 1000 \cdot z_f \cdot f_{anox \cdot f} \qquad (1)$$

where TD_f is the temperature dependence factor of the fibrous layer calculated as equation 5 in Walter and Heimann (2000), with mean soil temperature of that layer simulated by ED2.0-peat. $f_{labile\,f}$ is the relative liability of fibrous layer (1.0) based on Ise and Moorcroft (2006). R_0 is the CH_4 production rate [μM h^{-1}] at the reference temperature, as described in Walter and Heimann (2000). The depth of the fibrous layer z_f and the anoxic fraction of that layer $f_{anox \cdot f}$ are obtained from ED2.0-peat (Ise et al., 2008). Similarly, CH_4 production from humid layer is:

$$R_{prod \cdot h} = R_0 \cdot f_{labile \cdot h} \cdot f_{in} \cdot TD_h \cdot 1000 \cdot z_h \cdot f_{anox \cdot h} \tag{2}$$

TD_h, z_h , and $f_{anox \cdot h}$ are based on ED2.0-peat variables. $f_{labile\,f}$ is the relative liability of humid layer (0.044) based on Ise and Moorcroft (2006). Then, CH_4 oxidation rates of the fibrous layer $R_{oxid\,f}$ [μmol m^{-2} h^{-1}] and the humid layer $R_{oxid \cdot h}$ [μmol m^{-2} h^{-1}] is Michaelis-Menten type functions of CH_4 concentration of each layer $[CH_4]_f$ and $[CH_4]_h$, respectively:

$$R_{oxid \cdot f} = V_{\max} \cdot \frac{[CH_4]_f}{k_m + [CH_4]_f} \cdot TD_f \cdot 1000 \cdot z_f \cdot \left(1 - f_{anox \cdot f}\right) \tag{3}$$

$$R_{oxid \cdot h} = V_{\max} \cdot \frac{[CH_4]_h}{k_m + [CH_4]_h} \cdot TD_h \cdot 1000 \cdot z_h \cdot \left(1 - f_{anox \cdot h}\right) \tag{4}$$

where V_{max} and K_m are oxidation parameters described by Walter and Heimann (2000). Three modes of CH_4 transport (diffusion, bubbling, and vascular transport) were explicitly modeled. We assume that the plant rooting depth is at the boundary between fibrous and humid peat and root distribution is uniform. Thus, when the water table is above this boundary, the fast CH_4 transportation through ebullition (bubbling) occurs as described in Walter and Heimann (2000).

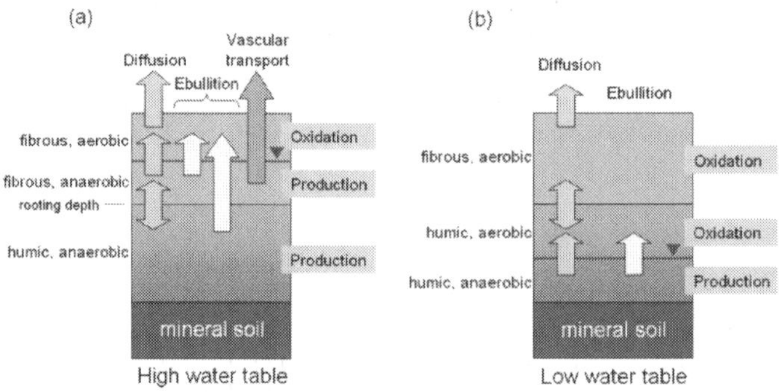

Fig. 1. Schematic diagram of ED2.0-peat, when water table is in fibrous layer (a) or in humic layer (b). Position of water table determines proportions of aerobic vs. anaerobic peat depth in each layer. CH_4 production occurs in anaerobic layers, and CH_4 oxidation occurs in aerobic layers. Three modes of CH_4 transportation are also shown. Diffusion is a slow, continuous process driven by CH_4 concentration gradient. Ebullition occurs when CH_4 concentration of an anaerobic layer exceeds a threshold (see section 2). When water table is above rooting depth, CH_4 is also transported through plant vascular system (see section 2).

To evaluate the model performance, we compared the simulation results by ED2.0-peat against field-observed data of water table depth, soil temperature at 10 cm below peat surface, and CH_4 emission rate from BOREAS NSA Fen site for 1994. Observed variables are gap-filled composites from 4 observation subsides and 32 measurement collars located in hummock, hollow, lawn, and moat. Gaps in the data were filled by linear interpolation. Each measurement time series are weighted equally to estimate mean and variance. Residuals were assumed to be normally distributed to generate 95-percent confidence intervals.

3. Results

The model reasonably reproduced water table and soil temperature dynamics observed at the site for 1994 (Fig. 2). ED2.0-peat generally captured seasonal patterns of soil thermal and moisture conditions, and the simulated trends were generally within the range of intra-site variations. There were some notable, systematic underestimate in the simulation of soil temperature in early summer because ED2.0-peat lacked treatment of lateral water flow that can bring a large amount of warm surface water into the site, which is typical for fens especially after soil melt. CH_4 emission rates were influenced by soil temperature and water table position and showed a similar seasonal pattern with these soil physical properties. The simulated CH_4 dynamics was generally consistent with the field observation. The SOC accumulation under the current climate was 116 kgC/m^2, (Ise et al., 2008), within the estimate range of 79-120 kgC/m^2 based on field studies (Trumbore and Harden, 1997).

Fen SOC was very sensitive to climate change. An experimental warming of 4°C caused an 86% loss of SOC (Fig. 3), because of the higher sensitivity caused by a positive feedback process between water loss and decomposition; an acceleration in SOC decomposition caused by the warming lowered the water table, resulting in further decomposition due to enhanced aerobic conditions (Fig. 3a and b). A reduction in summer-time soil insulation also accelerated decomposition (see Ise et al., 2008). Because of the radical changes in soil depth and hydrological and thermal environments, CH_4 emission responded to the climate change sensitively and showed a complex trajectory (Fig. 4). Due to the drier conditions caused by the climate change and loss in substrate, the equilibrium CH_4 emission rate under the warmer climate regime was significantly smaller than that under the current climate. However, the warming caused a transient increase in CH_4 emission due to the temperature dependency of methanogenic microorganisms (fast response: dominating in simulation years 2000-2100). Then, however, the SOC exposed to warmer and drier climate started to decompose and the reduction in soil depth and water holding capacity eventually decreased CH_4 emission (slow response). During the transition period after the climate change (simulation years 2200-2700), the water table was periodically in humid layer (Fig. 3b), and this caused dramatic reduction of CH_4 emission. After simulation year 2700, the new equilibrium of peat system was established and the mean position of water table is again in the fibrous layer, and thus the CH_4 emission rates were partially recovered.

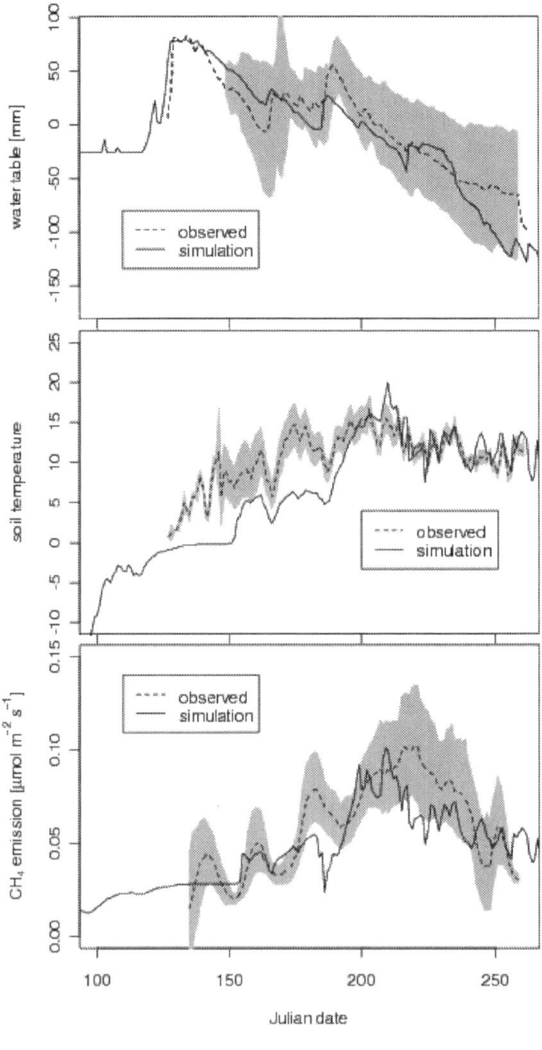

Fig. 2. (a) Water table depth, (b) soil temperature [°C] at 10 cm below peat surface, and (c) CH_4 emission dynamics for BOREAS Fen site in 1994. Observed variables are gap-filled composite datasets from 4 observation subsides and 32 measurement collars located in hummock, hollow, lawn, and moat. Each measurement time series are weighted equally to estimate mean and variance. Residuals were assumed to be normally distributed to generate 95-percent confidence intervals (grey shades).

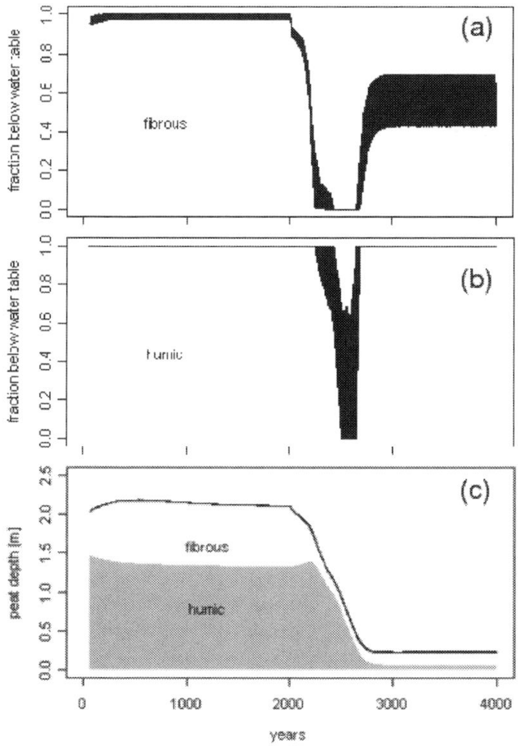

Fig. 3. 4000-year simulation of 12-year means of fraction below water table, (a) fibrous and (b) humid layers and (c) changes in peat depth. This simulation is applied to BOREAS Northern Study Area Fen site. 1994-2005 meteorological data is used repeatedly for this 4000-year simulation. A uniform rise of temperature by 4°C is applied after simulation year 2000.

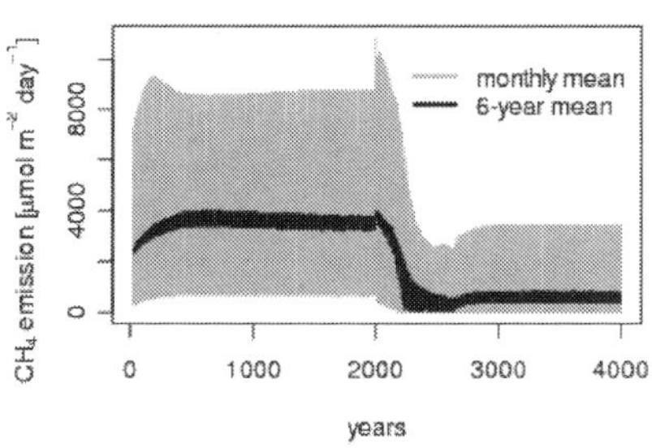

Fig. 4. 4000-year simulation of CH_4 emission from BOREAS Northern Study Area Fen site. 1994-2005 meteorological data is used repeatedly for this 4000-year simulation. A uniform rise of temperature by 4°C is applied after simulation year 2000.

4. Discussion

To our knowledge, this study is the first attempt to simulate CH_4 dynamics with a physics-based, fast-timescale model of soil temperature and moisture coupled to a long-term, biogeochemical model. ED2.0-peat captured the essential physical dynamics of BOREAS NSA Fen site. The short-term CH_4 dynamics with environmental dependency was reasonably reproduced. In contrast to the preexisting CH_4 models (e.g., Walter and Heimann, 2000), ED2.0-peat does not need soil moisture and temperature data as forcing variables. Thus, this model can be used in a predictive manner, under the future environmental changes. Moreover, ED2.0-peat explicitly considers long-term changes in soil properties such as peat depth. Thus, the CH_4 dynamics under a novel environment is estimated from an integrated system of soil temperature, moisture, and biogeochemistry. In addition, due to the characteristics of a mechanistic physics model, the model can reproduce a transient behavior in CH_4 dynamics under climate change. Together with a study for the shallow peat site (Ise et al., 2008) ED2.0-peat correctly reproduced soil biophysics of various peat lands in Northern Manitoba.

To appropriately study the future CH_4 in a predictive manner, ED2.0-peat explicitly simulated the dynamic interaction of soil temperature, moisture, and SOC in the peat land system. Our modeling scheme with mechanistically simulated soil hydrological and thermal processes has implications to future climate-soil interactions. An increase in temperature due to anthropogenic climate change will affect SOC decomposition rates because of their sensitivity to temperature. The Fen site lost 86% of SOC when subjected to a warming of 4°C, a rather conservative estimate of projected long-term climate change in boreal regions (IPCC, 2007). This significant collapse of peat land SOC is plausible when the global distribution of SOC is considered (Fig. 5). The transition from boreal to temperate biomes is currently characterized by a significant loss in stored SOC. If a temperature rise of 4°C at this site allows it to cross this threshold, a considerable shift in equilibrium SOC may occur. However, it should be noted that the Fen site also showed transient resistance against the collapse (Fig. 3c, simulation years 2000-2200); during the first few hundred years after the climate change, the decrease in SOC is rather slow, due to recalcitrance of slowly decomposing SOC.

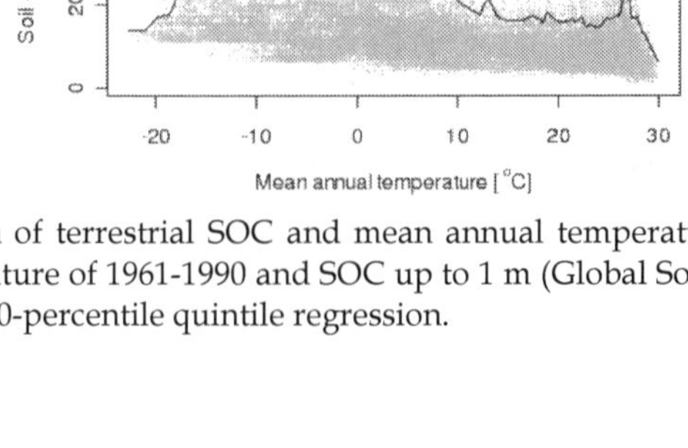

Fig. 5. Global pattern of terrestrial SOC and mean annual temperature. Data from 1° x 1° mean annual temperature of 1961-1990 and SOC up to 1 m (Global Soil Data Task 2000). The curve is a piecewise 90-percentile quintile regression.

For this study, as noted in Ise et al. (2008), we intentionally turned off the dynamic vegetation processes of ED2.0 (Medvigy et al., 2009) to construct the current version of ED2.0-peat. Responses of vegetation processes against climate change such as litter production may have a strong effect on CH_4 dynamics through substrate quantity and quality. However, as shown here, this study is the first attempt to explicitly simulate a long-term physical interplay between soil properties and biogeochemistry, and the resultant transient dynamics was complex and nonlinear. Our aim here was to show the inherent behavior in soil. Considering dynamic vegetation will increase the complexity of the system further, and a systematic analysis will be very difficult. Thus, we kept the model as simple as possible for this study, and an explicit treatment of dynamic vegetation will be our next project. In summary, this study has shown how the mechanistic linkages that exist between the physical and biogeochemical dynamics of peat lands have strong implications for the response of northern peat lands to climate change.

Acknowledgements

We appreciate helpful discussion with Profs. Steve Frolking, David Foster, and James McCarthy. The study of TI is supported by the James Mills Peirce Fellowship provided by the Department of Organismic and Evolutionary Biology at Harvard University and the Innovative Program of Climate Change Projection for the 21st Century (KAKUSHIN Program) of the Ministry of Education, Culture, Sports, Science and Technology, Japan.

5. References

Apps, M.J., Halliwell, D., 1999. BOREAS TE-13 biometry reports, data set. Available online at [http://www.daac.ornl.gov] from Oak Ridge National Laboratory Distributed Active Archive Center, Oak Ridge, TN, U.S.A.

Bond-Lamberty, B., Wang, C., Gower, S.T., 2004. Net primary production and net ecosystem production of a boreal black spruce wildfire chronosequence. Global Change Biology 10, 473-487.

Clymo, R.S., 1984. The limits to peat bog growth. Philosophical Transactions of the Royal Society of London, Series B 303, 605–654.

Dunn, A.L., Barford, C.C., Wofsy, S.C., Goulden, M.L., Daube, B.C., 2007. A long-term record of carbon exchange in a boreal black spruce forest: means, responses to interannual variability, and decadal trends. Global Change Biology 13, 577-590.

Friedlingstein, P., Cox, P., Betts, R., Bopp, L., Von Bloh, W., Brovkin, V., Cadule, P., Doney, S., Eby, M., Fung, I., Bala, G., John, J., Jones, C., Joos, F., Kato, T., Kawamiya, M., Knorr, W., Lindsay, K., Matthews, H.D., Raddatz, T., Rayner, P., Reick, C., Roeckner, E., Schnitzler, K.G., Schnur, R., Strassmann, K., Weaver, A.J., Yoshikawa, C., Zeng, N., 2006. Climate-carbon cycle feedback analysis: results from the (CMIP)-M-4 model intercomparison. Journal of Climate 19, 3337-3353.

Frolking, S., Roulet, N., Fuglestvedt, J., 2006. How northern peatlands influence the earth's radiative budget: sustained methane emission versus sustained carbon sequestration. Journal of Geophysical Research 111, G01008, doi:10.1029/2005JG000091.

Global Soil Data Task, 2000. Global soil data products CD-ROM (IGBP-DIS). International Geosphere-Biosphere Programme - data and information services. Available online at [http://www.daac.ornl.gov/] from Oak Ridge National Laboratory Distributed Active Archive Center, Oak Ridge, TN, U.S.A.

Holland, M.M., Bitz, C.M., 2003. Polar amplification of climate change in coupled models. Climate Dynamics 21, 221-232.

Intergovernmental Panel on Climate Change, 2007. Climate change 2007: the scientific basis. Available online at [http://www.grida.no/climate/ipcc_tar/].

Ise, T., Moorcroft, P.R., 2006. The global-scale temperature and moisture dependencies of soil organic carbon decomposition: an analysis using a mechanistic decomposition model. Biogeochemistry 80, 217–231.

Ise, T., Dunn, A.L., Wofsy, S.C., Moorcroft, P.R., 2008. High sensitivity of peat decomposition to climate change through water-table feedback. Nature Geoscience 1, 763-766.

Medvigy, D., Wofsy, S.C., Munger, J.W., Hollinger, D.Y., Moorcroft, P.R., 2009. Mechanistic scaling of ecosystem function and dynamics in space and time: Ecosystem Demography model version 2. Journal of Geophysical Research—Biogeosciences, 114, G01002, 10.1029/2008JG000812

Parton, W.J., Schimel, D.S., Cole, C.V., Ojima, D.S., 1987. Analysis of factors controlling soil organic matter levels in Great Plains Grasslands. Soil Science Society of America Journal 51, 1173-1179.

Pelletier, L., Moore, T.R., Roulet, N.T., Garneau, M., Beaulieu-Audy, V., 2007. Methane fluxes from three peatlands in the La Grande Riviere watershed, James Bay lowland, Canada. Journal of Geophysical Research—Biogeosciences 112, G01018, doi:10.1029/2006JG000216.

Schlesinger, W.H., 1997. Biogeochemistry: an Analysis of Global Change. Academic Press, San Diego, CA, 588pp.

Sellers, P., Hall, F., Margolis, H., Kelly, B., Baldocchi, D., Denhartog, G., Cihlar, J., Ryan, M.G., Goodison, B., Crill, P., Ranson, K.J., Lettenmaier, D., Wickland, D.E., 1995. The boreal ecosystem-atmosphere study (BOREAS): an overview and early results from the 1994 field year. Bulletin of American Meteorological Society 76, 1549-1577.

Trumbore, S.E., Harden, J.W., 1997. Accumulation and turnover of carbon in organic and mineral soils of the BOREAS northern study area. Journal of Geophysical Research—Atmospheres 102, 28817-28830.

Veldhuis, H., Eilers, R.G., Mills, G.F., 2002. Permafrost distribution and soil climate in the glacial Lake Agassiz basin in northcentral Manitoba, Canada. In: 17th World Conference on Soil Science, Bangkok, Thailand, 13-20 August.

Walter, B.P., Heimann, M. 2000. A process-based, climate-sensitive model to derive methane emissions from natural wetlands: application to five wetland sites, sensitivity to model parameters, and climate. Global Biogeochemical Cycles, 14, 745-765.

18

Towards a New Agriculture for the Climate Change Era in West Asia, Iran

Farzin Shahbazi[1] and Diego de la Rosa[2]
1- Soil Science Department, Faculty of Agriculture, University of Tabriz, Iran
2- Institute of Natural Resources and Agrobiology of Sevilla, CSIC, Spain

1. Introduction

Climate change means a change of climate which is attributed directly or indirectly to human activity that alters the composition of the global atmosphere and which is in addition to natural climate variability observed over comparable time periods. It will potentially lead to such eventualities as drought and famine, which some of the CWANA countries have already experienced. The capacity of national governments and communities to mitigate disasters will be limited in the short to medium term, rendering them still vulnerable to the adversities of climate change. Climate change is a global issue with regional implications. Many multilateral environmental agreements address these issues, and some countries of the region have ratified some such agreements (CWANA, 2009). Effects of climate change on land use refers to both how land use might be altered by climate change and what land management strategies would mitigate the negative effects of climate change (Dale, 1997). Asia is the most populous continent, population in 2002 was reported to be about 3,902 million, of which almost 61% is rural and 38.5% lives within 100 km of the coast (Duedall & Maul, 2005). Asia is divided into seven sub-regions, namely North Asia, Central Asia, West Asia, Tibetan Plateau, East Asia, South Asia and South-East Asia. All of Asia is very likely to warm during this century; the warming is likely to be well above the global mean in central Asia, the Tibetan Plateau and northern Asia, above the global mean in East and South Asia, and similar to the global mean in Southeast Asia. Extreme weather events in Asia were reported to provide evidence of increases in the intensity or frequency on regional scales throughout the 20th century. More investigations predicted that the area-averaged annual mean warming would be about 3°C in the decade of the 2050s and about 5°C in the decade of the 2080s over the land regions of Asia as a result of future increases in atmospheric concentration of greenhouse gases (Lal et al., 2001). In addition rainfall will be altered too. Rainfall in the Philippines would continue to be highly variable, as influenced by seasonal changes and climate extremes and be of higher intensity (Perez, 2008). Also, Changes in annual precipitation for Singapore would range from -2 to +15% with a median of +7%. Extreme rainfall and winds associated with tropical cyclones are likely to increase (Ho, 2008). Other investigations for west Asia has reported that long-term climatic changes of annual surface air temperature, surface wind and rainfall of the State of Qatar, Sultanate of Oman and the United Arab Emirates revealed that significant climate warming is taking place in entire three countries. However, there is no notable trend observed in the rainfall series at any of these places. There is a significant decrease in the mean wind speed at many locations in the region of investigation. The moisture deficit and ecologically fragile land is likely to have further water stress conditions. There has been a steady increase in the total emissions of carbon dioxide over all the three states (Govinda Rao et al., 2003). Some studies (Rosenzweig et al., 2001; FAO, 2004) agree that higher temperatures and longer growth seasons could result in increased pest populations in temperate regions of Asia where central and west Asia include several countries of predominantly arid and semi-arid region which have not been dedicated by these problems. On contrary, the stresses of climate change are likely to disrupt the ecology of mountain and highland systems in west Asia. The anthropogenic release of CO_2 has increased greatly since the industrial age began and fossil fuels began being intensively used as an energy source. Currently, 61% of the anthropogenic greenhouse forcing can be attributed to CO_2 increases (Shine et al. 1990). Research and assessment carried out during the Climate Change Enabling Activity Project, under the UN Framework Convention on Climate Change, predicts that if the CO_2 concentration doubles by the year 2100, the average temperature in Iran will increase by 1.5 - 4.5°C. As well as it has been reported in Kazakhstan by Dolgikh Kazakh (2003) where air temperature and the sum of precipitation are expected to be 6.9°C and -12%, respectively, under double CO_2 conditions. Following CO_2 enrichment and changes in temperature may also affect ecology, the evolution of weed species over time and the competitiveness of C3 v. C4 weed species (Ziska, 2003). In arid central and west Asia, changes in climate and its variability continue to challenge the ability of countries in the arid and semi-arid region to meet the growth demands for water (Abu-Taleb, 2000; UNEP, 2002; Bou-Zeid & El-Fadel, 2002; Ragab & Prudhomme, 2002). Decreasing precipitation and increasing temperature commonly associated with ENSO have been reported to increase water shortage, particularly in parts of Asia where water resources are already under stress from growing water demands and inefficiencies in water use (Manton et al., 2001). Crop simulation modelling studies based on future climate change scenarios indicate that substantial losses are likely in rainfed wheat in south and south-east Asia (Fischer et al., 2002). For example, a 0.5°C rise in winter temperature would reduce wheat yield by 0.45 tons per hectare in India (Lal et al., 1998; Kalra et al., 2003). Climate change can affect on land degradation risks in agricultural areas, soil erosion, and contamination corresponding to Mediterranean regions, too. Increased land degradation is one possible, and important, consequence of global climate change. Therefore the prediction of global environmental change impacts on these degradation risks is a priority (De la Rosa et al., 1996). Iran has located in desert belt where desertification, drought, water table reduction and flooding increment, vulnerability of land resources are the most relevant phenomena (Momeni, 2003). The impact of climate change in Iran includes changes in precipitation and temperature patterns and water resources, a rise in sea level, and an agricultural impact affecting food production, bioclimatic deficiency, land capability, agro-ecological field vulnerability and possibly more frequent droughts. The global demand for energy will increase in the coming decades, and this rising demand presents significant opportunities for our industry. As demand increases following population growth, however, the complexities of global climate change also pose serious questions for the energy industry and the broader society. During 1951 to 2003 several stations in different climatologically zones of Iran reported significant decrease in frost days due to rise in surface temperature. Also, some stations show a decreasing trend in precipitation (Anzali, Tabriz, Zahedan) while others (Mashad, Shiraz) have reported increasing trends (IRIMO, 2006 a & b; Rahimzadeh, 2006). Mean monthly weather data values from 1968 - 2000 for 12 major rainfed wheat production areas in north-west and western Iran have previously been used with a climate model, United Kingdom Meteorological Organization (UKMO), to predict the impact of climate change on rainfed wheat production for years 2025 and 2050. The crop simulation model, World Food Study (WOFOST, v 7.1), at CO_2 concentrations of 425 and 500 mg Kg^{-1} and rising air temperature of 2.7 - 4.7°C, projected a significant rainfed wheat yield reduction in 2025 and 2050. Average yield reduction was 18 and 24% for 2025 and 2050, respectively. The yield reduction was related to a rainfall deficit (8.3 - 17.7%) and shortening of the wheat growth period (8 - 36 d). Cultivated land used for rainfed wheat production under the climate change scenarios may be reduced by 15 - 40%. Potential improvements in wheat adaptation for climate change in Iran may include breeding new cultivars and changing agronomic practices like sowing dates (Nassiri et al., 2006). In a study conducted by the Office of Natural Resources & Environmental Policy and Planning (ONEP, 2008), negative impacts on corn productivity varied from 5–44%, depending on the location of production. The current research work for land evaluation therefore needs to be updated to reflect these newer concerns, some of which have been the focus of international conventions on climate change. The main objective is to introduce MicroLEIS, as a support system for agro-ecological land evaluations which can be used to assess soil quality and land use planning for selected time horizons.

2. MicroLEIS Agro-ecological Decision Support System

MicroLEIS, is an integrated system for land data transfer and agro-ecological land evaluation (De la Rosa et al., 1992). Decision support systems (DSS) are informatics systems that combine information from different sources; they help in the organization and analysis of information, and also, facilitate the evaluation (Sauter, 1997; Eom et al., 1998). MicroLEIS DSS provides a computer-based set of tools for an orderly arrangement and practical interpretation of land resources and agricultural management data. Its major components are: I) land evaluation using the following spatial units: place (climate), soil (site and soil), land (climate, site and soil) and field (climate, site, soil and management); II) data and knowledge engineering through the use of a variety of georeferenced database, computer programs, and boolean, statistical, expert system and neural network modelling techniques; III) monthly meteorological data and standard information as recorded in routine land surveys; IV) integrated agro-ecological approach, combining biophysical data with agricultural management experience; and V) generation of data output in a format readily accepted by GIS packages. Recently two components have been added in order to comply with rising environmental concerns (De la Rosa et al., 2001): prediction of global change impacts by creating hypothetical scenarios; and incorporating the land use sustainability concept through a set of tools to calculate current status; potentiality and risks; impacts; and responses. Thus, land evaluation requires information from different domains: soil, climate, crop and management. Soil surveys are the basic building blocks for developing the comprehensive data set needed to derive land evaluation which is normally based on data derived from soil survey, such as useful depth, soil texture, water capacity, drainage class, soil reaction or landscape (soil and site) attributes. The increasing pressure on natural resources leads to the erosion, physical degradation and chemical pollution of these resources, along with a reduction of their productive capacity. Computerized land evaluation techniques are a correct way to predict land productivity and land degradation, and to assess the consequences of changes such as climate. Therefore, other biophysical factors, mainly referred to monthly or daily climate parameters, are also considered as basic information or climate attributes (De la Rosa et al., 2004). There are various approaches to analyze the enormous complexity of land resource and its use and management from an agro-ecological perspective. It discusses the effectiveness of land evaluation for assessing land use changes in rural areas. Land evaluation analysis determines whether the requirements of land use and management are adequately met by the properties of the land. Within the new MicroLEIS DSS framework, land evaluation is considered as the only way to detect the environmental limits of land use sustainability (Shahbazi et al., 2010a). Today, MicroLEIS DSS is a set of useful tools for decision-making which in a wide range of agro-ecological schemes. The design philosophy follows a toolkit approach, integrating many software tools: databases, statistics, expert systems, neural networks, web and GIS applications, and other information technologies. It has divided to five packages: i) Inf & Kno; ii) Pro & Eco iii) Ero & Con; iv) Eng & Tec; and v) Imp & Res, while the packages related to climate observation and its perturbation were used to assessing the new agriculture for the climate change era in north-west of Iran. Diagrammatic scheme of the different packages and possibilities for using land evaluation models within the MicroLEIS framework and strategies supported by each model is presented in (Figure 1).

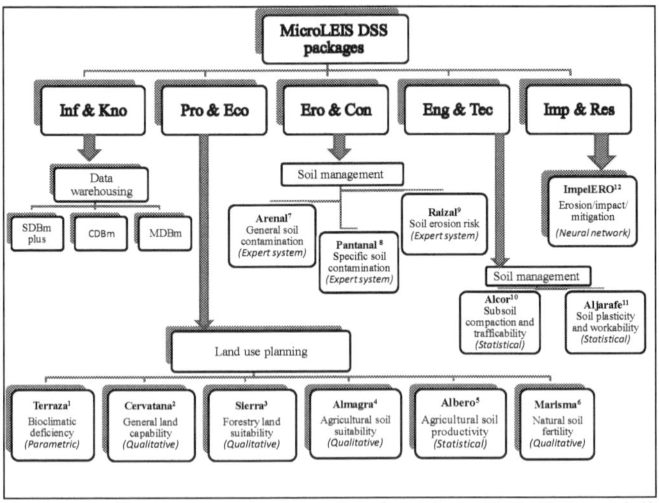

Fig. 1. General scheme of major components related to MicroLEIS DSS, modelling approach and supported strategies* (Shahbazi et al., 2010 a; Shahbazi & Jafarzadeh, 2010)

*Supported strategies by each model: [1]quantification of crop water supply and frost risk limitation; [2]segregation of best agricultural and marginal agricultural lands; [3]restoration of semi-natural habitats in marginal agricultural lands and selection of forest species; [4]diversification of crop rotation in best agricultural lands; [5]quantification of crop yields for wheat, maize and cotton; [6]identification of area with soil fertility problems and accommodation of fertilizer needs; [7]rationalization of total soil input application; [8]rationalization of specific soil input application such as N and P fertilizers, urban wastes, and pesticides; [9]identification of areas with soil erosion problems; [10]site-adjusted soil tillage machinery; [11]identification of soil workability timing; [12]formulating of management practices

3. GIS Spatialization

Geographic Information Systems have greatly improved spatial data handling (Burrough & McDonnell, 1998), broadened spatial data analysis (Bailey and Gatrell 1995) and enabled spatial modelling of terrain attributes through digital elevation models (Hutchinson 1989; Moore et al., 1991). The advent of GIS has brought about a whole set of new tools and enabled the use of methods that were not available at the time when the 1976 framework (FAO, 1976) was developed (FAO, 2006). Other systems, developed before the era of GIS, such as LESA, currently have been integrated with GIS (Hoobler et al., 2003). GIS and allows spatial monitoring and analyses where the knowledge of the stakeholders can be integrated. Tools related to environmental monitoring such as agroenvironmental indicators, soil-landscape relationships, land cover classification and analysis, land degradation assessment, estimation of agricultural biomass production potential and estimation of carbon sequestration all have their applications in land evaluation. Also risk assessment studies have grown in importance. The available GIS methods are usually combined with expert knowledge or production modelling to support studies such as land suitability assessment (Bouma et al., 1993; Bydekerke et al., 1998; Shahbazi et al., 2009a; Jafarzadeh et al., 2009) and risk analysis (Johnson & Cramb, 1996; Saunders et al., 1997; Shahbazi et al., 2009c).

4. Study Area

4.1. General Description

Iran, with an area of 1648000 km^2, is located between 25-40°N and 44-63 °E. The altitude varies from -40 to 5670 m, which has a pronounced influence on the diversity of the climate. Although, about 75% of total land area of Iran is dominated by an arid or semi-arid climate with annual precipitation rates from ~350 to less than 50 mm, Iran has a wide spectrum of climatic conditions. Lake sediments in western Iran and loess soil sequences in northern Iran have shown to be an excellent archive of climate change (Kehl, 2009). Total population inhabit 2004 was 69788000. Land area in 2002 was 163620000 ha where 17088000 ha and 15020000 ha were selected as permanent crops and arable land, respectively. Total forest area in 2005 was estimated 11075000 ha where 6.8% of them revealed as covered area (FAO, 2005). Natural renewable water resources in 2002 were 1900 m^3 capita^{-1}; Average production of cereals by 2005 was 21510000 T, while fish and fishery products in 2002 were estimated in average 5 Kg capita^{-1}. The average annual precipitation is 252 mm yr^{-1}. The northern and high altitude areas found in the west receive about 1600-2000 mm yr^{-1} (NCCO, 2003), while the central and eastern parts of the country receive less than 120 mm yr^{-1}. The per capita freshwater availability for the country was estimated at around 2000 m^3 capita^{-1} yr^{-1} in the year 2000 and expected to go below 1500 m^3 capita^{-1} yr^{-1} (the water scarcity threshold) by 2030 due to the population growth (Yang et al., 2003). Winter temperatures of -20 °C and below in high-altitude regions of much of the country and summer temperatures of more than 50 °C in the southern regions have been recorded (NCCO, 2003).

According to the national water planning report by the MOE (1998), Iran can be divided into eight main hydrologic regions (HR) comprising a total of 37 river basins where the case studied area included in this chapter are located in the north-west of Iran (Figure 2). As reported by MOE (1998), the second hydrologic region (HR_2) has covered a total of 131937 Km^2 where GRAS, SAVA, CRDY, CRWO, and SHRB are the most important land uses in the total of 54.22%, 17.53%, 14.2%, 11.3% and 2.61%, respectively. In HR_2, Urmia Lake is a permanent salt lake receiving several permanent and ephemeral rivers and also Aras, as an international river, has located in this region. It originates in Turkey and flows along the Turkish–Armenian border, the Iranian–Armenian border and the Iranian–Azerbaijan border before it finally meet with the Kura River, which flows into the Caspian Sea. This hydrologic region is important for agricultural activities, as the water resource availability and climatic conditions are suitable.

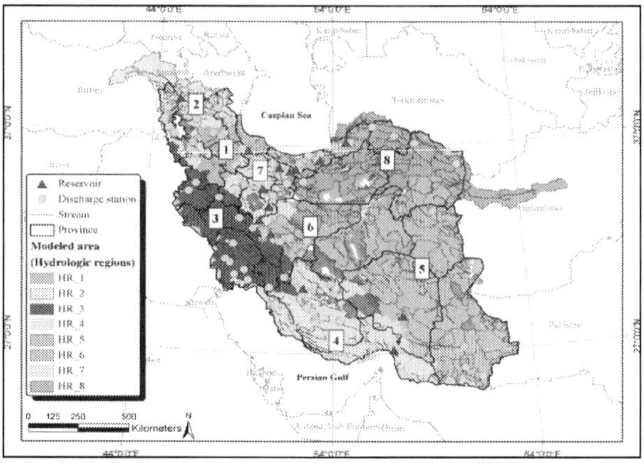

Fig. 2. Main hydrological divisions in Iran (Faramarzi et al., 2009)

4.2. Specific Description

Data required for this study were compiled from different sources belonged to the two major provinces, east and west Azerbaijan, where are located in the north-west of Iran. They include: Soil survey analyses for Ahar area where closed to Tabriz city in the east Azerbaijan province (Shahbazi et al., 2009a); Soil data extracted from the supported foundation by the university of Tabriz as an investigation for Souma area in the west Azerbaijan (Shahbazi et al., 2010 a); Climate data such as temperature for each month and total annual precipitation for last 20 consecutive years (1986-2006) from Ahar meteorological station and also 36 consecutive years (1966-2002) from Urmia meteorological station which is closed to Souma studied area according to Iran Meteorological Organization reports (IRIMO, 2006 b). IPCC refers to any change in climate over time, whether due to natural variability or as a result of human activity.

4.2.1. Site and Soil Information

Soil information is the engine of land evaluation process. Standard analyses, soluble salts and heavy metals, physical analyses, water content and hydraulic conductivity, and additional variables are the major laboratory works before land use planning or vulnerability assessment. Agriculture application is mainly related to site and soil information. Therefore, of course, only climate data will vary in this research work.

The first case study was performed in Ahar area which has located in the east Azerbaijan, Iran. It has different kinds of land use associated with soils of different parent material, such as limestone, old alluvium, and volcano-sedimentary rocks and covers about 9000 ha, between 47°00' to 47°07'30" east and 38°24' to 38°28'30" north. Its slopes range from < 2% to 30%, and the elevation is from 1300 to 1600m above sea level. Flat, alluvial plain, hillside, and mountain are the main physiographical units in the study area. A total of 44 soil profiles were characterized in the field and the lab, determining standard morphological, physical and chemical variables. According to the USDA Soil Taxonomy (USDA, 2006), the dominant soils are classified as Inceptisols, Entisols, and Alfisols. Additionally, 10 soil subgroups and 23 soil family were obtained. Typic Calcixerepts is the major subgroup more than 53%of total area (figure 3).

Fig. 3. Site and soil profile described in the study area
For example: Clayey, mixed, mesic, semiactive Typic Calcixerepts with soil horizons A, Bk1, Bk2, C of a dark greyish brown colour on topsoil); Location: 38° 24′31″ N and 47° 00′ 58″ E (Shahbazi, 2008).

The second studied area covers about 4100 ha, and includes natural regions of Havarsin, Kharghoush, Aghsaghghal, Johney and Bardouk in the west Azerbaijan province of Iran. It has located between 44°35' to 44°40' east longitude and 37°50' to 37°55' north latitude. Altitude varies from 1200 to 1400m with a mean of about 1300m, and slope gradients vary from flat to more than 9%. Thirty-five representative soil profiles were described while the nine benchmark soil families were selected between them to present the land characteristics correspond to the soil factors. Fluventic Haploxerepts and Typic Calcixerepts are dominant soils in the central and north-east of study area, respectively (Figure 4). Soil surveys generate large quantities of data from field description and laboratory analysis for both study area (Shahbazi, 2008; Shahbazi et al, 2008; Shahbazi et al., 2010 b) which these huge data were stored in SDBm plus.

4.2.2. Agro-climatic Indexes

4.2.2.1. Climate Observations

The projected temperature increase is widespread over the globe, and is greater at higher northern latitudes. In order to apply the land evaluation approaches due to climate change and perturbation, two scenarios were constructed. The first is defined as current situation extracted from the climate observations during the last 20 and 36 years for Ahar and Souma areas, respectively while the second one will be calculated based on projected changes in surface air temperature and precipitation for west Asia under the highest future emission trajectory (A1FI) for the 2080s (Christensen & Hewitson, 2007). Following the IPCC report, the mean temperature in this part of Asia will increase 5.1, 5.6, 6.3 and 5.7 °C in winter, spring, summer and autumn, respectively in the future scenario at the studied areas. On the other hand, total precipitation will decrease 11% and 25% in winter and spring, while it will be increased 32% and 52% in summer and autumn (Table 1).

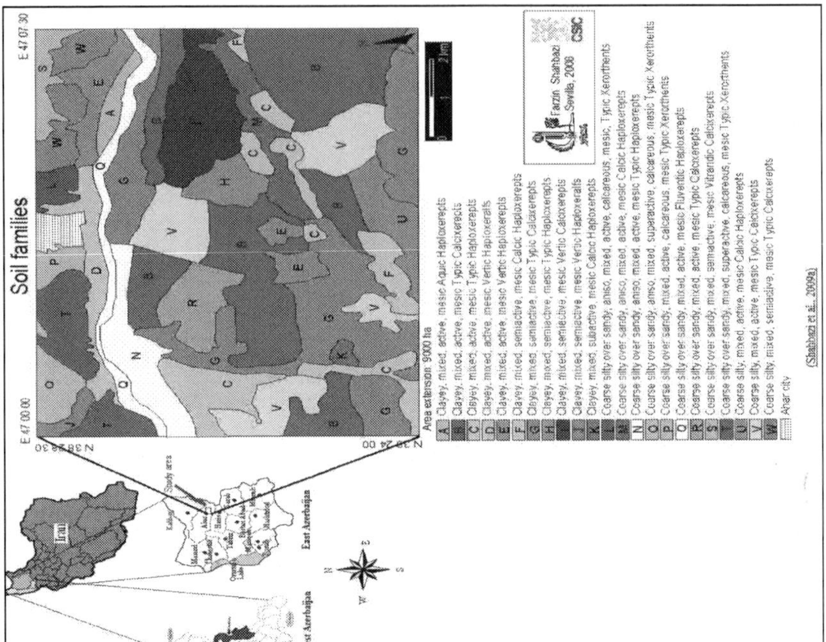

Fig. 4. Sites location and its soils covered in east and west Azerbaijan provinces, respectively (Shahbazi et al., 2009 a, 2010 a)

season	2010-2039				2040-2069				2070-2099			
	T(°C)		P (%)		T(°C)		P (%)		T(°C)		P (%)	
	A1FI	B1	A1FI	B1	A1FI	B1	A1FI	B1	A1FI	B1	A1FI	B1
DJF	1.26	1.06	-3	-4	3.1	2	-3	-5	5.1	2.8	-11	-4
MAM	1.29	1.24	-2	-8	3.2	2.2	-8	-9	5.6	3	-25	-11
JJA	1.55	1.53	13	5	3.7	2.5	13	20	6.1	2.7	32	13
SON	1.48	1.35	18	13	3.6	2.2	27	29	5.7	3.2	52	25

Table 1. Projected changes in surface air temperature and precipitation for west Asia, (12N-42N; 26E-63E) pathways for three time slices, namely 2020s, 2050s and 2080s (IPCC, 2007).
DJF= Dec., Jan., Feb.; MAM= Mar., Apr., May; JJA= Jun, Jul., Aug.; SON= Sep., Oct., Nov.;
T (°C)= Temperature; P(%)= Precipitation; A1FI= Highest future emission trajectory;
B1= Lowest future emission trajectory

4.2.2.2. Climate Perturbation

Future scenario in this chapter is now defined as climate data extracted from the pathway for the time slice 2080s using highest future emission trajectory (A1FI) according to Table 1. With the gradual reduction in rainfall during the growing season for grass, aridity in west Asia has increased in recent years, reducing growth of grasslands and increasing bareness of the ground surface (Bou-Zeid & El-Fadel, 2002). Increasing bareness has led to increased reflection of solar radiation, such that more soil moisture is evaporated and the ground has become increasingly drier in a feedback process, thus adding to the acceleration of grassland degradation (Zhang et al., 2003). Also, it is estimated that the agricultural irrigation demand in arid and semi-arid regions of Asia will increase by at least 10% for an increase in temperature of 1°C (Fischer et al., 2002; Liu, 2002). Paid attention to the literatures shows that towards a new agriculture for a climate change era in Iran (east and west Azerbaijan) will be visible in 2080s and must be attended. In this sense, estimated fresh climatic data are necessary to apply the land evaluation models for predicting coming events.

4.2.2.3. Calculated Climate Variables

Mean monthly values of a set of temperature and precipitation variables can be stored in a microcomputer-based tool named CDBm which includes software subroutines for calculating climate variables for use in agricultural land evaluation, organization, storage and manipulation of agro-climatic data. These interpretative procedures require large quantities of input data related to site, soil, climate, land use and management. The CDBm module has been developed mainly to help in the application of land use models, via their mechanization (e.g., De la Rosa and Crompvoets, 1998; De la Rosa et al., 1996; Shahbazi, 2008). Such models normally use monthly data from long periods of time. It is thus necessary to draw up climate summaries for such long periods. For periods longer than a year, the monthly data are mean values of the monthly dataset for the years under consideration. In this sense, evaporation and transpiration occur simultaneously and there is no easy way of distinguishing between the two processes. Apart from the water availability in the topsoil, the evaporation from a cropped soil is mainly determined by the fraction of the solar radiation reaching the soil surface. This fraction decreases over the growing period as the crop develops and the crop canopy shades more and more of the ground area. The evapotranspiration rate is normally expressed in millimeters (mm) per unit time which it expresses the amount of water lost from a cropped surface in units of water depth. Two main formula were considered within the CDBm to calculate it: By Thornthwaite (1948) and Hargreaves (Hargreaves et al., 1985) methods. The second one appears to give very good results in Mediterranean regions, and particularly in the Guadalquivir valley (Orgaz et al. 1996). For the Andalucian stations included in CDBm, the differences in results between this method and that of Thornthwaite are quite significant, above all for winter months. Calculated results taken by climatic observations from both station reports shows that total annual calculated evapotranspiration by using Hargreaves are higher than Thornthwaite method while it is going to increase for the climate change era (Table 2).

Season (months)		Current situation				Future scenario			
		EAT	EAH	WAT	WAH	EAT	EAH	WAT	WAH
winter	Dec.	2.5	46.7	5.3	46	9.3	59.2	13.1	57.9
	Jan.	0	43.7	0	42.1	2.8	52	6.2	53
	Feb.	0	47.6	0	46.3	5.2	61.2	6.7	59.6
spring	Mar.	16.1	64.9	14.6	61.4	25.3	80.7	24.2	77
	Apr.	42.4	83.3	43.2	82.9	55.8	99.5	55.9	99.3
	May	70	96.7	65.9	91.6	92.4	112.9	84.8	107.5
summer	Jun	95.1	111.5	89.2	104.4	134	129.4	121.7	121.9
	Jul.	122.5	123.9	109.7	109.1	158	142.1	139.5	125.9
	Aug.	119.8	132.1	110.4	115.7	155.4	151.6	139.5	133.5
Autumn	Sep.	89.5	126.1	84.5	112.8	121.7	145.5	111.4	130.7
	Oct.	57.3	98.3	56.9	91.3	75.8	116.3	73.8	108.3
	Nov.	22.3	67.6	24.2	63.9	32	82.3	32.2	73.5
Annual		**637.7**	**1042.5**	**603.9**	**967.5**	**868**	**1232.6**	**809**	**1148.1**

Table 2. Calculated potential evapotranspiration for two hypothetical scenarios
Calculated potential evapotranspiration for: EAT= East Azerbaijan using Thornthwaite method; EAH= East Azerbaijan using Hargreaves method; WAT= West Azerbaijan using Thornthwaite method; WAH= West Azerbaijan using Hargreaves method

Earlier investigations showed that there are the same differences in results for Ahar area (Shahbazi, 2008). Although, annual precipitation in east and west Azerbaijan during this era will be +3.4% and -3.6%, but total annual evapotranspiration will excess 230.3 and 205.1 mm, respectively. This emphasizes that before choosing one method or the other, it is essential to compare, in each case, with experimental measurements or those calculated using other, more exact procedures. However, all of other calculations for east and west Azerbaijan were performed according to Thornthwaite method. As crop evapotranspiration is directly affected by potential evapotranspiration, it seems that Humidity, Aridity, Precipitation concentration, Modified Fournier, and Arkley indexes will change which are dependant variables to potential evapotranspiratioin (Table 3). According to the results, Humidity and Precipitation concentration indexes will increase in both studied are. On contrary, Aridity and Arkley indexes will decrease. Therefore, effect of climate on degree of soil leaching will be monitored while it must carefully be paid attention to west Azerbaijan (Souma area) compared to east Azerbaijan (Ahar area). On the other hand irrigation effect and new methods can be assessed in east Azerbaijan. Although increment of growing seasons during this climate change era is certain, irrigation will be key role in this part of Asia. Graphical presentation for both studied area and climate change impact is shown in (Figure 5).

Variables	East Azerbaijan (Ahar station)		West Azerbaijan (Urmia station)	
	Current situation	Future scenario	Current situation	Future scenario
HUi	0.46	0.35	0.56	0.41
Ari	6	7	6	7
PCi	11	10	12	11
MFi	31	31	41	37
Aki	79.8	44.2	160	100
GS	9	11	8	11

Table 3. Calculated agro-climatic variables and climate change impact using CDBm

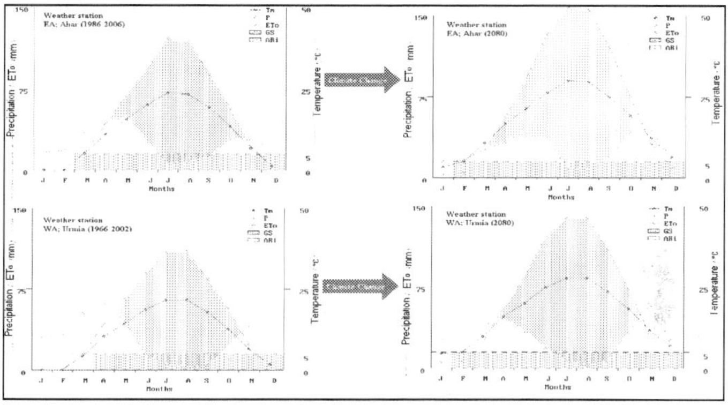

Fig. 5. Graphical presentation of some calculated parameters using CDBm
Tm = mean temperature; P = precipitation; Gs = growing period; ETo = potential evapotranspiration calculated by Thornthwaite method; Ari = aridity index; EA= East Azerbaijan; WA= West Azerbaijan

4.2.3. Agricultural Knowledge

The MDB database gives special attention to management/technological aspects at the field level combined with land characteristics. This database contains management information, which is described exclusively in technical terms and divided into two categories: crop properties and cultivation practices. It was used to capture, store, process, and transfer agricultural crop and management information obtained through interviews with farmers of Havarsin, Khargoush, Aghsaghghal, Johney and Bardouk natural regions related to Souma area. Also, water irrigation management for Ahar area where it is characterized by the seasonal distribution of precipitation, with summers more or less dry. This situation is not very suitable for crop growth. Therefore, most agricultural production systems depend basically on irrigation water as available water resource. The amount of water for irrigation of the selected crops in Ahar area varies between 3100 and 6800 m^3ha^{-1}, with 35% water use efficiency where The number of irrigations is 4-8 times in a growth period (Farshi et al., 1997). According to these extracted site, soil, climate and management data, bioclimatic deficiency and land capability evaluation in east Azerbaijan was being considered. In addition, land vulnerability evaluation due to water and wind erosion and contamination arising phosphorous, nitrogen, pesticides and heavy metals for the climate change era was examined.

5. Land Evaluation in Climate Change Scenarios

Bioclimatic deficiency, land capability, land vulnerability and finally in summary, land evaluation or land use planning will vary following the climate change impacts on the indexes. Thus, management will have an important role to achieve the sustainability.

5.1. Land Productivity Impact

5.1.1. Bioclimatic Deficiency in East Azerbaijan

While temperature conditions may be favorable for growing new types of crops, moisture deficits may preclude these new crops as an adaptation option. However, in order to adopt these new crops moisture deficits could be overcome through the use of irrigation (also an adaptive strategy). Decreasing availability of water for all users will lead to conflicts as producers compete with re-creationists, household users, electrical utilities, and the manufacturing and other industry for water for irrigation (Rosenberg, 1992; Wittrock & Wheaton, 1992). Moisture stress as affected by rainfed and irrigated conditions and impacts on yield reduction of production for wheat, alfalfa, sugar beet, potato, and maize as major crops in Ahar area was calculated applying the Terraza model (Figure 6).

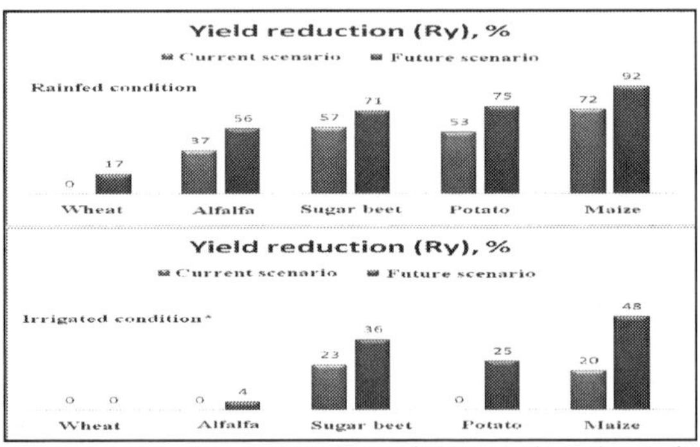

Fig. 6. Annual yield reduction for cultivation of irrigated and rainfed; comparing two scenarios (Shahbazi et al., 2009 a)

* Water irrigation supplement based on usual amount in the study area (see Table 6)

Bioclimatic classification; H1, 0-20%; H2, 20-40%; H3, 40-60%; H4, >60%

In the current situation, the Terraza modelling approach predicts that wheat has 0% (H1 class) of yield reduction in both rainfed and irrigated cultivations. The usual irrigation in the study area for potato and alfalfa is sufficient, increasing their bioclimatic classes from H3 and H2 to H1. Sugar beet and maize currently have 57% and 72% yield reduction of production, while this reduction will decrease to 23% and 20% respectively for the selected crops. Results reveal that usual irrigation, the amount of water is sufficient for wheat, alfalfa and sugar beet, but for potato and especially for maize is inadequate (Shahbazi et al., 2009 a; 2010 b). The Terraza model approach predicts that the currently high water deficit in Ahar area will be increased for the climate change era by the 2080s for all the crops except wheat. Although irrigation is indicated as very important in this semi-arid agriculture, results show that is possible cultivation of rainfed wheat in order to reduce the tillage operation costs.

Using new and classic irrigation methods can be recommended to increase the water use efficiency and decrease the yield reduction of production.

5.1.2. Bioclimatic Deficiency in West Azerbaijan

The predicted results of applying the Terraza model constituents of MicroLEIS DSS in Souma area showed that the annual yield reduction of maize is the highest amounts (74%) between the selected crops (Shahbazi et al., 2009 b) while it will increase up to 86% for the climate change era at rainfed condition in 2080s. Also, these annual reduction for wheat, alfalfa, potato and sugar beet is now calculated 0%, 39%, 55% and 60%, respectively where they are going to recalculated as 0%, 50% 61% and 70%. It means that in the current situation, west Azerbaijan has fewer limitations for wheat production and also it can be suggested as a rainfed cultivation because of its low stress.

5.1.3. Land Capability

Land comprises the physical environment, including climate, relief, soils, hydrology and vegetation, to the extent that these influence potential for land use. It includes the results of past and present human activity, e.g. reclamation from the sea, vegetation clearance, and also adverse results, e.g. soil salinization. The term "land capability" is used in a number of land classification systems notably that of the Soil Conservation Service of the U.S. Department of Agriculture (Klingebiel & Montgomery, 1961). In the USDA system, soil mapping units are grouped primarily on the basis of their capability to produce common cultivated crops and pasture plants without deterioration over a long period of time. Capability is viewed by some as the inherent capacity of land to perform at a given level for a general use, and suitability as a statement of the adaptability of a given area for a specific kind of land use; others see capability as a classification of land primarily in relation to degradation hazards, whilst some regard the terms "suitability" and "capability" as interchangeable. Capability units are soil groups within a subclass. The soils in a capability unit are enough alike to be suited to the same crops and pasture plants, to require similar management, and to have similar productivity. According to this preface, as climate observations have been included as a part of land characteristics, its change will impact on land capability and productivity. Given the potential changes in production variables, it is estimated that the average potential yields may fall by 10-30% (Williams et al., 1988). Across the prairies, crops yields will vary. For example, all crops in Manitoba may decrease by 1%, Alberta wheat, barley and canola may decrease by 7% and Saskatchewan wheat, barley and canola may increase by 2-8% (Arthur, 1988). Considering the type of soil loss impact in terms of productivity changes with time horizon (2020, 2050 and 2100) in southern Spain showed that the maximum impact according to the long-term productivity reduction (97%) for the 2100 time horizon (De la Rosa et al., 2000). The evaluation is based on the degree of limitation imposed on that land by a variety of physical factors which include erosion, soils, wetness and climate. Land is evaluated on the basis of the range of potential crops, productivity, and ease of management and risk of degradation. Therefore, the first step for land use planning to achieve sustainability is arable land identifications. Marginal agricultural land under any kind of farming system used to be the ideal scenario for soil erosion (De la Rosa & Sobral, 2008). For example, applying Terraza (bioclimatic deficiency) and Cervatana (land capability) models in the selected nine benchmark sites in Sevilla province of Spain showed that seven application sites are classified as arable or best agricultural lands, and another two as marginal or unsuitable lands. The Vega site (Typic Xerofluvent) and the Alcores site (Calcic Haploxeralf soil) present the highest capability for most agricultural crops; in contrast, the Sierra Norte site (Palexerult) and the Sierra Sur site (Vertic Xerorthent) show the most-unfavorable conditions (De la Rosa et al., 2009). Changes in land use from natural habitat to intensively tilled agricultural cultivation are one of the primary reasons for soil degradation. Deforestation for agricultural needs and overgrazing has led to severe erosion in the past. Usually, increasing agricultural land capability correlates with a decrease in the soil erosion process. In summary, a positive correlation between current land use and potential land capability would be necessary (De la Rosa & van Diepen, 2002).

Land use capability for a broad series of possible agricultural uses can be predicted by Cervatana model, as a component of MicroLEIS DSS (De la Rosa et al., 2004). The data requirements can be grouped in the following biophysical factors: relief, soil, climate, and current use or vegetation. This qualitative model works interactively, through different gradation matrixes, comparing the values of the input characteristics of the land unit to be evaluated with the generalisation levels established for each capability class. The first three classes - S1, S2, and S3 - include land considered able to support continuing, intensive agricultural use, while land of Class N is more appropriate for natural or forestry use. Studies in Suma area revealed that 80.49% of the total area was good capable for agricultural uses and 19.51% must be reforested and not dedicated to agriculture. Also, Sois of Typic Xerofluvents, Typic Calcixerepts with high carbonate percent and Fluventic Endaquepts with 812ha extension are not suitable for agricultural uses, while uses and must be reforested, while Typic Calcixerepts , Fluventic Haploxerepts with 3344 ha are mainly high suitable and in some cases optimum and moderately suitable (Jafarzadeh et al., 2009; Shahbazi & Jafarzadeh 2010). Following identification of agricultural land according to their limitations and ecological potentialities, prediction of land suitability for a specific crop or crop diversification (e.g. Figure 7; Shahbazi et al., 2009 d) over a long period of time is the subsequent option. In contrast, simplification of crop rotation as a relevant element of arable intensification has led to soil deterioration and other negative environmental impacts.

5.1.3.1. Case Study for the Climate Change Era

Agriculture has always been dependent on the variability of the climate for the growing season and the state of the land at the start of the growing season. The key for adaptation for crop production to climate change is the predictability of the conditions. What is required is an understanding of the effect on the changing climate on land, water and temperature. For instance, land evaluation analysis was developed for the current and future climate scenarios and for rainfed and irrigated conditions in east Azerbaijan province of Iran as follows: **I)** The land capability classification for irrigated cultivation using the normal water amount associated with 35% water use efficiency is divided in two sets: Dense cover (wheat and alfalfa) and moderate cover (sugar beet, potato, and maize). The first group presents similar capability classes to that for rainfed cultivation of wheat. Sugar beet cultivation showed no response to climate change concerning to constant bioclimatic deficiency class (H2), so 87.3% was good agricultural land but the rest was moderate agricultural land. The major limitation factors in classifying the capability of the area were bioclimatic and erosion risks, which were constant with climate change. The results showed that bioclimatic deficiency is the main agent in decreasing the capability classes in irrigated cultivation of potato and maize. **II)** For rainfed cultivation in both hypothetical scenarios (the current situation and the 2080s), model illustrated that wheat in all the simulated conditions has the same land capability classification. In summary, 41.7%, 45.6%, and 11.7% of the total area presents excellent (S1), well (S2), and moderate (S3) capability classes, respectively. Soil texture limitation was the main factor for converting the capability class from excellent to good. The bioclimatic limitation factor (b) was not determined in the cultivation of wheat. Therefore, the capability classes will not be changed in the long-term scenario. With climate change, 45.6% of the total area for alfalfa has been changed from good- to moderate-capability land. The same area for potato and sugar beet has been changed from good- to moderate-capability land. The whole area was not suitable in either the current situation or the 2080s for maize. Bioclimatic deficiency was the most-limiting factor. Concerning soil evaluation, eight application soil subgroups are classified as arable or best agricultural lands, and another two as moderate lands.

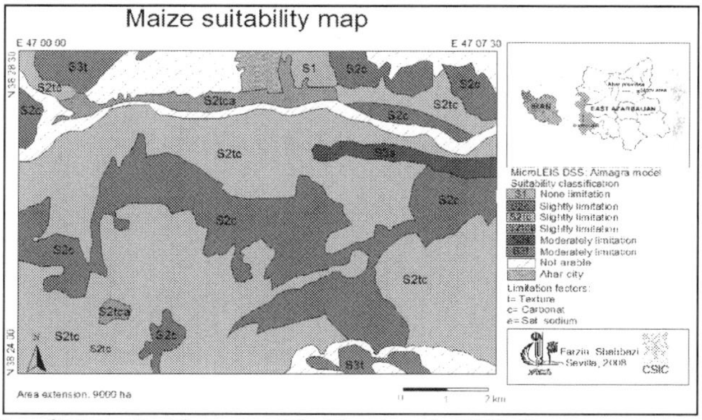

Fig. 7. Suitability of Maize in Ahar area (Shahbazi et al., 2009 d)

Typic Calcixerepts, Typic Haploxerepts, Vertic Calcixerepts, Vertic Haploxeralfs, Calcic Haploxerepts, and Vertic Haploxerepts present an extension of 22.8%, 7%, 5.6%, 3.1%, 1.83%, and 1.43%, respectively of S1 class for most of the crops. Soil and topography limitation are the two basic factors in classifying the Fluventic Haploxerept and Vitrandic Calcixerept subgroups as moderate lands that are currently dedicated to agricultural use. The change in these last two soil subgroups from natural habitat to intensively tilled agricultural cultivation is one of the primary reasons for soil degradation. Land use will be taken as optimum when considering the moderate arable lands as a natural habitat cultivation area. However, 45% of the study area is classified by the soil limitation factor as good-capability land (Shahbazi et al., 2009 a; Figure 8).

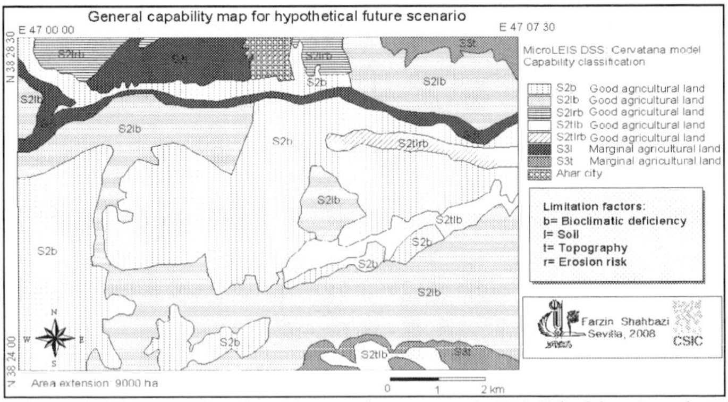

Fig. 8. General capability map for the climate change era in EA (Shahbazi et al., 2009 a)

5.2. Land Vulnerability Impact

The effects of agricultural and climate changes on the degradation of land resources are characterized not only by long-term perspectives, but also by diffuse incidence and large geographic areas impacted. The protection of these resources depends on the correct prediction of such effects (De la Rosa & Crompvoets, 1998). Land degradation is a global problem which involves climate, soil, vegetation, economic, and population conditions. It can be lifted by water and wind erosion or contaminants such as phosphorous, nitrogen, heavy metals and pesticides consumptions. When vulnerability is defined as the degree to which production and livelihood systems are susceptible to, or unable to cope with, adverse effect of climate change, including climate variability and extremes (IPCC, 2001), it is evident that rural poor will be the most vulnerable to these changes both in terms of risks to their production systems and infrastructures (e.g., houses and roads) because they have less assets to call upon in order to cope with extreme events such as prolonged droughts, intense storms and subsequent flooding (Thomas, 2008). Attempts to help the rural poor adapt to climate change must build on existing "coping strategies" that generally involve three elements: preparing for harsh climates by developing various types of insurances, actually coping with the stress when it happens and thirdly, adapting and recovering from the stress (Dietz & Verhagen, 2004). The third way in sustainable developing is the main goal which is completely related to management procedures versus natural variation and coming events. In Mediterranean Europe climatic variability and human pressure combine to produce soil sealing, erosion, salinization, fire risk, and landscape fragmentation, all regarded as important factors to start LD (Salvati & Zitti, 2009). Land vulnerability to degradation, environmental quality and management are all dynamic entities. Developing decision support systems appears as a promising tool to define trends and predict changes in land vulnerability and to promote efficient management of land degradation (Rubio & Bochet, 1998; Basso et al., 2000). It had been demonstrated that these systems could be used to predict for the climate change era. As reported by De la Rosa et al., (1996), two of the main desertification indices or land degradation risks in agricultural areas are soil erosion and contamination. Soil erosion by water is one of today's most important problems, in great part due to changes in agricultural land use and management (De la Rosa et al., 1999). Increased land degradation is one possible, and important, consequence of global climate change. Therefore, it is a priority to predict global environmental change impacts on these degradation risks. For this purpose, The Andalucia Region of Spain was used as the test region for applying Raizal and Pantanal models, based on the current climate and two climate change scenarios. The evaluation results show that 16% and 27% of the studied area is at elevated risk of soil rainfall erosion and contamination, respectively; and a further 58% and 33% at medium risk. For the present drought scenario, the modelling approach predicts that in 59% of land the erosion risk decreases, while for 24% of land this vulnerability increases. These values are 40% and 60%, respectively, for soil contamination vulnerability. The second scenario assumes the predicted climate change for 2050s for the Mediterranean area. This evaluation predicts that in 18% of land the erosion risk decreases, and increases in 47% of land. For the contamination vulnerability the predicted values are similar to those of the first scenario. Thus, change in rainfall amount affected erosion risks strongly, but this change proved to have little direct influence on contamination vulnerability. Pantanal model focuses on diffuse soil agro-contamination from agricultural substances. Tested case for hydrological change scenario in the province of Sevilla, 1 400 000 ha, within the Andalucia region correspond to six current agricultural change scenarios defined by the combination of several intensification production steps with three representative soil types, and with the major traditional crops showed that spatial variability in relation to soil and crop implies significant differences in vulnerability to the four types of soil contaminants considered. Ero&Con models evaluate the vulnerability risks of an agricultural field to land degradation, considering separately three types of vulnerability: attainable, management and actual; and for each degradation factor: water and wind erosion; and nitrogen, phosphorus, heavy metals (Cu, Zn, Cd, Hg, Pb) and pesticides (general, hydrophilic and hydrophobic) contamination. The attainable vulnerability considers the biophysical risk of the capability of the soil being harmed in one or more of its ecological functions. The management vulnerability considers the risk of a particular Field Utilization Type to land degradation. The actual vulnerability considers simultaneously the biophysical and management risk factors of a particular field unit.

5.2.1. Water and Wind Erosion

Ten soil erosion vulnerability classes established by Raizal for the attainable and actual Vulnerability risks (V1-V10). Increasing the number of classes equal with vulnerability risks increments and effect of management change on the vulnerability classes could be important. When class V10 (extreme) field units present an extremely high vulnerability to water or wind erosion. The field will erode until it has an intricate pattern of moderately deep or deep gullies. Soil profiles will be destroyed except in small areas between gullies. Such fields will not be useful for crops in this condition. Reclamation for crop production or for improved pasture is very difficult but will be practical if the other characteristics of the soil are favorable and erosion is controlled by soil conservation techniques, for example by construction of terraces. The assessment of the soil erosion management vulnerability is classified into four classes: V1-V4; very low, moderately low, moderately high, and very high. Three available states of risk types (attainable, management, and actual) for two hypothetical scenarios using Raizal model as point by point view in the whole studied area located in east Azerbaijan are completely summarized in (Table 4).

Natural regions	Current situation (1986-2006)						Future scenario (2008)					
	VAW	VAD	VMW	VMD	VCW	VCD	VAW	VAD	VMW	VMD	VCW	VCD
1-Kord Ahmad	V10	V6	V4u	V3o	V10e	V4	V10	V6	V4u	V3o	V10e	V4
4-Central Ahar	V9	V6	V3u	V4z	V10e	V2k	V9	V6	V3u	V4z	V10e	V2k
5-Dizaj Chalou	V10	V7	V4u	V3o	V10e	V5k	V10	V7	V4u	V3o	V10e	V5k
7-Kord Ahmad	V8	V3	V3u	V4z	V8e	V1	V9	V3	V3u	V4z	V9e	V1
8-Central Ahar	V8	V3	V3u	V4z	V8e	V1	V9	V3	V3u	V4z	V8e	V1
9-Central Ahar	V8	V3	V3u	V4z	V8e	V1	V9	V3	V3u	V4z	V9e	V1
10-Central	V10	V4	V4u	V3o	V8e	V2	V10	V4	V4u	V3o	V9e	V2
11-Central	V10	V4	V4u	V3o	V9e	V2	V10	V4	V4u	V3o	V10e	V2
12-KordAhmad	V8	V8	V3u	V4z	V8e	V4	V9	V8	V3u	V4z	V9e	V4
13-Dizbin	V10	V6	V4u	V3o	V8e	V4	V10	V6	V4u	V3o	V9e	V4
14-Dizbin	V10	V4	V4u	V3o	V8e	V2	V10	V4	V4u	V3o	V9e	V2
15-Mardehkatan	V10	V2	V4u	V3o	V8e	V1	V10	V2	V4u	V3o	V9e	V1
16-Garangah	V10	V8	V4u	V3o	V9e	V6	V10	V8	V4u	V3o	V10e	V6
18-Dizbin	V10	V2	V4u	V3o	V8e	V1	V10	V2	V4u	V3o	V9e	V1
19-Dehestan	V9	V6	V3u	V4z	V9e	V2	V9	V6	V3u	V4z	V10e	V2
20-Dizaj Talkhaj	V8	V3	V3u	V4z	V8e	V1	V9	V3	V3u	V4z	V9e	V1
21-Garangah	V10	V2	V4u	V3o	V8e	V1	V10	V2	V4u	V3o	V9e	V1
22-Garangah	V10	V4	V4u	V3o	V8e	V2	V10	V4	V4u	V3o	V9e	V2
23-Khonyagh	V10	V4	V4u	V3o	V8e	V2	V10	V4	V4u	V3o	V9e	V2
24-Dizbin	V10	V2	V4u	V3o	V8e	V1	V10	V2	V4u	V3o	V9e	V1
25-Dehestan	V10	V4	V4u	V3o	V8e	V2	V10	V4	V4u	V3o	V9e	V2
26-Mardehkatan	V10	V4	V4u	V3o	V8e	V2	V10	V4	V4u	V3o	V9e	V2
27-Garangah	V10	V8	V4u	V3o	V9e	V6	V10	V8	V4u	V3o	V10e	V6
28-Garangah	V10	V2	V4u	V3o	V8e	V1	V10	V2	V4u	V3o	V9e	V1
29-Khonyagh	V10	V2	V4u	V3o	V8e	V1	V10	V2	V4u	V3o	V9e	V1
30-kalhor	V10	V2	V4u	V3o	V8e	V1	V10	V2	V4u	V3o	V9e	V1
31-Dizaj Talkhaj	V10	V2	V4u	V3o	V8e	V1	V10	V2	V4u	V3o	V9e	V1
32-Mardehkatan	V10	V2	V4u	V3o	V8e	V1	V10	V2	V4u	V3o	V9e	V1
33-Garangah	V10	V6	V4u	V3o	V9e	V4	V10	V6	V4u	V3o	V10e	V4
34-Cheshmezan	V10	V4	V4u	V3o	V8e	V2	V10	V4	V4u	V3o	V9e	V2
35-kalhor	V10	V2	V4u	V3o	V8e	V1	V10	V2	V4u	V3o	V9e	V1
36-Dehestan	V10	V2	V4u	V3o	V8e	V1	V10	V2	V4u	V3o	V9e	V1
37-Kordlar	V10	V6	V4u	V3o	V8e	V4	V10	V6	V4u	V3o	V9e	V4
38-Kordlar	V9	V8	V3u	V4z	V8e	V4	V9	V8	V3u	V4z	V10e	V4
39-Garangah	V10	V6	V4u	V3o	V8e	V4	V10	V6	V4u	V3o	V9e	V4
40-Gorchi	V10	V2	V4u	V3o	V8e	V1	V10	V2	V4u	V3o	V9e	V1
41-Kalhor	V8	V3	V3u	V4z	V8e	V1	V8	V3	V3u	V4z	V8e	V1
42-Kordlar	V10	V2	V4u	V3o	V8e	V1	V10	V2	V4u	V3o	V9e	V1
43-Dehestan	V10	V4	V4u	V3o	V8e	V2	V10	V4	V4u	V3o	V9e	V2

Table 4. Summary of vulnerability classes due to water and wind erosion for the climate change era in east Azerbaijan using Raizal model (Shahbazi, 2008)

Natural regions (2, 3, 6, 17 and 44) were identified as marginal and not arable lands (12% of total area) by Cervatana model (see Figure 8);
Water erosion: VAW= attainable risk; VMW= Management risk; VCW= actual risk;
Wind erosion: VAD= attainable risk; VMD= Management risk; VCD= actual risk;
Vulnerability class: V1= none; V2= very low; V3= low; V4= moderately low; V5= slightly low;
V6= slightly high; V7= moderately high; V8= high; V9= very high; V10= extreme;
Land qualities: t= relief; k= soil erodibility; r= rainfall erosivity; e= wind erosion erodibility; Management qualities: o= crop properties to water erosion; z= cultivation practices to water erosion; c= crop properties to wind erosion; u= cultivation practices to wind erosion

Area extension for all mapping units and natural regions were calculated. According to the results, management vulnerability caused by current cultivation will be constant for the climate change era where wheat, alfalfa and apple garden were relevant land uses. In this sense, 73% and 15% of total area were distinguished as low and moderately low (V3&V4) vulnerable risk caused by water erosion while 9% and 79% of those areas rose with wind erosion. In summary attainable water erosion risk will not be affected by climate change (Figure 9), on contrary, attainable wind erosion is abruptly being increased.

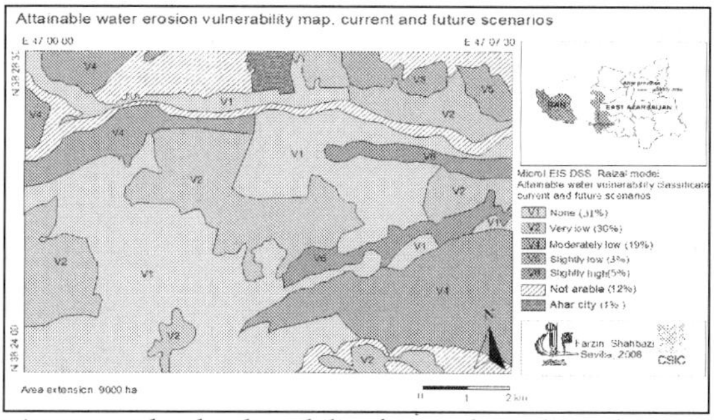

Fig. 9. Water erosion impact on land vulnerability for two hypothetical scenarios (Shahbazi, 2008)

5.2.2. Agricultural Management

Ero&Con models can also make hypothetical evaluations considering climate and management changes simultaneously. This option combines two of the changes: climate factors and management characteristics. Intensive cultivation of wheat, barley, alfalfa, maize, potato, and sugar beet as crop properties effect on water and wind erosions were examined. The order of these intensive cultivation impacts on decreasing land vulnerability raised by water erosion as follows: Sugar beet> alfalfa> wheat>. But there are not significant differences between other selected crops to reduce water erosion and vulnerability. Potato are now identified as the best land use to reduce wind erosion while wheat and maize are the worth one. Alfalfa, Barley and sugar beet have the same results versus wind erosion.

On the other hand, as reclamation for crop production or for improved pasture is very difficult but will be practical if the other characteristics of the soil are favorable and erosion is controlled by soil conservation techniques, for example by construction of terraces. Therefore, it is interested to assume cultivation practices (e.g., contouring and terraces) impact to control the movement of water over the soil surface and those effects on land vulnerability classes for the climate change era. The differences between two practices are shown in (Figure 10a & 10b) which will be achieved in the far future (Shahbazi, 2008). According to these results, terrace application without attention to economical condition and financial costs could be better than contouring to reduce risk of vulnerabilities. Also, the area covered with none level risk in the first examined item is 38% more than the second chosen one where 5% of a total are scattered near the Garangah and Mardekatan natural regions previously distinguished as low level risk will be altered to high level risk by selecting contouring practice instead of terrace procedure.

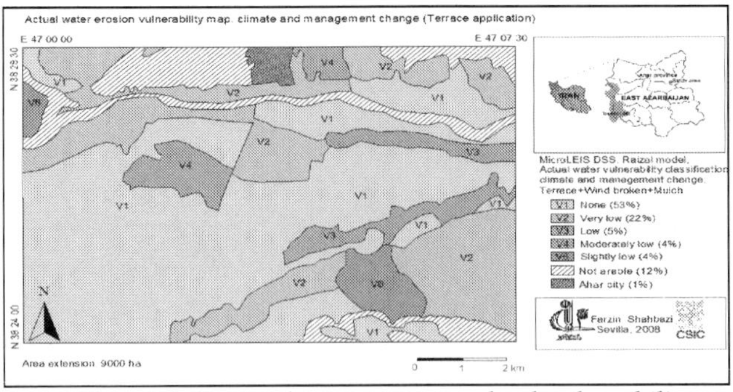

Fig. 10a. Terrace practice and climate change impact on land vulnerability caused by actual water erosion (Shahbazi, 2008)

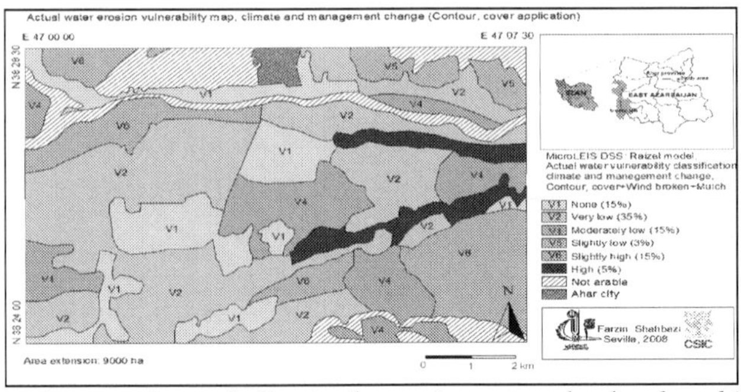

Fig. 10b. Contouring practice and climate change impact on land vulnerability caused by actual water erosion (Shahbazi, 2008)

5.2.3. Contaminants Risks

In general terms, the agrocontamination risk is considered to be directly related to the capacity of soils to store and immobilize toxic chemicals. The surface runoff transports high amounts of substances, such as phosphates in over-fertilized soils. Many biophysical and management factors control substance release from the soil to the water. The leaching of agricultural chemicals results from a complex interaction of physical, chemical and biological processes and attempts have been made to model these by equations based on classical mechanistic physics, and on a statistical or stochastic framework (De la Rosa & Crompvoets, 1998). However, models are not yet reliable enough to predict accurately the behavior of agrochemicals in the field. Soils are heterogeneous, climate and management factors vary, both in the short and long-terms. The development of land evaluation models is thus justified in terms of providing a tool with which to assess large amounts of soil information, such as that obtained from soil surveys, in order to yield the most practicable strategy for environmental protection (De la Rosa et al., 1993). The excesses of mineral nutrients and organic pesticides seem to be the most significant potential contaminants. However, impurities in fertilizers, manure and wastes can also be an important source of pollution especially with heavy metals. Therefore, the studied vulnerability types in west

Asia are: phosphorus, nitrogen, heavy metals and pesticides same as Mediterranean region. For Pantanal model establishment main following witticisms have been considered:
Phosphate substances are basically transported by runoff and constitute a possible source of eutrophication of waters. However, the phosphate fixation on clay minerals, along with its interaction with other soil components, was also estimated although the mobility of phosphate is usually very low in relation to other mineral nutrients. The amount of phosphate adsorbed by soil depends greatly on pH values, and also on particle size distribution and organic matter. Nitrate is the major nitrogen derived pollutant and the main source of groundwater contamination because of its high mobility. Along with land qualities associated with the rainfall partitioning, cation adsorption and denitrification are expected to predict this contamination risk. Retention of the heavy metals: copper, zinc and cadmium, by soils is analyzed considering the pH, as indicative of soil carbonate content, the main land characteristic controlling the different reactions. Organic matter content strongly affects adsorption-desorption and biodegradation of many pesticides, although other soil properties such as particle size distribution and CEC are also considered decision factors (De la Rosa & Crompvoets, 1998).

5.2.3.1. Case Study in East Azerbaijan

General contamination assessing in Ahar area revealed that only soil profiles under using of apple garden between the 44 studied profiles because of having artificial drainage has classified as moderate level risk (V2). Therefore, a total of 1560 ha (17.3%) are susceptible to contamination effect. In the current situation and without any climate and management changes risks of vulnerability raised by nitrogen and phosphorous (28% and 23% of studied area, respectively) are many times more than pesticides and heavy metals. It can be described as false management practices for using nitrogen fertilizers which are now presented in the whole are (88% area except not investigated lands where had been identified as marginal area by Cervatana model). Besides of that 57% area are distinguished as susceptible correspond to pesticides, correct management practices caused to be reduced the actual vulnerability compared with attainable one. Attainable and actual vulnerability classes for two hypothetical scenarios are summarized in (Table 5).

Vulnerability classes		Current and future scenarios (% of total area)			
		Phosphorous	Nitrogen	Heavy metals	Pesticides
V1	Attainable	32	55	57	1
V2		25	32	----	2
V3		4	1	31	49
V4		27	----	----	36
V5		----	----	----	----
V1	Actual	10	----	15	3
V2		29	----	47	11→12
V3		----	55	----	26→41
V4		26	32	26	48→32
V5		23	1	----	----

Table 5. Summary of Pantanal model application as a point by point view in Ahar area
* V1= none; V2= low; V3= moderate; V4= high; V5= extreme; → (impact of climate change)

According to the results, climate change will not effect on contamination vulnerabilities as well as water or wind erosion in part of Asia. The most important management practices accompany with climate change was examined as follows: Intensive wheat, barley, alfalfa, maize, potato, and sugar beet. Following orders present the best practice to decrease land vulnerability raised by: I) phosphorous; II) nitrogen; III) pesticides and IV) heavy metals, respectively.

I. Maize> Sugar beet> Barley> Wheat> Alfalfa> Potato
II. Alfalfa> Maize- Sugar beet- Wheat- Alfalfa- Potato
III. Potato> Maize> Barley> Sugar beet> Alfalfa> Wheat
IV. Maize> Barley- Sugar beet- Potato> Wheat

5.2.3.2. Case Study in West Azerbaijan for the Climate Change Era

Agro-ecological field vulnerability evaluation was compiled in Souma area where is closed to Urmia. Raizal model application resulted that for rainfall erosion, 72% of Souma lands are at none level of risk (ClassV1), and a further 28% at a very low and medium level. The medium risk area is more scattered in the north of study area which has established on plateau unit and characterised by a medium soil texture. In the simulated hypothetical scenario by long-term these results will be constant. Also, the study area is susceptible for wind vulnerability erosion and will increase in the future by climate change. The highest risk areas (V10) are located at the north-west and south-east of study area and refer to shallow Entisols. Soils No 2 and 6 areas will be altering from very high to extreme vulnerable land by climate change. Besides 10% extreme vulnerable land, 70% of the total area will be susceptible to vulnerability risks. A point-to-point application of Pantanal model results were summarized in (Table 6).

Soil	Phosphate		Nitrogen		Heavy metals		Pesticides	
No	current	future	current	future	current	future	current	future
1	V4	V4	V3	V3	V3	V3	V4	V4
2	V2	V2	V2	V1	V1	V1	V3	V3
3	V1	V1	V2	V1	V1	V1	V3	V2
4	V2	V2	V2	V1	V1	V1	V4	V3
5	V1	V1	V2	V1	V1	V1	V4	V3
6	V2	V2	V2	V1	V1	V1	V4	V3
7	V2	V2	V2	V1	V1	V1	V3	V3
8	V1	V1	V2	V1	V1	V1	V3	V3
9	V2	V2	V2	V1	V1	V1	V4	V3
Vulnerability classes*								
V1	18.63** (45%)	18.63 (45%)	0	37.53 (90%)	37.53 (90%)	37.53 (90%)	0	0
V2	18.9 (45%)	18.9 (45%)	37.53 (90%)	0	0	0	0	1.25 (3%)
V3	0	0	4.03 (10%)	4.03 (10%)	4.03 (10%)	4.03 (10%)	22.05 (53%)	36.28 (87%)
V4	4.03 (10%)	4.03 (10%)	0	0	0	0	19.51 (47%)	4.03 (10%)

Table 6. Summary of contamination vulnerability risk evaluation assessment in Souma (Shahbazi et al., 2009c)

* V1 = None; V2 = Low; V3 = Moderate; V4 = High, ** Area extention = km^2

According to obtained results, 10% of Souma area is at a high risk (Class V4) by phosphate while more than 45% is at a low level risk, and also 45% of the area presents no risk (ClassV1) of contamination. Reaction from local staff to the quality of the evaluation results for the current situation in Souma area was positive, although additional work on sensitivity and validation testing are needed in order to improve the prediction capacity of the risk evaluation approach (Shahbazi et al., 2009c).

6. Conclusion Remarks

Agro-ecological land evaluation appears to be a useful way to predict the potential index and/or general capability to distinguish the best agricultural land resulting from interactive changes in land use and climate. Due to bioclimatic deficiency is the most-sensitive factor affected by climate change; irrigation is indicated as very important in this semi-arid agriculture. However, the cultivation of rainfed wheat can be recommended instead of irrigated wheat in order to reduce the tillage operation costs. Also, the use of modern irrigation methods is recommended for the studied area in the future. Determining the impacts of climate change on land use systems involves also biophysical effects on agricultural management practices. Climate change might constrain or mandate particular land management strategies (e.g., irrigation); however, these options will be different for each particular site. In summary, the application of the land evaluation decision support system MicroLEIS DSS for planning the use and management of sustainable agriculture is suggested in west Asia region, for present and future climate conditions.

7. Abbreviations and Acronyms

AKi: Arkley index; **ARi:** Aridity index; **CDBm:** Monthly Climate database; **CRDY:** Dry land, Cropland, Pasture; **CRWO:** Cropland-Woodland mosaic; **CWANA:** Central and West Asia and North Africa; **ENSO:** El Niño-Southern Oscillation; **Eng & Tec:** Engineering and Technology Prediction; **Ero & Con:** Erosion and contamination modelling; **ETo:** Potential evapotranspiration; **GIS:** Geographic Information System; **GRAS:** Grassland; **GS:** Growth season; **HUi:** Humidity index; **ICCD:** Impacts of Climate Changes on Drylands; **Imp & Res:** Impact and Response simulation; **ImpelERO:** Integrated Model to Predict European Land use for erosion; **Inf & Kno:** Information and Knowledge databases; **IPCC:** Intergovernmental Panel on Climate Change; **LES:** Land Evaluation Systems; **LESA:** Land evaluation and site assessment; **LD:** Land degradation; **LUP:** Land use planning; **MDBm:** Management database; **MicroLEIS:** Mediterranean land evaluation information system; **MFi:** Modified Fournier index; **ONEP:** Office of Natural Resources & Environmental Policy and Planning; **p:** Monthly precipitation; **P:** Annual precipitation; **PCi:** precipitation concentration index; **Pro & Eco:** Production and Ecosystem modelling; **SAVA:** Savanna; **SDBm plus:** The multilingual soil database software; **SHRB:** Shrub land

8. References

Abu-Taleb, M.F. (2000). Impacts of global climate change scenarios on water supply and demand in Jordan. *Water International*, 25, 457-463

Arkley, R.J. (1963). Calculation of carbonate and water movement in soil from climatic data. *Soil Science*, 96, 239-248

Arnoldus, H.M.J. (1980). An approximation of the rainfall factor in the universal soil loss equation, In: *Assessment of Erosion*, Boodt, M. & Grabriels, D. (Eds.),. John Wiley & Sons, Inc., New York

Arthur, L.M. (1988). The implication of climate change for agriculture in the Prairie provinces, climate change digest 88-01. Downs view, ON: Atmospheric Environment Service, Toronto, Canada

Bailey, T.C. & Gatrell, A.C. (1995). *Interactive Spatial Data Analysis*. Longman, Harlow, UK.

Bouma, J.; Wagenet, R.J.; Hoosbeek, M.R. & Hutson, J.L. (1993). Using expert systems and simulation modeling for land evaluation at farm level - a case study from New York State. *Soil Use and Management*, 9, 131–139

Bou-Zeid, E. & El-Fadel, M. (2002). Climate change and water resources in Lebanon and the Middle East. *Water Res.* Pl.-ASCE, 128, 343-355

Burrough, P.A. & McDonnell, R.A. (1998). *Principles of Geographical Information Systems*. Oxford University Press. London

Bydekerke, L.; van Ranst, E.; Vanmechelen, L. & Groenemans, R. (1998). Land suitability assessment for cherimoya in southern Ecuador using expert knowledge and GIS. *Agriculture, Ecosystems and Environment*, 69, 89–98

CEC, (1992). CORINE; soil erosion risks and important land resources. *Commission of the European Communities*, DGXII. EUR, 13233 EN. Brussels

Christensen, J.; Hewitson, B.C.; Busuioc, A.; Chen, A.; Gao, X.; Jones, R.; Kwon, W.T.; Laprise, R.; Magana, V.; Mearns, L.; Menenedez, C.; Raisaene, J.; Rinke, A.; Kolli, R.K. & Sarr, A. (2007). Regional Climate Projections, In: *IPCC Fourth Assessment Report "Climate Change, The Scientific Basis"*, Cambridge University Press

CWANA, (2009). IASSTD Report. Agriculture at a crossroads, Vol 1: *Central and West Asia and North Africa*

Dale, V.H. (1997). The relationship between land use change and climate change. *Ecological Applications*, 7, 753–769

De la Rosa, D.; Anaya-Romero, M.; Diaz-Pereira, E.; Heredia, N. & Shahbazi, F. (2009). Soil-specific agro-ecological strategies for sustainable land use -A case study by using MicroLEIS DSS in Sevilla Province (Spain). *Land Use Policy*, 26, 1055-1065

De la Rosa, D. & Sobral, R. (2008). Soil quality and its assessment. In: *Land Use and Soil Resources*, Ademola, K.B. & Velk, L.G. (Eds.), 167-200, Springer

De la Rosa, D.; Mayol, F.; Diaz-Pereira, E.; Fernandez, M. & De la Rosa, Jr. D. (2004). A land evaluation decision support system (MicroLEIS DSS) for agricultural soil protection with special reference to the Mediterranean region. *Environmental Modelling & Software*, 19, 929-942

De la Rosa, D. & van Diepen, C. (2002). Qualitative and quantitative land evaluation. In: *Encyclopedia of Life Support System (EOLSS-UNESCO)*, Verheye, W. (Ed.), Section 1.5. Land use and land cover, Eolss, Oxford

De la Rosa, D.; Mayol, F.; Moreno, F.; Cabrera, F.; Diaz-Pereira; E. & Antoine, J. (2002). A multilingual soil profile database (SDBm plus) as an essential part of land resources information systems. *Environmental Modeling and Software*, 17, 721-731

De la Rosa, D.; Moreno, J.A.; Barros, J.; Mayol, F. & Rosales, A. (2001). MicroLEIS 4.1: exploring the agro-ecological limits of sustainability. Manual. Spanish National Research Council (CSIC), Institute of Natural Resources and Agrobiology (IRNAS), Sevilla, Spain

De la Rosa, D.; Moreno, J.A.; Mayol, F. & Bonson, T. (2000). Assessment of soil erosion vulnerability in western Europe and potential impact on crop productivity due to loss of soil depth using ImpelERO model. *Agriculture, Ecosystems and Environment*, 81, 179-190

De la Rosa, D.; Mayol, F.; Moreno, J.A.; Bonson, T. & Lozano, S. (1999). An expert system/neural network model (ImpelERO) for evaluating agricultural soil erosion in Andalucia region, southern Spain. *Agriculture, Ecosystems and Environment*, 73, 211-226

De la Rosa, D. & Crompvoets, J. (1998). Evaluating Mediterranean soil contamination risk in selected hydrological change scenarios. *Agriculture, Ecosystem and environment,* 67- 239-250

De la Rosa, D.; Crompvoets, J.; Mayol, F. & Moreno, J.A. (1996). Land vulnerability evaluation and climate change impacts in Andalucia, Spain: Soil Erosion and Contamination. *Agrophysics*, 10, 225-238

De la Rosa, D.; Moreno, J.A. & Garcia, L.V. (1993). Expert evaluation system for assessing field vulnerability to agrochemical compounds in Mediterranean regions. *Agric. Eng. Res.*, 56, 153–164

De la Rosa, D.; Moreno, J.A.; Garcia, L.V. & Almorza, J. (1992). MicroLEIS: a microcomputer based Mediterranean land evaluation information system. *Soil Use and Management*, 8, 89–96

Dietz, T. & Verhagen, J. (2004). The ICCD research, In: *The Impacts of Climate Changes on Drylands with a Focus on West Africa*, Dietz, A.J.; Ruben, R. & Verhagen, A. (Eds.), 403-408, Kluwer Academic Publisher, Dordrecht, The Netherlands

Dolgikh Kazakh, S. (2003). Current climate and climate change scenarios under global warming in Kazakhstan, *Proceeding of third regional & first national conference on climate change*, pp. 48-49, Isfahan-Iran

Duedall, I.W. & Maul, G.A. (2005). Demography of coastal populations, In: *Encyclopedia of Coastal Science*, Schwartz, M.L. (Ed.), 368-374, Springer, Dordrecht

Eom, S.B.; Lee, S.M.; Kim, E.B. & Somarajan, C. (1998). A survey of decision support system applications. *Journal of Operational Research*, 49, 109-120

FAO, (2006). Land evaluation. In: Towards a new framework. Land and Water Discussion FAO, 2005. Special event on impacts of climate change, pests and diseases on food security and poverty reduction. Background document 31st Session of the Committee on World Food Security, Rome, 10 pp

FAO, (2004). Data Base. Food and Agriculture Organization of the United Nations, Rome

FAO, (1976). A framework for land evaluation. Soils Bulletin 32. FAO, Rome

Faramarzi, M.; Abbaspour, K.C.; Schulin, R. & Yang, H. (2009). Modelling blue and green water resources availability in Iran. *Hydrological Progresses*, 23, 486-501

Farshi A.A.; Shariati, M.R.; Jarollahi, R.; Ghasemi, M.R.; Shahabifar, M. & Tolayi, M. (1997). *Water Requirement Estimating of Main Crops*. Part 1. Soil and Water Research Org., Karaj, IRAN

Fischer, G.; Shah, M. & van Velthuizen, H. (2002). Climate Change and Agricultural Vulnerability. International Institute for Applied Systems Analysis. Report prepared under UN Institutional Contract Agreement 1113 for World Summit on Sustainable Development. Laxenburg, Austria

Govinda Rao, P.; Ali Hamed, A.M. & Al-Sulaiti, M.H. (2003). Climatic changes and trends over Qatar, Oman and United Arab Emirates, *Proceeding of third regional & first national conference on climate change*, p. 43, Isfahan-Iran

Hargreaves, D.A.; Hargreaves, G.H. & Riley, J.P. (1985). Irrigation water requirements for Senegal River basin. *Journal of Irrigation and Drainage Division*, 3, 265-275

Ho, J. (2008). Singapore Country Report—A Regional Review on the Economics of Climate Change in Southeast Asia. Report submitted for RETA 6427: A Regional Review of the Economics of Climate Change in Southeast Asia. Asian Development Bank, Manila. Processed

Hoobler, B.M.; Vance, G.F.; Hamerlinck, J.D.; Munn, L.C. & Hayward, J.A. (2003). Applications of land evaluation and site assessment (LESA) and a geographical information system in east part county, Wyoming. *Journal of Soil and Water Conservation*, 58, 105–112

Hutchinson, M.F. (1989). A new procedure for gridding elevation and stream line data with automatic removal of spurious pits. *Journal of Hydrology*, 106, 211–232

IPCC, (2007). Impacts, Adaptation and Vulnerability. In: *Contribution of Working Group II to the Fourth Assessment Report of the Intergovernmental Panel on Climate Change*. Parry, M.L.; Canziani, O.F.; Palutikof, J.P.; van der Linden, P.J. & Hanson, C.E. (Eds.), 469-506, Cambridge University Press, Cambridge, UK

IRIMO, (2006a). Country Climate Analysis in year 2005, Islamic Republic of Iran Meteorological Organization, Tehran

IRIMO, (2006b), Country Climate Analysis in spring 2006, Islamic Republic of Iran Meteorological Organization, Tehran

Jafarzadeh, A.A.; Shahbazi, F. & Shahbazi, M.R. (2009). Suitability evaluation of some specific crops in Souma area (Iran), using Cervatana and Almagra models. *Biologia*, 64, 541-545

Johnson, A.K.L. & Cramb, R.A. (1996). Integrated land evaluation to generate risk-efficient land-use options in a coastal catchment. *Agricultural Systems*, 50, 287–305

Kalra, N.; Aggarwal, P.K.; Chander, S.; Pathak, H.; Choudhary, R.; Choudhary, A.; Mukesh, S.; Rai, H.K.; Soni, U.A.; Anil, S.; Jolly, M.; Singh, U.K.; Owrs, A. & Hussain, M.Z. (2003). Impacts of climate change on agriculture. In: *Climate Change and India. Vulnerability Assessment and Adaptatiopn*, Shukla P.R.; Sharma, S.K.; Ravindranath, A.; Garg, A. & Bhattacharya, S. (Eds.), 193-226, Orient Longman Private, Hyderabad,India

Kehl, M. (2009). Quaternary climate change in Iran- The state of knowledge. *Erdkunde*, 63, 1-17

Klingebiel, A.A. & Montgomery, P.H. (1961). Land capability classification. Agricultural Handbook 210, Soil Conservation Service. U.S. Govt. Printing Office, Washington, D.C. 21 p

Lal, M.; Harasawa, H.; Murdiyarso, D.; Adger, W.N.; Adhikary, S.; Ando, M.; Anokhin, Y. & Cruz, R.V. (2001). Asia. Climate Change 2001: Impacts, Adaptation, and Vulnerability. Contribution of Working Group II to the Chapter 10, Asia.

Lal, R. (1989). Land degradation and its impact on food and other resources. In: *Food and Natural Resources*, Pimentel, D. & Hall, C.V. (Eds.), 85-140, San Diego, C.A. Academic Press, USA

Liu, C.Z. (2002). Suggestion on water resources in China corresponding with global climate change. *China Water Resources*, 2, 36-37

Ministry of Energy of Iran (1998). An Overview of National Water Planning of Iran. Tehran, Iran (available in Persian)

Momeni, M. (2003). Climate change impact on ecological instability in iran. *Proceeding of third regional & first national conference on climate change*, pp. 149-154, Isfahan-Iran

Moore, I.D.; Grayson, R.B. & Ladson, A.R. (1991). Digital Terrain Modeling: A review of hydrological, geomorphological, and biological applications. *Hydrological Processes*, 5, 3–30

Nassiri, M.; Koocheki, A.; Kamali, G.A. & Shahandeh, H. (2006). Potential impact of climate change on rainfed wheat production in Iran. *Archives of Agronomy and Soil Science*, 52, 113-124

NCCO, (2003). Initial National Communication to United Nations Framework Convention on Climate Change. Published by Iranian National Climate Change Office at Department of Environment, Tehran, Iran

Oliver, J.E. (1980). Monthly precipitation distribution: A comparative index. *Professional Geographer*, 32, 300-309

ONEP, (2008). Climate Change National Strategy B.E. 2551-2555. Office of Natural Resources and Environmental Policy and Planning, Ministry of Natural Resources and Environment, Bangkok

Orgaz, F.; Pastor, M.; Vega, V. & Castro, J. (1996). Necesites de agua del olivar. Informaciones Técnicas 41/96, Pub. Junta de Andalucía, Sevilla (available in Spanish)

Perez, R. (2008). Philippines Country Report—a Regional Review on the Economics of Climate Change in Southeast Asia. Report submitted for RETA 6427: A Regional Review of the Economics of Climate Change in Southeast Asia. Asian Development Bank, Manila. Processed

Ragab, R. & Prudhomme, C. (2002). Climate change and water resources management in arid and semi-arid regions: prospective and challenges for the 21st century. *Biosystems Engineering*, 81, 3-34

Rahimzadeh, F. (2006). Study of precipitation variability in Iran, Research Climatology Institute, IRIMO, Tehran

Rosenzweig, C.; Iglesias, A.; Yang, X.B.; Epstein, P.R. & Chivian, E. (2001). Climate change and extreme weather events: implications for food production, plant diseases and pests. *Global Change and Human Health*, 2, 90-104

Salvati, L. & Zitti, M. (2009). Assessing the impact of ecological and economic factors on land degradation vulnerability through multiway analysis. *Ecological Indicators*, 9, 357-363

Sauter, V.L. (1997). *Decision Support System: an applied managerial approach*, John Wiley (Ed.), New York

Shahbazi, F. & Jafarzadeh, A.A. (2010). Integrated Assessment of Rural Lands for Sustainable Development Using MicroLEIS DSS in West Azerbaijan, Iran. *Geoderma*, 157, 175-184

Shahbazi, F.; Jafarzadeh, A.A. & Shahbazi, M.R. (2010a). Erosion and contamination impacts on land vulnerability in Souma area, using MicroLEIS DSS. *Final report, university of Tabriz, Tabriz, Iran*. 72pp

Shahbazi, F.; Jafarzadeh, A.A.; Sarmadian, F.; Neyshabouri, M.R.; Oustan, S.; Anaya-Romero, M. & De la Rosa, D. (2010b). Climate change impact on bioclimatic deficiency, Using MicroLEIS DSS in Ahar soils, Iran. *Journal of Agricultural Science and Technology, JAST*, 12, 191-201

Shahbazi, F.; Jafarzadeh, A.A.; Sarmadian, F.; Neyshabouri, M.R.; Oustan, S.; Anaya-Romero, M.; Lojo, M. & De la Rosa, D. (2009a). Climate change impact on land capability using MicroLEIS DSS. *Agrophysics*, 23, 277-286

Shahbazi, F.; Jafarzadeh, A.A. & Shahbazi, M.R. (2009b). Assessing sustainable agriculture development using the MicroLEIS DSS in Souma area, Iran. *Proceeding of International Conf. AgSAP*, pp. 304-305, Egmond aan Zee, The Netherlands

Shahbazi, F.; Jafarzadeh, A.A. & Shahbazi, M.R. (2009c). Agro-ecological field vulnerability evaluation and climate change impacts in Souma area (Iran), using MicroLEIS DSS. *Biologia*, 64, 555-559

Shahbazi, F.; Jafarzadeh, A.A.; Sarmadian, F.; Neyshabouri, M.R.; Oustan, S.; Anaya-Romero, M. & De la Rosa, D. (2009d). Suitability of wheat, maize, sugar beet and potato using MicroLEIS DSS software in Ahar area, north-west of Iran. *Journal of American and Environmental Science*, 5, 45-52

Shahbazi, F. (2008). Assessing MicroLEIS DSS application as a new method in land evaluation. *PhD Thesis*, Soil Science Department, Faculty of Agriculture, University of Tabriz, Iran, 347pp

Shahbazi, F.; De la Rosa, D.; Anaya-Romero, M.; Jafarzadeh, A.A.; Sarmadian, F.; Neyshabouri, M.R. & Oustan, S. (2008). Land use planning in Ahar area (Iran), using MicroLEIS DSS. *Agrophysics*, 22, 277-286

Shine, K.P.; Derwent, R.G.; Wuebbles, D.J. & Morcrette, J.J. (1990). Radiative forcing of climate. In: *Climate change: the IPCC Scientific Assessment*, Houghton, J.T.; Jenkins, G.J. & Ephraums, J.J. (Eds.), 40-68, Cambridge University Press, New York, New York, USA

Thomas, R.J. (2008). Opportunities to reduce the vulnerability of dryland farmers in central and west Asia and north Africa to climate change. *Agriculture, Ecosystems and Environment*, 126, 36-45

Thornthwaite, C.W. (1948). An approach toward a rational classification of climate. *The Geogr. Rev.*, 38, 55-94

UNEP, (2002). *Global Environment Outlook 3*, Earthscan, London, 426 pp

USDA, (2006). Keys to Soil Taxonomy. United State Department of Agriculture Natural Resources Conservation Service. Tenth Edition. The 18th World Congress of Soil Science. Philadelphia, Pennsylvania, USA

Williams, G.D.V.; Faultey, R.A.; Jones, K.H.; Stewart, R.B. & Wheaton, E.E. (1988). Estimating the effects of climatic change on agriculture in Saskatchewan. In: *The Impacts of Climatic Variations on Agriculture, Volume 1: Assessments in Cool, Temperate and Cold Regions*, Parry, M.L.; Carter, T.R. & Konjin, N.T. (Eds.), 219-379, Academic Publishers, Dordrecht

Yang, H.; Reichert, P.; Abbaspour, K.C. & Zehnder, A.J.B. (2003). A water resources threshold and its implications for food security. *Environmental Science and Technology*, 37, 3048–3054

Zhang, Y.; Chen, W. & Cihlar, J (2003). A process-based model for quantifying the impact of climate change on permafrost thermal regimes. *Journal of Geophys. Res.*, 108, 46-95

Ziska, L.H. (2003). Evaluation of the growth response of six invasive species to past, present and future atmospheric carbon dioxide. *Journal of Exp. Bot.*, 54, 395- 404

19

Simulated potato crop yield as an indicator of climate variability and changes in Estonia

Triin Saue[1,2] and Jüri Kadaja[1]
[1] *Estonian Research Institute of Agriculture*
[2]*Department of Geography, University of Tartu*
Estonia

1. Introduction

During the recent decades, global climate change has been at the centre of quite many scientific studies. Although the consensus is that climate is changing on a global scale, change on a regional or local scale is often more subtle and variable. Global climate change is mostly evaluated using the changes of annual average ambient temperature indicators, however, regional climate scenarios are not always consistent with global indicators. Consequently, the search for, and identification of, clear and unambiguous indicators of the impact of global climate change at a regional or local level is of vital importance.

Interactions between the biosphere and the atmosphere are obvious and have long been studied by several disciplines (e.g. Budyko, 1971, 1984; Fritts, 1976; Bolin, 1977; Tooming, 1977, 1984; Semenov and Porter, 1995; Scheifinger et al., 2002; Menzel, 2003; Aasa et al., 2004; McPherson, 2007). It has long been recognized that climate decides what can be cultivated, whereas soils indicate mainly to what extent climatic opportunities can be realized. The crops that continue to be grown in a particular location will primarily be determined by the changes in climate, and the seasonal distribution of rainfall and temperature that they experience. The main effect of temperature derives from the control of the growing period duration (Woodward, 1988), but also other processes linked with the accumulation of dry matter (leaf area expansion, photosynthesis, respiration, evapotranspiration etc.) are affected by temperature. Rainfall and soil water availability may affect the duration of growth through leaf area duration and the photosynthetic efficiency. These general climatic constraints on agricultural production are modified by local climatic constraints. In Northern countries the length of growing season, late spring and early autumn frost and solar radiation availability are typical climatic constraints, limiting the productivity of crops. For example, in Germany the growing season is one to three months longer than in Scandinavian countries (Mela, 1996).

Not surprisingly, also the reverse relation is true - biological and agricultural data can be used in climate assessments. Several biology-related indicators have been used by several scientists to assess past and present climate, its changes and variability, such as Palmer Drought Severity Index (e.g. Makra et al., 2002; Szep et al., 2005; Burke et al., 2006; Mpelasoka et al., 2007), growth season beginning and length (e.g. Menzel and Fabian, 1999;

Chmielewski & Köhn, 2000; Schwartz & Reiter, 2000; Sparks & Tryjanowski, 2007), dates of phenological phases (e.g. Ahas et al., 2004; Badeck et al, 2004; Chuine et al., 2004; Donnelly et al., 2004), etc. One of the complex variables, integrally describing summer weather conditions, is the biological production of plants and yield of agricultural crops. In this chapter, the potentiality of using the biological production and yield of agricultural crops as an indicator of summer climate variability and possible change is discussed. This approach is based on the postulate that the primary requirement for the success of a plant in a particular area is that its phenology would fit the environment. The signals of climate change usually occur more clearly in species growing at the borders of their distribution areas (Pensa et al., 2006) or whose growth is strongly influenced by climate, such as many arable crops (Hay & Porter, 2006).

Trends in individual climate variables or their combination into agro-climatic indicators show that there is an advance in phenology in large areas of North America and Europe, which has been attributed to recent regional warming. In temperate regions, there are clear signals of reduced risk of frost, longer growing season duration, increased biomass, insect expansion, and increased forest-fire occurrence that are in agreement with regional warming. Still, no detectable change in crop yield directly attributable to climate change has been reported for Europe (IPCC, 2007). Experimental studies of climate change through plant productivity are complicated indeed, as it is hard to distinguish the impact of climate variability or change from the effects of soil, landscape, and management. The worldwide trends in increasing productivity (yield per hectare) of most crops over the last 40 years, primarily due to technological improvements in breeding, pest and disease control, fertilisation and mechanisation, also make identifying climate-change signals difficult (Hafner, 2003). Thus, although the yield of agricultural crops is a quite commonly measured value, there is usually no long homogeneous time series of field crop yields. Therefore, the use of a simulated time series of crop yields, computed with dynamic plant production process models, is a more convenient and efficient way to draw climate estimations. These models are compiled from our knowledge of the different physiological processes in plants, and integrate different daily or more frequent weather data, calculating the development of plant production step-by-step. Traditionally, crop models are useful tools for translating climate forecasts and climate change scenarios into changes in yield, net returns, and other outcomes of different management practices. Additionally, those results can be turned backward and model-calculated yields can be used as an indicator to describe climate resources. In this chapter the concept of meteorologically possible yield (MPY) - the maximum yields under given meteorological conditions - is applied to derive qualitatively new information about climate variability. We will describe series of weather-reliant potato yields based on real existing meteorological series. Trends and variability changes within the series are assessed and compared to variability in the series of meteorological data. Probable range of temperature and precipitation in years 2050 and 2100 is applied to construct possible distribution of MPY in those years; future changes in mean values and variability are examined.

2. Material and methods

2.1 The model and the category of meteorologically possible yield

Plant productivity and thus the yields of field crops depend on many different closely interrelated factors. To introduce all of them into the model simultaneously is complicated. In our approach, the concept of the separation of factors, the principle of reference yields (Tooming, 1984; Kadaja & Tooming, 2004) was applied based on the principle of maximum plant productivity: such adaptation processes take place in a plant and plant community which are directed towards providing the maximum productivity of net photosynthesis possible under the existing environmental conditions (Tooming, 1967, 1970, 1977, 1984, 1988). Proceeding from this principle, maximum plant production is observed under different limiting factors, which can be divided into agroecological groups: biological, meteorological, soil, and agrotechnical groups. These groups of factors are included separately in the model, step by step, starting from the optimal conditions for the plant community (Tooming, 1993, 1998; Kadaja, 1994). Because the conditions specified as optimal involve no limitations, no input information regarding their optimal and limiting ranges is necessary. The corresponding categories of reference yields, as limits between the aforesaid groups, are in descending order: potential yield (PY), MPY, practically possible yield, and commercial yield (Fig. 1).

This concept is applied in the dynamic model POMOD to model the potato production process and yield (Sepp & Tooming, 1991; Kadaja & Tooming, 2004). In the present state, POMOD allows the computation of the PY and the MPY. The PY is the maximum yield of a given species or variety possible under the existing conditions of solar radiation, with all the other environmental and agricultural factors considered to be optimal. Therefore, PY is determined by the biological properties of the variety and the solar radiation available for utilization, and it expresses the radiation resources in units of biomass produced. The MPY is the maximum yield conceivable under the existing irradiance and meteorological conditions, with optimal soil fertility and agrotechnology, the levels of soil nutrients and the agrotechnology used do not limit production, and the effects of plant diseases, pests, and weeds are excluded. Only those soil properties related to the determination of the soil water content are applied.

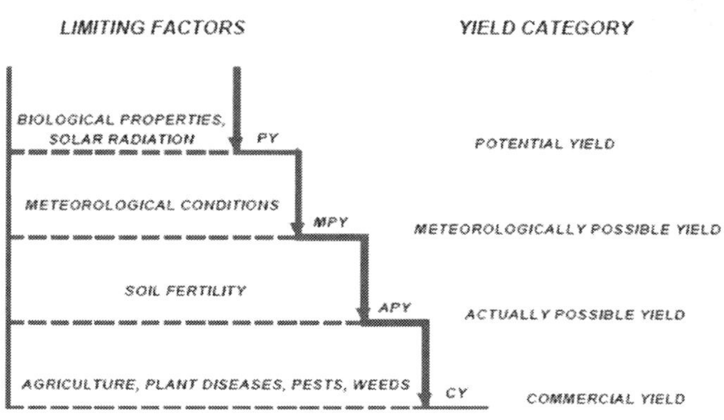

Fig. 1. The concept of yield limiting factors and corresponding reference yields (Zhukovsky et al., 1989).

As a result, MPY expresses agrometeorological resources, while its mean value and variability distribution over a long period characterize the agroclimatic resources in yield units. Using the category of MPY and the model of crop production, we can transform the complex of meteorological conditions into their yield equivalent and easily assess the agrometeorological resources of different years and the agroclimatic resources at different locations.

The underlying parameters of POMOD are the total biomass and the masses of plant organs (leaves, stems, roots, and tubers) per unit ground area (Kadaja & Tooming, 2004). The total growth of the plant biomass is calculated as the difference between the gross photosynthetic and respiration rates, integrated over time and leaf area index. The gross and net photosynthetic rates are expressed by equations derived from the principle of maximum plant productivity (Tooming, 1967). The meaning of parameters of gross and net photosynthesis irradiance curves are illustrated in Fig. 2. The initial slope a is the slope of tangent to the gross photosynthesis irradiance curve drawn from the origin of co-ordinates. R_a is the PAR flux density at the tangential point of net photosynthesis irradiance curve and its tangent drawn from the origin of co-ordinates. The intensity of photosynthetically active radiation (PAR) in the canopy is calculated from the total radiation and the leaf area above a particular level. The distribution of the total increase in biomass between different plant organs is determined using growth functions (Ross, 1966), which are given in the model as functions of accumulated positive temperatures. MPY is calculated taking into account the impact of meteorological factors on photosynthesis and respiration, and the influence of temperature on development rate.

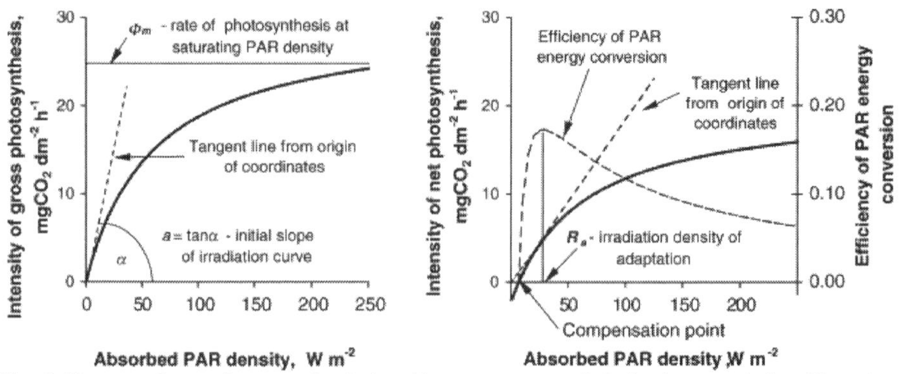

Fig. 2. Gross and net photosynthesis irradiance curves and their characteristics (Tooming, 1984).

The biological parameters of the potato varieties were determined on the basis of field experiments, not limited by nutrient deficiency, properly cultivated, weed and pest free, and regularly protected from late blight (Sepp & Tooming, 1991; Kadaja, 2004). The computed yields have proved similar to the real yields under these conditions, if the reduction in leaf area from late blight, not totally avoidable by protection, is included in the model. Differences in the real and computed yields did not exceed 5% in independent data collected under extremely good and bad growing conditions (Sepp & Tooming, 1991). Further verification of the model has been made on the basis of 20-year yield series at four stations of the Estonian Variety Control Network, with relatively stable cultivation and soils maintained during the period. Significant correlations between actual yields and calculated MPY were verified at three stations, whereas at the fourth, the correlation was not significant because of an increased level of plant diseases, grown without crop rotation.

2.2 Locations

To simulate time series of meteorologically possible yield, we compiled series of meteorological and agrometeorological data from the archives of the Estonian Meteorological and Hydrological Institute. We used the data from two stations: Tartu (58°15´N, 26°27´E) and Kuressaare (58°15´N, 22°29´E). These stations are located in regions with different local climates. Local climatic differences in Estonia result from, above all, the proximity of the Baltic Sea, which warms the coastal zone in winter and cools it especially in spring. According to the climatic classification of Estonia based on its air temperature regime, as proposed by Jaagus & Truu (2004), Tartu is located in the Mainland Estonia climatic region, characterized by a more continental climate and practically no climatic effect of the Baltic Sea, and Kuressaare is located in the Island Estonia region, with a much more maritime climate. Spring is much warmer in Tartu and summer starts earlier. In addition to different temperature regimes, there are considerable differences in precipitation between the two stations (Fig. 3). Furthermore, climate change effects appear to be different in the continental and coastal areas (Jaagus, 2006). For instance, because of the direct influence of the sea, the evident increase in annual mean temperature (1.0-1.7 °C at the different stations in Estonia during the second half of the 20th century) is less intense in spring in Kuressaare compared to that in Tartu. A significant increase in winter precipitation has also taken place in Estonia, but is much lower on the westernmost coast. In the same period, precipitation has increased remarkably in the coastal region in spring.

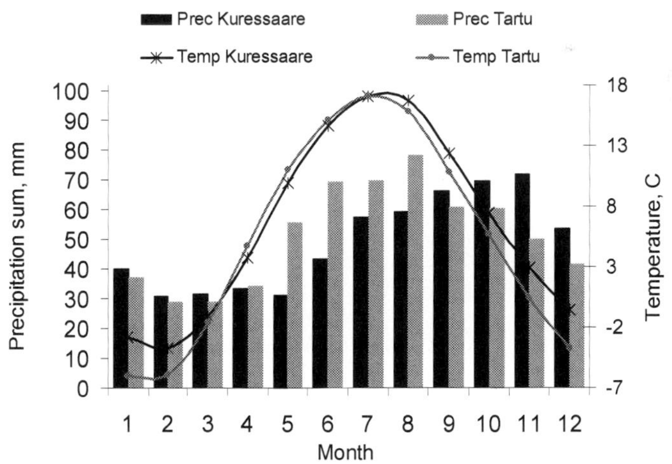

Fig. 3. Monthly mean temperatures (lines) and precipitation sums (bars) in Kuressaare and Tartu in 1965-2006.

2.3 Input data: calculations with current climate

The input information for the model can be divided into four groups: daily meteorological data, annual information, parameters of location, and biological parameters of the potato variety (Kadaja & Tooming, 2004).

The first group includes daily data on global radiation, air temperature, and precipitation for the growing period. For Tartu, meteorological data were available from 1901, for Kuressaare from 1923. Calculations were carried out up to 2006. As Kuressaare meteorological station was closed in 2001, the data for last years were calculated there on the basis of an adjacent station (Virtsu, Sõrve, Vilsandi, or Ristna, depending on which had the highest correlation for a particular factor or period). Direct measurements of global radiation have only been made since 1954 in Tartu. We computed the missing daily sums of global radiation from sunshine duration, using regression equations established separately for every month in Tartu.

Annual information included the year, the date and the value of the initial water storage in the soil (or the date when the soil moisture fell below the field capacity), the date of the permanent increase in temperature to above 8 °C in the spring, the dates of the last and first night frosts (≤ -2 °C), and the date of the permanent drop in temperature to below 7 °C in autumn. The initial soil moisture value is used as a basis for further calculations of soil moisture progression throughout the vegetation period. The dates of the temperature transitions are used as 'planting' and 'harvesting' dates for potatoes. We obtained the dates of night frosts and temperature transitions from the meteorological data sets of the stations. The data for the soil water status in spring was collected from the reports of the agrometeorological network using observations at Tartu-Erika (adjacent to Tartu) and at Karja on the island of Saaremaa (for Kuressaare). For the earlier period (up to the end of the 1940s) and for some later years when the agrometeorological network was not working, the data were derived from the meteorological data at the stations.

The locations are characterized by their geographical latitudes and the hydrological parameters of the soil, such as the wilting point, field capacity, and maximum water capacity. We used the parameters of the field soils (Kitse, 1978) prevalent at the locality. For Tartu, the parameters of a region with Albeluvisol (World Reference Base for Soil Resources) were used; for Kuressaare, the Skeletic Regosol prevails. All the soils are sandy silt loam, with quite similar hydrological parameters.

As parameters of variety, the model requires the parameters for photosynthesis, respiration, and the growth functions. We used the parameters of the early variety 'Maret' and the late variety 'Anti', both bred for Estonian conditions. The variety-specific photosynthesis variables, the initial slope of the photosynthesis irradiance curve a (kg CO_2 $s^{-1}W^{-1}$), the irradiation density of adaptation R_a (W m^{-2}), and the photosynthesis and respiration rates at the saturated PAR density given per unit mass of leaves, σ_1 and σ_2 respectively (kg CO_2 kg^{-1} s^{-1}), were estimated initially from the literature and adjusted for the specified varieties by a calibration method from experimental field data (Saue, 2006). Parameters σ_2 and α were considered constant throughout the vegetation period, while σ_1 and R_a were studied as variables. To associate parameters amongst each other, measured data of specific leaf weight of leaves, μ, were used. Specifically, different values were given to the maximum value of σ_1 and to the parameters describing its change within the temperature sums. The scope of change of σ_1 were first estimated by literature data (Tooming, 1977). R_a was calculated through σ_1, α and μ . To find the most optimal σ_1 value, relative errors between measured

and modelled data at different σ_1 values were calculated. Data of leaf area index and the biomass of all organs at all measurement dates were used.
Growth functions (Fig. 4) were determined on the basis of field experiments made from 2001 to 2006 (Kadaja, 2004, 2006).

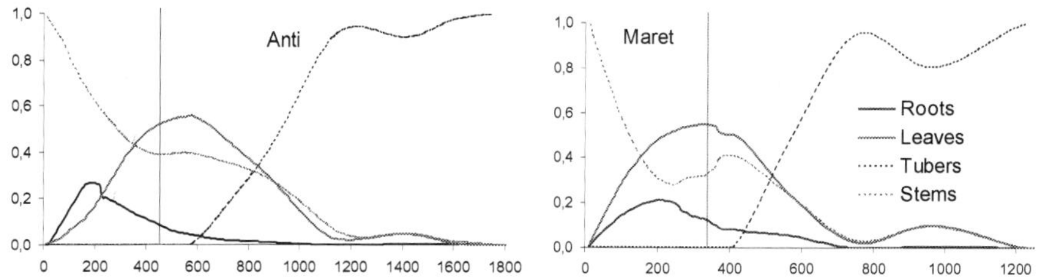

Fig. 4. Experimentally determined growth functions of late potato variety 'Anti' and early variety 'Maret'. Vertical lines denote the beginning of calculations.

2.4 Input data: calculations with future climate

Climate change could considerably affect the growth and yield of most crops (Adams et al., 1990; Easterling et al., 1992a, b). For model simulations of future potato production, future weather data were required. To achieve temperature and precipitation data for the years 2050 and 2100, climate change scenarios were generated for Estonia using a simple coupled gas-cycle/climate model MAGICC (Model for the Assessment of Greenhouse-gas Induced Climate Change) that drives a spatial climate-change scenario generator (SCENGEN). MAGICC has been one of the primary models used by IPCC since 1990 to produce projections of future global-mean temperature and sea level rise; we used the 5.3 version of the software, which is consistent with the IPCC Fourth Assessment Report (http://www.cgd.ucar.edu/cas/wigley/magicc/UserMan5.3.v2.pdf).
Because projections of climate change depend heavily upon future human activity, climate models are run against scenarios. There are over 40 different scenarios, each making different assumptions for future greenhouse gas pollution, land-use and other driving forces. Assumptions about future technological development as well as the future economic development are thus made for each scenario. Four alternative illustrative emission scenarios were used in our study to generate climate change scenarios for Estonia: A1B, a scenario of an integrated world with rapid economic growth, slowing population increase and a quick spread of new and efficient technologies with a balanced emphasis on all energy sources; A2, a scenario of a more divided world with continuously increasing population and an emphasis on family values and local traditions; B1, scenario of a world of "dematerialization" and introduction of clean technologies with rapid economic growth and increasing population; B2, a scenario of a world with an emphasis on local solutions to economic and environmental sustainability, with moderate economic growth and slowed population increase (Nakićenović & Swart, 2000). The highest climate warming is projected by A2; the lowest by B1. The year 1990 is used as the reference year in MAGICC/SCENGEN, all the climatic changes are calculated with respect to this year.
Data of changes in mean monthly air temperature and precipitation, averaged over 18 GCM experiments available on SCENGEN were applied. The idea of averaging more than one

GCM experiment and constructing a composite pattern for future climate change was first introduced by Santer *et al.* (1990); later Hulme *et al.* (2000) reported the clear supremacy of the technique over just only one model. The data are displayed in MAGICC/SCENGEN in a grid resolution of 2.5° latitude/longitude, thus the Estonian territory is covered by three grid boxes, with medium coordinates 58.8°N/21.3°E, 58.8°N/23.8°E and 58.8°N/26.3°E. Kuressaare and Tartu fall into two outermost boxes. However, the direct use of the SCENGEN output is not possible, because these predictions are available as changes in monthly means, but the crop model depends on daily time-series of weather as one of its main inputs. To calculate the future values of MPY, we used observed daily weather data in those stations during the baseline period 1965-2006. This shorter period is applied instead of previously used longer periods, since in climate change calculations it is necessary to use data outside the heretofore growing period. Global radiation was assumed not to change. Future daily temperatures and precipitation were calculated by adding the predicted monthly corrections to the observed series of daily data. This way, not just the one average predicted future value for temperature and precipitation, but 41 possible series of those meteorological elements were obtained for the two target years, suggesting the possible future weather distribution. Such setup also leads to the variability in the future climates being almost identical to the variability of the historical climate. Although the variability of climate in the future may alter (Rind et al., 1989; Mearns, 2000), inducing possible decrease in mean crop yields (Semenov & Porter, 1995; Semenov et al., 1996), some researchers (Barrow et al., 2000; Wolf, 2002) have reported that for potato, changes in climatic variability in northern Europe generally resulted in no changes in mean yields and its coefficient of variation.

Thus converted future weather data series are employed to calculate the date and the value of the initial water storage in the soil (or the date when the soil moisture falls below the field capacity), the date of the permanent increase in temperature to above 8 °C in the spring, the dates of the last and first night frosts (≤ -2 °C), and the date of the permanent drop in temperature to below 7 °C in autumn for each individual year of the new series. For determination of the soil water status in spring a relationship between radiation balance R_{fc} from permanent transition of temperature over 0° C to soil moisture fall below the field capacity, and meteorological data was derived using 30-year data of 13 stations of the Estonian Agrometeorological Network. To calculate R_{fc} , incoming global radiation and evaporative energy of precipitation (precipitation multiplied by latent evaporative heat) were accounted. The strongest correlations of R_{fc} were achieved with temperature sums from March to April $T_{3\text{-}4}$ and precipitation sums from February to April $U_{2\text{-}4}$:

$$R_{fc} = 468.2 - 1.587\, T_{3\text{-}4} - 0.517\, U_{2\text{-}4} \qquad r = 0.66 \tag{1}$$

To apply relationship (1) into the future dataset, a submodel calculates R_{fc} as well as permanent date of temperature rise over 0° C for each year of the new weather data series for 2050 and 2100. Next, from that date, the running radiation balance is summarized day-by-day. The date when the running radiation balance exceeds R_{fc} is counted as the date of achieving the soil field capacity and it is considered as the 'first possible' planting date. Additionally, 'optimal planting date' is applied – the date achieved by postponing the day of planting in model calculations day-by-day until the maximum yield is obtained. To prevent staying to a side maximum this postponing is conducted until the MPY drops below 70% of its maximum value, or until the date of summer equinox.

The dates of last and first night frosts in the future series are found on the basis of the earlier determined relationships between mean daily air temperature and ground level minimum temperature, dependent on the radiation sum of previous day.

3. Results

3.1 Time series of meteorological resources: current climate

Series of meteorologically possible yield were compiled for early and late maturing potato varieties in two different Estonian localities. In Table 1 we present long-term mean yields calculated with existing meteorological data series, using real and computed (both first possible and optimal) planting dates; the yields thus describe real, possible and optimal climatic resources for plant growth during given period.

With real planting dates, there was practically no difference in average values of the MPY between long and short (from 1965) series. As expected, the late variety produced higher yields at all locations. Overall, the MPY series showed only weak and insignificant trends (Fig. 5), although reliable trends are apparent for some shorter periods. The longest period with a significant ($P < 0.05$) decreasing trend was observed in Kuressaare from 1977 to 2006. Generally, 'Anti' demonstrated higher variance in yields. For both varieties, the variability reached higher in Kuressaare. Variability increases in all cases when using computed planting dates instead of real dates.

Closer investigation of the MPY variability showed a significant increase in variance in Tartu since the early 1980s. In the MPY calculations contrived with real meteorological data, the standard deviation of MPY was significantly lower for 'Maret' in 1901-1980 compared to 1981-2006 ($P = 0.006$, according to F test); for 'Anti', the change was smaller yet significant ($P = 0.046$). When using shorter time series and optimal planting times, the same difference in yield variance was detected both for 'Maret' ($P = 0.002$) and 'Anti' ($P = 0.015$). The meteorological elements series revealed no similar changes in climate variability. Reliable dispersion differences were detected only in the precipitation series, but their significance was lower than that of the yields.

	ANTI				MARET			
	Tartu		Kuressaare		Tartu		Kuressaare	
	MPY	Var. coeff.	MPY	Var. coeff.	MPY	Var. coeff.	MPY	Var. coeff.
Real dates								
Long series to 2006	55.5	0.20	50.3	0.27	45.0	0.16	37.8	0.21
1965-2006	54.5	0.21	50.3	0.28	45.1	0.19	37.7	0.22
1901-1980	56.1	0.18			45.5	0.14		
1981-2006	53.9	0.25			43.5	0.22		
1923-1938			51.0	0.16				
1939-2006			50.1	0.29				
Computed dates								
1965-2006, first planting date	58.8	0.24	49.8	0.33	42.4	0.18	38.2	0.27
1965-2006, optimum planting date	58.9	0.23	50.2	0.32	44.0	0.19	39.3	0.26

Table 1. Mean values of MPY and corresponding coefficient of variation for different periods.

Therefore, the separate meteorological elements did not reflect the influence of their combined effect on the variability of biological production. Significant differences in yield variability, not identified in the meteorological series, were also observed for 'Anti' at Kuressaare, where the standard deviation was approximately two times lower before 1939 than in later periods ($P < 0.017$).

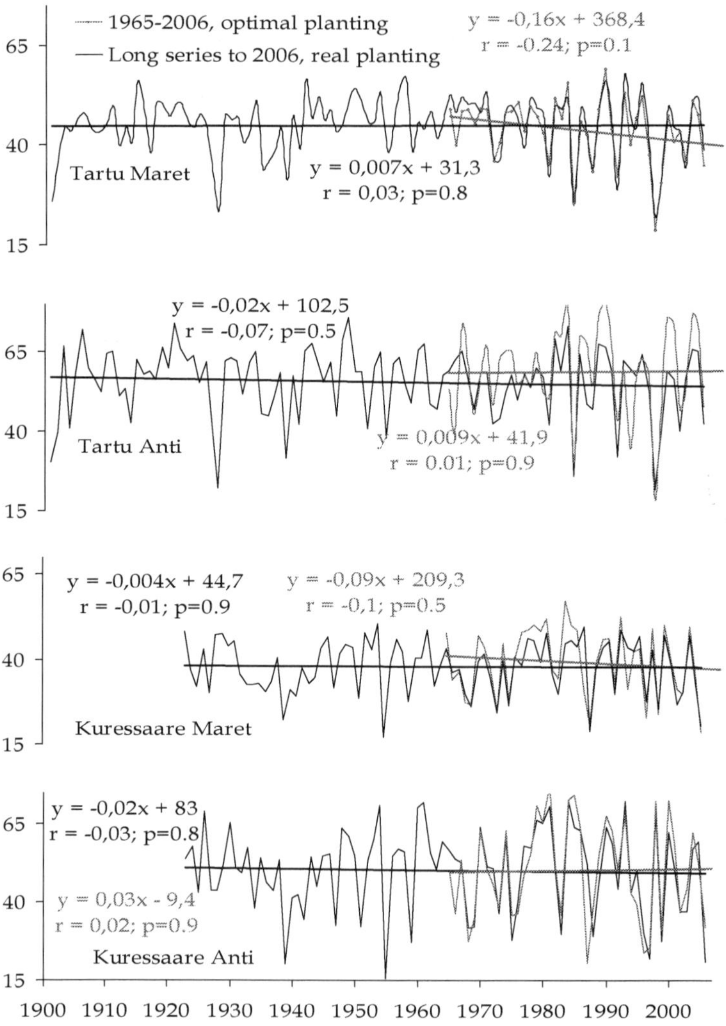

Fig. 5. Series of MPY of the early potato variety 'Maret' and the late potato variety 'Anti' in Tallinn and Kuressaare.

3.2 Relationships between MPY and other indicators

In Estonia, like elsewhere in temperate zone, crop yield variation is highly influenced by weather conditions (Carter, 1996; Karing et al.,1999). When using real, measured potato yield data, potato yield variance was found to be mostly dependent on weather conditions, while the impact of fertilization and soil management proved less significant and in interaction with weather (Saue et al., 2010). Of meteorological conditions, potato proved the most susceptible to spring temperatures, yielding higher in years with a warm spring; negative linear relation between yields and precipitation during the same period concurred. The positive influence of precipitation was expressed after flowering.

In this paragraph, we will compare simulated yields and direct meteorological series of precipitation, temperature and solar radiation, using accumulated values for those meteorological elements over different periods, in order to explain the extent to which individual factors allow us to describe the whole complex. Correlation analyses (linear and second-order polynomial) were performed.

In Tartu , linear correlations between MPY and the accumulated meteorological factors were weak, although they were significant in some cases since the series were long (Table 2). The correlations with temperature were slightly higher, but only for the early variety.

In Kuressaare, significant ($P < 0.01$) linear correlations were identified between MPY and all the accumulated meteorological factors in the selected periods: positive for precipitation and negative for solar radiation and temperature. In general, the period with the highest correlations began earlier for precipitation (from May for 'Maret' and from June for 'Anti'), and later for temperature and radiation (from June and July, respectively). The results for Kuressaare are quite different from those for Tartu because its location on the island of Saaremaa in the western part of Estonia is characterised by a mild marine climate and dry summers. Low precipitation at the beginning of summer causes dry conditions, so water deficit is the main limiting factor there. The relationships between MPY and solar radiation and temperatures are largely indirect, and these factors correlate negatively with precipitation.

As a rule, if a curve with a maximum describable by a second-order polynomial is applied, better correlation will be apparent between MPY and the accumulated meteorological elements. This means that for all factors, the limitation derives from both deficit and excess. Again, the highest correlations occurred in Kuressaare: for 'Anti' with precipitation (June-August: r = -0.77, May-August: r = -0.76), and for 'Maret' with temperature from June to September (r = -0.71). The only exception, where the correlations are almost equal on the linear and polynomial curves, is the early variety in Kuressaare. There, the conditions are dry, especially in the first half of summer, so the limiting factor for the early variety in most years is a deficit of precipitation. For the late variety, the decrease in yield is occasionally caused by an excess of water. However, the latter is much more common in inland regions, represented by Tartu, where intense rainy periods produce soil moisture near its maximum content in June and July, causing the loss of soil aeration and a very significant reduction in yield.

The limiting from two sides and high variances between MPY and the cumulative meteorological elements allow us to conclude that, under our conditions, MPY gives qualitatively new information about climate variability in summer, especially regarding climatic favourableness, by integrating the effects of different weather factors. In conditions with one very dominant limiting factor, there is no need for such an indicator, e.g., near the

Polar Circle, where MPY correlates very well with temperature (Sepp et al., 1989) or in arid regions, where the dominant factor is water deficit. For the stations analyzed in our work, Kuressaare is the most likely to be affected by a single dominant limiting factor, but the variance is still quite high there.

Station	Meteo-element	Relation-ship	Early variety 'Maret'			'Late variety Anti'		
			May-Aug	June-Aug	May-Sept	May-Aug	June-Aug	May-Sept
Tartu	R	LIN	0,03	0,02	0.03	0,01	-0,03	0.02
		POL	**--0,36**	**--0,41**	**--0.31**	**--0,47**	**--0,52**	**--0.43**
	P	LIN	--0,07	--0,02	--0.13	--0,06	--0,12	--0.03
		POL	**--0,53**	**--0,40**	**--0.49**	**--0,64**	**--0,56**	**--0.40**
	T	LIN	**--0,26**	**--0,37**	--0.24	--0,04	--0,20	--0.03
		POL	**--0,35**	**--0,50**	--0.29	**--0,41**	**--0,55**	**--0.35**
		POL	--0,25	**--0,32**	**--0.26**	**--0,34**	**--0,35**	**0.34**
	P	LIN	0,19	**0,27**	0.05	**0,26**	**0,34**	0.10
		POL	**--0,31**	**--0,33**	**--0.34**	**--0,42**	**--0,46**	**--0.42**
	T	LIN	--0,17	**--0,41**	--0.24	0,14	--0,09	0.08
		POL	**--0,41**	**--0,52**	**--0.34**	**--0,46**	**--0,44**	**--0.41**
Kuressaare	R	LIN	**--0,50**	**--0,55**	**--0.51**	**--0,46**	**--0,56**	**--0.45**
		POL	**--0,50**	**--0,55**	**--0.51**	**--0,47**	**--0,57**	**--0.47**
	P	LIN	**0,65**	**0,61**	**0.64**	**0,65**	**0,72**	**0.61**
		POL	**--0,68**	**--0,66**	**-0.65**	**--0,76**	**--0,77**	**--0.69**
	T	LIN	**--0,56**	**--0,68**	**--0.61**	**--0,30**	**--0,44**	**--0.35**
		POL	**--0,58**	**--0,69**	**--0.62**	**--0,48**	**--0,57**	**--0.51**

Table 2. Correlation coefficients *r* for the linear (LIN) and polynomial (POL) relationships between meteorologically possible yield (MPY) and accumulated solar radiation (R), precipitation (P), and temperature (T) at two stations. Bold indicates significance levels of $P < 0.01$.

3.3 Climate Change

Most climate change scenarios project that greenhouse gas concentrations will increase through 2100 with a continued increase in average global temperatures (IPCC, 2007). Results of the four emission scenarios, each containing 18 General Circulation Models (GCM) experiments used in SCENGEN provide a wide variety of possible climate change scenarios (Table 3). In this paragraph we will look at the results by four illustrative emission scenarios, achieved by using the multi-model average for two locations in Estonia. All scenarios project the increase in annual mean temperature, the highest warming is supposed to take place during the cold part of the year (Fig. 6). During the plant-growth period (April to September), the increase of air temperature will be lower. Average annual precipitation is also predicted to increase (Fig. 7), however, changes in the annual range of monthly precipitation vary highly between models and scenarios and are less certain than changes in temperature. On average, the highest change in precipitation is predicted for January and

November; August and September are predicted a small increase or even a slight decrease. All the projected climatic tendencies have already been noted during the last century (Jaagus, 2006), indicating evident climate warming in Estonia. In previous analogous works (Keevallik, 1998; Karing et al., 1999; Kont et al., 2003), temperature rise has been predicted higher; however we believe that moderate warming is more realistic.

		Temperature change, ° C		Precipitation change, %	
Year	Scenario	Tartu	Kures-saare	Tartu	Kures-saare
2050	A1B	2.40	2.37	8.5	8.1
	A2	2.60	2.54	10.0	8.8
	B1	1.73	1.71	6.2	5.8
	B2	2.25	2.24	8.1	8.0
2100	A1B	4.65	4.64	16.2	16.3
	A2	5.78	5.72	20.7	19.5
	B1	3.11	3.14	10.7	11.2
	B2	4.13	4.13	14.7	14.4

Table 3. Changes in annual air temperature and precipitation calculated as a mean of experiments by 18 different GCM for four different emission scenarios.

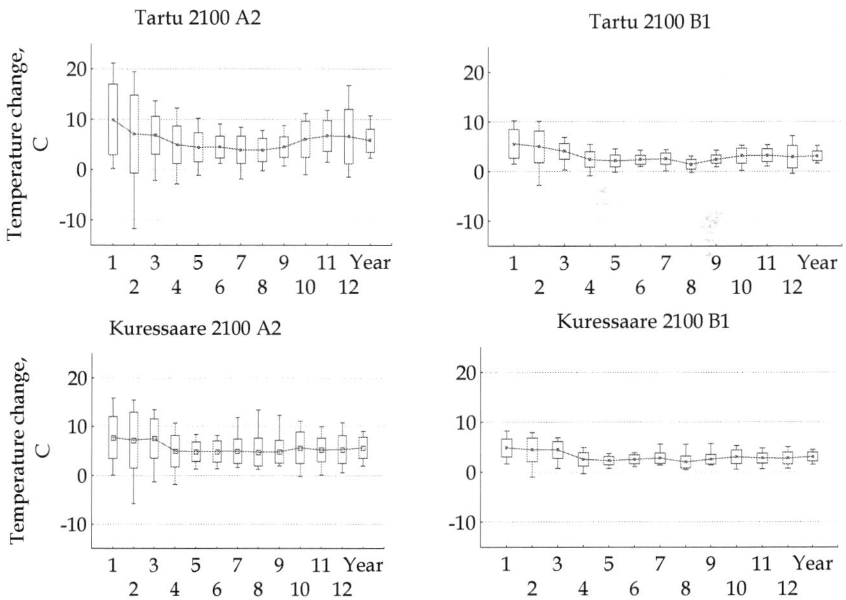

Fig. 6. Changes in monthly mean temperature (° C) predicted by 18 global climate models for the A2 and B1 emissions scenarios for year 2100 compared to the baseline period (1961–1990) at two Estonian sites. Lines connect the values of monthly mean change, boxes mark mean change ± standard deviation and whiskers mark the range of all models.

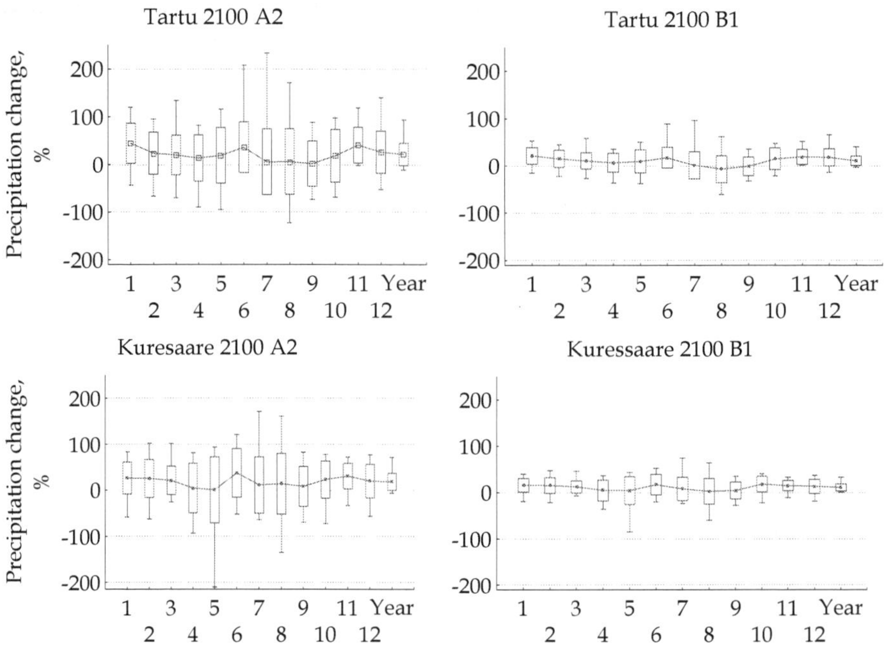

Fig. 7. Changes in monthly sum of precipitation (%) predicted by 18 global climate models for the A2 and B1 emissions scenarios for year 2100 compared to the baseline period (1961–1990) at two Estonian sites. Lines connect the values of monthly mean change, boxes mark mean change ± standard deviation and whiskers mark the range of all models.

3.4 MPY in the future

From now on, all changes in MPY are referred as compared to baseline period (1965-2006) and we will discuss the yields achieved with optimal planting time. The productivity and yield changes related to the rise of CO_2 in the atmosphere rise are not considered.

For the late variety 'Anti', the long-term mean MPY values, calculated by using historical climate data of 1965-2006 with computed optimal planting time, describing the optimal climatic resources for plant growth, are 58.9 t ha-1 in Tartu and 50.2 in Kuressaare (see Table 1). For the early variety 'Maret' the values are 44.0 and 39.3, respectively.

For early variety, all four considered scenarios predict losses in all given localities (Fig. 8). Stronger scenarios cause higher losses, up to 37% in Tartu and 32% in Kuressaare by 2100. In Kuressaare, the change in mean MPY is statistically significant for the year 2050 only by the strongest, A2 scenario (p=0.03); for the year 2100 all scenarios predict significant loss (p<0,001). In Tartu, for the year 2050 the change in MPY is significant by A2 (p=0.002), A1B (p=0.01) and B2 (p=0.03) scenarios; for the year 2100, the loss in MPY is significant by all scenarios (p<0.001).

For late variety, remote rise in yields is predicted for year 2050. Lower temperature rise through milder scenarios is more favourable for potatoes – B1 scenario predicts 5.5% yield rise in Tartu and 5% in Kuressaare, while for A2 scenario the rise is 2.5 and 2%. For year 2100, all scenarios predict yield losses, stronger scenarios up to 15% in Tartu, up to 19% in Kuressaare for 2100 as compared to present climate. The changes in 'Anti' MPY are however not statistically significant for any location, year or scenario.

Compared to yield variability in baseline climate, the predicted yield variability of 'Anti' turned to be significantly ($p<0.05$) lower in Kuressaare in case of the strongest climate change (A2 scenario for the year 2100) (standard deviation 11.6 compared to 15.8 t ha^{-1}). The 'Maret' MPY variability is also lower in Kuressaare in 2100 by scenarios A1B ($p<0.001$), A2 ($p<0.001$) and B2 ($p=0.02$), standard deviation declining from 10.1 to 6.3, 5.7 and 7.7 t ha^{-1}, respectively. In Tartu, the change in variability was only significant ($p=0.009$) for A2 in 2100 (standard deviation 7.8 to 5.4 t ha^{-1}).

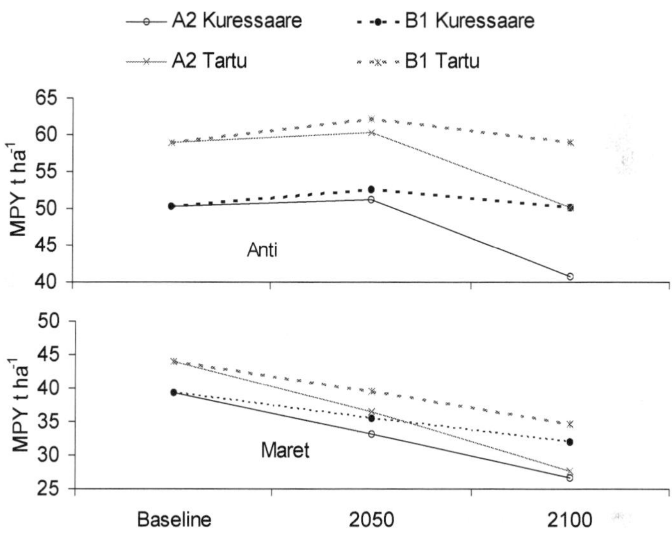

Fig. 8. Mean values of the meteorologically possible yield (MPY) of late potato variety 'Anti' and early potato variety 'Maret' for baseline period (1965-2006), years 2050 and 2100 by the two scenarios predicting the strongest (A2) and weakest (B1) warming.

3.5 Cumulative distribution of MPY

An applicable method for comparing the extent of MPY variability among different varieties and locations is based on their cumulative distributions, which expresses the probabilistic climatic yield forecast (Zhukovsky et al., 1990). For the baseline climate, the late variety 'Anti' produced higher yields across the entire range of probabilities and the distribution of the yield is not a symmetric one. Low yields, corresponding to extreme meteorological conditions and forming deep deviations in time series (Fig. 5), stretch the cumulative distribution out in the left part (Fig. 9 & 10). For the current climate, the decline in the cumulative distribution is quite steep after the mean value of MPY. High MPY values correspond to the years in which the different meteorological resources are well balanced throughout the summer period. As a rule, these are climatically similar to the climatic norms for all the factors in Estonia. The MPY distribution for 'Anti' is lower in Kuressaare, predominantly in the range of lower and central MPY values, resulting in a smoother decline in the range of the highest yields. Even larger inequalities in mean values as well as in their distributions appear between two locations for the early variety 'Maret'. We can conclude that the differences in climatic conditions during the first half of summer have a greater effect on early varieties. The shape of the distribution curve is more symmetric for the early variety.

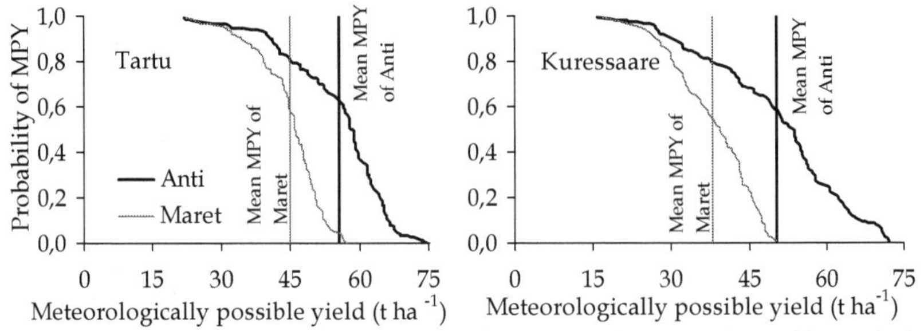

Fig. 9. Cumulative distribution of the MPY for the current climate, achieved by real planting dates.

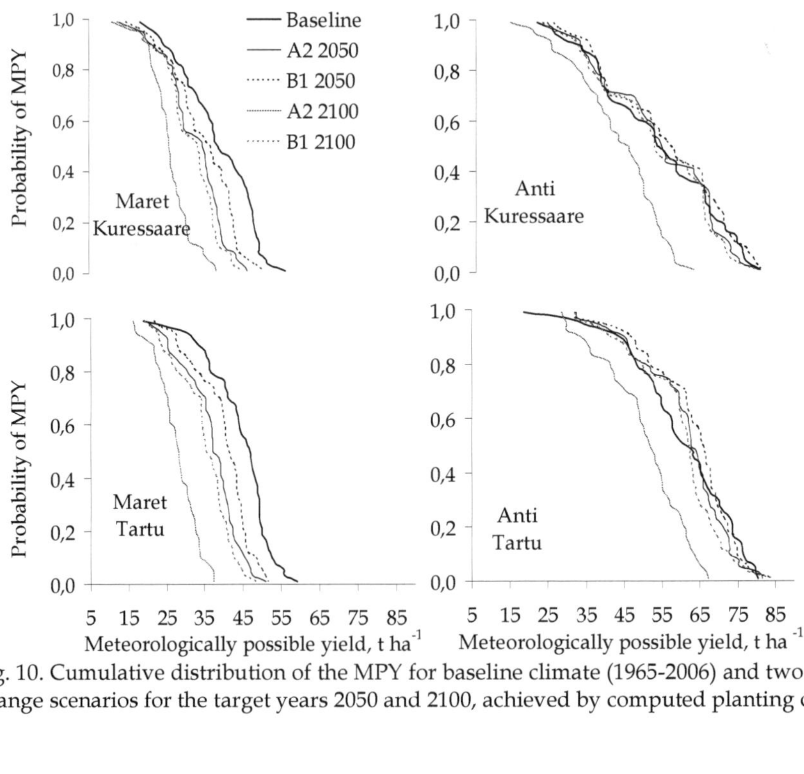

Fig. 10. Cumulative distribution of the MPY for baseline climate (1965-2006) and two climate change scenarios for the target years 2050 and 2100, achieved by computed planting dates.

Cumulative distribution of the future MPY values (Fig. 10) shows greater differences between scenarios and target years for 'Maret', witnessing the higher weather sensitivity of early variety. For all cases, A2 scenario certifies definite disadvantage of strong warming modelled for the year 2100. For 'Anti', the cumulative yield differences between scenarios and target years are not very stark, enabling to conclude the advantage of longer maturing varieties for future climate warming.

4. Conclusions and discussion

The main objective of this chapter was to show that computed yields give additional information about climatic variability compared with the traditional use of individual meteorological elements. Our results indicate that none of the observed separate meteorological factors sufficiently reflects the variations in the computed MPY series. We found significant linear correlations for only the western Estonian coastal zone, represented by the station at Kuressaare, because of the dominant limiting factor, the water deficit during the first half of summer in most years. Although the polynomial correlations were higher, indicating a dual influence of the factors, there was still high variance. The significant changes in MPY variability, as observed in Tartu in the second half of the period, were only weakly expressed in the precipitation series and were absent from the temperature and radiation data. Evidently, the combined effects of weather conditions on plant production processes have a more complex character than can be measured with long-term statistics for individual meteorological elements. Consequently, the use of MPY to express the agrometeorological resources available for plant production in yield units introduces additional information about the impact of climatic variability. The changes in MPY and their statistical distribution are better indicators of the impact of climate change on plant production than are changes in the time series of any individual meteorological elements. This holds particularly true if simulations for species adapted to local climatic conditions are used. If species are located at the borders of their distribution areas, some meteorological factors will predominantly limit their growth and will describe the climatic resources without being combined with other factors. The MPY series collected through 83-106 years revealed no significant trends. However, significant trends do exist in terms of shorter periods. The variability of MPY has been increasing in the island regions of Estonia since the 1940s and in the continental areas since the 1980s.

The above-described results have been further expanded into the future and future values of meteorologically possible potato crop yield have been generated. This allows to estimate the influence of climate change on agrometeorological resources for potato growth in Estonia. All of the four climate change scenarios projected the increase in annual mean temperature for Estonia, the highest warming during the cold part of the year. Average annual precipitation was also predicted to increase, however, changes in the annual range of monthly precipitation vary highly between models and scenarios and are less certain than changes in temperature. All the projected climatic tendencies have already been noted in observations during the last century (Jaagus, 2006), indicating evident climate warming in Estonia.

Changes in MPY were calculated using historical weather variability and projected changes in mean monthly values. For early potato variety, all scenarios predict losses in potato yields, while the scenarios of more notable warming cause higher losses. For late variety, a slight rise in yields is predicted for 2050, which turns to loss by 2100. However, the changes are not statistically significant for the late variety. This result is a development from previous results with the same model (Kadaja & Tooming, 1998; Karing et al., 1999; Kadaja, 2006), which predicted yield rise with moderate scenarios for late variety and loss only occurs with strong warming scenarios.

There have been several researches in different regions about possible climate-change-related variation in potato growth. Peiris et al. (1996) calculated increases in tuber yield by temperature rise for potato in Scotland due to faster crop emergence and canopy expansion and thus a longer growth period. Wolf (1999 a, 2002) has reported small to considerable increases in a mean tuber yield with climate change in the Northern Europe, being caused by the higher CO_2 concentration and by the temperature rise. Wolf and van Oijen (2002) showed yield increase for the year 2050 in all regions of the EU, mainly due to the positive yield response to increased CO_2. Such disagreement with our results likely derives from the fact that in our study no effect of CO_2 rise on potato growth has been considered. There is clear evidence since 1950s (Keeling et al., 1995) that atmospheric CO_2 is increasing, and plant physiologists have repeatedly demonstrated that such increases likely have already caused substantial increases in leaf photosynthesis of C_3 species (Sage, 1994). The presence of large sinks for assimilates in tubers makes potato crop a good candidate for large growth and yield responses to rising CO_2; this effect tends to be smaller for late cultivars (Miglietta et al., 2000). However, since the optimal temperature range for tuber growth (between 16 and 22 °C) is small (Kooman, 1995), and since with climate change the prevailing temperature during tuber growth will likely be different, the positive effect of CO_2 may be counteracted by the effect of a concominant temperature rise. Wolf (1999a; 2002) has shown such effect for central and southern Europe, where the negative effect of temperature rise was expected sometimes to exceed the positive effect of CO_2 enrichment. Under hotter and wetter scenarios for Great Britain, Wolf (1999b) demonstrated tuber yields to become lower, caused by the temperature rise, which speeded the phenological development of the crop and reduced the time for growth and biomass production. At the same time, under the smaller temperature rise the yield had mainly increased at the same locations. Rosenzweig et al. (1996) have also calculated decreases in tuber yield for most sites in the USA due to the negative effect of temperature rise on yield that was stronger than the positive effect of CO_2 enrichment. Miglietta et al (2000) have described a model experiment for Dutch weather conditions, where the elevated temperature reduced the positive effect of elevated CO_2. For predicted future temperature rise (without an increase in atmospheric CO_2) over England and Wales, Davies et al. (1997) calculated variable and little changes in tuber yield of potato.

Based on this knowledge and our current research result, we can thus say that the climatic resources for potato growth are predicted to become worse under climatic change because of increased temperature and variable rainfall; however in higher latitudes this effect may be altered and turned positive by the change in plants photosynthetic activity and production.

The variability of potato yields is predicted to decrease slightly due to climate change. This is however not a plausible result, since the change in meteorological variability has not been counted in. Further investigation need rises in this area. Also Wolf (1999a) has shown the variability of non-irrigated tuber yield to essentially zero to moderately decrease in Northern Europe.

Acknowledgements

Financial support from the Estonian Science Foundation grants No 6092 and 7526 is appreciated.

5. References

Aasa, A.; Jaagus, J.; Ahas, R. & Sepp, M. (2004). The influence of atmospheric circulation on plant phenological phases in central and eastern Europe. *International Journal of Climatology*, 24 (12), 1551-1564

Adams, R.M.; Rosenzweig, C.; Peart, R.M.; Ritchie, J.T. & 6 others (1990). Global climate change and US agriculture. *Nature*, 345, 219-223

Ahas, R. ; Jaagus, J. & Aasa, A. (2000). The phenological calendar of Estonia and its correlation with mean air temperature. International Journal of Biometeorology, *44 (4), 159 - 166*

Badeck, F.-W., Bondeau, A., Böttcher, K., Doktor, D., Lucht, W., Schaber, J. & Sitch, S. (2004). Responses of spring phenology to climate change. *New Phytologist*, 162, 295-309

Barrow, E. M.; Hulme, M.; Semenov, M. A. & Brooks, R. J. (2000). Climate change scenarios. In: *Climate Change, Climatic Variability and Agriculture in Europe: an integrated assessment*, Downing, T. E.; Harrison, P. A.; Butterfield, R. E. & Londsdale, K. G.(Eds.), 11-27, Environmental Change Institute, University of Oxford, UK

Bolin, B. (1977). Changes of Land Biota and Their Importance for the Carbon Cycle. *Science*, 196, 613-615

Budyko, M.I. (1971). *Climate and life*. Gidrometeoizdat. Leningrad. 471 pp. [in Russian, with English abstract]

Budyko, M.I. (1974). *Evolution of biosphere*. Gidrometeoizdat. Leningrad. 488 pp. [in Russian, with English abstract]

Burke, E. J.; Brown, S.J. & Christidis, N. (2006). Modeling the Recent Evolution of Global Drought and Projections for the Twenty-First Century with the Hadley Centre Climate Model. *Journal of Hydrometeorology*, 7, 1113-1125

Carter, T. R. (1996). Global climate change and agriculture in the North. *Agric Food Sci Finland*, 5, 222-385

Chmielewski, F.-M. & Köhn, W. (2000). Impact of weather on yield and yield components of winter rye. *Agric. Forest Meteorol*, 102, 253-261

Chuine, I.; Yiou, P.; Viovy, N; Seguin, B.; Daux, V. & Ladurie E.L.R. (2004). Historical phenology: Grape ripening as a past climate indicator. *Nature*, 432, 289-290

Davies, A.; Jenkins, T.; Pike, A.; Shao, J.; Carson, I.; Pollock, C.J. & Parry, M.L. (1997) Modelling the predicted geographic and economic response of UK cropping systems to climate change scenarios: the case of potatoes. *Ann Appl Biol*, 130, 167-178

Donnelly, A., Jones, M.B., Sweeney, J. 2004. A review of indicators of climate change for use in Ireland. *International Journal of Biometeorology*, 49, 1-12

Easterling, W.E.; McKenney, M.S.; Rosenberg, N.J. & Lemon, K.M. (1992a). Simulations of crop response to climate change: effects with present technology and no adjustments (the 'dumb farmer' scenario). *Agric For Meteorol*, 59, 53-73

Easterling, W.E.; Rosenberg, N.J.; Lemon, K.M. & McKenney, M.S. (1992b). Simulations of crop responses to climate change: effects with present technology and currently available adjustment (the 'smart farmer' scenario). *Agric For Meteorol*, 59, 75-102

Fritts, H.C. (1976). *Tree Rings and Climate*. Academic Press, London

Hafner, S. (2003). Trends in maize, rice, and wheat yields for 188 nations over the past 40 years: a prevalence of linear growth, *Agriculture, Ecosystems & Environment*, 97, 275 - 283

Hay, R.K.M. & Porter J.R. (2006). *The physiology of crop yield. 2nd ed.*, Blackwell Publishing, UK

Hulme, M.; Wigley, T.M.L.; Barrow, E.M.; Raper, S.C.B.; Centella, A.; Smith, S.J. & Chipanshi, A.C. (2000). *Using a Climate Scenario Generator for Vulnerability and Adaptation Assessments: MAGICC and SCENGEN Version 2.4 Workbook*. Climatic Research Unit, Norwich UK

Hurrell, J.W. & van Loon H. (1997). Decadal variations in climate associated with the North Atlantic Oscillation. *Clim Change*, 36, 301–326

IPCC (2007). Climate Change 2007: The Physical Science Basis. In: *Contribution of Working Group I to the Fourth Assessment Report of the Intergovernmental Panel on Climate Change*. Solomon, S., D. Qin, M. Manning, Z. Chen, M. Marquis, K.B. Averyt, M. Tignor & H.L. Miller (Eds.). Cambridge University Press, Cambridge, UK and New York

Jaagus, J. (2006). Climatic changes in Estonia during the second half of the 20th century in relationship with changes in large-scale atmospheric circulation. *Theor. Appl. Climatol*, 83, 77–88

Jaagus, J. & Truu J. (2004). Climatic regionalisation of Estonia based on multivariate exploratory techniques. *Estonia. Geographical studies*, 9, 41–55

Kadaja, J. (1994). Agrometeorological resources for a concrete agricultural crop expressed in the yield units and their territorial distribution for potato. In: *Proceedings of the GIS – Baltic Sea States 93*, Vilu H., Vilu R. (Eds.), 139–149, Tallinn Technical University, Tallinn

Kadaja, J. (2004). Influence of fertilisation on potato growth functions. *Agronomy Research*, 2(1), 49-55

Kadaja, J. (2006). Reaction of potato yield to possible climate change in Estonia. *Book of proceedings. IX ESA Congress. Part I. Bibliotheca Fragmenta Agronomica 11*, Fotyma, M. & Kaminska B. (eds.), 297–298, Pulawy–Warszawa

Kadaja, J. & Tooming, H. (1998). Climate change scenarios and agricultural crop yields. In: *Country case study on climate change impacts and adaptation assessments in the Republic of Estonia*, Tarand, A. & Kallaste, T. (Eds.), 39–41, Ministry of the Environment Republic of Estonia, SEI, CEF, UNEP, Tallinn

Kadaja, J. & Tooming, H. (2004) Potato production model based on principle of maximum plant productivity, *Agric. For. Meteorology*, 127 (1–2), 17–33

Karing, P.; Kallis, A & Tooming, H. (1999). Adaptation principles of agriculture to climate change. *Clim Res*, 12, 175–183

Keeling, C.D.; Whorf, T.P.; Whalen, M. & van der Plicht, J. (1995). Interannual extremes in the rate of rise of atmospheric carbon dioxide since 1980. *Nature*, 375, 666–670

Keevallik, S. (1998). Climate change scenarios for Estonia. In: *Country case study on climate change impacts and adaptation assessments in the Republic of Estonia*, Tarand, A. & Kallaste, T. (Eds.), 1–6, Ministry of the Environment Republic of Estonia, SEI, CEF, UNEP, Tallinn

Kitse, E. (1978). *Mullavesi [Soil water]*. Tallinn, Valgus [in Estonian]

Kont, A.; Jaagus, J. & Aunap, R. (2003). Climate change scenarios and the effect of sea-level rise for Estonia. *Global and Planetary Change*, 36, 1 -15

Kooman, P.L. (1995).*Yielding ability of potato crops as influenced by temperature and daylength.* PhD thesis, Wageningen Agricultural University, Wageningen, The Netherlands

Makra, L.; Horváth, S.; Pongrácz R.& Mika, J. (2002). Long term climate deviations: an alternative approach and application on the Palmer drought severity index in Hungary, *Physics and Chemistry of the Earth*, 27, 1063–1071

McPherson, R. (2007). A review of vegetation—atmosphere interactions and their influences on mesoscale phenomena. *Progress in Physical Geography*, 31, 261-285

Mearns, L.O. (2000). Climate change and variability. In: *Climate Change and Global Crop Productivity*. Reddy, K.R. & Hodges, H.F. (Eds.), 7-35, CAB International Publishing

Mela, T. (1996). Northern agriculture: constraints and responses to global climate change. *Agricultural and Food Science in Finland*, 5 (3), 229–234

Menzel, A. (2003). Plant Phenological Anomalies in Germany and their Relation to Air Temperature and NAO. Climatic Change, *57 (3), 243–263*

Menzel, A. & Fabian, P. (1999). Growing season extended in Europe. *Nature*, 397, 659

Miglietta, F.; Bindi, M.; Vaccari, F.P.; Schapendonk, A.H.C.M.; Wolf, J.; Butterfield, R.E. (2000). Crop Ecosystem Responses to Climatic Change: Root and Tuberous Crops. In: *Climate Change and Global Crop productivity*. Reddy, K.R; Hodges, H.F (Eds.). CAB International Publishing

Mpelasoka, F.; Hennessy. K., Jones R. & Bates B. (2007). Comparison of suitable drought indices for climate change impacts assessment over Australia towards resource management. *International Journal of Climatology*, 28, 1283–1292

Nakićenović, N. & Swart, R. (Eds.). (2000). *Special Report on Emissions Scenarios*. Cambridge University Press, Cambridge, UK, 570 pp.

Peiris, D.R.; Crawford, J.W.; Grashoff, C.; Jefferies, R.A.; Porter, J.R. & Marshall, B. (1996). A simulation study of crop growth and development under climate change. *Agric For Meteorol*, 79, 271–287

Pensa, M.; Sepp, M. & Jalkanen, R. (2006). Connections between climatic variables and the growth and needle dynamics of Scots pine (Pinus sylvestris L.) in Estonia and Lapland. *International Journal of Biometeorology*, 50 (4), 205–214

Rind, D.; Goldberg, R. & Ruedy, R. (1989). Change in climate variability in the 21st century. *Clim Change*, 14, 5–37

Rosenzweig, C.; Phillips, J.; Goldberg, R.; Carroll, J. & Hodges, T. (1996). Potential impacts of climate change on citrus and potato production in the US. *Agric Syst*, 52, 455–479

Ross J. (1966). About the mathematical description of plant growth. *DAN SSSR* 171 (2b), 481–483 [In Russian]

Sage, R.F. (1994). Acclimation of photosynthesis to increasing atmospheric CO_2. The gas exchange perspective. *Photosynthesis Research*, 39, 351–368

Santer, B.D.; Wigley, T.M.L.; Schlesinger, M.E. & Mitchell, J.F.B. (1990). *Developing Climate Scenarios from Equilibrium GCM Results.* Max-Planck-Institut für Meteorologie Report No. 47, Hamburg, Germany

Saue, T. (2006). *Site-specific information and determination of parameters for a plant production process model.* Masters thesis. Eurouniversity, Tallinn.

Saue, T.; Kadaja, J. (2009). Simulated crop yield - an indicator of climate variability. *Boreal Environment Research*, 14(1), 132–142.

Saue, T.; Viil, P.; Kadaja, J. (2010). Do different tillage and fertilization methods influence weather risk on potato yield? *Agronomy Research*, xx-xx [in press]

Scheifinger, H.; Menzel, A.; Koch, E.; Peter, C.; Ahas, R. (2002). Atmospheric mechanisms governing the spatial and temporal variability of phenological phases in central Europe. *International Journal of Climatology*, 22, 1739–1755

Schwarz, M.D.; Reiter, B.E. (2000). Changes in North American spring. *Int. J. Biometeorol*, 20, 929–932

Semenov, M.A.; Porter, J.R. (1995). Climatic variability and the modelling of crop yields. *Journal of Agricultural and Forest Meteorology*, 73, 265–283

Semenov, M.A.; Wolf, J.; Evans, L.G.; Eckersten, H.; Iglesias, A. (1996). Comparison of wheat simulation models under climate change. II. Application of climate change scenarios. *Climate Research*, 7, 271–281

Sepp J.; Tooming, H. & Shvetsova, V.M. (1989). Comparative assessment of potato productivity in the Komi A.S.S.R. and in Baltic republics by the method of dynamic modeling. *Fiziologija Rastenii (Plant Physiology)* 36 (1), 68–75 [In Russian with English summary]

Sepp J. & Tooming, H. (1991). *Productivity resources of potato*, Gidrometeoizdat, Leningrad [in Russian with English abstract]

Sparks, T. & Tryjanowski, P. (2007). Patterns of spring arrival dates differ in two hirundines. *Climate Research*, 35, 159–164

Sparks, T.H., Jeffree, E.P. & Jeffree, C.E., (2000). An examination of the relationship between flowering times and temperature at the national scale using long-term phonological records from the UK. *Int. J. Biometeorol.* 44, 82–87

Szep I.J.; Mika J.; Dunkel Z. (2005). Palmer drought severity index as soil moisture indicator: Physical interpretation, statistical behaviour and relation to global climate. *Physics and Chemistry of the Earth*, 30 (1-3 SPEC. ISS.), 231–243

Tooming, H. (1967). Mathematical model of plant photosynthesis considering adaptation. *Photosynthetica*, 1 (3 - 4), 233–240

Tooming, H. (1970). Mathematical description of net photosynthesis and adaptation processes in the photosynthetic apparatus of plant communities. In: *Prediction and Measurement of Photosynthetic Productivity*. Setlik I. (Ed.), 103–114, Pudoc, Wageningen

Tooming, H. (1977). *Solar radiation and yield formation*. Gidrometeoizdat, Leningrad [In Russian with English abstract]

Tooming, H. (1984). *Ecological principles of maximum crop productivity*. Gidrometeoizdat, Leningrad [in Russian with English summary].

Tooming, H. (1988). Principle of maximum plant productivity. In: *Lectures in Theoretical Biology*. Kull K., Tiivel T. (Eds.),. 129–138, Valgus, Tallinn

Tooming, H. (1993). Evaluation of agrometeorological resources based on the potential productivity of crops. *Journ. Agric. Met.* (Jap.) 48 (5), 501–507

Tooming, H. (1998). Climate change and estimation of ecologically founded yields. In: *Climate change studies in Estonia*, Kallaste, T., Kuldna, P. (Eds). 141–152, Stockholm Environment Institute Tallinn Centre - Ministry of environment, Republic of Estonia, Tallinn

Tooming, H. & Kadaja, J. (1999). Climate changes indicated by trends in snow cover duration and surface albedo in Estonia. *Meteorol Zeitschrift*, 8, 16–21

Tooming, H. & Kadaja, J. (2006). Relationships of snow cover in Estonian climate - relations from winter to spring. In: *Handbook of Estonian snow cover.* Tooming, H. & Kadaja, J. (Compilers), Kallis A. (Ed.),. 112–133. Estonian Meteorological and Hydrological Institute - Estonian Research Institute of Agriculture, Tallinn - Saku

Wolf, J. (1999a). Modelling climate change impacts at the site scale on potato. In: *Climate Change, Climate variability and Agriculture in Europe: an integrated assessment.* Report No. 21. Downing, T.E.; Harrison, P.A.; Butterfield, R.E.; Lonsdale, K.G. (Eds.). 135–154, Environmental Change Unit, University of Oxford, Oxford, UK

Wolf, J. (1999b). Modelling climate change impacts on potato in central England. In: *Climate Change, Climate variability and Agriculture in Europe: an integrated assessment.* Report No. 21. Downing, T.E.; Harrison, P.A.; Butterfield, R.E.; Lonsdale, K.G. (Eds.), 239–261, Environmental Change Unit, University of Oxford, Oxford, UK

Wolf J. (2002) Comparison of two potato simulation models under climate change. II. Application of climate change scenarios. *Clim Res,* 21, 187–198

Wolf, J. & van Oijen, M. (2002). Modelling the dependence of European potato yields on changes in climate and CO_2. *Agricultural and Forest Meteorology,* 112 (3-4), 217-231

Woodward, F.I. (1988). Temperature and the distribution of plant species and vegetation. In: *Plants and Temperature,* Long S.P and Woodward F.I. (Eds), 59 -75, Society of Experimental Biology by The Company of Biologists Limited, Cambridge

Zhukovsky, E.E., Sepp J. & Tooming, H. (1989). On the possibility of the yield calculation forecasting calculation. *Vestn. S.-H. Nauki (Messenger of agricultural sciences)* (5), 68-79 [In Russian with English abstract]

Zhukovsky, E.E., Sepp J. & Tooming, H. (1990). Probabilistic forecasts of possible yield. *Meteorologija i Gidrologiya (Meteorology and Hydrology)* (1), 95-102 [In Russian with English abstract].

20

Determining the relationship between climate variations and wine quality: the WEBSOM approach

Subana Shanmuganathan and Philip Sallis
Auckland University of Technology
New Zealand

1. Introduction

Climate change has the potential to impact on all forms of agriculture and vegetation and the impact is predicted to be inconsistent across the globe. Thus the polarising debates on climate change, the phenomenon that has become to be famously known as 'global warming' or 'global climate change', has increased scientific and commercial interest immensely in this topic and predictions relating to it. The potential influence of climate variation on viticulture and enology is considered to be *dramatic* because grapevine varieties are among the most sensitive cultivated crops that thrive only under niche climate and environmental conditions. Historical viticulture records evidence the fact that the winemaker ability to produce premium quality wine is highly influenced by the environment and climate apart from the grapevine variety itself as described by a famous Mediterranean concept "*Terrior x cultiva*". Major variations in climate can even force shifts in whole wine regions. Minor or seasonal variations generally result in differences in quality among vintages. The historical data on the factors relating to climate and wine quality is proving invaluable in comparing contemporary data with the past and it is essential for any forecasting or prediction over time (Hansen, et al., 2001: Jones, 2005).
Literature reviewed for this research reveals that in the past viticulturists and winemakers introduced and developed subtle changes to cultivation practices and winemaking processes to turn the annual (or vintage-to-vintage) climate change effects, favourable for winemaking. This has been occurring over centuries to turn climate variation effects to grapevine growing advantage in an endeavour to produce finer wine labels. On the other hand, major shifts in cultivation methods in whole wine-producing regions have occurred in the past and it appears that all these happened in order to produce grapes with a higher percentage of sugar but without comprising the other aroma and colour pro-protein compounds in the berry ripening process. These characteristics of wine quality are considered to be the principal factors relating to climate for viticulture regions throughout the world. Irrigation, frost, wind and solar influences are also major factors in grape production and therefore, primary determinants of wine quality (Jones, 2005).
In view of the above facts, approaches based on modern knowledge discovery methods are now being increasingly investigated to improving our understanding on climate and environmental influences on wine quality. Sections 2 and 3 illustrate on some of the literature and our related

research in modelling wine quality using sensory data in both text and numeric formats. Section 3 details on the other novel methods being experimented at the Geoinformatics Research Centre (GRC), Auckland University of Technology (AUT) in New Zealand, to modelling the blended quantitative and qualitative grapevine phenology data that determines the sugar and protein components in grapes and its influence on the final wine product. Initial results of this research are outlined in this section. Consequently, the WEBSOM approach to mining unstructured data in other major problem domains is outlined. Section 4 illustrates the WEBSOM approach being investigated in the Centre to analysing sommelier comments in free text format with sample data set, extracted from a web magazine (Wine enthusiast, 2009), with an ultimate aim of modelling the climate change effects on grapevine phenology and wine quality. The final section proposes future research to model the effects of climate change in greater detail with larger data sets from more grape growing regions within New Zealand and Chile to study the climate change effects on the world's major wine regions in the southern hemisphere. From this analysis we expect to be able to predict the wine quality, style and appellations suitable for future climate change, short-/long-term, with data from climate models already developed.

2. The effects of climate change

Climate change is predicted to bring about significant modifications to all forms of agriculture and vegetation on earth at varying degrees (Atkins, et al., 2006). Its potential impact on Viticulture, the world's old and most expensive cultivated crop and *enology* based on the science that underpins it, is considered by many observers to be *dramatic* (Jones, et al., 2005). The recent model predictions relating to future climate change suggest that the effects to be inconsistent across the globe (severe in the northern hemisphere and mild in the southern) and also, to have a variable effect on different grapevine varieties. Grapevine phonology, such as crop budburst, *floraison, veraison,* and harvest, greatly depends on local weather conditions in different regions, and this is a major factor in determining wine vintage quality. For example, even over a single degree centigrade change in temperature is predicted to make the production of the world's famous Mediterranean wine appellations impossible. Grape varieties thrive under significantly narrow or niche climate and environmental conditions, and historical evidence as well suggests that the production of premium quality wine labels as highly prone to any change in current climate, annual as well as long term. Research findings with Australian grapevines and wines (Web, 2006) suggest that a change of grapevine varieties could be a way to overcome the effects of future climate change in that country's major wine regions. This would of course, be an extremely expensive exercise (Gutierrez, 2005), which is why objective scientific analysis for scenario building and prediction is of great significance at the moment.

The research discussed in this paper relates to the overarching research project, *Eno-Humanas* (www.geo-informatics.org) that is aimed at building models based on correlations between dependent and independent sets of diffident combinatorial data collected on the environment, climate and atmosphere, soil, terrain, moisture and plant responses in association with sensory perception data relating to flavour, odour and fruit robustness (Sallis, et al., 2008 : Shanmuganathan, et al., 2008). Hence, the title *Eno-Humanas,* is about the combination of precise ecological data and the less precise qualitative opinion data that comes from wine consumers. This paper relates to the imprecise data analysis aspect in that it analyses the descriptions of wine quality coming from experts; Master Wine Sommeliers.

3. Modelling grape wine using text based comments

This section of the chapter outlines some major methodologies from the literature on this topic and then discusses GRC research in wine quality and sensory data analysis. Finally, the WEBSOM approach and its applications are discussed.

3.1 Wine sensory analysis

The wine quality literature relating to sensory and chemical data analysis can be broadly categorised into the following:

1) Wine characterisation and discrimination using chemical and sensory properties: Most of the papers reviewed fall into this category. Wine of all main appellations, for example, champagne, chardonnay, and pinot noir, have been analysed with chemical sensory analyses, in search of better ways to identify the differences between the wines for use in classifying sub-appellations within the main ones. For example, in (Parr, et al., 2007) a distinctive New Zealand wine style Marlborough Sauvignon Blanc was classified by sensory characterisation and chemical analysis for selected aroma compounds for that wine type. In (Kontkanen, 2005) the differences that might be supportive for designating three sub-appellations of red Niagara Peninsula Bordeaux style were investigated using chemical and sensory analysis on forty-one commercially available wines. Similarly, (Vannier, et al., 1999) and (Gawel, 2001) looked at strategies a) to control champagne wine quality based on sensory and b) red table wine quality characterised by pleasing and complex mouth-feel sensations respectively.
2) Professional versus novice taster abilities. There are many studies in this area cited in the literature and another project within *Eno-Humanas* is considering this from an audio-mining perspective to elicit the degree of emphasis (passion) expressed about wine quality in recording of wine tasting by both professionals and novices.
3) Wine ratings, favourable climatic conditions and price fluctuations. Research on this subject described by Jones, et al., (2005) looked at climate and global wine quality factors and discussed a year-to-year comparison over a ten year period. It includes a description of wine quality factors in juxtaposition with prices and vintage ratings. Citing many earlier studies the authors of this work pointed out that the analysis of the relationships between climate variables and wine prices to be based on an underlying hypothesis that beneficial climate conditions would improve the wine quality and that in the past, these had in turn led to short-term price hikes. They also reflected that the unavailability of consistent price data for multiple regions and with different styles over many years was a shortcoming for any analysis on the study of long-term effects. They also pointed out that the vintage ratings to be a strong determinant of the annual economic success of a wine region based on analysis by (Nemani, 2001) but then went on to say that the ratings could be determinants of wine quality not necessarily a predictor based on (Ashenfelter, 2000) where ratings were described to be reflective qualitatively of the wine quality; they had same weather factors documented to be the determinants of the same wine quality.
4) Analysis of wine taster descriptions in free-text. There are not many studies of this kind cited in the literature. Of the studies revealed under this category, two major approaches are outlined herein. In (Brochet, 2001) authors analysed taster comments

by different wine experts using a software product called ALCESTE to study the structure of the language used by the experts in describing wines they tasted. In this study, the analysts grouped the word categories within different expert corpuses by calculating the χ^2 of co-occurrences of words and classified the categories into different classes, such as idealistic, odour, colour, somesthesic, taste and hedonistic. The study concluded that the language structure used by wine experts as not organised along their sensory dimensions instead with prototypes. When describing wine taste, experts tend to relate it to a prototype rather than stating its properties.

In another interesting piece of work (Be´cue-Bertaut, et al., 2008), the researchers calculated synthetic liking scores by studying the correlations between pairs of original scores and word groups/counts in free text comments and then comparing these synthetic scores with the original ratings for the sample set of wines studied. The authors used multiple factor analysis to establish the correlations between each pair of comments and their respective liking scores.

Research conducted at University of California Davis (Frøst, 2002) found that only 25% of wine liking ratings was linked to wine sensory descriptive data in a map created with statistical analysis results of the latter on y axis and ratings on x axis. The authors as well found some descriptors, such as "leather" and "sour", as having a negative effect on preference and wine tasters liked some descriptors, such as "vanilla/oak" and "canned vegetable", and noted even though 75% of the variations in liking could not be explained, the results should be read with caution.

3.2 GRC research: modelling New Zealand wine sensory properties

The section summarises previous GRC research conducted in text mining wine comments for studying the vintage variability in wine quality with sommelier comments that could be eventually extended for modelling the year-to-year climate change effects on wine quality.

3.2.1 New Zealand wine regions and wine styles

New Zealand's (NZ) wine industry continues to grow rapidly in total grapevine cultivation area as well as fine wine production for both domestic and export markets. With extremely diverse climate and environmental conditions combined with incredible enology skills, NZ wineries are able to produce finer quality wine with extraordinary flavours in a wide range of appellations exceeding global market standards. Even though NZ wine industry has a chequered history, in recent times the industry has been witnessing a rapid growth and this has increased the interest in scientifically understanding the link between the country's climatic/ environmental conditions (site specific attributes) and berry component formation, the overall impact on the ultimate end product, wine and its quality. The major NZ wine varieties, wine regions (fig. 2), varieties cultivated and ecological niche (climate, environmental, soil and topographic factors represented by vegetation) are presented in figs. 1a-d.

1a: Chardonnay 1b: Sauvignon Blanc 1c: Pinot Noir

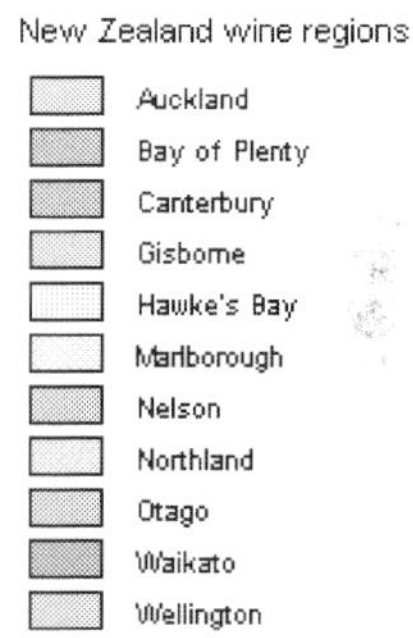

Fig 1a-c: Maps showing major New Zealand wine regions and three major styles from the 95 NZ wine (comments) being analysed in this research. As seen in the maps Chardonnay crops are not cultivated in southern and similarly, Pinot Noir and Sauvignon Blanc are rare in northern New Zealand.

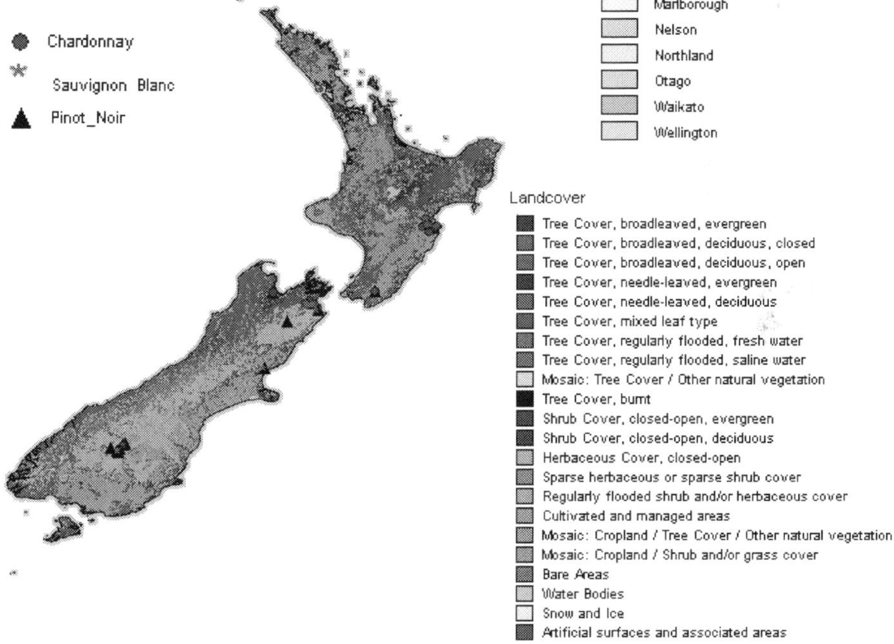

Fig 1d. Map of 3 major wine styles and NZ regional vegetation

NZ wine region	Appellations
Northland	Cabernet Sauvignon, Merlot and Chardonnay
Auckland	Cabernet Sauvignon
Waikato	Chardonnay, Riesling and Cabernet Sauvignon
Gisborne	Muller Thurgau, Chardonnay and Gewurztraminer
Hawkes Bay	Sauvignon Blanc, Chardonnay, Cabernet Sauvignon and merlot
Wellington	Shardonnays, Rieslings and Pinot Noir
Nelson	Rieslings and Chardonnay
Marlborough	Sauvignon Blanc, Chardonnay, Pinot Noirand Riesling, Pinot Gris, Gewurztraminer, Merlot and Cabernet Sauvignon
Waipara	Pinot Noir, Chardonnay Riesling and Sauvignon Blanc
Canterbury	Pinot Noir, Chardonnay, Riesling and Pinot Gris
Otago	Pinot Noir, Riesling and Chardonnay

Fig. 2. New Zealand wine regions and appellations

3.2.2 text mining wine comments with SOM[1] techniques

Every wine label consists of vintage rating, wine details and comments provided by sommeliers that describe the wine colour, aroma, mouth feel and after taste in text format or in some case in audio clips. Both audio and text data (structured and unstructured) is also made available via web based wine catalogues and wine comments of 95 New Zealand vintages from a web magazine called *Wine Enthusiast* (Wine enthusiast, 2009) are analysed to model the vintage variability in the wine quality of these NZ wines. The following are the information generally found in wine vintage labels/ descriptions:

1. Name of the winery
2. Wine style:
3. Wine region
4. Vintage:
5. Rating

Initially, the 95 New Zealand wine comments obtained from *Wine Enthusiast* were separated into structured and text data. The text data was then pre-processed to remove stop words (such as a, the, is) and a matrix of 95 wines into 117 words (*lemmas*) was created. Of the 117 words very common and rare words were removed. Finally, using formula (1) weights (*wi*) for the selected words (wine descriptors) were calculated (fig. 3) The formula is from a well-known information retrieval system called Slaton's vector space model, which has been successfully applied to information storage and retrieval, such as a) local information from individual documents and b) global information from the collection of documents.

---

[1] Self-organising maps (SOMs) are single layered artificial neural networks (ANNs) that use an unsupervised training algorithm developed by Tuevo Kohonen based on the functioning of the cortex cells of the human brain (www.cis.hut.fi/research/som-research/teuvo.html). SOMs are very useful in exploratory data analysis where prior knowledge on the problem domain is limited. SOMs can project multi-D data onto low D displays with most of the topological details of the complex datasets preserved thus the SOM maps enhance analyst ability to visualise new knowledge embedded in the original data in the form of patterns and correlations.

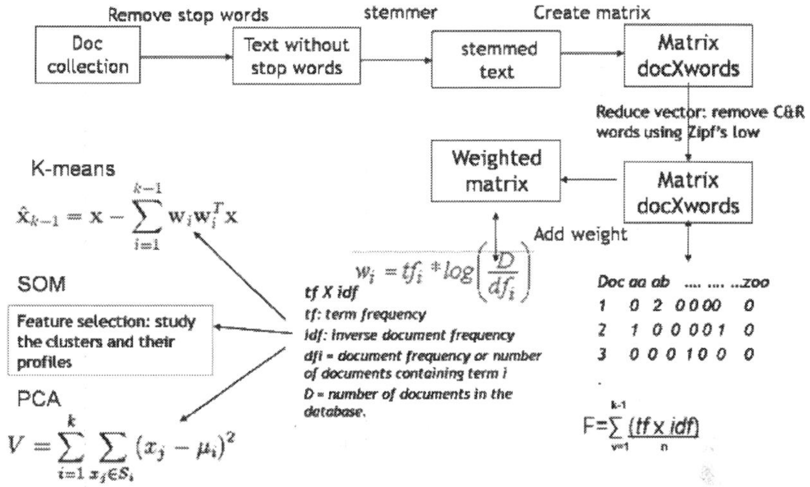

Fig. 3. Schematic diagram showing the steps followed in text mining 95 NZ wine comments.

$$w_i = tf_i * log\left(\frac{D}{df_i}\right) \tag{1}$$

Where,
tfi = term frequency (counts) or number of times a term *i* occurs in a document.
dfi = document frequency or number of documents containing term *i*
D = number of documents in the collection/database

A SOM (figs. 4 a and b) was created using 85 wine descriptors to see the vintage groupings within the 95 New Zealand wines analysed and are outlined here onwards.
The nine SOM clusters of 95 New Zealand wine descriptors show the wine groupings based on the descriptors used by sommeliers to describe the wines. The cluster profiles show the distribution of 85 wine descriptors and their relationships with wine style, region, rating and vintage. For example, cluster 7 wines (Figs. 4 a, b and 5a) are described using *ag_2* (age), *black_9* (black) *Cherri_12* (cherry), *chocol_13* (chocolate), *cinnamon_14* (cinnamon), *cola_17* (cola), *dri_21* (dried), *plum_63* (plum), *spice_74* (spice) and *structure_75* (structure). This cluster consists of Pinot Noir and Bordeaux Blend (BB) wines from Hawke's Bay, Waipara, Marlborough, Martinborough and Central Otago (Fig 5 b). Of these wines, Pinot Noir from Martinborough 2005 rated 93 (wine 8) fetched $ 60 per bottle and was described as "Since its debut in 2003 this has been one of New Zealand-s top Pinot Noirs combining power structure and complexity. Smoky and richly peppery at first it turns more floral with aeration and while it's big in the mouth it is also silky in texture. The black cherry plum vanilla and spice flavors fan out on the long layered finish. Drink now-2015". This shows that the correlations between wine ratings (descriptors) and regional aspects, such as climate and environmental of "*terroir*" as the French say, could be established however, currently there are not any conventional methods or approaches for this purpose.

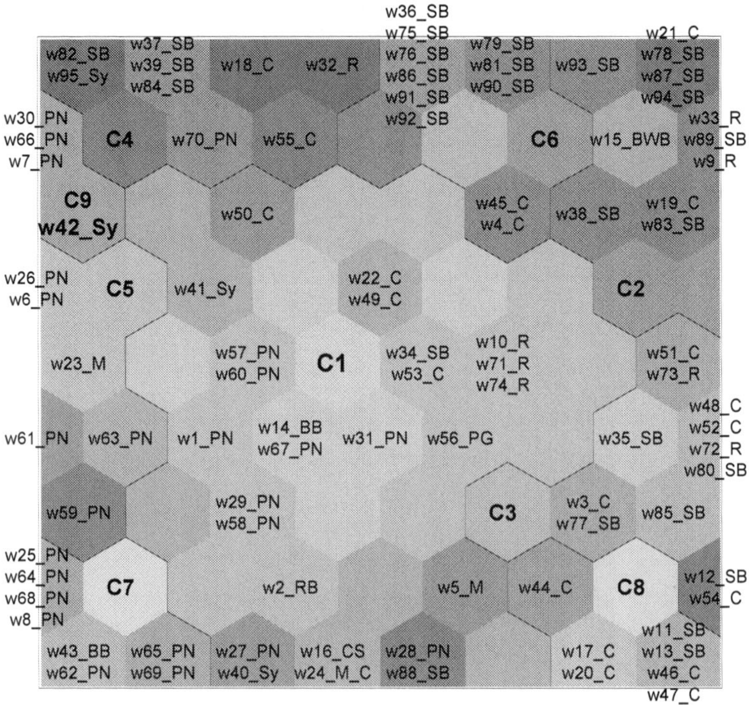

Fig. 4. a: SOM of 85 wine descriptors with 9 clusters. The descriptors obtained from sommelier comments provided for 95 NZ wines illustrate the quality of the vintage in terms of colour, aroma and flavour that are dependent on the grapes that is in turn dependent upon the weather that ripened them apart from winemaker talent and experience.

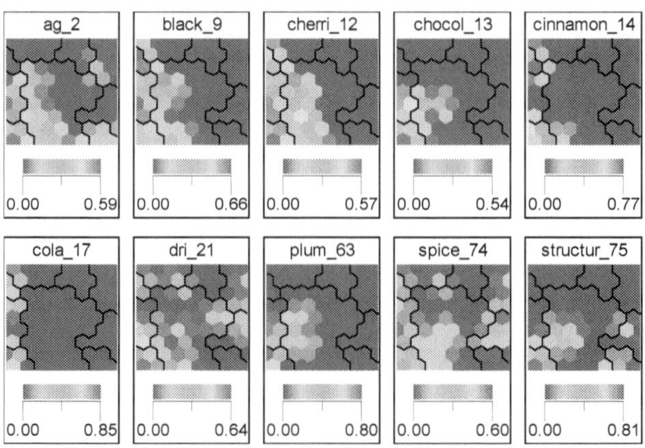

Fig 4. b: SOM components of selected wine descriptors that are related to cluster 7 Pinot Noir vintages

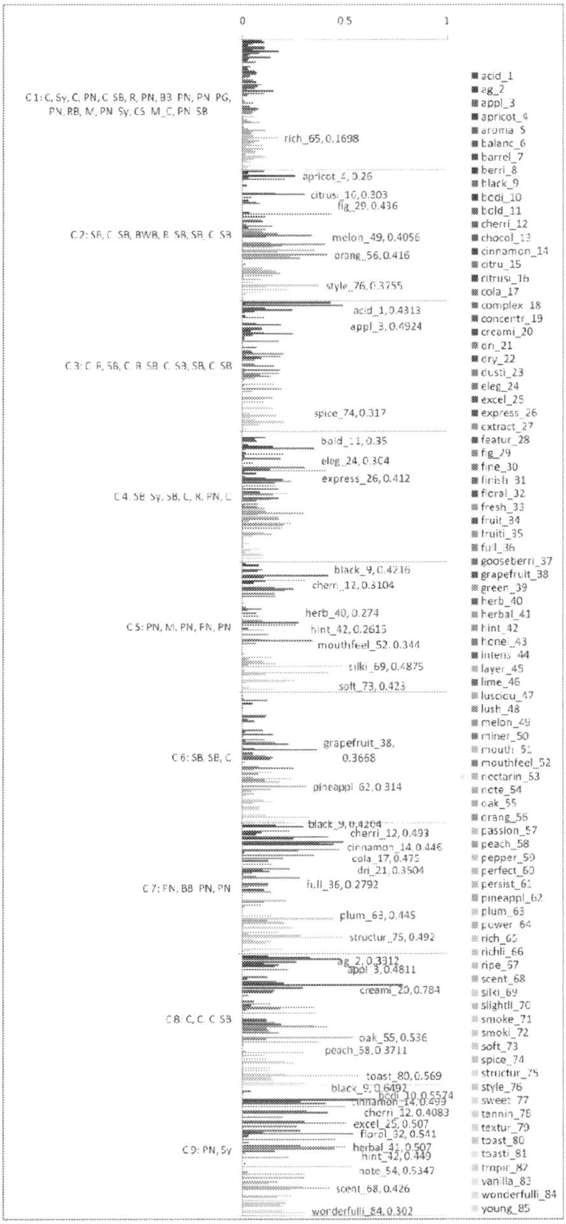

Fig. 5. a. Graph showing 85 wine descriptor distribution within the 9 SOM clusters. The SOM was created with 85 descriptors extracted from wine comments made by sommeliers for 95 New Zealand wines.

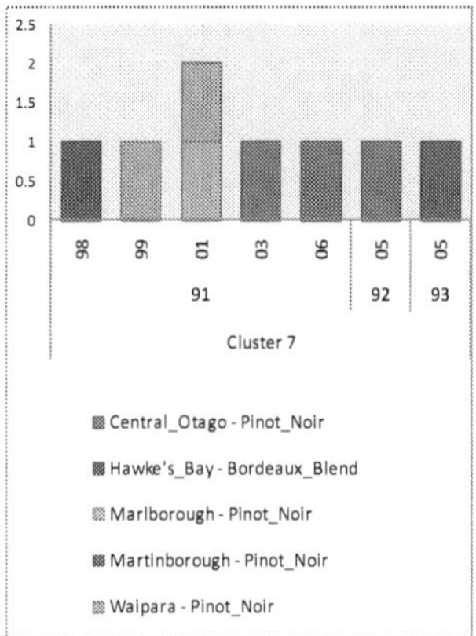

Fig. 5. b. Graph showing cluster 7 vintage, rating (x axis), count of wines (y axis), wine style and regional details.

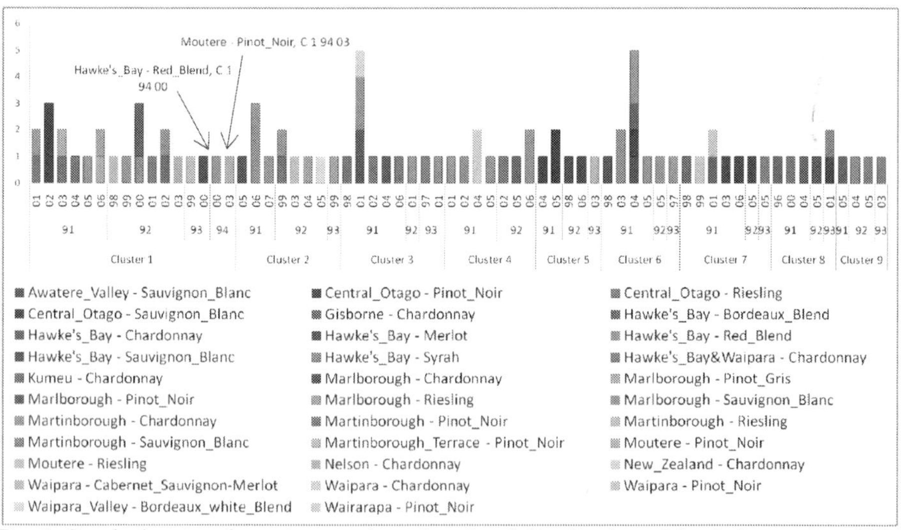

Fig. 6. Graph showing the ratings, style and regional distribution of the 9 clusters in a SOM created with 85 taster descriptors of the 95 New Zealand wines. X axis consists of (from top to bottom) year, rating and cluster no. y axis consists of the count of wines.

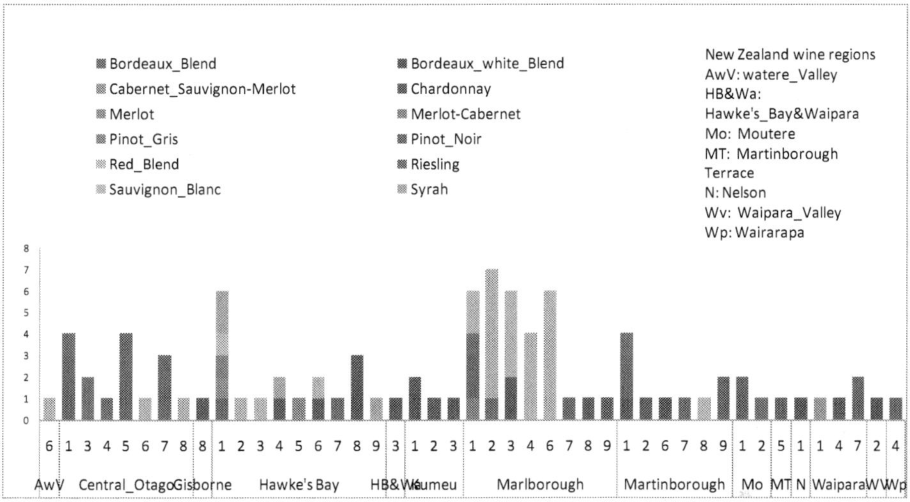

Fig 7. Graph showing the SOM cluster number and wine style distribution (x axis) within the regions relating to the 95 New Zealand wines. The SOM was created using 85 wine descriptor weights calculated based on formula 1 and procedures in Fig 1. Marlborough region consists of more Sauvignon Blanc (SB) wines. All of SB come from the central and southern regions of NZ, mostly from the central Marlborough region.

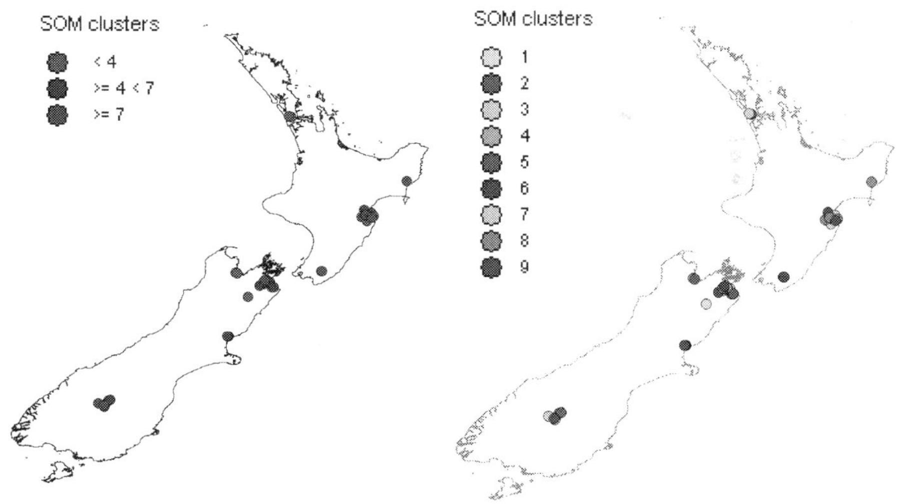

Fig. 8. Winery locations projected on NZ map based on a (left) : three and b (right) : nine clustering of the SOM created with 85 wine descriptors extracted from 95 NZ wine comments. The wine descriptor clustering interestingly reflects more of the aspects relating to winery's location rather than the style.

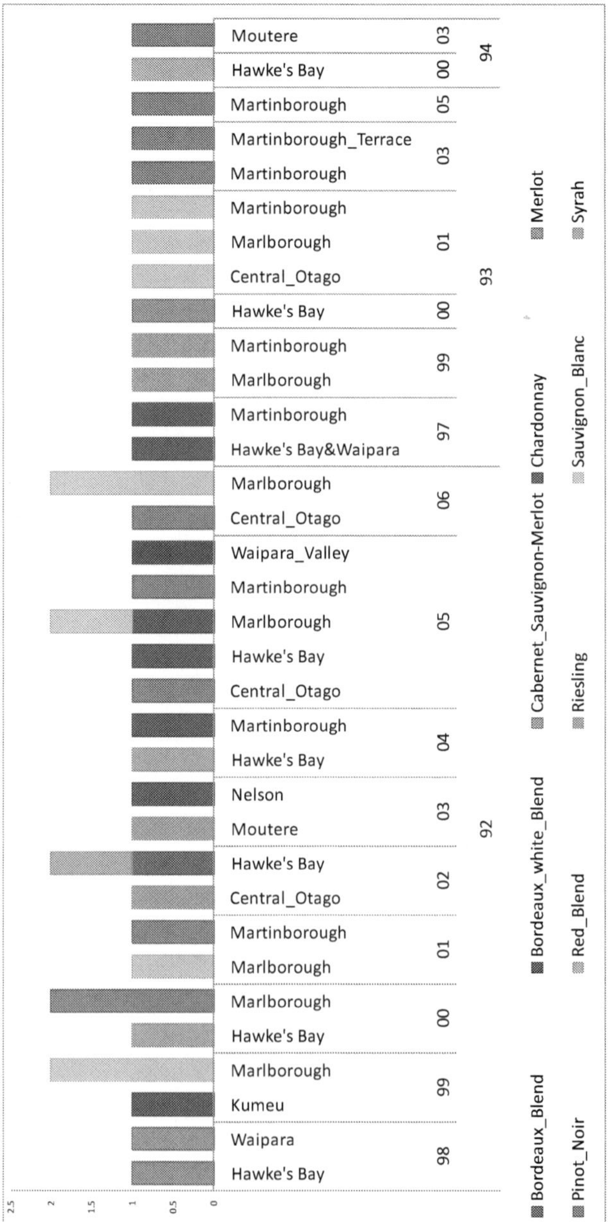

Fig. 9. Graph showing winery region, vintage, rate 92-94 (x axis) and wine style. Pinot Noirs from Hawke's Bay (2000) and Moutere (2003) were rated 94, both clustered in SOM cluster 1 (fig. 6).

Even though the SOM clustering of the 95 New Zealand wines based on the use of wine descriptors show the correlations between wine ratings, vintage and wine type, the clustering is more influenced by the regional aspects as seen in fig. 8.

3.4 WEBSOM approach and its applications

WEBSOM is an approach that provides an efficient methodology for full-text information retrieval and exploration of large collections of documents generally from websites. It uses SOM techniques to statistically analyse the relations between collections of words based on their co-occurrence in documents being analysed, and then based on the relationships, creates document maps. As the word co-occurrences are used as basic components for clustering, similar documents get clustered close to each other on the document map. The WEBSOM approach was first developed as a map of documents providing a good basis for search and exploration of documents (Lagus, 1996). WEBSOM applications include unsupervised (Kaski, 1996), partially supervised processing of newsgroups (Honkela, 1996), browsing interface for web pages for the exploration of document collections (Lagus, 1996) and as a method/ tool for data mining in textual databases (Lagus, 1996).
In this research, WEBSOM is used to group wine descriptors that co-occur in sommelier comments to study the year-to-year variations that may exist between the descriptor-groups that best describe different features relating to vintage quality/ wine sub appellations in terms of appearance, aroma, flavour and mouth-feel.

4. The WEBSOM approach to analysing wine quality data

A WEBSOM was created to analyse the variability in the 95 NZ wine vintages used in the earlier GRC research and the results are discussed here onwards.

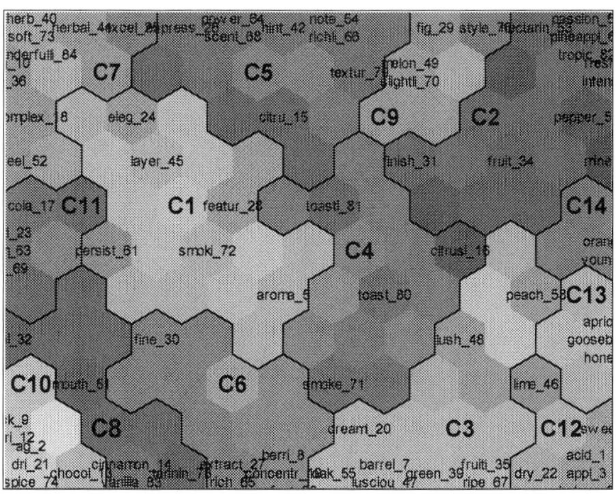

Fig. 10. WEBSOM of 85 wine descriptors (same as used for SOM clustering of wines) extracted from sommelier comments of 95 New Zealand wines.

Chardonnay 1998-2005

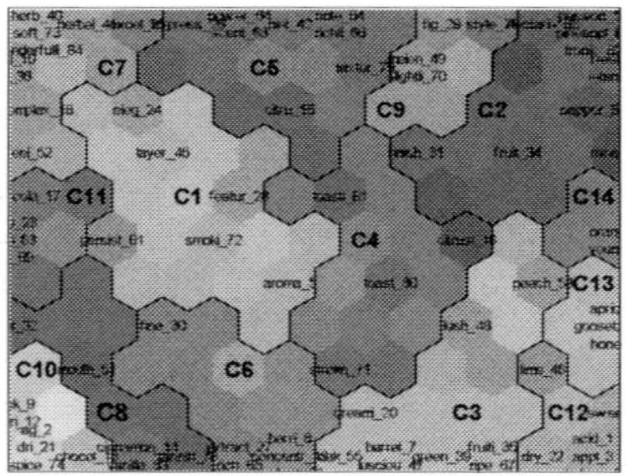

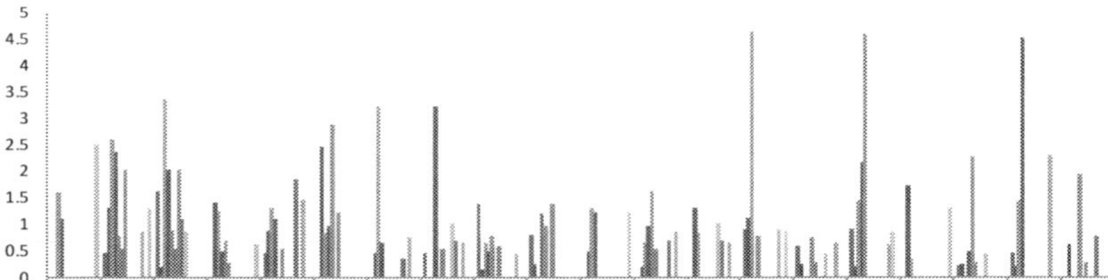

	w44	w46	w47	w45	w17	w18	w3	w48	w49	w50	w19	w52	w51	w20	w53	w21	w4	w22	w54	w55
	2000	1999		2005	2003		2001	2001	2002	2005	2001	2005	unknown	1998	2004	2005	1999	2003	1998	unknown
	91	91			92		93	91			92	91		92	91	92	93	92	91	91
	Gisborne	Hawke's_Bay					Hawke's_Bay&Waipara	Kumeu				Marlborough			Martinborough			Nelson	New_Zealand	Waipara
Sum of C1 aroma_5 eleg_24 featur_28 layer_45 persist_61 smoki_72	0	0.45	1.58	0	0.45	0	0	0.45	1.36	0.79	0	0	0	0.9	0.57	0.9	0	0.22	0.45	0
Sum of C2 finish_31 fresh_33 fruit_34 grapefruit_38 intens_44 miner_50 nectarin_53 passion_57 pepper_59 pineappl_62 tropic_82	0	1.27	0.19	1.39	0.86	2.44	0.45	0	0.14	0.24	0.48	0.19	1.27	1.09	0.24	0.19	1.7	0.24	0.19	0.59
Sum of C3 barrel_7 creami_20 fruiti_35 green_39 lusciou_47 lush_48 oak_55 peach_58 ripe_67	1.57	2.57	3.34	1.24	1.28	0.8	3.2	0	0.64	0	1.28	0.64	0.8	4.62	0	1.41	0.32	0	1.41	0
Sum of C4 citrusi_16 smoke_71 toast_80 toasti_81	1.07	2.35	2.01	0.47	1.07	0.94	0.64	3.21	0.47	1.18	1.2	0.94	0	0	0	2.14	0	0.47	4.49	0
Sum of C5 citru_15 express_26 hint_42 note_54 power_64 richli_66 scent_68 textur_79	0	0.75	0.86	0.68	0	2.86	0	0	0.75	0.94	0	1.58	0	0.75	0.73	4.55	0	2.23	0	1.9
Sum of C6 berri_8 concentr_19 extract_27 fine_30 perfect_60 rich_65	0	0.53	0.53	0.27	0.53	0	0	0.53	0	0	0	0.53	0	0	0.27	0	0	0.27	0	0
Sum of C7 bodi_10 bold_11 complex_18 excel_25 full_36 herb_40 herbal_41 mouthfeel_52 soft_73 wonderfulli_84	0	2.01	2.01	0	0	1.2	0	0	0.57	1.37	0	0	0	0	0	0	0	0	0	0.27
Sum of C8 cinnamon_14 floral_32 mouth_51 tannin_78 vanilla_83	0	0	1.07	0	0	0	0	0	0	0	0	0	0	0	0	0	0	0	0	0
Sum of C9 fig_29 melon_49 slightli_70 style_76	0	0	0.83	0	0	0	0	0.98	0	0	0	0	0.98	0	0.42	0	0	0.42	0	0
Sum of C10 ag_2 black_9 cherri_12 chocol_13 dri_21 spice_74 structur_75	0	0	0	0	1.82	0	0.34	0.68	0	0	0	0.68	0.68	0	0	0	0	0	0	0.76
Sum of C11 cola_17 dusti_23 plum_63 silki_69	0	0	0	0	0	0	0	0	0	0	0	0	0	0.9	0	0	0	0	0	0
Sum of C12 acid_1 appl_3 balanc_6 dry_22 lime_46 sweet_77	0	0.83	0	0	1.43	0	0.74	0.64	0.42	0	0	0.83	0.64	0	0.62	0.6	0	0	2.26	0
Sum of C13 apricot_4 gooseberri_37 honei_43	0	0	0	0	0	0	0	0	0	0	0	0	0	0.83	0	0.83	0	0	0	0
Sum of C14 orang_56 young_85	2.48	1.28	0	0.6	0	0	0	0	0	0	1.2	0	0	0	0	0	1.28	0	0	0

Fig. 11. WEBSOM cluster profile of 20 Chardonnays within the 95 NZ wines. Higher frequencies of C2 & C3 show high rating (92/3).

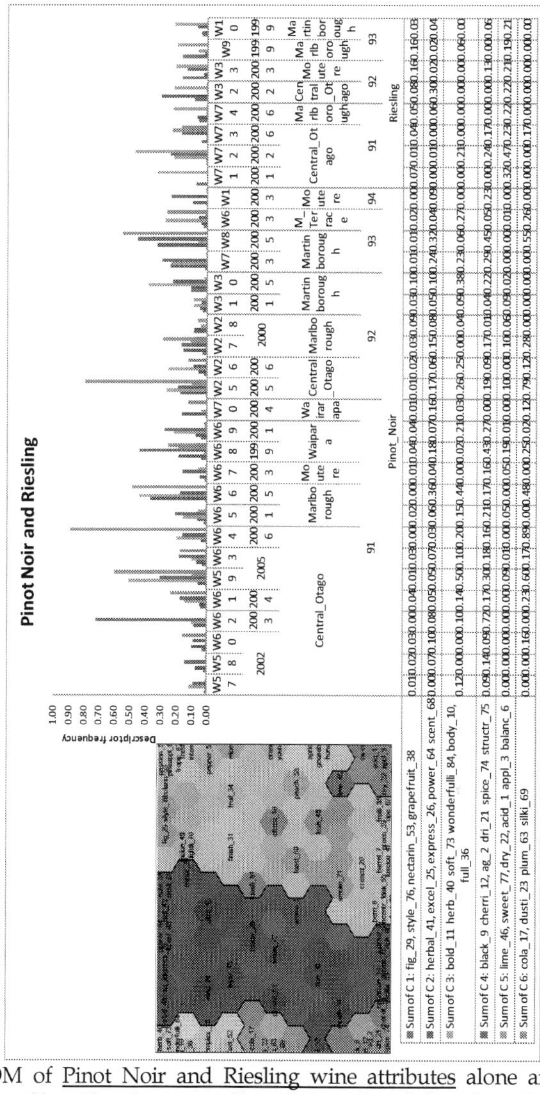

Fig. 12. a: WEBSOM of Pinot Noir and Riesling wine attributes alone and b: graph of the WEBSOM cluster profiles. The cluster profiles of the selected "attributes" clearly indicate the descriptor frequencies that can be used for the characterisation of the vintages within and among the wine appellations. Pinots have more of C3 and C6 descriptors meanwhile Riesling wines have more of C5. The frequencies as well relate to ratings. W1 Pinot Noir (2003) rated 94 from Moutere consists of C2 and C4 descriptors alone. Pinots from other regions even though have higher frequencies for C2 and C4 descriptors they also have more of C3 and C6 descriptors. Hence, W1 is rated high for its sharpness or C3 and C6 could be imposing a negative impact on the overall rating.

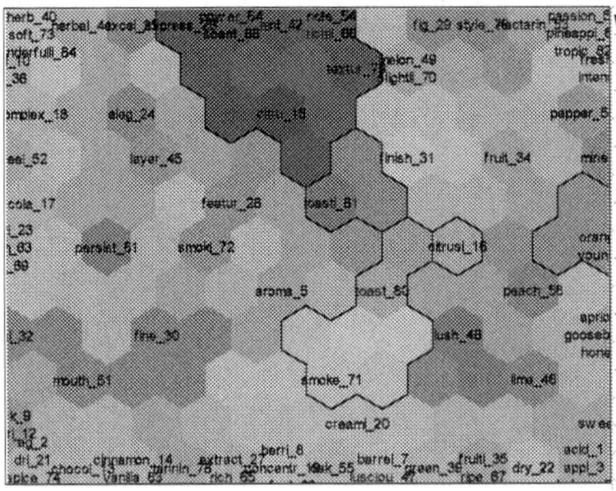

WineNc	Region	C 1: bold_	C 2: expre	C 3: toast_	C 4: orang	C 5: toasti	C 6: citrusi
44	Gisborne	0.023	0	0.468	1.225	0.002	0.006
17	Hawke's_Bay	0.087	0	0.465	0	0.003	0.004
18	Hawke's_Bay	0.064	0.333	0.004	0	0.932	0.001
45	Hawke's_Bay	0.0428	0.0828	0.0014	0.2972	0.4651	0.0022
46	Hawke's_Bay	0.108	0.074	1.172	0.633	0.004	0.009
47	Hawke's_Bay	0.134	0.116	0.471	0.001	0.935	0.009
3	Hawke's_Bay&	0.0584	0	0.3523	0.0018	0.0004	0.0019
19	Kumeu	0.026	0	0.002	0.594	0	1.183
48	Kumeu	0.046	0	0.468	0	0.933	1.184
49	Kumeu	0.0444	0.0891	0.002	0	0.4664	0.0003
50	Kumeu	0.0344	0.1216	0.583	0	0.0027	0.0034
20	Marlborough	0.12	0.074	0.008	0.005	0	0.005
51	Marlborough	0.0619	0	0.0037	0	0	0.0005
52	Marlborough	0.0404	0.195	0.0007	0.0015	0.9301	0.0035
4	Martinboroug	0.029	0	0	0.634	0	0.002
21	Martinboroug	0.057	0.569	0.003	0.003	0.932	1.184
53	Martinboroug	0.0294	0.0888	0	0	0	0.0006
22	Nelson	0.0167	0.2832	0.0009	0	0.4663	0.0007
54	New_Zealand	0.064	0	1.169	0.001	0.934	1.19
55	Waipara	0.0251	0.2222	0	0	0.0008	0.0004

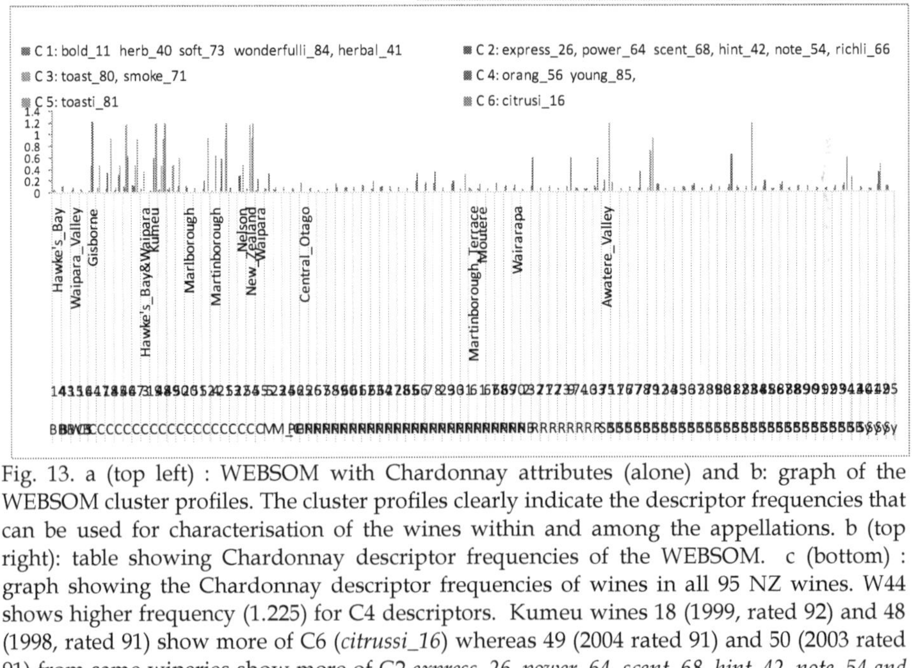

Fig. 13. a (top left) : WEBSOM with Chardonnay attributes (alone) and b: graph of the WEBSOM cluster profiles. The cluster profiles clearly indicate the descriptor frequencies that can be used for characterisation of the wines within and among the appellations. b (top right): table showing Chardonnay descriptor frequencies of the WEBSOM. c (bottom) : graph showing the Chardonnay descriptor frequencies of wines in all 95 NZ wines. W44 shows higher frequency (1.225) for C4 descriptors. Kumeu wines 18 (1999, rated 92) and 48 (1998, rated 91) show more of C6 (*citrussi_16*) whereas 49 (2004 rated 91) and 50 (2003 rated 91) from same wineries show more of C2 *express_26, power_64, scent_68, hint_42, note_54 and richli_66.*

The WEBSOM of all 85 words (fig. 9) give interesting patterns in the use of the descriptors by sommeliers:

1) Descriptors of clusters C8 (*cinnamon_14 floral_32 mouth_51 tannin_78 vanilla_83*) and C11 (*cola_17 dusti_23 plum_63 silki_69*) are not generally used to describe Chardonnay wines, they are used for Pinot noirs instead

2) Similarly, descriptors of C13 (*apricot_4 gooseberri_37 honei_43*) and C14 (*orang_56 young_85*) are not used for Pinots at all however sparingly used for Chardonnays.
3) Higher frequencies of C5 (*citru_15 express_26 hint_42 note_54 power_64 richli_66 scent_68 textur_79*) high ratings (92/3).
4) C 10 descriptors (*ag_2 black_9 cherri_12 chocol_13 dri_21 spice_74 structur_75*) are commonly used for Pinots
5) C4 descriptors (*citrusi_16 smoke_71 toast_80 toasti_81*) are not used at all to describe Pinots.

Based on the results of SOM technique and then the WEBSOM clustering of wines, both approaches investigated with the same set of 85 descriptors extracted from sommelier comments of 95 New Zealand wines, the latter seems be a better tool for modelling the year-to-year (vintage) variability within the wine descriptors. Even though with both methods it is possible to subdivide the main clusters and look for patterns for establishing the correlations between the descriptor frequency and different wine styles, vintages and ratings the WEBSOM approach gives more flexibility to choose wine attributes and drill down the analysis to within and among regions, appellations or vintages.

5. Conclusions

The paper looked at the SOM and then WEBSOM based clustering approaches to modelling the year-to-year (vintage) variability within 95 New Zealand wines using 85 taster descriptors extracted from web based sommelier comments provided for the NZ wines. The results of this initial research favour WEBSOM approach for establishing the vintage variability within wines from New Zealand's major regions as well as within a region with all or selected attributes (relating to a single wine style , such as Chardonnay or Pinot Noir). Further research is already underway to investigate the variability among vintages from New Zealand and Chilean wine regions with WEBSOM approach using larger sample sets of sommelier comments. This will enhance a broader understanding on the climate effects on different wine styles and vintages from both countries' wine regions and from a comparative perspective as well on the effects in two major wine producing regions in the southern hemisphere.

6. Acknowledgements

The authors wish to thank Professors Ajit Narayanan and Jeffery Hunter for advice on multi dimensional data analysis. Winemaker Michael Brajkovich is thanked for his expertise and time spent with the authors. GRC research partners are acknowledged for their continued support. Peter Sumich is thanked for his efforts in organising discussions and funding for the main *Eno-Humanas* as well as its sub projects.

7. References

Ashenfelter, O., & Jones, G.V., (2000). The demand for expert opinion: BordeauxWine. VDQSAnnual Meeting, d'Ajaccio, Corsica, France. October, 1998. Published in Cahiers Scientifique from the Observatoire des Conjonctures Vinicoles Europeenes, Faculte des Sciences

Atkins, T. A., & Morgan, E. R., (2006). Modelling the effects of possible climate change scenarios on the phenology of New Zealand fruit crops ISHS Acta Horticulturae 276: II International Symposium on Computer Modelling in Fruit Research and Orchard Management ???

Be´cue-Bertaut , M., A´ lvarez-Esteban, R., and Page` S, (2008). Rating of products through scores and free-text assertions: Comparing and combining both. Food Quality and Preference vol. 19 pp. 122–134. 133

Brochet F., and Dubourdieu, D., (2001). Wine Descriptive Language Supports Cognitive Specificity of Chemical Senses. Brain and Language Vol. 77, Issue 2, May 2001: 187-196 .

Frøst M. B., and Noble, A., (2002). Preliminary study of the effect of knowledge and sensory expertise on liking for red wines. American Journal of Enology and Viticulture. vol. 53(4) :275-284

Gawel, R. I. P., Characterizing the astringency of red wine: a case study (2001) Food Quality and Preference vol 12: 83-94.

Gutierrez, A. P., Luigi, P., Ellis, C. K., and d'Oultremont, T., (2005). Analysis of climate effects on agricultural systems. Report published by California Climate Change Center CEC-500-2005-188-SD pp 28 + appendices A1-7

Hansen, A., and Dale V., (2001). Biodiversity in US Forests under Global Climate Change. Ecosystems vol. 4 161–163

Honkela, T., Kaski, S., Lagus, K., and Kohonen, T., (1996). Newsgroup exploration with WEBSOM method and browsing interface. Technical Report A32, Helsinki University of Technology, Laboratory of Computer and Information Science, Espoo. WEBSOM home page. http://websom.hut.fi/websom/

Jones , G. V., [et al.] (2005). Climate and Global Wine Quality. Climatic Change by Springer vol. 73:319–343.

Jones, G. V., "How Hot Is Too Hot?" (2005). Wine Business Monthly, February 2005. pp 1-4.

Jones, G.V., Making Wine in a Changing Climate (2004) Geotimes, August 2004, Vol. 50, No. 7, 22-27

Kaski S., Honkela, T., Lagus, K. and Kohonen, T. (1996). Creating an order in digital libraries with self-organizing maps. [Conference] // Proceedings of World Congress on Neural Networks (WCNN-96) http://websom.hut.fi/websom/

Kontkanen D R Canadian terroir: sensory characterization of Bordeaux-style red wine varieties in the Niagara Peninsula (2005). Food Research International vol. 38: 417–425.

Lagus, K., Honkela, T., Kaski, S. and Kohonen, T. (1996). Self-organizing maps of document collections: A new approach to interactive exploration [Conference] // Knowledge Discovery and Data Mining (KDD-96) https://eprints.kfupm.edu.sa/64289/

Lagus, K., Honkela, T., Kaski, S., Kohonen, T., (1996). Creating an order in digital libraries with self-organizing maps [Conference] // proceedings of STeP'96. Jarmo Alander, Timo Honkela and Matti Jakobsson (eds.) , Publications of the Finnish Artificial Intelligence Society: 73-78.. -

Lagus, K., Kaski, S., Honkela, T. and Kohonen, T. (1996). Browsing digital libraries with the aid of self-organizing maps. [Conference] // In Hopgood, B., editor, in proceedings of Fifth International World Wide Web Conference, Paris, volume posters, 71-79.. -

Nemani, R. R., White, M. A., Cayan, D. R., Jones, G. V., Running, S.W., and Coughlan, J. C. (2001). Asymmetric climatic warming improves California vintages. - Clim. Res. Vol 19:25–34.

Parr, W. V., Green, J. A., White, K. G., and Sherlock, R. R., (2007). The distinctive flavour of New Zealand Sauvignon blanc: Sensory characterisation by wine professionals. Food Quality and Pref vol. 18:849–861.

Sallis, P.J., Shanmuganathan, S., Pavesi, L., and Jarur, M. (2008). A system architecture for collaborative environmental modelling research [eds] McQuay Waleed W. Samari and William. - A publication of the IEEE, New Jersey, USA. ISBN: 978-1-4244-2248-7, Irvine, California, May 19-23 2008 pp 39-47 :

Shanmuganathan, S., Ghobakhlou, A., and Sallis, P., Sensor data acquisition for climate change modelling (2008). WSEAS Circuits & Systems, Issue 11, Volume 7, Nov. 2008. ISSN: 1109-2734 pp 942-952.

Vannier, A., Bruna, O. X., and Feinberg, M. H., Application of sensory analysis to champagne wine (1999). Characterisation and discrimination. Food Quality and Preference vol. 10:101-107.

Web, L. B., (2006). The impact of projected greenhouse gas-induced climate change on the Australian wine industry, PhD thesis, School of Agriculture and Food Systems, University of Melbourne pp 277.

Wine enthusiast (2009). Wine enthusiast magazine. www.winemag.com/buyingguide/search.asp?db= [Online]

21

Ecological modernisation and the politics of (un)sustainability in the Finnish climate policy debate

Tuula Teräväinen
University of Helsinki
Finland

1. Introduction

The recent policy developments at the international level (IPCC, 2007; EC, 2008; OECD, 2009) have strengthened the politicisation of climate change and created an unparalleled international unanimity about the need for urgent action to tackle global warming. At the same time, the market and innovation-led ideas of the knowledge-based economy have penetrated into national and supranational policies, becoming normative principles for all societal development. They have highlighted the role of science and technology in the global politico-economic order and raised high technologies into the core of modern knowledge production and diffusion (Häyrinen-Alestalo & Kallerud, 2004; Godin, 2006). The recognition that the recent global economic and environmental crises derive from the same origins and cannot therefore be solved separately has resulted in the increasing extension of the ideas of the knowledge-based economy into the sphere of climate and energy policies. Supranational organisations have launched new political concepts, such as 'green growth', 'new sustainable social market economy' and 'Green New Deal', referring to a global transformation towards a 'knowledge-based, greener, competitive and more inclusive economy' that would be 'based on knowledge and new environmental technologies' (EC, 2009; OECD, 2009; UNEP, 2009).

Although concern for the environment has in this way become a new ideological 'master frame' (Eder, 1996), little has been done to change the prevailing politico-economic structures towards a more sustainable social order. A broadly adopted environmental policy approach in many industrialised countries and the EU has been ecological modernisation, whose key tenets include the aim of transcending the traditional division between economic growth and environmental protection through the development and diffusion of technological innovations (Mol & Sonnenfeld, 2000; Jänicke, 2008). Suggesting a win-win solution between these conflicting goals has fitted well to the framework of the knowledge-based economy and created new trust among national governments and supranational organisations in the ability of technology to solve societal problems. Yet the capability of ecological modernisation to solve societal problems has recently been challenged by new economic and technological uncertainties, the rebound effects of technological solutions, increased competition for environmental resources, and the dynamics between the global

North and South (e.g. Baker, 2007; Olsen, 2007; Jänicke, 2008; Jänicke & Lindemann, 2010). This has also been interpreted as an indication of a paradigm change in eco-politics towards post-ecologism and its unsustainable politics (Blühdorn & Welsh, 2007).

Despite supranational climate initiatives, specified emission reduction targets are assigned to individual countries, and meeting climate policy objectives importantly depends on the formulation and implementation of national level policies. Climate policy thus appears as a compelling field where the high-technology and market-oriented tenets of the knowledge-based economy confront global responsibility and regulation on the one hand and national concerns, such as self-sufficiency, the energy supply security, and undisturbed energy distribution, on the other. This article scrutinises the political side of technology by analysing the recent climate policy debate at the national level through the example of Finland. It focuses on the use of eco-modernist strategy in legitimating a technology-led climate policy and scrutinises the related tensions that emerge between the national and international levels. Finland provides a good example in this respect. It has pursued long-term growth-oriented economic and industrial policies along with strong investments in R&D (Cabinet Programmes, 1987–2007). At the same time, the post-war period's centralised steering of national economy has gradually been replaced by the deregulation of financial markets and the opening of national borders for international competition. In comparison to many countries and the EU on average, Finland's R&D investments have increased rapidly, particularly in the late-1990s, and have thereafter continuously remained at a relatively high level (Statistics Finland, 2010). A special characteristic of Finland has also been its strong emphasis on technology and innovation in creating societal well-being and the adoption of growth-oriented technology policy as a guiding principle for other national policies (Pelkonen et al., 2008). These transformations have been characterised, for example, through a shift from a natural resource-intensive to a knowledge-intensive economy (Schienstock, 2007) and from a welfare state to a neoliberal/competition state (Alestalo, 1997; Heiskala & Luhtakallio, 2006; Pelkonen, 2008). Recently, Finland has also been placed among the leading economies in the world in several competitiveness rankings (EC, 2007; WEF, 2009; WEF, 2010; cf., however, IMD, 2010), which has further encouraged the knowledge-intensive and market-led approach among national decision-makers (See also Häyrinen-Alestalo et al., 2005). Yet energy policy has been strongly directed by negotiations between the state and the economically important heavy industry, both not only as key actors in the national innovation system but also as major shareholders in large energy companies.

The following questions will be asked:

1. How is technology-led climate policy legitimised in the national policy debate?
2. What kinds of tensions emerge between national interests and international climate policy targets in this framework?
3. To what degree does Finnish climate policy reflect eco-modernist visions of green growth and what kinds of implications does this have in terms of the global dimensions of climate policy?

The data consist of the government of Finland's official documents on climate, energy, environmental, technology, and innovation policies; strategies, reports and documents of key ministries; parliamentary proceedings (plenary sessions and committee statements) (2005–2009); and stakeholder statements, documents, reports, and press releases (2008–2010). In addition, key strategies, initiatives, and policy documents of relevant supranational organisations and think-tanks (the EU, IPCC, OECD, UN, WEF, and NEF) are utilised in the analysis. The analysis was based on a data-oriented approach. Utilising qualitative inductive content analysis (e.g. Hsieh & Shannon, 2005), the documentary data were first coded at the level of individual statements and sentences. The data were then elaborated by organising the initial codings into broader thematic categories, and complementing and expanding on them with the parliamentary and the stakeholder data. The findings were then specified by re-reading the data and scrutinising the ways different arguments were used in specific contexts. Finally, the analysis was deepened by identifying and analysing the main rhetorical strategies and related tensions (Cf. Palonen & Summa, 1996) through which various claims were justified in the policy debate.

2. Ecological modernisation and the demand for sustainability

At the core of national and international climate policy debates has been the question about an optimal social order that would foster environmentally and economically sound development. During the past few decades, many countries have adopted an eco-modernist environmental policy strategy, which seeks to overcome the division between economic and environmental objectives by seeing them as complementary (Hajer, 1995; Christoff, 1996; Mol & Sonnenfeld, 2000). While the idea of sustainable development has been to harmonise economic growth and environmental protection along the principles of intra- and intergenerational justice (WCED, 1987), a central assumption in ecological modernisation is that technology can contribute to both ecological and environmental progress and that it is possible to combine environmental protection with economic growth through the development and diffusion of cleaner technologies (Jänicke, 2008). Other key elements of ecological modernisation include integrating environmental policy to other policy fields, developing new policy instruments, such as voluntary agreements and eco-audit, and emphasising the role of economic actors in ecological transformations (Mol & Sonnenfeld, 2000; Baker, 2007). At the national level, it can be characterised as a government-led action programme that aims at cleaner economic growth through eco-innovation and energy efficiency (Murphy, 2000).

As an environmental policy strategy, ecological modernisation can be characterised as reformist rather than radical, as it aims to redefine the structures of contemporary society to some extent but does not suggest any radical changes to the prevailing status quo (Dryzek, 2005). Reflecting the tradition of modernist eco-political thought, it also suggests *a* human rationality as the way towards achieving *a* global politico-economic equilibrium (Blühdorn, 2000). Ecological modernisation thus tends to attach itself more or less to the key premises of liberal capitalist society, where the increasingly market-driven policy orientation (Larner, 2000; Jessop, 2002; Brenner & Theodore, 2002) has strongly highlighted economic and technological aspects in combating climate change. In many industrialised countries and the EU, investing in new technologies has been recently emphasised as the most innovative and cost-efficient way to contribute to emission abatement through producing low-carbon energy solutions and energy efficiency (Cabinet Programme of Finland, 2007; EC, 2008).

The attractiveness of ecological modernisation has undoubtedly related to its compatibility with the key tenets of the knowledge-based economy. As an umbrella concept for all kinds of developments and policies related to science, technology, and innovation, the knowledge-based economy has combined fashionable ideas from new growth theories, the National Innovation System, the New Economy, and the information society, producing a network of concepts that have reinforced each other (Godin, 2006, 23). The OECD has defined the concept as referring to 'trends in advanced economies towards greater dependence on knowledge, information and high skill levels, and the increasing need for ready access to all of these by the business and public sectors' (OECD, 2005). In a similar vein, ecological modernisation emphasises an innovation-oriented approach and highlights the critical importance of the production, dissemination, and utilisation of knowledge and innovation. Yet this market and high technology-led political strategy has so far not been able to balance various dimensions of sustainability, thus leaving the controversy between the environment and the economy ultimately unsolved. According to Blühdorn and Welsh (2007), it has become evident that the key parameters determining the ways in which environmental issues are perceived have changed. The established ecologist values, diagnoses, and strategies have become outdated, resulting in a paradigm change in the realm of eco-politics. The new phase of post-ecologism is signalled, for example, by neo-materialism commanding late modern societies, the de-ideologisation of politics, the loss of identity of eco-politics as a specific policy field, and the decline of ecologism's political actors. In this framework, key societal and environmental concepts are redefined as economic and efficiency issues in a way that they simulate continuity in relation to modernist values but are at the same time compatible with the perpetuation of the capitalist growth economy (Blühdorn, 2002). Yet manifestations of post-ecologism tend to vary across different politico-economic and cultural contexts. The following sections scrutinise the ways in which they are reflected in the Finnish policy debate.

3. Reorientations in Finnish climate and energy policies

Finnish energy policy has traditionally been based on strong state-led governance and regulation, where state-owned energy companies have dominated the production and distribution of energy. Characteristic of the management of the energy system has also been corporatist negotiations among the state and large energy producers, distributors, and buyers. Since the 1980s, however, the energy sector has experienced transformations through deregulation and market liberalisation. Early reforms in this respect included transforming the distributive price control of petrol and diesel oil, liberalising the refinery prices of a large oil company, *Neste*, demolishing the integrated price system of oil products at the national level, and removing the import licensing system of raw oil (Ruostetsaari, 2005). In 1995, the Electricity Market Act meant a change from a centralised state-led steering towards a market-based governance of the energy economy. At the same time, the national energy markets were opened for international competition and Finland became a member of the EU, which transformed national policies towards a more international orientation.

Peculiar for Finland has also been certain continuity in energy policy. Since the 1970s, national energy policy has been largely built upon the concern over the relatively high import dependency, which has raised versatility, self-sufficiency, and security of energy supply among key political questions. A related feature directing energy policy has been the relatively high share of fossil fuels in the total energy supply. In 2005, over half of Finland's total primary energy supply came from fossil fuels, out of which the share of oil was 30%, natural gas 10%, and coal and peat together 14% (IEA, 2008a). Nuclear power (17%) and biomass (20%) have also formed an important part of the national energy mix, which has gained attention in international comparisons on renewable energy and national fuel mix (IEA, 2008b). Finland's relative share of renewable energy has been among the highest in the EU and the OECD countries. However, while the production of renewable energy has mostly consisted of wood-based biomass generated from by-products and wood residuals of the forest industry, the shares of other renewable energy sources like hydropower (3%) and geothermal, solar, and wind power (together less than 0.3%) (IEA, 2008a) have remained almost insignificant. Finland's ranking has thus been far from the leading European countries like Germany and Spain when measured, for instance, through installed wind and solar power capacity (EurObserv'ER, 2010). In fact, Finland, together with some Eastern European countries like Hungary, Lithuania, Latvia, and Romania, has been placed at the bottom in these comparisons.

The importance of forestry and wood-based fuels in domestic energy production and consumption can be largely explained by the lack of oil, gas, and coal reserves as well as the national economic production structure's reliance on energy-intensive heavy industry (in particular pulp, paper, and metal industries). This has directed national energy policy along the principles of securing the operational conditions of economically important national industry and providing inexpensive electricity for industrial use (Cabinet Programmes, 1987–2007). An important feature in the Finnish energy system has also been the centralisation of energy production, as a small number of energy companies, such as *Fortum* and *Neste Oyj*, account for a large share of Finland's energy production. As the state holds a majority ownership in these energy companies, its interest as a shareholder in maximising share values has, together with the objective of supporting national heavy industry, biased energy policy towards favouring nuclear power and coal over new renewable energies (Ruostetsaari, 2009). Nuclear power has also been an attractive energy production option in Finland because the power plants are largely owned by the state and heavy industry. Two current nuclear power plants are operated by *Fortum*, a public-listed energy company of which the Finnish government owns 51%, and two by *Teollisuuden Voima Oyj*, a public-private partnership company that was founded in 1969 to supply electricity to its shareholders (mostly heavy industry) at cost. While many other European countries committed themselves to phase out nuclear power in the 1990s and 2000s, the Finnish Parliament approved *Teollisuuden Voima*'s application to build a fifth nuclear power plant in 2002. In 2008 and 2009, three new nuclear power construction applications were filed at the Ministry of Employment and the Economy, and the Parliament accepted two application licences (Teollisuuden Voima and Fennovoima) in July 2010.

Guided by the key energy policy principles of self-sufficiency, diversified production, and the security of supply as well as broader growth-oriented economic development policies, Finnish climate and energy policies have been gradually directed towards a technology and innovation-driven approach accompanied by substantial financial allocations to developing new energy technologies. Foundations for this approach were built in the 1960s and 1970s, when the construction of nuclear power plants, together with the international oil crisis, accelerated the development of technology policy and directed R&D activities towards nuclear energy, energy saving, and energy efficiency research (Murto et al., 2006). Yet these efforts were long seen as part of general industrial and technology policies primarily designed to enhance Finland's competitiveness and economic growth (KTM-81 Committee, 1981; Committee of Technology Policy, 1985). In the 1990s, the development of the national innovation system 'to support the economy, entrepreneurial activities and employment'

(Academy of Finland, 1997) became a key national policy target that was fostered by the government's decision to aggressively increase the GDP share of R&D investment (Cabinet Programme, 1995; Ministry of Education, 1996). The growth of public R&D funding was largely covered by privatising state-owned companies, and most of the additional funding was directed towards technology programs of the Finnish Funding Agency for Technology and Innovation (Tekes), research projects strengthening the technology base in Finland, and R&D activities of the selected industrial 'clusters', one of which was energy and environment (Academy of Finland, 1997). The growing importance of energy technologies was also indicated by several Tekes-funded technology programmes, ministries' renewable energy programmes, and the Academy of Finland's research programmes thereafter. A recent development in this respect has been the establishment of six Strategic Centres for Science, Technology and Innovation (SHOKs) in selected technological fields, one of which is energy and the environment. Co-funded by the state and private companies, they are expected to renew industrial clusters and create radical innovations through public-private partnerships and industry-academia collaboration.

The recent national energy and climate strategies (MTI, 2001; MTI, 2005; MEE, 2008) further develop these ideas. They connect the development of energy technologies more closely to national and international climate policy objectives and highlight the need to move from the import-dependent energy system and fossil energy sources towards the exploitation of low-carbon energy. The strategies are based on 'the government's target-oriented economy policy that supports growth in employment' (MTI, 2001, 40), where strong R&D investments in selected technological fields are expected to increase the productivity and competitiveness of the national economy (Cabinet Programme, 2007; MEE, 2008). Following the ideas of ecological modernisation and recent EU policy developments, Finnish strategies build upon the idea that it is possible to combine the objectives of environmental sustainability, energy supply security, and economic competitiveness through fostering technological advancement (e.g., MEE, 2008, 30). They also imply rather high expectations about high technology-led development. Technological innovations and respective financing are seen to be focal tools not only in attaining Finland's climate policy objectives but also in fostering international technological cooperation, improving Finnish companies' innovation capacity, and promoting export activities. Government funding is increasingly allocated to encourage innovations, especially in areas that develop novel energy technologies or have a high technological risk. The aim is to double the investment in the research and development, implementation, and commercialisation of new technologies by 2020 (MEE, 2008, 54). This technology-driven climate policy orientation reflects the broader political project of the growth-oriented knowledge-based economy (Jessop, 2002; Godin, 2006). At the same time, it serves as an illustrative example of a political field where ideas of technology policy are confronting and partly overlapping with traditional corporatist regulation, resulting in struggles over power between different spheres and levels of decision-making.

4. Legitimising high technology-led climate policy

In the Finnish policy debate, high technology-led policy orientation is justified by three rather powerful forms of persuasive speech, namely the rhetoric of responsibility, the rhetoric of possibility, and the rhetoric of necessity. Through these discursive devices, high technology-led and market-oriented climate policy is represented as having a broad societal acceptance and political legitimacy. It is, however, also in these points where the weaknesses of the eco-modernist approach in particular and the knowledge-based economy in general become visible, creating tensions between the national and global levels and pointing to an underlying tendency of trying to 'sustain something that is known to be unsustainable' (Blühdorn, 2000).

The rhetoric of responsibility

In terms of responsibility, many documents suggest redirecting national climate policy towards greater responsiveness to the international politico-economic framework. The commitments to international, particularly EU, climate policy targets together with uncertainties related to the current politico-economic situation are seen to bind Finland to carry its part of the global responsibility and to take action in preventing climate change. The policy documents place Finland as part of the global energy system, bearing responsibility for climate change not only as a single country but also as a member of a 'global community' (MEE, 2008; Prime Minister's Office, 2009). They refer to national climate policy as a broader project that extends to a supranational scale through market-based mechanisms, such as the emissions trade and technology transfer. Tackling global warming is seen to require 'immediate, comprehensive and unprecedentedly strong international cooperation' (Prime Minister's Office, 2009, 29), where the primary role of industrialised countries is to create climate mitigation tools for developing countries, that is, developing new energy technologies and high-technology components that can be utilised in low-carbon energy production.

Referring to common but differentiated responsibilities based on the available technologies and financial resources in different countries, however, this techno-optimist ecological modernisation argument not only assumes technological development as inherently natural for industrialised countries like Finland but also naturalises uneven economic and technological development on a global scale. The assumed technological superiority of industrialised countries also implicitly suggests their supremacy in making judgments on the needed forms of action, thus clearly departing from key principles of sustainability, such as global equity and distributive justice. This becomes visible also in the documents on the Clean Development Mechanism, a flexible mechanism under the Kyoto Protocol, that tend to focus on technological, judicial, and economic aspects of technology transfer with few references to different dimensions of sustainable development (See Teräväinen, 2009). This interpretation of responsibility thus seems to reflect more Finland's aspirations of new market opportunities and emission credits under the Kyoto Protocol than objectives of balanced sustainable development. Moreover, it retains a national orientation while narrowing Finland's global responsibility to the technological sphere and largely bypassing the world's changing power relations, particularly the strengthening role of developing countries in global techno-economic development. The policy documents' 'global community' thus seems to become rather a politically persuasive rhetorical construction reflecting a world order dominated by industrialised countries than a real collective effort or a sense of global responsibility.

Aside from the global level, the rhetoric of responsibility has a strong grounding in national interests. In terms of Finland's responsibility to tighten its own emission reduction targets, the policy documents selectively use some elements of a weak, techno-corporatist variant of ecological modernisation (Christoff, 1996) to foster rather traditional national interests. In particular, they prioritise the interests of the economically important national heavy industry over international (technology-led) climate policy objectives and point to the responsibility of the national political system to ensure the competitiveness of national industry (MTI, 2005; MEE, 2008). Characteristic of the policy documents has also been the technological selectivity illustrated by the government's highly selective resource allocation policies for certain technological fields and open scepticism towards those renewable energy technologies that would cause additional costs to national industry (Cabinet Programme, 2007; Prime Minister's Office, 2008). In the negotiations concerning the EU climate policy package (EC, 2008), for instance, Finnish representatives actively opposed policy proposals that suggested substantial increases to the share of renewable energy by claiming that Finnish industry is already a forerunner in this respect. While *in principle* emphasising the idea of know-how, knowledge, and innovation as the cornerstones of climate policy, this stance clearly reflected national heavy industry's interests, prioritising voluntary energy efficiency agreements and market-based measures like technology export over additional renewable energy investments and binding climate policy targets (cf. EK, 2009). Given the persistence of corporatism and the traditionally strong state in Finland (Alestalo, 1997; Arter, 1999), these arguments highlight the prevailing importance of the old state-industry negotiations. Despite the recent tendency to outsource industrial production to developing countries and the reduction of domestic manufacturing capacity, a considerable part (81% in 2007) of Finnish renewable energy production consists of by-products of the forest, pulp, and paper industries (Motiva, 2009). Active state intervention and technological development are thus seen to be needed to support the development of the forest industry, as Finland's capability of meeting its renewable energy obligations is claimed to importantly depend on the forest industry's production capacity (MEE, 2008; Prime Minister's Office, 2009). These rather protectionist tones indicate a tension that emerges between international climate commitments and the functioning of global markets on the one hand and national industrial interests on the other.

The rhetoric of possibility

The rhetoric of possibility refers to technology-led climate policy as a self-evident continuum for innovation and knowledge-oriented economic growth policies. Pointing to Finland's long-term R&D investments within the energy sector, political documents highlight the need to utilise the allegedly reliable and high level Finnish research and technical know-how in developing renewable energies, energy efficiency, and energy-saving solutions. New energy technologies are seen to have a huge potential both domestically and internationally. According to the national climate and energy strategy (MEE, 2008), for instance, environmental and energy technologies contribute to fostering domestic low-carbon energy sources, improving energy efficiency, and strengthening national self-sufficiency in energy production. The aim is also to take full advantage of international market opportunities: investments in energy technologies are expected to enhance domestic production and energy efficiency, but, more importantly, create commercial innovations and goods for 'the expanding global market' (MTI, 2005; MEE, 2008; Prime Minister's Office, 2010). The economic possibilities are highlighted by setting rather high expectations to new technologies' growth potential. The policy documents emphasise, in particular, the possibility of utilising the growing demand of new energy technologies in developing countries (MEE, 2008). In this respect, they call for strengthening technology transfer and trade liberalisation in order to maximise Finland's export opportunities in the world market. Some have even claimed that the clean technology sector has the possibility of creating 'green *Nokias*' (Ministry of Education, 2005; Anttila, 2007; Hassi, 2009), and, alongside with metal, forest, and the ICT industries, become a 'new supporting pillar' for the national economy (Prime Minister's Office, 2009).

The strong influence of technology policy in national politics (Häyrinen-Alestalo et al., 2005; Pelkonen, 2008), together with the emphasis on energy technologies' market potential, indicates that the policy documents largely articulate climate policy in technological and economic terms. It serves as part of broader economic and industrial policies in which new technologies are primarily seen as an engine for economic growth and export revenues (Prime Minister's Office, 2008). In this frame, investing in new energy technologies is not primarily motivated by preventing environmental catastrophes but indeed by promoting Finland's technological and economic competitiveness. The policy documents thus subordinate climate concerns to economic rationale and shift political power from the nation-state to the international market.

The effectiveness of eco-modernist political strategy depends, however, on not only the radicalness of environmental innovations but also the degree of their diffusion (Jänicke, 2008). This issue has been taken up both by international comparisons of techno-scientific performance (Naumanen, 2005; Lehtoranta et al., 2007) and by national competitiveness strategies and evaluations (Prime Minister's Office, 2004). The recent evaluation of the national innovation system (Veugelers et al., 2009), for instance, illustrates that Finnish policy has been too focused on input-oriented activities and calls for more attention to the demand side to correct this imbalance. It suggests active state intervention to provide incentives for the development of technological innovations. In the area of energy, climate, and environment, proposed policy measures include the public procurement of innovation and the establishment of norms and regulations favouring clean technologies. Similar suggestions have been made at the international level in recent visions of green growth (Aghion & Howitt, 2009; OECD, 2009). In practice, though, expectations of technology-led climate policy have been overly optimistic. Despite Finland's strong input activities in R&D and good rankings in international technological and competitiveness comparisons, it has been weak in the utilisation and diffusion of technology. Although Finland has plenty of low-carbon technologies available, their utilisation has remained at a relatively low level. For instance, the production of wind power technologies and components has increased notably in recent years but much of them are produced for export, and the share of wind power in domestic energy consumption has been marginal (0.3% of total energy consumption in 2008). A shortcoming of this kind of a climate policy orientation is thus its weakness in terms of policy implementation and outcomes, which questions the validity of the 'new green economy' thesis.

The rhetoric of necessity

The rhetoric of necessity reflects the commitments to global climate change mitigation and the EU's renewable energy obligations, which are seen to 'impact Finland's short-term economic growth potential' (Prime Minister's Office, 2008) and 'bring along unforeseeable pressures that weaken the public economy' (Prime Minister's Office, 2010, 49). Increasing world population and consumption, together with the tightening global competition over raw materials, are seen to require renewals in regional, national, and global production and consumption structures. The National Innovation Strategy, for instance, identifies globalisation, new technologies, the ageing of the population, and sustainable development as key drivers of economic and societal change, arguing that 'the increased awareness of climate change and the threats related to it have created pressure to move towards ecologically sound production and consumption. This pressure is strengthened by the scarcity and the sharply rising prices of raw materials' (Prime Minister's Office, 2008, 3). Along the lines of ecological modernisation, technology is considered to be the key to simultaneously promoting economic growth and environmental and societal well-being. Technological breakthroughs are seen to contribute to not only managing scarce natural resources and tightening emission reduction targets but also creating 'a more sustained economic growth and societal well-being' (Prime Minister's Office, 2009; Sitra, 2009). They are also represented as a necessary means to achieve a greener development path. The Government Foresight Report on Climate and Energy Policy (Prime Minister's Office, 2009), for instance, claims that Finland is 'obliged to develop technologies that are crucial to global emission reductions', such as carbon capture and storage technology that is seen to be needed in the future especially in countries that have large coal reserves, such as China.

Besides developing technological innovations, however, it remains unclear what is meant by the 'economic and societal change' that is called for in recent policy documents. In fact, there are few signs of structural reforms that would point to changes in the prevailing patterns of energy production and consumption. Instead, Finnish climate and energy policies are largely based on the very same premises as in the 1970s and 1980s. The policy documents replicate the assumption that the consumption of energy is continuously increasing and suggest building additional energy production capacity to fulfil this growing energy demand. As in the 1970s, the main response to the additional need of energy is building more nuclear power plants (MTI, 2005; Cabinet Programme, 2007; MEE, 2008). The rhetoric of necessity is thus used strategically not only to promote technology-driven development but also to foster the acceptability of additional nuclear power.

The rhetoric of necessity also points to the characterisation of Finland as a Nordic corporate state and a consensual democracy (Arter, 1999), where the successive coalition governments have since the late-1970s enabled a rather consensual policy style among the main political parties, decision-makers, industry, and labour market organisations. Political documents refer to high technology-led climate policy as having a broad political acceptance at the national level due to a number of stakeholder consultations and the broad-based parliamentary committee work in the preparatory phases of the national climate strategies. Moreover, the consensus-seeking nature of the energy policy decision-making processes has been strengthened by the state ownership of large national energy companies, which has provided an indirect means of regulating the energy sector and reduced open political conflicts. The continuity of the Finnish model of corporatist governance has also been visible in that since the late-1980s and 1990s, there have been few changes in the composition of the Finnish energy elite (Ruostetsaari, 2009). Redirecting climate policy towards an innovation and market-led approach is constructed as a joint national project and a shared goal that requires a profound commitment of all political actors. In this sense, political consensus becomes both a justification and a policy objective.

Although there seems to be a broad consensus on the necessity of the national technology-led policy vision, its realisation seems to remain somewhat problematic. One problem in turning the rhetorical construction into a coherent political strategy has been the fragmented division of responsibilities between different sectoral ministries on environmental issues. While the coordination of international climate policy is placed under the Ministry of the Environment, national climate and energy policies belongs to the Ministry of Employment and the Economy, and the flexible mechanisms to the Ministry of Foreign Affairs. Other ministries are also involved with climate and energy policy preparations in their respective fields, such as agricultural and forest policies (the Ministry of Agriculture and Forestry) and knowledge infrastructure (the Ministry of Education). The unclear division of responsibilities, together with the traditionally strong boundaries between sectoral ministries, has recently raised discussion about reforming policy coordination structures through enhancing cross-sectoral cooperation across administrative, organisational, and political borders and/or establishing a separate climate and energy ministry (Vapaavuori, 2010). Another problematic issue has been the concentration of power within the energy sector. In this respect, an emerging question has been the multiple role of the Ministry of Employment and the Economy in governing both energy policy and nuclear power security issues and acting as permission authority, while, at the same time, being a substantial owner of companies applying permissions for new nuclear power construction projects, which has been claimed to have resulted in an interest controversy (Lampinen, 2009, 41). Discussion has also emerged about the limited role of the Ministry of the Environment, which is responsible for international climate policy but not for national level climate policy. Despite the politically powerful consensus rhetoric, it might therefore be difficult to speak of *a* governmental rationality and *a* (national) political will in governing such a complex and multi-layered policy field.

5. Towards a green economy?

Ecological modernisation, at least in the weak, techno-corporatist version (Christoff, 1996), has become a broadly accepted political strategy in Finland. It has reinforced the ideas of the knowledge- and innovation-driven economic development model and provided politically persuasive 'green' elements for national technology policy. In addition, ecological modernisation has neither required radical interventions to the established patterns of production and consumption nor challenged high-technology and market-led economic growth policies. Emphasising the knowledge-based economy's principles of flexibility, openness, and fluidity (Beerkens, 2008), particularly in relation to the international market, this approach highlights the development and commercialisation of environmental innovations and new technologies as a key solution to the problem of climate change. At the same time, Finland's technological and economic competitiveness in the world market are raised as important national policy objectives, as the rapidly growing world market is seen to favour 'continuously renewing' and 'efficient' economies with 'proactive adaptation capability' (STPC, 2006; Cabinet Programme, 2007). This is also reflected in the understanding of the environment, which has changed from a cost burden to an opportunity for economic profit.

The policy documents' selective use of ecological modernisation also signifies the interconnectedness of the knowledge-based economy and unsustainability. While emphasising a knowledge- and high technology-driven development path, this approach suggests the continuance of almost unlimited economic growth. It subordinates climate, environmental, and broader societal objectives to the techno-economic rationale, leading to preferring short-term political choices in a situation where long-term solutions would be needed. Moreover, it has had little to say about the distribution of costs and benefits on a global scale, thus largely ignoring the uneven development impacts of global economic growth. In this respect, the Finnish approach seems to follow Blühdorn's (2007, 263–264) characterisation of post-ecologism by prioritising 'economic competitiveness and growth, the security of Northern life-styles and the preservation of established global power relations', where technological advancement, economic growth, and continuous global competition have become non-negotiable imperatives.

Moreover, the knowledge- and technology-driven development path reflects a rather narrow understanding of sustainability. The policy documents contain only generic references to sustainable development, and the relationship between economic growth and environmental degradation remains undefined. In addition, they place climate change high on the national political agenda and treat it as a political field of its own, leaving other environmental concerns, such as environmental equity and biodiversity, largely outside the mainstream political debate. Guided by a nation state-centric perspective aiming primarily to respond to an individual industrialised country's interests, Finnish climate policy is thus incapable of adequately taking into account a broader global development perspective. Characteristic for the Finnish approach is also, as in Blühdorn's 'politics of unsustainability', that key societal and environmental concepts are being reinterpreted to legitimise the technology-led economic development path. Environmental concepts become translated into economic language, as illustrated, for instance, by the policy documents' strong emphasis on the cost efficiency and profit opportunities of new environmental technologies. They are also strategically used to justify certain controversial political choices, such as nuclear power that has recently been reframed as a 'clean', even 'green' technology. It seems indeed that in the framework of the knowledge-based economy, all climate policy measures become conditioned by Finland's technological and economic competitiveness. Yet many authors, such as Paul Krugman (1994), have argued that it is irrational to assume that nation-states could compete with one another like private enterprises.

The adaptation of new policy concepts, such as 'green economy', 'green growth', and 'Green New Deal', indicate an effort to integrate environmental issues into broader growth policies. At the same time, however, they illustrate the prevailing supremacy of consumerist market economy's values in relation to ecologically sound development. Redefining the knowledge-based economy as being driven by eco-innovation does not, as such, bring much new to the assumptions of market-oriented new growth theories. In fact, suggesting a state of equilibrium through the objective of 'sustainable economic growth', it follows the ideas of neo-classical economics. The green growth, green economy, and Green New Deal models thus contain a built-in controversy. As Schumpeter has argued (See Lundvall, 1999), innovation processes (whether green or not) cannot be based on a state of harmony and equilibrium because they always involve constant creation of new ideas, products, services, and needs. Furthermore, the very idea of knowledge and innovation as the key drivers of economic progress is by no means new. Most of the ideas of the knowledge-based economy and the new growth theories can be traced back to the work of Moses Abramovitz (1989/1952) and Fritz Machlup (1962), which highlighted the significance of knowledge production and innovation for economic growth already in the 1950s and 1960s.

In both scientific and public climate policy debates, questions about the optimal scale for the economy and the environmental implications of economic growth have recently been revived. This has resulted in mainstream economic theories and the knowledge-driven growth paradigm as a whole having been increasingly questioned. One vision of a sustainable society has been derived from John Stuart Mill's (1965/1848) concept of the 'stationary state', which has been used to refer to a zero-growth society and 'a continuous state of dynamic equilibrium', in which technological advancements would be directed towards a more equitable redistribution of wealth (Winch, 2004; Lin, 2006). Alternative suggestions based on critiques of unlimited economic growth have also been developed within the degrowth movement (See, for example, Fournier, 2008; Latouche, 2009). Central to this approach has been an emphasis on 'escaping from the economy', that is, challenging the ways in which economy and its practices are thought of and conceptualising forms of social organisation that do not rely on an economic vocabulary (Fournier, 2008). Common to these critiques has been the idea that economic growth is not an answer but part of the problem of environmental degradation and needs therefore be called into question (cf. Meadows et al., 1972). A report of a British think-tank, the New Economic Foundation (2010, 1-2), for instance, argues, 'Just as the laws of thermodynamics constrain the maximum efficiency of a heat engine, economic growth is constrained by the finite nature of our planet's natural resources (biocapacity)'. Yet in the mainstream policy debates, these voices have so far remained relatively marginal. The Finnish experience illustrates that while advocating an integrative (innovation) policy approach with greater inclusiveness towards environmental concerns, the green economy thesis persistently reproduces a basis of legitimacy for the continuance of the knowledge-based economy's growth model. At the same time, it actually narrows the space for any alternative suggestions that might challenge the prevailing politico-economic status quo.

6. References

Abramovitz, M. (1989/1952). Economics of Growth. In: *Thinking about Growth and other Essays on Economic Growth and Welfare* (1989), Cambridge University Press, Cambridge.

Academy of Finland (1997). *Kansallinen tutkimuksen huippuyksikköstrategia.* [A National Strategy for Centres of Excellence in Research.] Suomen Akatemian julkaisuja 5/97, Edita, Helsinki.

Aghion, P, & Howitt, P. (2009). *The Economics of Growth,* MIT Press, Cambridge, MA.

Alestalo M. (1997). Variations in State Responsiveness. The Science System and Competing Theories of the State. *International Sociology,* 12, 1, 73-92.

Anttila, S.-L. (2007). Suomesta ympäristöteknologian huippumaa [Finland to the Leading Countries of Environmental Technology], Speech of the Finnish Parliement's First Deputy Speaker, 4 February, Helsinki, http://web.eduskunta.fi/Resource.phx/pubman/templates/24.htx?id=705 (visited 20 February 2010).

Arter D. (1999). *Scandinavian Politics Today,* Manchester University Press, Manchester.

Baker, S. (2007). Sustainable Development as Symbolic Commitment: Declaratory Politics and the Seductive Appeal of Ecological Modernisation in the European Union. *Environmental Politics,* 16, 2, 297-317.

Beerkens, E. (2008). University Policies for the Knowledge Society: Global Standardization, Local Reinvention. *Perspectives on Global Development and Technology,* 7, 15–36.

Blühdorn, I. (2000). *Post-Ecologist Politics. Social Theory and the Abdication of the Ecologist Paradigm,* Routledge, London/New York.

Blühdorn, I. (2002). Unsustainability as a Frame of Mind - and How We Disguise It: The Silent Counter-Revolution and the Politics of Simulation. *The Trumpeter* 18, 1, 59–69.

Blühdorn, I. (2007). Sustaining the Unsustainable: Symbolic Politics and the Politics of Simulation. *Environmental Politics,* 16, 2, 251–275.

Blühdorn, I. & Welsh, I. (2007). Eco-Politics beyond the Paradigm of Sustainability: A Conceptual Framework and Research Agenda. *Environmental Politics,* 16, 2, 185–205.

Brenner, N. & Theodore, N. (Eds.) (2002). *Spaces of Neoliberalism: Urban Restructuring in North America and Western Europe.* Blackwell, Oxford.

Cabinet Programmes of Finland 1987; 1991; 1995; 1999; 2003 & 2007.

Christoff, P. (1996). Ecological Modernisation, Ecological Modernities. *Environmental Politics,* 5, 3, 476–500.

Committee of Technology Policy (1985). *Opetusministeriön hallinnonalan teknologiapolitiikkaa koordinoivan työryhmän muistio* [A Memorandum of a Coordinative Committee of Technology Policy under the Ministry of Education], Prime Minister's Office, Helsinki.

Dryzek, J. S. (2005). *The Politics of the Earth. Environmental Discourses,* 2nd Edition, Oxford University Press, Oxford.

Eder, K. (1996). *The Social Construction of Nature,* Sage, London.

EK, Confederation of Finnish Industries (2009). *Oikeudenmukaista ja tuloksellista ilmastopolitiikkaa* [Fair and Successful Climate Policy], Confederation of Finnish Industries, Helsinki.

EC, European Commission (2007). *European Innovation Scoreboard (EIS) 2006: Comparative Analysis of Innovation Performance,* Commission of the European Communities, Brussels.

EC, European Commission (2008). *20 20 by 2020. Europe's Climate Change Opportunity,* COM (2008)30 final, Commission of the European Communities, Brussels.

EC, European Commission (2009). *Consultation on the 'EU 2020' Strategy,* Commission Working Document, COM(2009)647 final, http://ec.europa.eu/eu2020/pdf/eu2020_en.pdf (visited 20 February 2010).

EurObserv'ER (2010). *The State of Renewable Energies in Europe,* 9th EurObserv'ER Report, http://www.eurobserv-er.org/pdf/bilan9.asp (visited 16 February 2010).

Fournier, V. (2008). Escaping from the Economy: The Politics of Degrowth. *International Journal of Sociology and Social Policy,* 28, 11/12, 528–545.

Godin B. (2006). The Knowledge-Based Economy: Conceptual Framework or Buzzword. *Journal of Technology Transfer,* 31, 1, 17–30.

Hajer, M. (1995). *The Politics of Environmental Discourse,* Oxford University Press, Oxford.

Hassi, S. (2009). Uusi energiatekniikka on nyt vihreää kultaa [New Energy Technology Is Now Green Gold]. Opinion. *Savon Sanomat,* 28 May 2009.

Heiskala R, Luhtakallio E. (Eds) (2006). *Uusi jako: miten Suomesta tuli kilpailukyky-yhteiskunta* [The New Division: How Finland Became a Competition Society], Gaudeamus, Helsinki.

Häyrinen-Alestalo, M.; Pelkonen, A.; Teräväinen, T. & Villanen, S. (2005). Changing Governance for Innovation Policy Integration in Finland. In: *Governance of Innovation Systems: Vol. 2. Case Studies in Innovation Policy*, Remoe, S.-O. (Ed), pp. 111–138, OECD, Paris.

Häyrinen-Alestalo, M. & Kallerud, E. (2004). Introduction: Towards a Biotech Society - Nordic Perspectives. *In: Mediating Public Concern in Biotechnology. A Map of Sites, Actors and Issues in Denmark, Finland, Norway and Sweden.* Häyrinen-Alestalo, M. & Kallerud, E. (Eds.), pp. 7–21, NIFU, Oslo.

IEA, International Energy Agency (2008a). *Energy Policies of IEA Countries: Finland 2007 Review*, OECD/IEA, Paris.

IEA, International Energy Agency (2008b). *Renewables Information: 2008 Edition*, OECD, Paris.

IMD, International Institute for Management Development (2010). *IMD 2010 World Competitiveness Yearbook,* International Institute for Management Development, Lausanne.

IPCC, the Intergovernmental Panel on Climate Change (2007). *The Fourth Assessment Report (AR4)*, IPCC, Geneva.

Jänicke, M. & Lindemann, S. (2010). Governing Environmental Innovations. *Environmental Politics*, 19, 1, 127–141.

Jessop B. (2002). *The Future of the Capitalist State*, Polity Press, Cambridge.

Jänicke, M. (2008). Ecological Modernisation: New Perspectives. *Journal of Cleaner Production*, 16, 5, 557–565.

Krugman, P. (1994). Competitiveness: A Dangerous Obsession. *Foreign Affairs*, 73, 2, 28–44.

KTM-81 Committee (1981). *KTM-81-komitean mietintö* [KTM-81 Committee Report], Komiteanmietintö 1981:59, Helsinki.

Lampinen, A. (2009). An Analysis of the Justification Argumentsin the Application for the New Nuclear Reactor in Finland. In: *The Renewal of Nuclear Power in Finland*, Kojo, M. & Litmanen, T. (Eds.), pp. 41–68, Palgrave Macmillan, Basingstoke.

Larner W. (2000). Neo-liberalism: Policy, Ideology, Governmentality. *Studies in Political Economy*, 63, 5–26.

Latouche, S. (2009). *Farewell to Growth*, Polity Press, Cambridge.

Lehtoranta, O.; Pesonen, P.; Mononen, E.; Ahlqvist, T. & Loikkanen, T. (2007). *TEKbaro. Teknologiabarometri kansalaisten asenteista ja kansakunnan suuntautumisesta tietoon perustuvaan yhteiskuntaan* [Technology Barometer on Citizens' Attitudes and Finland's orientation towards the Knowledge-Based Society], Tekniikan Akateemisten Liitto TEK, Helsinki.

Lin, B. C.-A. (2006). A Sustainable Perspective on the Knowledge Economy: A Critique of Austrian and Mainstream Views. *Ecological Economics*, 60, 1, 324–332.

Lundvall, B.-Å. (1999). Innovation Policy and Economic Theory. In: *Transformation Towards a Learning Economy*, Schienstock, G. & Kuusi, O. (Eds.), pp. 397-419, Hakapaino, Helsinki.

Machlup, F. (1962).*The Production and Distribution of Knowledge in the United States*, Princeton University Press, Princeton.

Meadows, D.; Meadows, D. & Randes, J. (1972). *Limits to Growth*, Universe Books, New York.

MEE, Ministry of Employment and the Economy (2008). *Pitkän aikavälin ilmasto- ja energiastrategia. Valtioneuvoston selonteko eduskunnalle 6. päivänä marraskuuta 2008* [Long-term Climate and Energy Strategy. Government Report to Parliament 6 November 2008], Ministry of Employment and the Economy, Helsinki.

Mill, J. S. (1965/1848). *The Collected Works of John Stuart Mill, Volume III - The Principles of Political Economy with Some of Their Applications to Social Philosophy,* Robson, J. M. (Ed.), Routledge and Kegan Paul, London.

Ministry of Education (1996). *Koulutuksen ja korkeakouluissa harjoitetun tutkimuksen kehittämissuunnitelma vuosille 1995–2000* [Government's Decision in Principle on the Development Plan of Education and Academic Research in 1995–2000], valtioneuvoston periaatepäätös 21.12.1995, Ministry of Education, Helsinki.

Ministry of Education (2005). *Kolme puheenvuoroa luovuuden edistämisestä. Luovuusstrategian osatyöryhmien raportit* [Three Perspectives for Promoting Creativity. Reports of the Creativity Strategy Working Groups], Publications 35, Ministry of Education, Helsinki.

Mol, A. & Sonnenfeld, D. (Eds.) (2000). *Ecological Modernisation around the World: Perspectives and Critical Debates,* Frank Cass, London.

Motiva (2009). *Uusiutuvan energian käyttö Suomessa* [Renewable Energy in Finland], http://www.motiva.fi/toimialueet/uusiutuva_energia/uusiutuvan_energian_kaytto_suomessa (visited 20 February 2010).

MTI, Ministry of Trade and Industry (2001). *Kansallinen ilmastostrategia. Valtioneuvoston selonteko eduskunnalle* [National Climate Strategy. Government Report to Parliament], VNS 1/2005 vp, Ministry of Trade and Industry, Helsinki.

MTI, Ministry of Trade and Industry (2005). *Outline of Energy and Climate Policy for the Near Future: National Strategy to Implement the Kyoto Protocol,* MTI Publications 27, Ministry of Trade and Industry, Helsinki.

Murphy, J. (2000). Ecological Modernisation. *Geoforum,* 3, 1, 1–8.

Murto, E.; Niemelä, M. & Laamanen, T. (2006). *Finnish Technology Policcy from the 1960s to the Present Day – Public Sector Activities Promoting Research and Development.* Ministry of Trade and Industry, Helsinki.

Naumanen, M. (2005). *TEKbaro. Teknologiabarometri kansalaisten asenteista ja kansakunnan suuntautumisesta tietoon perustuvaan yhteiskuntaan* [Technology Barometer on Citizens' Attitudes and Finland's Orientation towards the Knowledge-Based Society], Tekniikan Akateemisten Liitto TEK, Helsinki.

NEF, New Economic Foundation (2010). *Growth Isn't Possible. Why We Need a New Economic Direction,* New Economic Foundation, London.

OECD (2005). *The Measurement of Scientific and Technological Activities: Guidelines for Collecting and Interpreting Innovation Data,* Oslo Manual, 3rd Edition, prepared by the Working Party of National Experts on Scientific and Technology Indicators, OECD, Paris.

OECD (2009). *Declaration on Green Growth,* adopted at the Council Meeting at Ministerial level on 25 June 2009, http://www.olis.oecd.org/olis/2009doc.nsf/ (visited 15 March 2010).

Olsen, K. (2007). The Clean Development Mechanism's Contribution to Sustainable Development: a Review of the Literature. *Climatic Change,* 84, 1, 59–73.

Palonen, K. & Summa, H. (Eds.) (1996). Pelkkää retoriikkaa [Just Rhetoric], Vastapaino, Tampere.

Pelkonen A. (2008). *The Finnish Competition State and Entrepreneurial Policies in the Helsinki Region,* Research Reports 254, Department of Sociology, University of Helsinki.

Pelkonen, A.; Teräväinen, T. & Waltari, S.-T. (2008). Assessing Policy Coordination Capacity: Higher Education, Science, and Technology Policies in Finland. *Science and Public Policy,* 35, 4, 241–252.

Prime Minister's Office (2004). *Osaava, avautuva ja uudistuva Suomi* [Finland's Competence, Openness and Renewability], Prime Minister's Office Publications 19/2004, Edita, Helsinki.

Prime Minister's Office (2008). *Innovaatiostrategia. Valtioneuvoston innovaatiopoliittinen selonteko eduskunnalle* [Innovation Strategy. Government Report to Parliament], Prime Minister's Office, Helsinki.

Prime Minister's Office (2009). *Government Foresight Report on Long-Term Climate and Energy Policy: Towards a Low-Carbon Finland,* Publications 30/2009, Prime Minister's Office, Helsinki.

Prime Minister's Office (2010). *Kestävästä kasvusta hyvinvointia ja elämänlaatua.* Kasvutyöryhmän väliraportti [Sustainable Growth for Well-being and Quality of Life. Interim Report of the Growth Project Working Group], Reports 1/2010, Prime Minister's Office, Helsinki.

Hsieh, H.-F. & Shannon, S. E. (2005). Three Approaches to Qualitative Content Analysis. Qualitative Health Research, 15, 9, 1277–1288.

Ruostetsaari, I. (2005). *Energiahuollosta kilpailullisiin energiamarkkinoihin: toimijakentän muutokset* [From Energy Management to Competitive Energy Market], presentation at the 10th anniversary celebration of the Energy Market Authority of Finland, 1 June 2005, Helsinki, www.energiamarkkinavirasto.fi/files/IlkkaRuostetsaari_20050601.pdf (visited 15 March 2010).

Ruostetsaari, I. (2009). Governance and Political Consumerism in Finnish Energy Policy-Making. *Energy Policy,* 37, 1, 102–110.

Schienstock G. (2007). From Path Dependency to Path Creation: Finland on Its Way to the Knowledge-Based Economy. *Current Sociology,* 55, 1, 92–109.

Sitra, the Finnish Innovation Fund (2009). *Kansallisen luonnonvarastrategian taustaraportti: luonnonvaroissa muutoksen mahdollisuus* [Background Report for the National Natural Resource Strategy], the Finnish Innovation Fund Sitra, Helsinki.

Statistics Finland (2010) *Tutkimus- ja kehittämistoiminnan menot rahoituslähteen mukaan sektoreittain 1971–2008* [Research and Development Expenditure by Funding Source and Sector, 1971–2008], http://www.stat.fi (visited 15 March 2010).

STPC, Science and Technology Policy Council of Finland (2006). *Tiede, teknologia, innovaatiot* [Science, Technology, Innovations], Science and Technology Policy Council, Helsinki.

Teräväinen, T. (2009). The Challenge of Sustainability in the Politics of Climate Change: A Finnish Perspective on the Clean Development Mechanism. *Politics,* 29, 3, 173–182.

UNEP, the United Nations Environment Programme (2009). *Global Green New Deal: An Update for the G20 Pittsburgh Summit,* http://www.unep.org/pdf/G20_policy_brief_Final.pdf (visited 15 March 2010).

Vapaavuori, J. (2010). Perustetaan Suomeen ilmasto- ja energiaministeriö [Finland Needs a Climate and Energy Ministry], Sunday Debate, *Helsingin Sanomat,* 14 February 2010.

Veugelers et al. (2009). *Evaluation of the Finnish National Innovation System – Policy Report,* prepared by an evaluation panel chaired by R. Veugelers. Taloustieto, on behalf of the Ministry of Education and the Ministry of Employment and the Economy, Helsinki.

WCED, World Commission on Environment and Development (1987). *Our Common Future,* Oxford University Press, Oxford.

Winch, D. (2004). Thinking Green, Nineteenth-Century Style: John Stuart Mill and John Ruskin. In: *Markets in Historical Contexts: Ideas and Politics in the Modern World,* Bevir, M. & Trentmann, F. (Eds.), pp. 105–128, Cambridge University Press, New York.

WEF, World Economic Forum (2009). The Global Competitiveness Report 2009–2010, World Economic Forum, Geneva.

WEF, World Economic Forum (2010). The Lisbon Review. Towards a More Competitive Europe, World Economic Forum, Cologny/Geneva.

22

Transportation and climate change

Yuri Yevdokimov
University of New Brunswick
Canada

In its fourth assessment report published in spring 2007, leading scientists on the Intergovernmental Panel for Climate Change (IPCC) reached consensus that human activity is responsible for many observed climate changes, particularly the warming temperatures of the last several decades, and concluded that there is a need for far more extensive adaptation than is currently occurring to reduce vulnerability to future climate changes.

However, while human activities cause significant climate changes on the one hand, these changes affect humans and their activities in return on the other. Therefore, it is natural to talk about two-directional link between human activities and climate change, and transportation is a good example of this approach.

Numerous studies have examined the link between climate change and the transportation sector. These studies have been conducted primarily from the perspective of transportation's contribution to global warming through the burning of fossil fuels, which releases carbon dioxide (CO2) and other greenhouse gases (GHGs) into the atmosphere. Far less attention has been paid to the consequences of potential climate changes for transportation infrastructure and operations. For example, projected rising sea levels, flooding, and storm surges could swamp marine terminal facilities, airport runways near coastlines, subway and railroad tunnel entrances, and roads and bridges in low-lying coastal areas.

There are also likely to be many indirect effects of climate changes on transportation. Potential climate-caused shifts in demographics as well as redistribution of production and consumption in agricultural sector, manufacturing sector, forestry, fisheries and others can affect the existing transportation patterns. Transportation patterns can also shift as the tourism industry responds to changes in ecologically or recreationally interesting destinations. Eventually it may lead to significant redistribution of transportation flows since all of the above from an economic standpoint are consumers of transportation. Therefore, it appears to be that while direct impacts of climate change affect supply of transportation, indirect impacts affect demand for transportation, and both types of impacts should be taken into account.

Here we analyze the link between transportation and climate in both directions - contribution of transportation to climate change as well as impacts of climate change on transportation.

1. Contribution of Transportation to Climate Change

Transportation sector is the primary consumer of petroleum products. It accounts for a growing fraction of global carbon dioxide (CO2) emissions - one of the main greenhouse gases (GHG). Transportation currently accounts for approximately 21% of global CO2 emissions. Even with projected improvements in efficiency, emissions from transportation are going to increase to 23% of total CO2 by 2030 (IEA WEO 2009). In some studies dedicated to the future of the GHG emissions, transportation emissions are estimated to account for an even larger fraction of future emissions, with greater reductions in emissions coming from other sectors. This result is not surprising if one considers the continuing growth of transportation all over the world.

Carbon dioxide is not the only GHG emitted by transportation. Methane (CH4) and nitrous oxide (N2O) are the other two. As an example of transportation GHG emissions, let us take a closer look at Canadian transportation sector.

1.1 Transportation sector in Canada as an example of greenhouse emissions

Canada has the world's 13th largest economy and is the world's 8th largest emitter of GHG. The latest 2009 GHG inventory from Environment Canada shows that after a slight decrease in 2004-2006, Canada's total GHG emissions increased again. According to this report, it is transportation and energy production that has driven emissions up. Between 1990 and 2007, emissions from energy industries such as the oil sands and transportation increased by about 143 million tons, or most of the overall increase of 155 million tones, the report says.

In general, transportation is responsible for more than 26% of Canada's total GHG emissions which means that in Canada transportation is the largest single contributor to GHG. 63% of these GHG emissions come from passenger travel and 37% from freight transportation. 50% of all personal GHG emissions in Canada are from transportation. The following graph shows dynamics of the GHG emissions by transportation in Canada over period of 1990-2007.

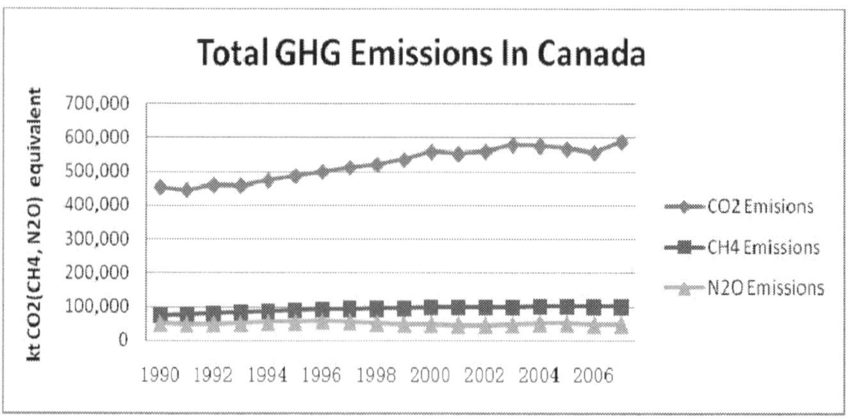

The above described situation is typical for developed economies, and it points to a significant contribution of transportation emissions to the climate change via the so-called greenhouse effect. The greenhouse effect is the process by which absorption and emission of infrared radiation by gases in the atmosphere warm a planet's lower atmosphere and surface. It was discovered by Joseph Fourier in 1824 and was first investigated quantitatively by Svante Arrhenius in 1896 (Weart, 2008).

The question is: Why should we care about climate changes at all? There are many reasons for that. However, we are going to emphasize only economic ones. From an economic standpoint, changes in climate directly affect our infrastructure and shift our consumption and production patterns. Moreover, these changes are associated with a lot of uncertainty which increases risk of future economic development. Investors in many fields are continually making decisions in the face of uncertain information about risks and outcomes. Wrong decisions can lead to some negative macroeconomic and microeconomic consequences. That is why addressing climate change in terms of quantitative assessments, such as, for example, the development of probabilistic climate change scenarios, may reduce this risk and the costs associated with wrong decisions. In terms of transportation, the results of such quantitative studies can be directly incorporated into planning forecasts and engineering design guidelines and standards. And since transportation professionals typically plan 20 to 30 years into the future, many decisions taken today, particularly about the location of infrastructure, will help shape economic development patterns and markets that endure far beyond these planning horizons

1.2 Analysing the link: climate-transportation emissions-climate change

There are two general views on the relationship between economic development on the one hand and the state of our environment on the other. The first one is associated with the so-called neo-Malthusian approach started with the study "The Limits to Growth" (Meadows et al, 1972). This study produced two basic conclusions: (i) under existing consumption rates the world would run out of resources in the next 100 years; and (ii) to prevent this from happening it is necessary to reduce economic growth, population growth and the level of pollution. According to Daly and Cobb (1989), "further growth beyond the present scale is overwhelmingly likely to increase costs more rapidly than it increases benefits thus ushering in a new era of "uneconomic growth" that impoverishes rather than enriches"

In contrary, the second view states that economic growth and development would eventually lead to a decrease in pollution and cleaner environment. In this regard, numerous studies have been conducted to prove the existence of a positive link between economic growth and environmental quality.

In 1990, this debate between two views led to the appearance of the so-called Environmental Kuznets Curve - an inverted U-shape relationship between the level of pollution and per capital income: With an increase in per capital income, the level of pollution increases initially but decreases eventually. In order to explain this phenomenon economists proposed 3 models: (i) overlapping generations' models; (ii) production/consumption models, and (iii) political economy models.

Grossman and Krueger (1991) conducted the first empirical study of the EKC relationship as part of a wider study to assess the environmental impacts of the North American Free Trade Agreement (NAFTA). Their analysis confirmed that ambient concentrations of SO2 and dust exhibit the EKC relationship with turning points between $4,000 and $5,000 in 1985 US dollars. After that numerous studies in various countries have been conducted to detect and estimate the EKC relationship.

Since then there have been enormous number of studies dedicated to the existence of the EKC. These studies tested the EKC hypothesis for various pollutants for different countries, for regions within a country and for different economic sectors of a country. Unfortunately, transportation sector has not been a primary target of these studies. That is why we decided to test the relationship between GHG emissions by transportation and Canadian standard of living statistically. In particular, a more specific question was addressed: Do GHG emissions by transportation in Canada and per capita income follow the inverse U-shape relationship known in the literature as Environmental Kuznets Curve (EKC) and how climate enters this relationship?

The relationship between environment and economy, expressed by the EKC, is affected by the climate in two ways:

1. Climate-pollution link. Many factors govern the severity and timing of pollution. Meteorological factors are among them. Various studies have been conducted by researchers in this area. For example, Aw and Kleeman (2003) modeled the link between temperature and pollution, Mickley et al.(2004) tested the effect of changing wind patterns on pollutant concentrations, Hogrefe et al. (2004) focused on the effect of climate change on surface ozone. Overall these studied proved the existence of the link between climate and the level of pollution. Moreover, it is a fact that environmental adaptive capacity - the capacity of a environment to adapt to pollution over time – depends on climate, and if the latter changes so does the adaptive capacity which means a change in the relationship between the environment and the level of pollution.
2. Climate-economy link. From an economic viewpoint, climate change leads to a change in consumption and production patterns that eventually affect the standard of living expressed in the EKC by per capita income.

Therefore, climate affects the EKC relationship on both sides – on pollution side and on income side – and therefore, it should be explicitly included into this relationship if a study period is long enough. In our study a period of 18 years was chosen with annual temperature as the most important indicator of the climate change.

There is a broad variety of the EKC specifications in the literature. In some studies it is expressed in levels while in others in logarithms. In some studies concentrations of pollutants are used while in others emissions. In some studies quadratic relationship is estimated while in others cubic. In this study we decided to combine all of these approaches. Of course, since transportation is associated with mobile sources of pollution, only GHG emissions make sense to detect the EKC. That is why three transportation GHG emissions were used as dependent variables. However, we estimated the EKC relationship with annual temperature in both levels and logarithms as well as for quadratic and cubic functional forms.

The data set used in that study included three blocks: (i) the income block; (ii) the pollution block, and (iii) the climate block Sample period was from 1990 to 2007.

The income block included GDP data at provincial level - 10 Canadian provinces and 2 territories – and population data. *The pollution block* consisted of three GHG emissions by transportation namely CO_2, CH_4 and N_2O. Annual emission data on 10 Canadian provinces and 2 territories was taken from the database of Environment Canada. *The climate block* consisted of annual average temperatures from 1990 to 2007 taken from Environment Canada. One major city data from each province/territory was chosen.

Estimation of the EKC relationship between transportation GHG emissions and per capita GDP in Canada has been performed at two levels: (i) federal level, and (ii) provincial level.

At federal level, in general, evidence in favour of the EKC relationship was found in 55% of all our statistical experiments. The strongest evidence for this relationship was found for methane CH4. This relationship was found to hold in levels as well as in logarithms for all provinces and territories. The value of turning point varied in range from \$297 to \$16,623 in regressions in levels and from \$19,930 to \$24,338 in regressions in logs which is consistent with findings by other researchers. Temperature appeared to be significant at 10% level in regressions in logarithms only.

For CO2, turning point was defined in the range from \$86,000 to \$98,000 in the EKC in levels. The EKC relationship in logarithms was found in all regressions. However, the value of the turning point ranged from \$189,874 to \$719,795. It is somewhat high value. One reason for that is sensitivity of the logarithmic transformation to both types of errors - measurement and estimation. However, more importantly CO2 appears to be the most sensitive transportation emission to climate change. Annual temperature is significant at 5% in logs in all regressions and at 10% significance level in regression in levels.

Finally, N2O does not seem to exhibit the EKC relationship in levels. Only weak cubic relationship between N2O emissions and per capita GDP was found. However, the EKC relationship was supported by regressions in logs with turning points in the range from \$23,181 to \$44,112 which is also consistent with the existing literature including Canadian study by Day and Grafton (2001). Annual temperature plays no role in regressions in levels but it is significant at 10% level in almost all regressions in logs.

Overall, results for the EKC relationship in logs exhibit higher degree of consistency compared to the results in levels. With respect to annual temperature, it seems to be insignificant in levels but significant in logs.

What conclusions can be drawn from these results? Since we are interested in climate change impacts it appears to be that mostly warmer climate leads to a higher level of pollution by transportation. On the other hand, higher level of pollution contributes to a warmer climate. In particular it is true with respect to CO2 emissions by transportation - the largest contributor to climate change. When we estimated this relationship at provincial level, we ended up with slightly better results. It looks like the link climate-pollution-climate change in the case of transportation is even stronger at this level.

2. Climate change impacts on transportation

Climate change will affect transportation primarily through increases in several types of weather and climate extremes, such as very hot days; intense precipitation events; intense hurricanes; drought; and rising sea levels, coupled with storm surges and land subsidence. The impacts will vary by mode of transportation and region, but they will be widespread and costly in both human and economic terms and will require significant changes in the planning, design, construction, operation, and maintenance of transportation systems. In 2008 report of U.S. Transportation Research Board the following statement was made:

"Potentially, the greatest impact of climate change for North America's transportation systems will be flooding of coastal roads, railways, transit systems, and runways because of global rising sea levels, coupled with storm surges and exacerbated in some locations by land subsidence... The Atlantic and Gulf Coasts are particularly vulnerable because they have already experienced high levels of erosion, land subsidence, and loss of wetlands... "

That is why we decided to study transportation in Atlantic Canada as an example of climate change impacts on transportation.

2.1 Atlantic Canada transportation sector as an example of regional approach

Transportation sector in Canada accounts for about 4% of Canada's gross domestic product in terms of value added, employs more than 800,000 people, and is associated with about 13% of total expenditures by Canadians. It has been estimated that the road system alone has an asset value approaching $100 billion (Richardson, 1996). As a matter of fact, in 2001 Canadian transportation system was rated as one of the best in the world (World Economic Forum, 2001).

However, it appears to be that transportation in Canada remains sensitive to a number of weather-related hazards. The first general assessment of climate change impacts on transportation in Canada was undertaken in the late 1980s (BI Group, 1990), and focused mainly on sensitivities and expert opinion. In the late 1990s, Andrey and Snow (1998) conducted a more comprehensive analysis. They concluded that it is difficult to generalize about the effects of climate change on Canada's transportation system since impacts are certain to vary by region and mode.

Why these impacts are so important? Climate and weather affect the planning, design, construction, maintenance and performance of transportation fixed facilities such as roads, railways, airport runways, shipping terminals, canals and bridges throughout their service life. Major concern is that future weather conditions may reach or exceed the limits of tolerance for some parts of the transportation system with all negative consequences.

Of course, climate change impacts on transportation infrastructure will vary regionally, reflecting differences both in the magnitude of climate changes, and in environmental conditions. As well, climate changes will have secondary impacts on demand for transportation since many sectors of modern economies are consumers of various transportation services. Climate change impacts in these sectors will affect their demand for transportation causing redistribution of transportation flows. It implies that detailed studies are needed to evaluate the consequences of the climate change impacts at regional level. And as an example of such regional approach, we decided to analyze climate change impacts on ground freight transportation in Atlantic Canada.

There are approximately 67,000 kilometers of highways in the Atlantic Canada of which 2,880 kilometers have been designated as part of the National Highway System (Transport Canada, 2008). Freight rail transportation in Atlantic Canada accounts for over 11 percent of all Canadian national railways' total traffic (Transport Canada, 2008). In terms of infrastructure, there are approximately 2,400 kilometers of main railway tracks in the Atlantic Region, of which 925 kilometers are operated by Canadian National (CN), while the rest is owned and operated by five provincial short-line railways. In addition, there are approximately 950 kilometers of spur and yard track age (Transport Canada, 2008). Sea ports and harbors in Atlantic Canada play an important role in economic activities that enable more and more commodities to be transported to inland markets. Intra-, interregional and international trades are active since there is a linkage between shipping and trucking/rail.

The above facts show the importance of freight transportation in Atlantic Canada for regional economy. In our opinion, Atlantic Canada is a nice representative for regional studies on climate change impacts on transportation. Moreover, assessing the vulnerability of transportation system in Atlantic Canada to climate changes is an important step toward ensuring a safe, efficient and resilient transportation system in general.

2.2 Climate change in Atlantic Canada

As the first step in our analysis, we addressed the following questions: Is there a climate change in Atlantic Canada? In general, the existing literature suggests that Atlantic Canada may not experience as much warming as central, western, and northern Canada, however, the region may be particularly hard hit by secondary effects such as

- rising sea level
- extreme weather events
- coastal erosion
- wetter winters
- drier summers
- reduced freshwater resources
- drought on farms
- exotic pests bringing new diseases and threats to farms and forests with infestation
- increased forest fires
- plant and animal communities may not be able to adapt fast enough

Based on the existing literature and expert opinion, it was hypothesized in our study that climate change in Atlantic Canada came about sometime in the period between 1940 and 2006. In this regard, time series data was collected on climate related variables over this period to primarily test the following stylized facts:

(i) An increase in mean annual temperature
(ii) An increase in sea level rise
(iii) Wetter winters
(iv) Drier summers

The data was obtained from the National Climate Archive of the Environment Canada. The main idea of our statistical experiments was to detect potential regime changes in the following time series:

- Annual mean temperature;
- Sea level;
- Annual total precipitations;
- Annual snowfall;
- Annual rainfall.

According to our statistical analysis, in the 1990s on average mean annual temperature in Atlantic Canada has been 0.8°C higher. As well during that period there has been a 3.2 cm increase in the overall trend in the sea level with an average rise of 0.2 cm per year. In addition, snowfall has decreased by 94 cm per year during the 1990s while rainfall has increased by 30 mm/year over the same period. This dynamics allows us to make the following conclusion: Due to an increase in average temperature in the 1990s by 0.8°C, some snowfall was converted into rainfall; however, the decrease in snowfall exceeded the increase in rainfall, and that is why total precipitations decreased

Based on comprehensive time series analysis eventually we came to the following conclusions:

- There was a definite climate change in the 1990s in Atlantic Canada. Two series - mean temperature and sea level rise - unambiguously point to a regime change in the mid 1990s.
- Regime change with respect to precipitation is not so obvious. Break points associated with the 1990s are not dominant but statistically significant.
- There was a regime change in the 1950-1960s with respect to precipitations
- Statistical evidence supports "the drier summers" stylized fact while rejecting "the wetter winters" fact.

More importantly, our analysis produced marginal impacts of the climate change on regional economy. These impacts can be summarized as follows:

- a one-degree increase in annual temperature increases regional GDP by $142.2 million;
- a one-centimetre increase in the sea level decreases regional GDP by $155.8 million;
- an increase in total precipitation by 100 millimetres increases regional GDP by $79.1 million;
- a decrease in total precipitation by 100 millimetres decreases regional GDP by $79.1 million.

2.3 Estimating consequences of climate change impacts

According to the existing literature, resource based economies are likely to be harder hit by climate change than industrial economies. Therefore, Atlantic Canada may suffer proportionately more economic hardship than, for example, central Canada. If the warming trend continues in Atlantic Canada, ice break-up and flooding on the rivers will become more severe and less predictable. This can cause increasing damage to public and private property, including transportation infrastructure such as highways and bridges.

According to the Government of Canada (2004), the Atlantic Canada can also anticipate an increased risk of trees blowing down as storms become more frequent and intense as a result of climate change. For example, a massive blow-down in 1994 caused 30 million trees to be felled which cost $100 million in damages.

Agricultural sector of the regional economy, which is one of the largest consumers of freight transportation, will be the primary beneficiary of climate change. Due to warming, the growing season for such crops as corn and other cereals will be prolonged producing larger yields. However, the probability of droughts is going to increase, thus raising the issue of supplementary irrigation. In addition, warmer winters will boost insect reproduction, forcing local farmers to apply bigger amounts of pesticides. Some other natural phenomena such as floods and hail can substantially damage crops as well as livestock.

Forestry is another large consumer of freight transportation. Climate change may increase the risk to forests in Atlantic Canada as well. According to the Government of Canada report *"warmer winter temperatures may allow invasive insects, such as the gypsy moth, to become more pervasive, while warmer, drier summers would increase the threat of forest fires in the Atlantic Provinces"* (Government of Canada, 2004).

Analysis of the cost structure of the major industrial producers in Atlantic Canada shows that adjustment to the global climate change will result in the following (Government of Canada, 2004):

- The price of pulp and paper will rise by 0.06 percent or by about 59 cents per ton;
- The price of electricity (coal) will rise by 1.94 percent or by 0.14 cents per KWH;
- The price of electricity (gas) will rise by 0.60 percent or by 0.04 cents per KWH;
- The price of steel (conventional) will rise by 0.29 percent or by $2.10 per ton;
- The price of aluminum will rise by 0.23 percent or by $4.73 per ton.
- The price of natural gas will rise by 0.14 percent or by 0.5 cents/million cubic feet.

All these changes will inevitably affect the allocation of production, consumption and trade flows in the regional economy which, in turn, will affect allocation of freight transportation flows.

In order to address this issue at aggregate level, the following three scenarios were assumed in order to trace consequences of the climate change impacts on regional economy and its transportation system:

	Expected	**Best**	**Worst**
Sea level rise	0.5 cm/year	0.2 cm/year	1.0cm/year
Temperature	0.03/year	0.01/year	0.05/year
Precipitations	0	0	-50 mm/year

Table 1. Scenarios of climate changes in Atlantic Canada

These scenarios were formed on the basis of the forecasts made by the Government of Canada for Atlantic Canada. Using these assumptions, computer simulation was performed. The results of the computer simulation are summarized in the following tables:

Year	Best Case Scenario	Expected Scenario	Worst Case Scenario
2010	-0.06%	-0.14%	-0.36%
2050	-1.05%	-2.60%	- 6.66%

Table 2. Change in the volume of ground freight transportation in Atlantic Canada

Year	Best Case Scenario	Expected Scenario	Worst Case Scenario
2010	-0.17%	-0.41%	-1.05%
2050	-1.04%	-2.58%	- 6.57%

Table 3. Change in Atlantic Canada GDP

In general, the results are consistent with the Government of Canada forecasts with respect to regional GDP. Government of Canada (2004) predicted a 0.3% decrease in Atlantic Canada GDP by the year of 2010. According to our simulation, except for the worst case scenario, a decrease in Atlantic Canada GDP is expected to be in the range of 0.17-0.41%.

Therefore, it is possible to argue that a longer forecast until 2050 is also valid as a first approximation as well as long-term impacts on transportation.
A distinguishing future of our result is: While in the short-term climate change has a stronger impact on economy as a whole compared to freight transportation, in the longer term the consequences of the climate change for the regional economy and freight transportation converge. It is not surprising since in the short-run economy is unable to quickly adjust to the climate change impacts while in the long-run it will adjust through redistribution of production, consumption, trade and transportation flows.

3. Conclusion

As stated at the beginning of this chapter, the relationship between transportation and climate is two-directional. Based on our statistical analysis performed for Canada, we can make some general conclusions about this relationship. On the one hand, transportation is one of the largest contributors to GHG emissions which, in turn, cause various changes in climate. On the other hand, these climate changes negatively affect transportation in terms of its infrastructure and operations. Therefore, this study sends a clear signal to policy makers: "Do-nothing" alternative with respect to climate changes and transportation is no longer viable, and the interaction between the two should be explicitly taken into account by transportation professionals.
In terms of contribution of transportation to climate change, any global warming that will be experienced during the next several decades will largely be the result of GHG emissions. As already noted, transportation sector in general is a significant source of GHG emissions. In order to limit future warming, GHG concentrations in the atmosphere must be stabilized. This will require reducing GHG emissions by transportation not merely to below what they might otherwise be if present trends were to continue but to well below current levels. Some decisive steps must be taken now to begin to implement different approaches. For example, U.S. Transportation Research Board proposes the following approaches:
- Utilization of carbon neutral fuels;
- Improved traffic flow and other changes in transport activity resulting from better integration of transport systems, enabled by information technology;
- Use of power trains that are highly energy efficient;
- Change in the historical mix-shifting trend to larger vehicle categories;

And some others.
In terms of climate change impacts on transportation, first of all, transportation agencies and service providers should inventory critical infrastructures according to the existing climate change projections. As well, transportation agencies and service providers should incorporate potential climate changes into their long-term capital improvement plans, facility designs, maintenance practices, operations, and emergency response plans. It means that from now on climate change should become a mandatory factor in any design of transportation facilities and operations.

4. References

Andrey, J. and Snow, A. (1998), Transportation Sector in Canada Country Study: Climate Impacts and Adaptations, Volume VII, National Sectoral Volume, Chapter 8, Environment Canada, p. 405-447.

Aw, J., and M.J. Kleeman, Evaluating the First-Order Effect of Intra-annual Air Pollution on Urban Air Pollution, *Journal of Geophysics Research*, 108, 4365, 10, 2003

BI Group (1990), The Implications of Long-term Climatic Changes on Transportation in Canada, Environment Canada, Downsview, Ontario, Climate Change Digest, CCD90-02

Canada's 2007 Greenhouse Gas Inventory, Environment Canada, 2009, http://www.ec.gc.ca/pdb/ghg/inventory_report/2007/som-sum_eng.cfm

Climate Change 2007: Impacts, Adaptation and Vulnerability, IPCC Fourth Assessment Report, Cambridge, UK: Cambridge University Press, 976 pages

Daly, H. and Cobb Jr., J., Redirecting the Economy toward Community, the Environment, and a Sustainable Future, Boston: Beacon Press, 1989

Day K. and Grafton, Q., Growth and the Environment in Canada: An Empirical Analysis, *Canadian Journal of Agricultural Economics*, 51, 2003, 197-216

Government of Canada, Modeling the Economic Impacts of Addressing Climate Change. Report. Available on-line at *http://www.climatechange.gc.ca/english/publications*

Grossman, G.M. and Krueger, A.B. (1991) "Environmental Impacts of a North-American Free Trade Agreement", NBER Working Paper 3914

Hogrefe, C., B. Lynn, K. Civerolo, J.-Y. Ku, J. Rosenthal, C. Rosenzweig, S. Gaffin, K. Knowlton, and P.L. Kinney, Simulating Changes in Regional Air pollution over the Eastern United States due to Changes in Global and Regional Climate and Emissions, *Journal of Geophysical Research*, 109, 2004

Meadows, D.H., Meadows, D.L., Randers, J. and Behrens III, W., The Limits to Growth, New York: Universe Books, 1972

Mickley, L. J., D. J. Jacob, B. D. Field, and D. Rind, Effects of Future Climate Change on Regional Air Pollution Episodes in the United States, *Journal of Geophysical Research*, Let., 30, L24103, 2004

Potential Impacts of Climate Change on U.S. Transportation, Transportation Research Board Special Report 290, Washington, D.C., 2008

Richardson, S. (1996), Valuation of the Canadian road and highway system, Transport Canada, TP 1279E, 20 p.

Transport Canada (2008), Transportation in Canada 2008, Transport Canada Annual Report. Available on-line at http://www.tc.gc.ca/policy/report/aca/anre2008/index.html

Weart, Spencer (2008), "The Carbon Dioxide Greenhouse Effect". *The Discovery of Global Warming*, American Institute of Physics

World Economic Forum (2001), The Global competitiveness report 2001-2002, World Economic Forum 2001; executive opinion survey produced in collaboration with Center for International Development at Harvard University and Institute for Strategy and Competitiveness, Harvard Business School, CD-ROM

World Energy Outlook, 2007, International Energy Agency, http://www.iea.org/weo/2009.asp

23

Cost-optimal technology and fuel choices in the transport sector under a stringent climate stabilization target

Takayuki Takeshita
Transdisciplinary Initiative for Global Sustainability, The University of Tokyo
Japan

1. Introduction

Climate change is one of the most serious challenges in the 21st century. To avoid dangerous climate change, a variety of greenhouse gas (GHG) mitigation actions have increasingly been taken in all sectors of the global energy system. The International Energy Agency (IEA) indicated that the transport sector accounted for about 23% of energy-related CO_2 emissions in 2005 and is likely to have a higher share in the future unless strong action is taken (IEA, 2008). Furthermore, the IEA showed that if a halving of 2005 energy-related CO_2 emissions is to be achieved by 2050, the transport sector must make a significant contribution, despite the fact that transport's central economic role and its deep influence on daily life have made rapid changes difficult to achieve (IEA, 2000, 2008). It is, therefore, critically important to find a long-term, cost-effective strategy for reducing CO_2 emissions from the transport sector.

So far, several studies have been carried out to address this issue using long-term global technology-rich bottom-up energy system models, with notable examples being Azar et al. (2003), Turton (2006), IEA (2008, 2009), and Grahn et al. (2009). Although these studies investigated the future role of alternative propulsion systems and fuels in the light-duty vehicle sector under CO_2 constraints, all of these studies except IEA (2008, 2009) did not place sufficient focus on the other modes of transport. The IEA (2008, 2009) derived the results for energy use and CO_2 emissions in the transport sector from a number of scenarios using the model covering all modes of transport. However, these scenario results are substantially affected by arbitrary assumptions about the diffusion rates of alternative propulsion systems and fuels. Moreover, these IEA scenarios have a time horizon until 2050, rather than a time horizon until 2100 adopted in the other three previous studies cited above, which makes it difficult to assess the very long-term prospects for radically new transport technologies.

In this context, the objective of this chapter is to examine the cost-optimal choice of propulsion systems and fuels for each of 13 transport modes over the 21st century under a constraint that the long-term global mean temperature rise would be limited to 2.0 to 2.4 degrees Celsius. This chapter also presents the results of the sensitivity analysis with respect to three important factors: (1) the climate stabilization target; (2) the cost of a proton exchange membrane (PEM) fuel cell stack and a hydrogen storage tank; and (3) the demand for supersonic air travel. These analyses are done by using a global energy system model called REgionally Disaggregated Global Energy Model with 70 regions (REDGEM70), which describes the transport sector in detail.

The rest of the chapter proceeds as follows. Section 2 outlines the structure of the REDGEM70 model and describes how to model the transport sector. Section 3 gives key input data and assumptions for the model. The model simulation results and discussion are presented in Section 4. Section 5 concludes the chapter.

2. Model Descriptions

2.1 Overview of REDGEM70

REDGEM70 is a bottom-up type, global energy systems optimization model with a detailed technological representation, which is formulated as an intertemporal linear programming problem. With a 5% discount rate, the model is designed to determine the optimal energy strategy for each of 70 world regions from 2000 to 2100 at 10-year time steps so that total discounted energy system costs are minimized under constraints on the satisfaction of exogenously given useful energy and energy service demands, the availability of primary energy resources, the maximum market penetration rate of new technologies, the atmospheric CO_2 concentration, etc. The model has a full flexibility in when and where CO_2 emissions reductions are achieved to stabilize the atmospheric CO_2 concentration at a given level.

The 70 regions of REDGEM70 are categorized into "energy production and consumption regions" and "energy production regions." The whole world is first divided into the 48 energy production and consumption regions to which future useful energy and energy service demands are allocated. The 22 energy production regions, which are defined as geographical points, are then distinguished from the energy production and consumption regions to represent the geographical characteristics of the areas endowed with large amounts of primary energy resources. Such a detailed regional disaggregation enables the explicit consideration of the regional characteristics in terms of energy resource supply, energy demands, and geography.

REDGEM70 is also characterized by a detailed description of the whole energy system, from primary energy supply through energy conversion to final energy consumption, as illustrated in Fig. 1. In the downstream part of the model, future useful energy demand trajectories are given for each of the industrial and residential/commercial sectors and decomposed by demand category, whereas future energy service demand trajectories (expressed in passenger-km (pkm) or tonne-km (tkm)) given for each of 13 transport modes. For each end-use demand category, the possibility of price-induced demand reductions, substitutability among final energy carriers (for example, high-quality energy carriers can be used for a wide range of applications), and changes in efficiency and costs associated with final energy substitution are considered in the model. All transport technologies, which refer to possible combinations of propulsion systems and transport fuels in this chapter, are characterized by parameters such as energy intensity, capital cost, and operating and maintenance (O&M) cost, and their cost-optimal mix is endogenously determined for each transport mode in the model.

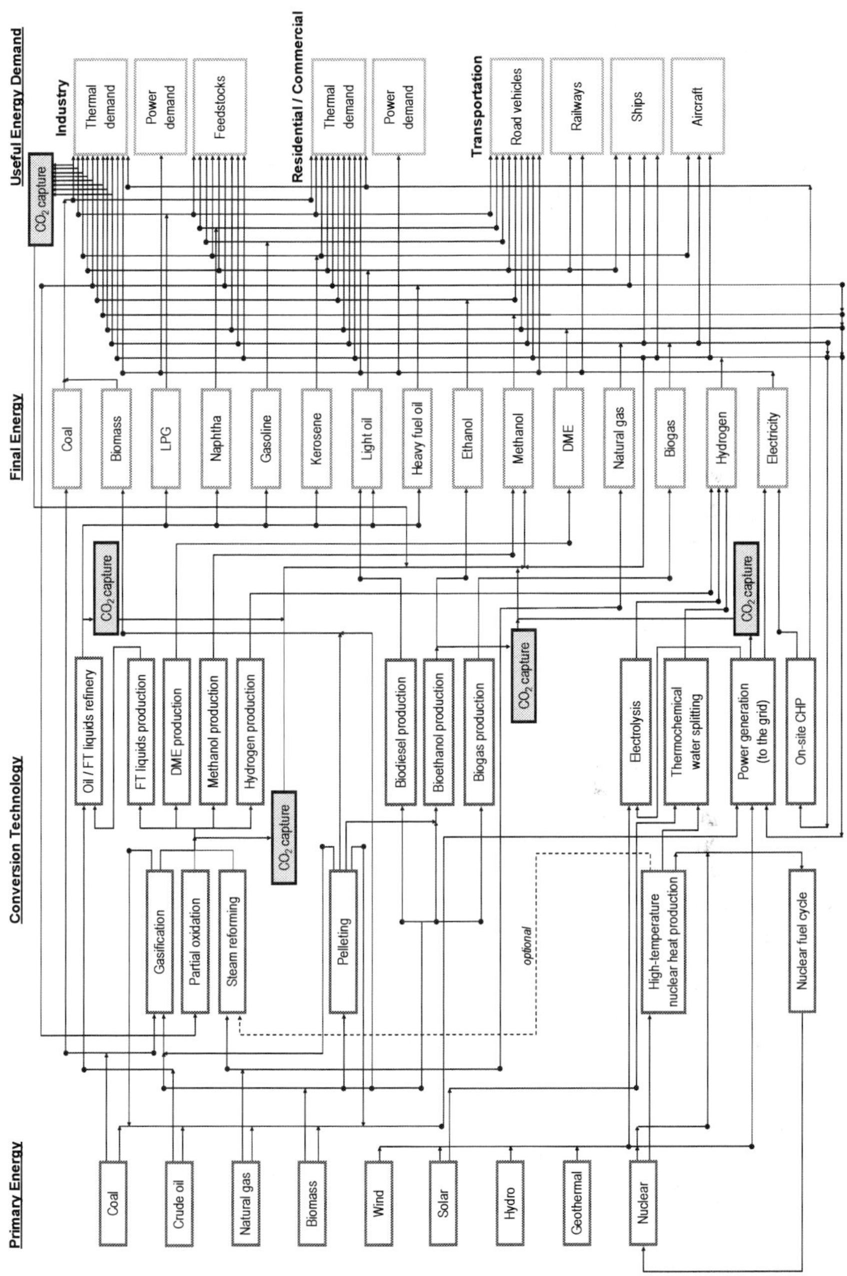

Fig. 1. Schematic representation of the structure of REDGEM70

On the supply side, REDGEM70 considers the entire supply chain of final energy carriers, which includes primary energy production, interregional energy transportation, coastal storage, conversion into secondary energy, intraregional secondary energy distribution, and final energy supply at retail sites (e.g., refuelling). To represent the economics of each of these final energy supply chain stages in a realistic manner, the model considers the capital and O&M costs separately at each stage of the fuel supply chain (excluding resource extraction) by treating the corresponding infrastructure explicitly. Note that final energy carriers are not always supplied in this order: a wide variety of final energy supply patterns can be selected in the model. The model treats the interregional transportation of 10 types of energy carriers and CO_2 between representative cities/sites in the 70 model regions and is able to identify its cost-optimal evolution path. Furthermore, the model considers the difference in the cost of local secondary energy distribution not only by energy carrier, but also by time point, region, and end-use sector. To make such modelling possible, the spatial structure of energy production and consumption regions is represented in detail in the model by consideration of the distribution of energy system components in this type of model regions, as illustrated in Fig. 2. The inclusion of the entire supply chain of final energy carriers, the separate consideration of capital and O&M costs across their entire supply chain, and the differentiation of intraregional secondary energy distribution costs (as described above) are three key features to help the model better represent the economics of transport fuels.

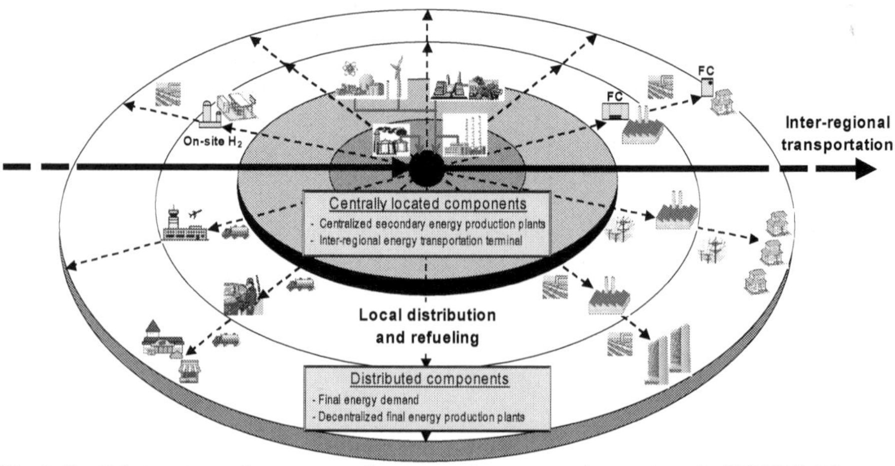

Fig. 2. Spatial structure of energy production and consumption regions in REDGEM70

REDGEM70 considers a number of promising energy conversion technologies. In particular, the model comprehensively includes technologies for producing alternative energy carriers such as synthetic fuels (i.e., hydrogen, methanol, dimethyl ether (DME), and Fischer-Tropsch (FT) synfuels) and conventional biofuels (i.e., bioethanol, biodiesel, and biogas). For biomass resources, the model considers not only plantation biomass such as energy crops (which are defined as fast-growing trees, e.g., hybrid poplars and willows, in the model), modern fuelwood, sugar crops, grain crops, and oilseed crops, but also waste biomass.

Given the amount of excess cropland that can be used for energy purposes, the model determines its optimal allocation among different plantation-based crop biomass productions based on crop yields per hectare of land, crop supply costs, and characteristics of conversion technologies available. The model also describes in detail the refinery process streams for crude oil and raw FT liquids, which consist of a lot of refinery processes. In the model, the CO_2 generated from power plants (excluding those used for on-site combined heat and power production and biomass-fired steam cycle power production), synthetic fuels production plants (excluding those used for converting stranded gas and decentralized small-scale hydrogen production), ethanol production plants, oil/FT refinery plants, and industrial processes can be captured for subsequent sequestration in geologic formations or methanol synthesis.

2.2 Transport sector submodel

In REDGEM70, passenger transport modes included are motorized two-wheelers, light-duty vehicles, buses, ordinary rail, high-speed rail, subsonic aircraft, and supersonic aircraft, whereas medium-duty trucks, heavy-duty trucks, freight rail, domestic shipping, international shipping, and freight air distinguished for freight transport. To take into account the inertia of each transport mode, its capital vintage structure (i.e., age structure) is represented in the model, where vehicles other than motorized two-wheelers and light-duty vehicles produced at a certain time period exist at the next time period.

In the model, energy requirements in the transport sector are derived from transport activity (measured in pkm and tkm) and actual in-use energy intensity (measured in MJ/pkm and MJ/tkm). The actual in-use energy intensities of road vehicles are calculated by dividing their respective on-road fuel economy (measured in MJ per vehicle-km) by their respective average occupancy rate (measured in passenger per vehicle and tonne per vehicle), whereas those of non-road transport modes are exogenous inputs to the model. The model allows for price-induced transport activity demand reductions by incorporating the long-run price elasticity of transport activity demand.

The road traffic supply-demand constraints are given by:

$$Ract(m,i,t) \leq \sum_s \sum_v LF(m,i,t)^* ADT(m,i,t)^* vin(m,s,t)^* V(m,v,i,s) + S(m,i,t) \tag{1}$$

where $Ract(m,i,t)$ is the demand for road transport (in pkm/tkm) carried by mode m in region i at time period t; $LF(m,i,t)$ is the load factor (i.e., vehicle occupancy rate) for mode m in region i at time period t; $ADT(m,i,t)$ is the annual distance travelled per vehicle (i.e., annual mileage per vehicle) for mode m in region i at time period t; $vin(m,s,t)$ is the remaining rate of transport technologies of vintage s available for mode m in their fleet stocks at time period t; $V(m,v,i,s)$ is the number of transport technologies v available for mode m produced in region i at time period s (which is endogenously determined in the model); and $S(m,i,t)$ is the price-induced transport activity demand reductions in mode m in region i at time period t.

On the other hand, the non-road traffic supply-demand constraints are given by:

$$NRact(m,i,t) \leq \sum_s \sum_v LF(m)^* vin(m,s,t)^* CAP(m,v,i,s) + S(m,i,t) \tag{2}$$

where $NRact(m,i,t)$ is the demand for non-road transport (in pkm/tkm) carried by mode m in region i at time period t and $CAP(m,v,i,s)$ is the capacity of transport technology v available for mode m produced in region i at time period s, which is defined in terms of pkm per year or tkm per year and is endogenously determined in the model. In this equation, domestic shipping is classified into two modes: large ships and small ships.

3. Data and Assumptions

3.1 Scenario driving forces

Future trajectories for scenario driving forces such as population, gross domestic product measured in purchasing power parities (GDP_{PPP}), and end-use demands are based on the "Middle Course" case B developed by the International Institute for Applied Systems Analysis (IIASA) and the World Energy Council (WEC) (Nakicenovic et al., 1998). End-use demand projections were first made for each of 11 world regions used in the IIASA/WEC study (Nakicenovic et al., 1998). They were then disaggregated into the 48 energy production and consumption regions of REDGEM70 by using country- and state-level statistics/estimates (and projections if available) on population, GDP_{PPP}, geography, energy use by type, and transport activity by mode, and by taking into account the underlying storyline of the case B that regional diversity might be somewhat preserved throughout the 21st century. Note that throughout this chapter, an 11-region classification is identical to that of the joint IIASA/WEC study (Nakicenovic et al., 1998).

Future transport activity demands were projected for each of the 13 transport modes and each of the 11 world regions mainly based on Victor (1990), Azar et al. (2000, 2003), Schafer & Victor (2000), and Fulton & Eads (2004). Fig. 3 shows the resulting passenger and freight transport activity demand projection by mode at the global level. Domestic ship transport is carried out by large and small ships. The share of each ship type in total domestic shipping activity was set for each of the 11 world regions based on Fulton & Eads (2004).

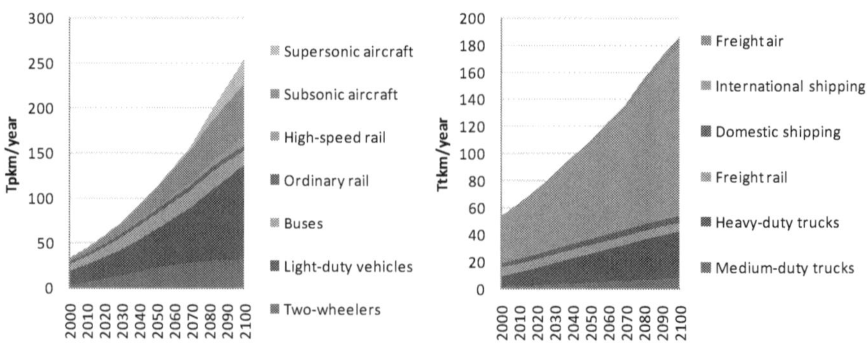

Fig. 3. Projected global passenger (left) and freight (right) transport activity demand

3.2 Delivered costs for transport fuels

This section focuses on the data and assumptions for the intraregional distribution and refuelling of transport fuels. A detailed description of the data and assumptions for the other stages of the final energy supply chain is given in Takeshita & Yamaji (2008) and Takeshita (2009, 2010). Table 1 shows the intraregional distribution and refuelling costs for each transport fuel. It is implicitly assumed that the intraregional distribution of CNG and GH_2 is made by pipeline and that liquid transport fuels are distributed intraregionally by truck, except that the distribution of LNG and LH_2 to airports is by rail. For the supply of LNG or LH_2 to aircraft, two possible pathways are considered: (1) the receipt of CNG/GH_2 via pipeline at an airport boundary followed by the liquefaction of natural gas/hydrogen and the supply of LNG/LH_2 to aircraft; and (2) the receipt of LNG/LH_2 via rail at an airport boundary followed by the supply of LNG/LH_2 to aircraft (Brewer, 1991).

Transport fuel	Intraregional distribution cost (USD/GJ)	Refuelling cost (USD/GJ)
Petroleum and FT products	0.8	1.3
Liquefied petroleum gas (LPG)	1.1	2.1
Ethanol	1.0	1.9
DME	1.5	3.0
Liquefied natural gas (LNG)		
LNG supply to international ocean-going ships	0	4.8
LNG supply to aircraft	1.8	4.8
Liquid hydrogen (LH_2)		
LH_2 delivery and GH_2 refuelling	2.5	5.6
LH_2 delivery and LH_2 refuelling		
LH_2 supply to medium-duty trucks	2.5	5.0
LH_2 supply to aircraft	2.5	6.7
Compressed natural gas (CNG)		
CNG supply to light-duty vehicles and heavy-duty trucks	3.3	3.3
CNG supply to buses and medium-duty trucks	2.0	3.3
CNG supply to aircraft	1.3	-
Gaseous hydrogen (GH_2)		
Centralized H_2 production		
GH_2 supply to light-duty vehicles	4.7	4.7
GH_2 supply to buses and medium-duty trucks	2.8	4.7
GH_2 supply to domestic freight ships	1.9	6.3
GH_2 supply to international ocean-going ships	0	6.3
GH_2 supply to aircraft	1.9	-
Decentralized H_2 production	-	3.9
Electricity		
Electricity supply to two-wheelers and light-duty vehicles	5.1	5.0
Electricity supply to buses and medium-duty trucks	3.1	5.0

Table 1. Intraregional distribution and refuelling costs for transport fuels

In addition to their temporal development, REDGEM70 takes into account the site-specific feature of the intraregional distribution costs of transport fuels, in particular gaseous fuels (Azar et al., 2000). Following the approach proposed by Ogden (1999a), the intraregional distribution costs of CNG, GH_2, and electricity are assumed to vary depending on the density of final energy demands. They are estimated to be lower for urban areas where a geographically concentrated demand exists (Ogden, 1999a; van Ruijven et al., 2007). It is assumed that there is a high correlation between the density of final energy demands and the level of urbanization (i.e., the percentage of the population living in urban areas). By using this relationship and the local GH_2 distribution cost function proposed by Ogden (1999a, p.252), the intraregional distribution cost of GH_2 was estimated for each world region and each time period as a function of the level of urbanization. The intraregional distribution costs of CNG and electricity were estimated similarly with their world average values for the year 2000 taken into account.

In the light of the degree of spatial distribution of refuelling points for each transport mode, ranging from centralized to completely decentralized, the model considers the difference in the intraregional distribution costs of CNG, GH_2, and electricity by transport mode: costs of distributing them to aircraft and domestic freight ships are assumed to be 60% lower than, costs of distributing them to buses and medium-duty trucks are assumed to be 40% lower than, and costs of distributing them to motorized two-wheelers and heavy-duty trucks are assumed to be the same as those of distributing them to light-duty vehicles, whereas the intraregional distribution of transport fuels to international ocean-going ships is assumed to be unnecessary. These assumptions are based on the fact that delivery trucks and buses are usually centrally refuelled, and that long-haul heavy-duty trucks must be able to refuel at reasonable distances (IEA, 2008). The intraregional distribution costs of liquid transport fuels are assumed to be the same across all transport modes because the distribution distance has a small impact on them (Amos, 1998; Simbeck & Chang, 2002).

The share of capital costs in total costs is assumed to be 85% for pipeline distribution and electric power transmission, whereas the corresponding estimate is 33% for truck distribution and 75% for refuelling (Amos, 1998; Simbeck & Chang, 2002). Considering that the major expense is not the pipeline cost itself but installing the pipeline (Amos, 1998) and that installed pipeline capital costs are site specific (Ogden, 1999a), installed capital costs of pipelines and power transmission lines by world region were calculated by applying a region-specific location factor.

3.3 Techno-economic data and assumptions for transport technologies

It is assumed that the average lifetime is 10 years for motorized two-wheelers and light-duty vehicles, 15 years for buses and trucks, and 20 years for trains, ships, and aircraft. Based on data from Landwehr & Marie-Lilliu (2002), the long-run price elasticity of transport activity demand was set at -0.17 for motorized two-wheelers and light-duty vehicles, -0.18 for aircraft, -0.20 for trucks, and 0 for the other transport modes.

Fig. 4 shows the actual in-use energy intensity of a conventional reference transport technology by transport mode for the years 2000, 2050, and 2100. For the definition of a conventional reference transport technology, see footnote in Fig. 4. Note that the actual in-use energy intensity of transport technologies of the vintages of the same year as that in which they are operated is shown in these figures.

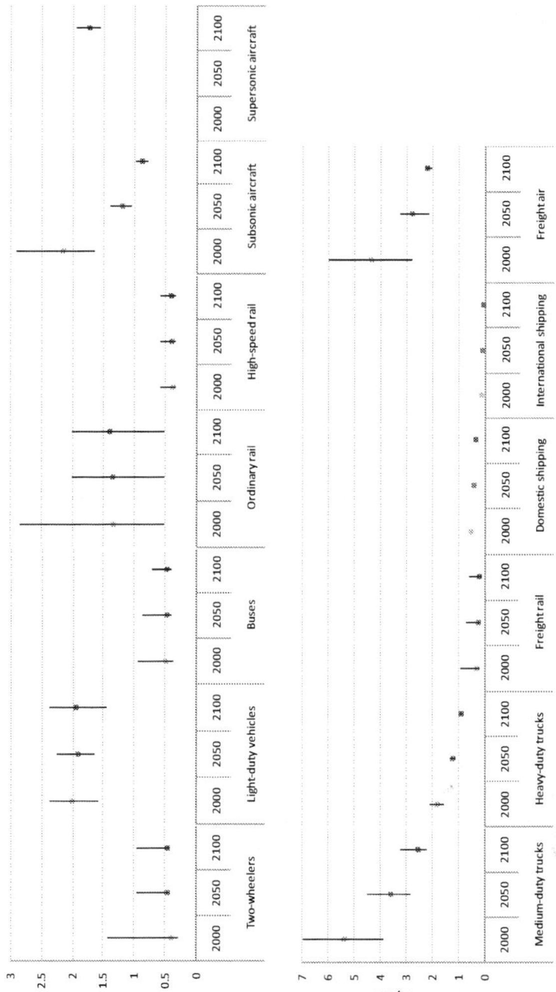

Fig. 4. Projected actual in-use energy intensities of passenger (upper) and freight (lower) transport modes[a,b]

[a] These figures show the actual in-use energy intensities of reference transport technologies. It is assumed that the reference transport technology is a gasoline internal combustion engine (ICE) vehicle for motorized two-wheelers and light-duty vehicles, a diesel ICE vehicle for buses, trucks, non-high-speed rail, and domestic shipping, a heavy fuel oil (HFO) ICE vehicle for international shipping, and a kerosene ICE vehicle for aircraft.

[b] The world average shown as squares in these figures is calculated as the activity-weighted average of the actual in-use energy intensity of each transport mode. The range denotes the difference by world region.

The on-road fuel economy of conventional gasoline ICE light-duty vehicles was projected for each of the 11 world regions by taking into account future improvements in their test-based fuel economy due to technical progress, recent trends (e.g., towards larger and more powerful vehicles), current and future expected policies, and the gap between their test and on-road fuel economy. Except for high-speed rail and aircraft, improved fuel efficiencies of passenger transport technologies would be offset to some small or large degree by declining vehicle occupancy rates (Schafer & Victor, 1999; Azar et al., 2000). For high-speed rail, it is assumed that a development towards faster speeds would offset technical efficiency gains (Azar et al., 2000). In contrast, it is indicated that large reductions in the actual in-use energy intensity of aircraft are possible (Schafer & Victor, 1999).

By conducting a comprehensive survey of literature and interviewing experts, possible combinations of propulsion systems and transport fuels were defined for each transport mode and techno-economic parameters were set for each transport technology. As an example, Table 2 shows the assumed possible combinations of propulsion systems and transport fuels for road vehicles. A hybrid propulsion system is not considered for long-haul heavy-duty trucks because they operate primarily on highways at near to maximum rated power and because hybrids are estimated to provide virtually no efficiency benefits on highway driving cycles (Fulton & Eads, 2004). Durability is a key issue for fuel cell propulsion systems, so they are not considered for long-haul heavy-duty trucks that often travel over 100,000 km/year (IEA, 2008).

Transport technologies available for non-high-speed rail are assumed to be diesel and electric trains, while those available for high-speed rail are assumed to be high-speed electric trains and magnetic levitation (maglev) systems. Contrary to IEA (2008) and Electris et al. (2009), fuel cell propulsion systems are not considered for the non-high-speed rail sector for the same reason as in the case of heavy-duty trucks. Because the two transport technologies available for high-speed rail are powered by electricity and because the actual in-use energy intensity of the maglev systems is estimated to fall to that of high-speed electric trains (Azar et al., 2000), the electricity consumption of the high-speed rail sector is given exogenously to the model and each of the two transport technologies is not characterized in the model. As regards the freight shipping sector, transport technologies available for small ships are assumed to be diesel ICEs, diesel ICEs with electric motors, and GH_2 fuel cell hybrids, while those available for large ships are assumed to be HFO ICEs, LNG ICEs with electric motors, and HFO ICEs with a GH_2 fuel cell auxiliary power unit (APU).

Based on Victor (1990) and IEA (2005), it is assumed that not only kerosene-fuelled aircraft but also LNG- and LH_2-fuelled aircraft are available for the subsonic aviation sector. In contrast, the supersonic aviation sector is assumed to have no CO_2 mitigation options other than biomass-derived FT kerosene. This is because supersonic aircraft fly in the stratosphere 80-85% of the time, where water vapour has a far more powerful greenhouse effect than in the troposphere (Penner et al., 1999), and because the intensity of water vapour emissions, expressed as amount of emissions per unit of transport activity, is much higher for LNG- and LH_2-fuelled aircraft than for kerosene-fuelled aircraft (more than three times higher for LH_2-fuelled aircraft than for kerosene-fuelled aircraft). Supersonic aircraft are assumed to be consistently half as energy efficient as subsonic aircraft (Victor, 1990).

Transport technology	Two-wheelers	Light-duty vehicles	Buses	Medium-duty trucks	Heavy-duty trucks
Gasoline ICEVs	+	+	+	+	
Diesel ICEVs		+	+	+	+
LPG ICEVs		+	+	+	
Gasohol ICEVs	+	+	+	+	
Ethanol ICEVs	+	+	+	+	+
DME ICEVs		+	+	+	+
CNG ICEVs		+	+	+	+
GH_2 ICEVs		+	+		
LH_2 ICEVs				+	
Gasoline HEVs		+	+	+	
Diesel HEVs		+	+	+	
LPG HEVs		+	+	+	
Gasohol HEVs		+	+	+	
Ethanol HEVs		+	+	+	
DME HEVs		+	+	+	
CNG HEVs		+	+	+	
GH_2 HEVs		+	+	+	
Gasoline PHEVs		+		+	
Diesel PHEVs		+		+	
Gasohol PHEVs		+		+	
Ethanol PHEVs		+		+	
Gasoline FCHVs		+	+	+	
DME FCHVs		+	+	+	
GH_2 FCHVs		+	+	+	
BEVs	+	+	+	+	

Table 2. Possible combinations of propulsion systems and transport fuels for road vehicles[a,b]
[a] Possible combinations of propulsion systems and transport fuels are marked by pluses (+).
[b] ICEVs=internal combustion engine vehicles; HEVs=hybrid electric vehicles; PHEVs=plug-in hybrid electric vehicles; FCHVs=fuel cell hybrid vehicles; BEVs=battery electric vehicles. Gasohol is defined as a 10% ethanol to 90% gasoline volumetric blend.

Except for pure electric vehicles, the capital cost of light-duty vehicles was estimated for all alternative transport technologies that have a consumer performance (such as range, acceleration, passenger and cargo capacity) comparable to that of their conventional gasoline ICE counterpart. Based on Grahn et al. (2009) and IEA (2009), pure electric light-duty vehicles are assumed to have a driving range of 200 km, whereas all other transport technologies available for light-duty vehicles are assumed to have a driving range of 500 km. To compensate for such reduced driving range, pure electric vehicles are likely to require fast charging stations in cities and/or along certain corridors (IEA, 2009). Following the method of Simbeck & Chang (2002), they were estimated to add USD 5/GJ to the delivered cost of electricity (see Table 1). Similar to Grahn et al. (2009), plug-in hybrid vehicles are assumed to operate as electric vehicles for 65% of their daily driving.

The assumptions about the specific cost of batteries (in USD/kWh) designed for road vehicles are based on IEA (2009). Li-ion batteries for pure electric light-duty vehicles with a 200 km range were estimated to cost USD 478/kWh in 2020, and their specific cost was expected to decline to USD 330/kWh by 2030. The specific cost of Li-ion batteries for pure electric buses and pure electric medium-duty trucks can be estimated from the relationship between the energy (kWh) and specific cost of Li-ion batteries: the specific cost of Li-ion batteries for pure electric vehicles was estimated to be 13% and 10% lower for buses and medium-duty trucks, respectively, than for light-duty vehicles. Specific battery costs differ by vehicle type. For light-duty vehicles, the specific cost of Li-ion batteries was estimated to eventually drop to USD 460/kWh for conventional hybrids and USD 420/kWh for plug-in hybrids, respectively.

On the other hand, the specific cost of a PEM fuel cell stack (in USD/kW) was estimated to drop to USD 500/kW in 2030 and to eventually reach USD 95/kW in 2050 (IEA, 2008; Grahn et al., 2009). For hydrogen storage, the specific cost of a GH_2 storage tank at a pressure of 700 bar (in USD/kg) was estimated to drop to USD 447/kg in 2030 and to eventually reach USD 313/kg in 2050 (IEA, 2005; Grahn et al., 2009), and the specific cost of a LH_2 storage tank (in USD/kg) is assumed to drop to USD 313/kg in 2050 (WBCSD, 2004). For the purpose of sensitivity analysis, two different values were considered for the future costs of these technologies. Under optimistic assumptions, the specific cost in 2050 was estimated to be USD 65/kW for a PEM fuel cell stack and USD 179/kg for a GH_2/LH_2 storage tank. Under pessimistic assumptions, the specific cost in 2050 was estimated to be USD 125/kW for a PEM fuel cell stack and USD 447/kg for a GH_2/LH_2 storage tank. These assumptions were made based on IEA (2008) and Grahn et al. (2009).

3.4 Climate policy scenario

Unless otherwise noted, REDGEM70 is run under the constraint that the atmospheric concentration of CO_2 will be stabilized at 400 ppmv in 2100, which has been assumed to assure stabilization of climate change at 2.0 to 2.4 degrees Celsius by 2100 (Metz et al., 2007). The reason for the choice of this constraint is because the Intergovernmental Panel on Climate Change (IPCC) Fourth Assessment Report (Metz et al., 2007) states that avoidance of many key vulnerabilities requires temperature change in 2100 to be below 2.6 degrees Celsius above pre-industrial levels and estimates that achieving the CO_2 stabilization target of 400 ppmv would be a sufficient condition for limiting the global mean temperature change below 2.6 degrees Celsius above pre-industrial levels, using a best estimate climate sensitivity of 3.0 degrees Celsius. Overshoots are allowed before 2100 in model simulations.

4. Simulation Results and Discussion

4.1 Definition of simulation cases

The five cases as defined in Table 3 are simulated with REDGEM70 to examine (1) the cost-optimal choice of transport technologies under the 400 ppmv CO_2 stabilization constraint, (2) the effect of future costs of hydrogen-fuelled transport technologies on the cost-competitiveness of hydrogen in the transport sector under the 400 ppmv CO_2 stabilization constraint, and (3) the effect of the appearance of supersonic aircraft on the cost-optimal technology strategy for the transport sector under the 400 ppmv CO_2 stabilization constraint.

Case	Climate policy	Costs of a PEM FC stack and a H_2 storage tank	Demand for supersonic aviation
No CO_2 constraint case	No policy intervention	Reference values	Reference values
400 ppmv case	CO_2 stabilization at 400 ppmv	Reference values	Reference values
400 ppmv case with OPT assumptions on hydrogen vehicles	CO_2 stabilization at 400 ppmv	Optimistic values	Reference values
400 ppmv case with PESS assumptions on hydrogen vehicles	CO_2 stabilization at 400 ppmv	Pessimistic values	Reference values
400 ppmv case without the demand for supersonic aviation	CO_2 stabilization at 400 ppmv	Reference values	Assumed not to occur

Table 3. Cases considered for simulation

4.2 Results for the entire transport sector

Fig. 5 shows the cost-optimal mix of transport fuels at the global level. In this figure, the consumption of each transport fuel is shown for each transport mode to examine the cost-optimal choice of transport technologies by transport mode. If the climate stabilization constraint is not imposed, petroleum products continue to dominate the global transport fuel consumption and the contribution of CO_2-neutral transport fuels to it is very small. In contrast, the global final-energy mix of the transport sector becomes diversified in the CO_2 400 ppmv stabilization cases. Comparing the results of the no CO_2 constraint and 400 ppmv cases shows that hydrogen, electricity, biomass-derived FT synfuels, and natural gas are promising transport fuels contributing substantially to the reduction of CO_2 emissions from the transport sector.

As an alternative fuel for diesel engines, FT diesel is preferred to DME because FT synfuels have an advantage over DME in that they are largely compatible with current vehicles and existing infrastructure for petroleum fuels. In all regions, biodiesel is produced from all the available amount of waste grease and oil and used in the transport sector from 2020, but its small resource potential makes the share of biodiesel negligible.

Total global transport fuel consumption in the CO_2 400 ppmv stabilization cases is smaller than that in the no CO_2 constraint case. This is mainly due to the deployment of highly efficient transport technologies such as conventional and plug-in hybrids in the former cases. This trend is especially evident from 2040 onward because of technical progress and discounting. However, even in these CO_2 400 ppmv stabilization cases, total global transport fuel consumption begins to increase sharply from around 2070, which is caused by the increasing demand for supersonic aviation. The lack of CO_2 mitigation options other than biomass-derived FT kerosene in the supersonic aviation sector and insufficient biomass supply potential are the reasons for this.

As expected, the assumptions on the costs of a PEM fuel cell stack and a GH_2/LH_2 storage tank have an evident impact on the total global hydrogen consumption of the transport sector under the 400 ppmv CO_2 stabilization constraint.

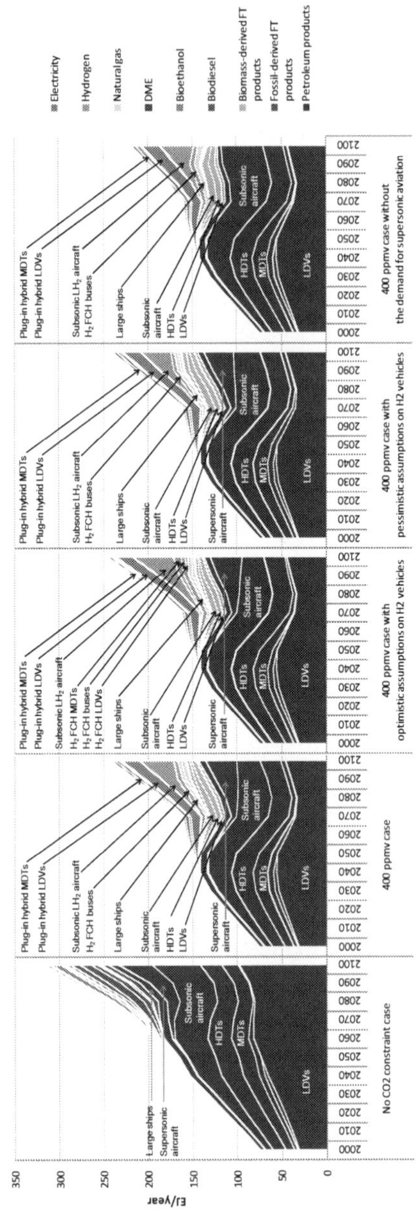

Fig. 5. Cost-optimal mix of transport fuels[a]

[a] LDVs = light-duty vehicles; MDTs = medium-duty trucks; HDTs = heavy-duty trucks; FCH = fuel cell hybrid.

4.3 Mode-specific results

It should be noted that the relative share of each transport fuel in total transport fuel consumption expressed in energy terms (as shown in Fig. 5) does not reflect its actual contribution to meeting the demand for transport, which can be evaluated by looking at the relative share of transport technologies running on each transport fuel in total transport activity. In the light of the relative importance of each transport mode in terms of transport activity and transport fuel consumption, Figs. 6-10 show the cost-optimal choice of transport technologies, at the global level, for the light-duty vehicle, medium-duty truck, heavy-duty truck, large ship, and passenger aviation sectors, respectively.

It can be seen from Figs. 5-10 that the cost-optimal choice of transport technologies under the 400 ppmv CO_2 stabilization constraint differs significantly by transport mode. In the light-duty vehicle sector, plug-in hybrids and biomass-derived FT gasoline are selected as cost-effective CO_2 mitigation options (see Figs. 5 and 6). Gasohol as well as gasoline (including FT gasoline) is the main fuel for light-duty vehicles in all cases. Regardless of the simulation cases, there is a trend in this sector that propulsion systems of choice shift from ICEs to conventional hybrids and ultimately to plug-in hybrids. In all cases, plug-in hybrids account for a large share of the total global activity of light-duty vehicles in the second half of the century. This implies that although pure electric vehicles are not selected in any case (which is also true for all other transport modes), promoting the electrification of light-duty vehicles is a robust future technology strategy. The cost-competitiveness of hydrogen fuel cell hybrid light-duty vehicles is weak, judging from the result that they could account for a tiny share toward the end of the century only in the 400 ppmv case with optimistic assumptions on hydrogen vehicles.

Fig. 5 shows that buses are a key niche market for hydrogen fuel cell (hybrid) vehicles. The first reason for this is that, as pointed out by Ogden (1999b) and IEA (2005), in the bus market cost goals for hydrogen fuel cell (hybrid) vehicles which if successfully met can insure high marketability are not as stringent as for light-duty vehicles.

In general, the share of capital costs in total lifetime costs is smaller for large commercial vehicles such as buses and trucks than it is for light-duty vehicles because the annual mileage and average lifetime of the former are longer than those of the latter. Hence, the benefits of high energy efficiency and low maintenance costs are more pronounced for large commercial vehicles. In other words, such characteristics of large commercial vehicles allow the additional capital cost to be spread over a longer period of time, which results in more favourable conditions for highly efficient, capital-intensive transport technologies such as hydrogen fuel cell (hybrid) vehicles. The second reason is that buses are centrally refuelled, which helps to overcome fuel infrastructure obstacles (IEA, 2008).

Contrary to the prior expectation, the conditions found in the bus market are not applied to the medium-duty truck market. The results in Figs. 5 and 7 reveal that the medium-duty truck sector makes the choice of transport technologies similar to that of the light-duty vehicle sector under the 400 ppmv CO_2 stabilization constraint, except that diesel engines are preferred by the former. One explanation might be that unlike buses, the actual in-use energy intensity of diesel engine medium-duty trucks exhibits a consistently declining trend (see Fig. 4), which does not provide an adequate incentive for the uptake of highly efficient, capital-intensive hydrogen fuel cell hybrid vehicles to the medium-duty truck market.

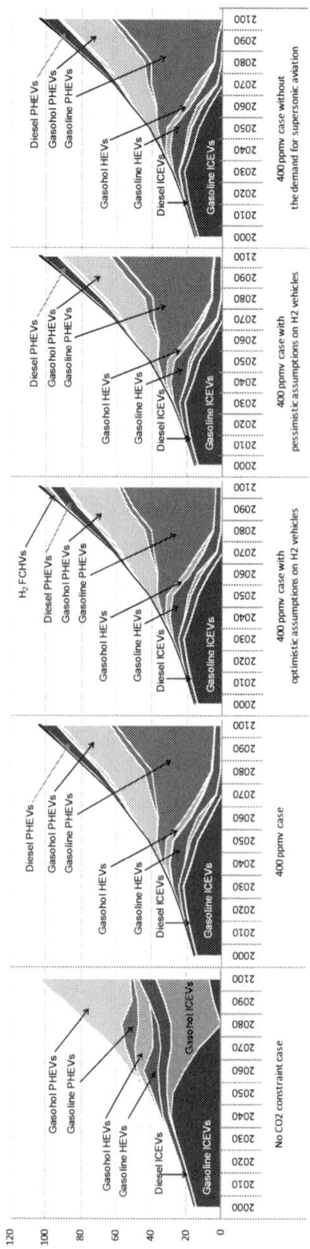

Fig. 6. Cost-optimal choice of transport technologies in the light-duty vehicle sector

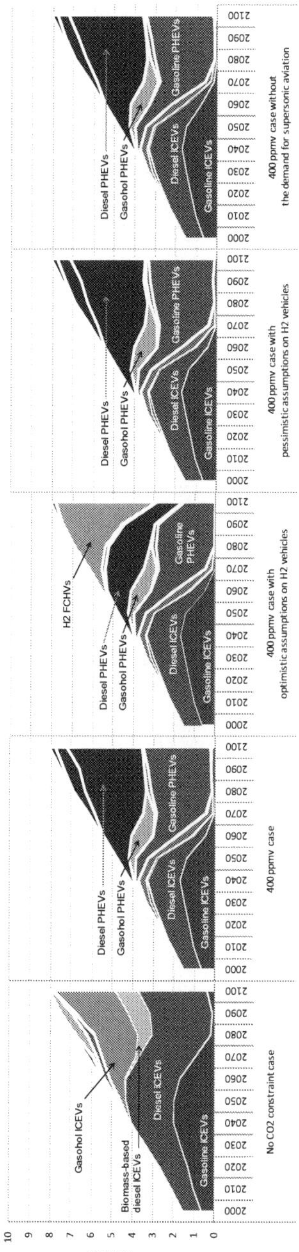

Fig. 7. Cost-optimal choice of transport technologies in the medium-duty truck sector

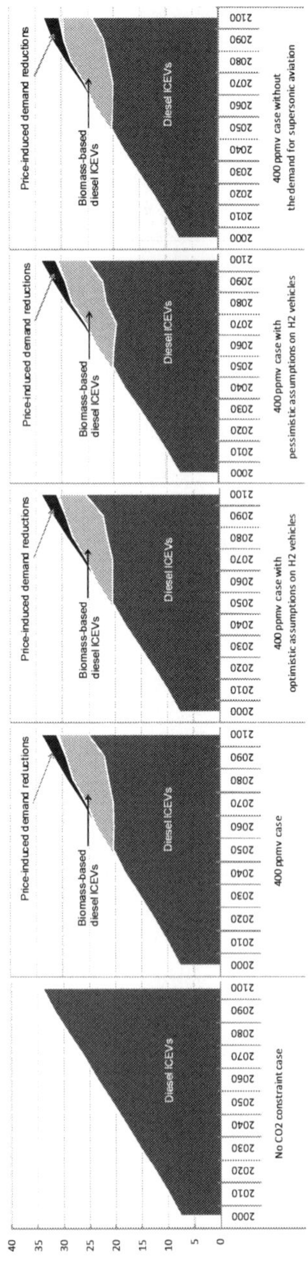

Fig. 8. Cost-optimal choice of transport technologies in the heavy-duty truck sector

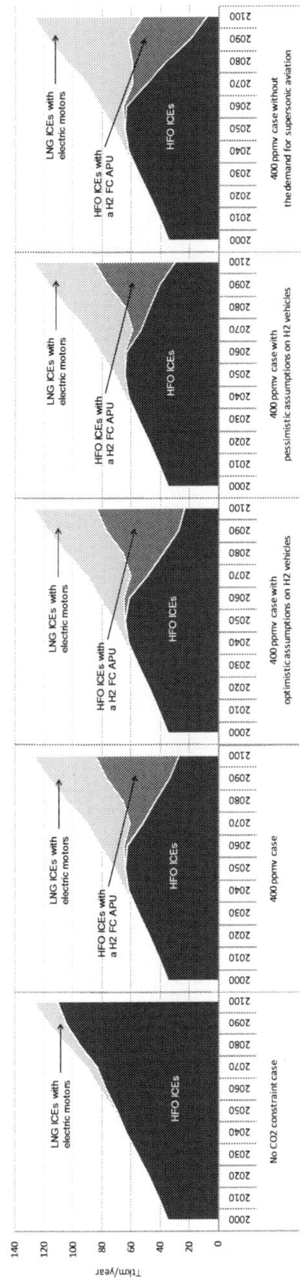

Fig. 9. Cost-optimal choice of transport technologies in the large ship sector

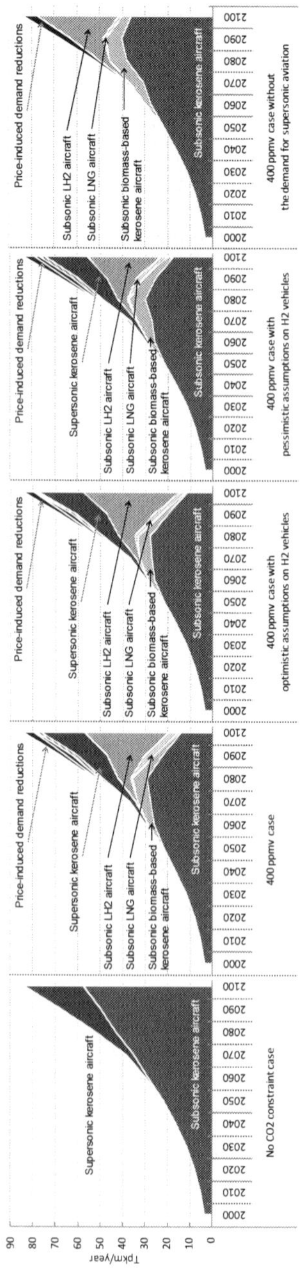

Fig. 10. Cost-optimal choice of transport technologies in the passenger aviation sector

On the other hand, biomass-derived FT diesel is introduced into the heavy-duty truck market as a cost-effective CO_2 mitigation option (see Figs. 5 and 8). Even if the 400 ppmv CO_2 stabilization constraint is imposed, petroleum diesel ICEs have a dominant share of the total global activity of heavy-duty trucks. This is also mainly due to the consistently declining trend of the actual in-use energy intensity of diesel engine heavy-duty trucks. The assumption that the heavy-duty truck sector has a limited range of CO_2 mitigation options is another reason for this. In the CO_2 400 ppmv stabilization cases except the 400 ppmv case without the demand for supersonic aviation, the share of petroleum diesel increases toward the end of the century, which is a by-product of refinery operations to provide petroleum kerosene for the supersonic aviation sector as described below.

It is interesting to note that the results for the motorized two-wheeler sector are similar to those for the ordinary rail sector. First, energy consumption of the motorized two-wheeler and rail sectors (including the high-speed rail sector) is relatively small compared to other transport sectors because of their low actual in-use energy intensity and low levels of transport activity. Second, the share of electricity in the total global final energy consumption of each of the two sectors increases over time to significantly high levels in all cases. It should be emphasized, however, that there is a stark contrast between the cost-optimal fuel mix of power generation in the CO_2 400 ppmv stabilization cases and that in the no CO_2 constraint case. This finding for the motorized two-wheeler sector can be regarded as reasonable because electric scooters are already very popular in China and because they are usually powered by small electric motors driven by lead-acid batteries making them the cheapest form of motorized transport in China (IEA, 2009). This finding for the ordinary rail sector suggests that the assumed autonomous trend towards the electrification of the rail system combined with the progressive decarbonisation of power generation cost-effectively reduces CO_2 emissions from the ordinary rail sector on a well-to-wheel basis.

Despite very high levels of the activity of large ships, their considerably low actual in-use energy intensity resulting from the high efficiency of large, slow-speed, two-stroke engines (as much as 50% on a lower heating value basis) makes their total global final energy consumption very small. The total global final energy consumption of small ships is negligible because of their low actual in-use energy intensity and low levels of transport activity. As shown in Fig. 9, an electric propulsion system composed of LNG-fuelled ICE generators and electric motors serves as a cost-effective CO_2 mitigation option for large ships.

According to the results shown in Figs. 5 and 10, biomass-derived FT kerosene-fuelled and LH_2-fuelled aircraft enter the subsonic passenger aviation market as cost-effective CO_2 mitigation options. In the CO_2 400 ppmv stabilization cases, subsonic aviation becomes the largest user of hydrogen in the transport sector toward the end of the century. The cost-competitiveness of hydrogen in the subsonic aviation sector is significantly strong toward the end of the century in the CO_2 400 ppmv stabilization cases: subsonic LH_2-fuelled aircraft hold a share of 65.4% in the total global activity of subsonic passenger aircraft in 2100 in the 400 ppmv case. Even when the future cost of a LH_2 storage tank takes its upper bound value, subsonic LH_2-fuelled aircraft account for a large share toward the end of the century under the 400 ppmv CO_2 stabilization constraint. The first reason for the strong competitiveness of subsonic LH_2-fuelled aircraft is that they do not suffer from the high cost of a fuel cell propulsion system. The second reason is that aircraft are centrally refuelled. The third reason is that under the stringent climate stabilization constraint, hydrogen production from biomass with CO_2 capture and storage (CCS) is preferred to FT synfuels production from biomass with CCS because the former can sequester larger amounts of CO_2. In contrast, the supersonic passenger aviation sector relies heavily on petroleum kerosene in the CO_2 400 ppmv stabilization cases. This is because of the assumed unavailability of LNG and LH_2 in the supersonic aviation sector, the low efficiency of supersonic aircraft, and the severe supply constraint on biomass.

5. Conclusions

This chapter has examined the cost-optimal choice of propulsion systems and fuels for each of the 13 transport modes over the 21st century under a constraint that the long-term global mean temperature rise would be limited to 2.0 to 2.4 degrees Celsius. It has been shown that the cost-optimal choice of transport technologies under the 400 ppmv CO_2 stabilization constraint differs significantly by transport mode. Such a detailed analysis by transport mode is a new contribution. A future study will consider the impact of the specific cost of batteries on the cost-optimal choice of transport technologies under the stringent climate stabilization target and include non-CO_2 GHGs (e.g., water vapour) in the model.

6. References

Amos, W.A. (1998). Costs of Storing and Transporting Hydrogen, National Renewable Energy Laboratory, NREL/TP-570-25106, Golden, CO.

Azar, C.; Lindgren, K. & Andersson, B.A. (2000). *Hydrogen or methanol in the transportation sector ?*, KFB, ISBN: 91-88371-90-5, Stockholm.

Azar, C.; Lindgren, K. & Andersson, B.A. (2003). Global energy scenarios meeting stringent CO_2 constraints - cost-effective fuel choices in the transportation sector. *Energy Policy*, Vol. 31, No. 10, 961-976, ISSN: 0301-4215.

Brewer, G.D. (1991). *Hydrogen Aircraft Technology*, CRC Press, ISBN: 0-8493-5838-8, Boca Raton, FL.

Electris, C.; Raskin, P.; Rosen, R. & Stutz, J. (2009). *The Century Ahead : Four Global Scenarios*, Tellus Institute, Technical Documentation, Boston, MA.

Fulton, L. & Eads, G. (2004). *IEA/SMP Model Documentation and Reference Case Projection*, International Energy Agency/World Business Council for Sustainable Development, Available at: <http://www.wbcsd.org/web/publications/mobility/smp-model-document.pdf>.

Grahn, M.; Azar, C.; Williander, M.I.; Anderson, J.E.; Mueller, S.A. & Wallington, T.J. (2009). Fuel and vehicle technology choices for passenger vehicles in achieving stringent CO_2 targets: connections between transportation and other energy sectors. *Environmental Science & Technology*, Vol. 43, No. 9, 3365-3371, ISSN: 0013-936X.

IEA (International Energy Agency) (2000). *World Energy Outlook 2000*, IEA, ISBN: 92-64-18513-5, Paris.

IEA (2005). *Prospects for Hydrogen and Fuel Cells*, IEA, ISBN: 92-64-109-579, Paris.

IEA (2008). *Energy Technology Perspectives 2008*, IEA, ISBN: 92-64-04142-4, Paris.

IEA (2009). *Transport, Energy and CO_2*, IEA, ISBN: 978-92-64-07316-6, Paris.

Landwehr, M. & Marie-Lilliu, C. (2002). *Transportation Projections in OECD Regions*, IEA, Paris.

Metz, B.; Davidson, O.R.; Bosch, P.R.; Dave, R. & Meyer, L.A., (Ed.) (2007). *Climate Change 2007: Mitigation of Climate Change*, Cambridge University Press, ISBN: 978-0521-70598-1, Cambridge.

Nakicenovic, N.; Grubler, A. & McDonald, A., (Ed.) (1998). *Global Energy: Perspectives*, Cambridge University Press, ISBN: 0521-64569-7, Cambridge.

Ogden, J.M. (1999a). Prospects for building a hydrogen energy infrastructure. *Annual Review of Energy and the Environment*, Vol. 24, 227-279, ISSN: 1056-3466.

Ogden, J.M. (1999b). Developing an infrastructure for hydrogen vehicles: a Southern California case study. *International Journal of Hydrogen Energy*, Vol. 24, No. 8, 709-730, ISSN: 0360-3199.

Penner, J.E.; Lister, D.H.; Griggs, D.J.; Dokken, D.J. & McFarland, M. (1999). *Aviation and the Global Atmosphere*, Cambridge University Press, ISBN: 0521-66404-7, Cambridge.

Schafer, A. & Victor, D.G. (1999). Global passenger travel: implications for carbon dioxide emissions. *Energy*, Vol. 24, No. 8, 657-679, ISSN: 0360-5442.

Schafer, A. & Victor, D.G. (2000). The future mobility of the world population. *Transportation Research Part A*, Vol. 34, No. 3, 171-205, ISSN: 0965-8564.

Simbeck, D.R. & Chang, E. (2002). *Hydrogen Supply: Cost Estimate for Hydrogen Pathways – Scoping Analysis*, National Renewable Energy Laboratory, NREL/SR-540-32525, Golden, CO.

Takeshita, T. (2009). A strategy for introducing modern bioenergy into developing Asia to avoid dangerous climate change. *Applied Energy*, Vol. 86, Supplement 1, S222-S232, ISSN: 0306-2619.

Takeshita, T. (2010). Cost-optimal use of bioenergy under a stringent climate stabilization target, In: *Green Energy & Technology*, Machrafi, H., (Ed.), Bentham Science Publishers, Bussum, The Netherlands (forthcoming).

Takeshita, T. & Yamaji, K. (2008). Important roles of Fischer-Tropsch synfuels in the global energy future. *Energy Policy*, Vol. 36, No. 8, 2791-2802, ISSN: 0301-4215.

Turton, H. (2006). Sustainable global automobile transport in the 21st century: an integrated scenario analysis. *Technological Forecasting & Social Change*, Vol. 73, No. 6, 607-629, ISSN: 0040-1625.

van Ruijven, B.; van Vuuren, D.P. & de Vries, B. (2007). The potential role of hydrogen in energy systems with and without climate policy. *International Journal of Hydrogen Energy*, Vol. 32, No. 12, 1655-1672, ISSN: 0360-3199.

Victor, D.G. (1990). Liquid hydrogen aircraft and the greenhouse effect. *International Journal of Hydrogen Energy*, Vol. 15, No. 5, 357-367, ISSN: 0360-3199.

WBCSD (World Business Council for Sustainable Development) (2004). *Mobility 2030: meeting the challenges to sustainability*, WBCSD, ISBN: 2-940240-57-4, Geneva, Switzerland.

Impact of climate change on health and disease in Latin America

Alfonso J. Rodríguez-Morales, Alejandro Risquez and Luis Echezuria
Department of Preventive and Social Medicine, Luis Razetti Medical School, Faculty of Medicine, Universidad Central de Venezuela
Caracas, Venezuela

1. Introduction

Climate change is a widely known problem and consequence of multiple interacting phenomena. Nobody can deny it, and several issues are identified as cause and implications of this global problem. Among these causes there is a basic important element, its anthropogenic origin. Many styles and quality of modern life, explained by the use and abuse of contaminating energy sources, including the forms of generating, producing and distributing it. Additionally, demographical changes of the World have intensified this climate change crisis, including the increase in the magnitude of the global population, its utilization of more equipment and electrical devices.

Climate change has multiple effects on society, including direct and indirect influences on human health and is one of the spheres that has been recently highlighted by multiple research reports in regard to the importance of climatic change for global public health (Martens et al, 1997).

General effects of climate change have been already detected; physical sciences have shown it clearly, the earth is warming up. If the trend of global warming continues, man will face many threats, diseases and deaths related to natural disasters: hurricanes, torrential rains, heat waves and other climate anomalies. In some regions of the world numerous populations will be displaced by the increase of the level of the sea or will be seriously affected by droughts and famines, decrease of suitable lands for the agriculture, increase of the food-borne diseases, water-borne diseases, vector-borne diseases as well as an increase of premature deaths and diseases related to the air pollution (Mills, 2009; PAHO, 2008; United Nations, 2006).

Global warming and climate change are products of many physical imbalances but also from biological and social ones in the way of evolution of mankind. Apparent modern quality of life, assumed in the last years and particularly during the last century, as well as population growth and overpopulation in some areas of the World, have lead to modern societies that have significantly increased the consumption of energy and waste production (Mills, 2009; PAHO, 2008; United Nations, 2006; Diaz, 2006).

2. Basics about Environmental Changes and Health

The elements and inputs for other acquaintances and accepted as traditional and necessary for a good health are: drinkable, constant and safe water, safe and sufficient food, immunizations, epidemiological alertness of the morbidity and response to this one, safe and effective fight against disease vectors and the sufficient preparation against disasters. All of them are indispensable and decisive components of the practices of public health that also constitute the best adjustments and adaptation to climatic change (PAHO, 2008; United Nations Development Programme, 2008; United Nations, 2006).

Population segments with higher poverty will be the most affected by climate change. The physical effects of the climatic change will be different in diverse geographical localities, and social conditions will make the poorest areas more vulnerable to climate change. Repercussions, influences or impacts in human health will be determined and modified by such conditions. Herein we should include the level of development (importantly measured by the Human Development Index, HDI), poverty (currently measured by many indicators such as the Poverty Index and the Unsatisfied Basic Needs index), education (included as a variable of the HDI,) the infrastructure of public health, the landscape use and the political structure, among others. Initially, developing countries will be the most vulnerable and affected. Countries with high levels of poverty and malnutrition, fragile health infrastructures and/or political instability will be the most vulnerable in facing the multiple impacts of climate change (PAHO, 2003; United Nations Development Programme, 2008).

It is vital to prepare ourselves and face these phenomena as well as the elements that contribute to the problem. One of the most important consequences, global warms, as well the climate variability and environmental effects influence human health. Already, multiple research reports have described the effects of climate change on multiple biological imbalances within in many diverse forms of life. As many diseases are consequences of a complex interplay of biological organisms (microbial organisms, insects or vectors, animals and human beings), significant environmental changes can originate displacements of serious endemic and epidemic diseases. These include malaria, leishmaniasis, hemorrhagic transmissible fevers and dengue, among many others, from the original endemic areas in South East Asia, neo-tropical Americas and Africa, to southward and northward. Additionally, increase in global temporary and definitive migration will also contribute to the increase of such diseases in endemic areas and will be seen in non-endemic areas in other, even developed, countries (PAHO, 2003; United Nations Development Programme, 2008).

Many communicable or infectious diseases are sensitive to the climate variability. Climate change represents an additional risk which must be dealt with in order to take the corresponding preventive measurements. In the last 20 years significant evidences have been generated from multiple science fields demonstrating how the climate change affects, directly and indirectly, disease vectors (particularly mosquitoes) (Diaz, 2006; Parry et al, 2007). Climate change can accelerate biological development and increase vectors population available to transmit pathogens and diseases. This is a consequence of climate change on the environment, altitude, cold and heat, and water reservoirs and, particularly, wetlands.

With a more spread and greater population of vectors, disease risk spectrum is a consequence of more time of exposition. In some affected areas of the World climates have become more suitable for disease vectors survival. Until approximately twenty years ago

(1990), slightly more than 30% of the World's population was living in regions at risk for dengue and malaria. Today it is known that with the climate change this percentage has been increased, and it is foreseen that between 50 and 60% of the World's population will live in zones at risk for transmission of dengue in future years. This translates to mean, among other things, that the poorest populations will be less able to resist than others. This is mainly due to the lack of drinkable water and the need to collect water, which in the case of dengue is particularly suitable for the vector (*Aedes aegypti*) development (Gubler et al, 1981; Sukri et al, 2003; Rifakis et al, 2005; Halstead, 2006).

Other infectious diseases discussed later are also significantly influenced by the climate change. Many of them can change patterns to come into levels of epidemics when significant climate anomalies occur. These include torrential rains, water-courses, hurricanes, which originate floods, collapse of structures and faults in the basic services, with severe consequences in the populations. When these phenomena occur in poorer regions, other problems are generated and explained by the losses of activities and the climate change. Finally this will generate major direct losses to the health, to the economy and to the family and social structure of the affected regions and countries (Parry et al, 2007; PAHO, 2008).

As previously stated, major impacts will be felt in developing countries. This is explained due to the high levels of poverty and the few response capacities of the health systems and the lack of drinkable, constant and regular water supply services, as well as waste disposal systems (United Nations Development Programme, 2008).

It becomes imperative and of utmost relevance to study, the impacts of climate change on health. This knowledge must be used for monitoring and preparing for diseases, early alert and response, spreading knowledge, educating, strengthening the population, and concentrating efforts on the formal education of the populations more at risk in the basic education and students of health sciences. These new approaches in preventive medicine, tropical medicine, travel medicine, epidemiology, among other medical and health disciplines, should be considered and taught, in order to prepare for damages and diseases related to the climate change (United Nations Development Programme, 2008; PAHO, 2008).

We must improve the preparedness and all the systems of early alertness, to be more effective and to better approach the possible increase in the number of environmentally related disasters like hurricanes, torrential or copious rains, floods, fade, among others (PAHO, 2008). Additionally, climate change can originate emotional and psychiatric personal and collective conflicts related to the fight for water resources, the availability and access to drinkable water, food production, and in the generation of diseases related to nutritional deficits. In the long term, the climate change will alter natural economic and social systems that help to support acceptable levels of health (Ortega Garcia, 2007; Ebi & Paulson, 2007).

To protect the health and reduce the risks related to the climate change, we must intensify and promote innovative measures of education as well the epidemiological alertness and control of infectious diseases, the employment of safe sources of water and the efficient coordination of the health teams' response to the related emergencies (PAHO, 2008).

Challenges before the imminent natural dangers and the subsequent impact in the countries and populations who need major protection, will always exist. We must work to assure that the reduction of risks due to initiatives of health, including medical primary assistance, safety of patients, women, children and workers, should be sustainable in time (PAHO, 2008).

In the future years, climate change will increase the risks before disasters, making them not only more frequent, intense and risky, but also population vulnerability will be greater than currently exists. More frequent and intense storms, floods and long-lasting droughts can concern the communities, the attitude to be prepared, answer and reconstruct after events. Other adverse impacts on the public health, ecosystems, food safety, migrations and the most vulnerable groups like children, the elderly and women, will increase the vulnerability of communities to natural dangers of all the types (PAHO, 2008).

In the last 20 years, the number of disasters has doubled from 200 to more than 400 per year. Floods are more frequent (of approximately 50 in 1985, it increased to more than 200 in 2005). Damage areas are increasingly more extensive than twenty years ago. These trends indicate a future where the variability of extreme climate and its consequences probably will become the norm. The human implications are significant, with more rains, hurricanes, and increasing sea levels which will increase the risk of floods, and vulnerability will be higher (Liverman, 2009).

Between the years 1991 and 2005, 3,470 million persons were affected by disasters, 960,000 persons died, and the economic losses were estimated in US $1,193 billion. The poorest countries have been more affected, which is due to their internal vulnerabilities, capacity of response with effective measures in reducing risks. Small developing countries are indeed more vulnerable. For example: losses in Granada from hurricane Ivan in 2004 totalled US $919 million. That amount equalled 2.5 times Granada's gross domestic product (GDP). Similar figures were seen in other Central America countries such as Honduras, Nicaragua, Guatemala, El Salvador and Belize, in which hurricane Mitch, in 1998, caused around 9,550 deaths, destroyed 13,785 homes and affected a population of 3,174,700 people in those countries (Liverman, 2009; McMichael et al, 2003). Although developing countries in Latin America are most vulnerable to the impacts of natural disasters (related, or not, to climate change), there are a clear internal differences between the development of its countries that make them very different in their vulnerability, consequences (in morbidity, mortality and incapacity) and responses against them. For example: recently, two significant earthquakes in Latin America affected two countries, Haiti (January 2010) and Chile (February 2010), respectively. However, effects were remarkably different. In the Haitian earthquake, general and health consequences were practically devastating. In the Chilean earthquake, although it affected a large geographical area with a stronger intensity, general and health consequences were much lower.

The risk for conflicts may increase, especially in already fragile social environments. The risk of discrimination and violation of the most fundamental rights: human, economic, social and cultural, would need special attention. Many challenges will be raised, and as a consequence of this, forced migrations from the changeable climate areas will occur (PAHO, 2008).

Adaptation and mitigation of the effects of climatic change will need special analysis of vulnerability. Development of risk maps, contingency plans, and other measures will be needed. Better communication is needed with societies and governments from academia and research organizations regarding the impact of climate change on public health. Additional financial and human resources will be needed to fight against the effects of climate change on health and disease (PAHO, 2008; Lapola et al, 2008).

Multiple strategies and mechanisms for the reduction of risks should be implemented before the disasters and the human consequences of the climate change affect the World. In order to protect the human safety, this must include social and economic development, the preparation before the emergencies, and the proper response and mechanisms of recovery at all the levels (United Nations Development Programme, 2008).

It is also important to have materials, human resources and reliable financial resources to address the risks related to disasters and the management of them in this setting of climate change (PAHO, 2008).

As previously stated climate change will affect health and disease in multiple ways and is currently impacting communicable and non-communicable disease epidemiology in many areas of the World. This includes Latin America. In this important Neotropical area, multidisciplinary biomedical research, particularly from the public health perspective, has shown that climate change has significantly impacted the epidemiology of tropical, and in general, of communicable diseases.

3. Climate change and Communicable diseases in Latin America

3.1 Communicable diseases endemic in Latin America

As previously described, it is well established that climate is an important determinant of the distribution of vectors and pathogens, such as those of malaria (*Anopheles spp.-Plasmodium falciparum, P. vivax, P. ovale, P. knowlesii*), dengue (*Ae. aegypti*-Dengue viruses), and leishmaniasis (sandflies *Phlebotomus spp. and Lutzomyia spp.-Leishmania spp.*), among many others in different areas of the World, including Latin America. This is particularly true in its tropical regions but also in its subtropical areas (Rodriguez-Morales, 2005).

In this Neotropical region, recent contributions in the field have demonstrated strong and significant links between climate variability, climate change, emerging and re-emerging infectious diseases that represent public health issues for the region. Many diseases have varied their morbidity, mortality and even chronic consequences, such as disabilities, as a consequence of climate variability and change (Rodriguez-Morales, 2009).

Countries such as Venezuela, Colombia, Mexico, Ecuador, Brazil, Argentina, Peru and Bolivia, among others, have evidenced and suffered the impacts of climate change in the socioeconomic systems, such as agriculture and fishing. These impacts are a consequence of the phases of the El Niño Southern Oscillation (ENSO) phenomena, but also in specific health conditions, such as tropical and infectious diseases including malaria, dengue, leishmaniasis (cutaneous and visceral), yellow fever, cholera and diarrhoeas, and probably Chagas disease (Figure 1) (Rodriguez-Morales, 2005; Araújo et al, 2009) are a result of climate change. Tropical areas of Latin America have been suitable for those diseases; these are endemic, and climate change is now triggering its increase, persistence, re-emergence in non-previous endemic areas or in areas where they were eliminated, eradicated or controlled (Figure 1).

3.2 Approaches to the Study of Communicable diseases in Relation to Climate

Different statistical analyses, most based on linear regressions, have linked extreme climatic anomalies with significant alterations in the epidemiological patterns of diseases and are sometimes coupled, directly and indirectly, on time and space (McMichael et al, 2003). Additionally to statistic techniques, geographical information systems (GIS) and remote sensing (spatial epidemiology) have supported these observations and are helping in the development of systems for predicting and forecasting such diseases based on climate variability and climate change (McMichael et al, 2003). Now availability of data, images and software, and new technologies for the region (including satellites) allows better defining of the impact of climate change on health and disease (Rodriguez-Morales, 2008). In Argentina, there is now an aero-spatial institution with an area dedicated to satellite epidemiology, use of data from remote sensing (satellites) applied to the study of diseases (Rodriguez-Morales, 2005; Beck et al, 2000).

3.3 Evidences regarding Climate Change and its Potential Effect on Disease: Cutaneous and Visceral Leishmaniasis

In this regard, evidences from Latin America have accumulated useful qualitative and quantitative information that indicates how climate variability and change influenced particular tropical diseases (McMichael et al, 2003; Arria et al, 2005). The impact of El Niño Southern Oscillation climatic fluctuations during 1985–2002 in the occurrence of leishmaniasis in two north-eastern provinces of Colombia (North Santander and Santander) was reported. During that period, it was identified that during El Niño, cases of leishmaniasis increased up to 15.7% in disease incidence in North Santander and 7.74% in Santander, whereas during La Niña phases, leishmaniasis cases decreased 12.3% in Santander and 6.8% in North Santander. When mean annual leishmaniasis cases were compared between La Niña and El Niño years, significant differences were found for North Santander ($p<0.05$) but not for Santander ($p=0.05$) (Cárdenas et al, 2006). During the same study period in southern provinces, effects of climate variability and change were also studied regarding leishmaniasis incidences. In this study, 11 southern departments of Colombia were analyzed: Amazonas, Caquetá, Cauca, Huila, Meta, Nariño, Putumayo, Tolima, Valle, Vaupes and Vichada. Climatic data were obtained by satellite and epidemiologic data were obtained from the Health Ministry. National Oceanographic and Atmospheric Administration (NOAA) climatic classification and SOI (Southern Oscillation Index)/ONI (Oceanic Niño Index) indexes were used as indicators of global climate variability. Yearly variation comparisons and median trend deviations were made for disease incidence and climatic variability. During this period there was considerable climatic variability, with a strong El Niño for six years and a strong La Niña for eight years. During this period, 19,212 cases of leishmaniasis were registered, for a mean of 4,757 cases/year. Disease in the whole region increased (mean of 4.98%) during the El Niño years in comparison to the La Niña years, but there were differences between departments with increases during El Niño (Meta 6.95%, Vaupes 4.84%). The remainder showed an increase during La Niña (between 1.61% and 64.41%). Differences were significant in Valle ($p<0.01$), Putumayo ($p<0.001$), Cauca ($p=0.03$), and for the whole region ($p<0.01$), but not in the remaining departments (Cárdenas et al, 2008). This information shows how climatic changes influence the occurrence of leishmaniasis in north-eastern and southern Colombia.

Similar results have been described in Venezuela. Between 1994 and 2003, a study in 2,212 cutaneous leishmaniasis cases also linked climate variability to disease incidence in an endemic area of the country, Sucre state. During that period, three important El Niño phases were observed: 1994-1995, 1997-1998, and 2001-2003. The 1997-1998 phase was the most relevant one and was followed by a chilly and rainy season in 1999 (La Niña). During 1999-2000, 360 cutaneous leishmaniasis cases were recorded in Sucre, with an important variability within a year, and a 66.7% increase in cutaneous leishmaniasis cases ($F=10.06$, $p<0.01$) associated with the presence of a weak La Niña phenomenon (not too cold and rainy). Models showed that with higher Southern Oscillation Index (SOI) values, there was a reduced incidence of cutaneous leishmaniasis (r^2=0.3308; p=0.05). The increase with respect to the average trend in rain was associated with increases in trends for cutaneous leishmaniasis in the period from 1994 to 2003 (p=0.036) (Cabaniel et al, 2005).

Although not described in such detail, the Suriname cutaneous leishmaniasis is a seasonal disease. The rainy seasons are from November to January and from May to July. In a recent study (2008), most patients with this disease were registered during the short dry season in March (35%) (van der Meide et al, 2008). In Brazil, studies made on leishmaniasis vector have characterized spatial distribution of them. In Mato Grosso, the vector sandfly *Lu. whitmani* s.l. have been positively correlated with deforestation rates and negatively correlated with the Brazilian index of gross net production (IGNP), a primary indicator of socio-economic development. Authors found that favourable habitats occur in municipalities with weaker economic development. This confirms that vector occurrence is linked to precarious living conditions found either in rural settlement of the Brazilian government's agrarian reform program, or in municipalities with intense migratory flows of people from lower social levels (Zeilhofer et al, 2008). In Colombia, another entomological study in 5,079 sand flies collected (*Lu. spinicrassa* represented 95.2% of them) have linked population densities to climate. The climatic period where the collection of vectors was done corresponded to a dry season of El Niño (highest Oscillation Niño Index in the last 2006 trimester). In general, the main components analyses evidenced a significant inverse relation between *Lu. spinicrassa* abundance and the relative humidity ($p<0.05$) and rainfall ($p<0.05$), but not for the average temperature ($p>0.05$) (Galvis et al, 2009). In Costa Rica and Bolivia, recent studies have also linked social and climate changes with cutaneous leishmaniasis (Chaves and Pascual, 2006; Gomez et al, 2006).

In the case of visceral leishmaniasis, other studies in Latin America have linked its incidence to climate. Prolonged droughts in semi-arid north-eastern Brazil have provoked rural-urban migration of subsistence farmers and a re-emergence of visceral leishmaniasis (Confalonieri, 2003). A significant increase in visceral leishmaniasis in Bahia State (Brazil) after the El Niño years of 1989 and 1995 has also been reported (Franke et al, 2002).

3.4 Evidences regarding Climate Change and its Potential Effect on Disease: Malaria

For malaria, many studies in the region have linked climate to disease. Classically, after the onset of El Niño (dry/hot) there has been described a risk of epidemic malaria in coastal regions of Colombia and Venezuela (Poveda et al., 2001). Even new patterns of disease have been described in association with climate variability, as in the so-called phenomena of highland malaria described in Venezuela and Bolivia. In November 1999, in an Andean area of Venezuela, there was an epidemic outbreak of highland malaria in the Parish of Guaramacal, Trujillo state. This was an area historically classified as without malaria, with altitudes up to 2200 meters above the sea level (masl). Nine cases of malaria were reported from this area: two of these were classified as introduced and seven were classified as imported. Four species of mosquitoes of the genus *Anopheles*, subgenus *Kerteszia*, probably implicated in this outbreak were collected; they were identified as *Anopheles homunculus* (n=27; 65.9%), *Anopheles lepidotus* (n=9; 21.9%), *Anopheles neivai* (n=3; 7.3%) and *Anopheles pholidotus* (n=2; 4.9%). These mosquitoes were not previously reported as vectors of malaria in Venezuela or from Trujillo State. The most important breeding sites were the epiphytic bromeliads (*Tillandsia spp*). The presence of introduced cases was probably brought about by frequent migrations of people to and from La Laguneta, La Fernandera, Agua Fría in Guaramacal Parish, and the village of San Juan de Dios in Portuguesa state. These people worked in culturing corn and yucca during an epidemic outbreak of malaria in that region; however, these social issues coupled with intense climate changes and particularly intense rainfall during the year of the outbreak would explain this occurrence of highland malaria (Benitez et al, 2004).

In Peru, recent studies have described the relation between climate and disease. This has been explored in Loreto, a north-eastern Amazon jungle area of Peru, during a 13 year period. In this ecological study conducted with data from the monthly average temperature (ºC), relative humidity (%), precipitation (mm) and level of the Amazon River (meters), malaria was linked to climate variables. Authors found significant negative correlation between temperature and cases of malaria for five years: 1997, 1999, 2003, 2005 and 2006; river level for four years: 1997, 1998, 2003 and 2005; and humidity for three years: 1996, 2005, 2006. No association was found for any years with rainfall. The multiple regression models were significant in three years (1999, 2003 and 2006) with r^2 values between 0.870 and 0.937 (Ramal et al, 2009). In Brazil and Ecuador, malaria has been studied in regard to the influence of climate variability (Kelly-Hope & Thomson, 2008).

3.5 Evidences regarding Climate Change and its Potential Effect on Disease: Other Parasitic Diseases

In Brazil, other parasitic diseases such as schistosomiasis have been linked to climate variability (Kelly-Hope & Thomson, 2008). In Venezuela, some evidences suggested that onchocerciasis (river blindness) would also be associated to climate (Botto et al, 2005). In this country, ascariasis has been linked to climate (Benitez et al, 2005). Chagas disease probably will be influenced by climate change; however, there is no significant number of reports that have shown relevant evidence supporting this theory. Other trematode infections different to schistosomiasis, such as fascioliasis and paragonimiasis would be susceptible to the impacts of climate change given their complex parasite life cycles. Regard cestodes few studies on taeniasis and cysticercosis, hydatidosis, hymenolepiasis, among others have also been fewly studied in relation to climate change and climate variability.

3.6 Evidences regarding Climate Change and its Potential Effect on Disease: Dengue

Dengue, as described before, has been significantly linked to climate change, including evidences generated from Latin America. Many countries in Latin America are endemic for this disease which is particularly important in urban centres, such as Caracas, the capital city of Venezuela. In this location between 1998 and 2004, a study found significant associations between dengue hospital morbidity and climate variability. This study used microclimatic data such as rainfall and maximal and minimal monthly temperatures. Macroclimatic indexes such as NAO (North Atlantic Oscillation), SOI (Southern Oscillation Index) and ONI (Oceanic Niño Index), were used. Seasons were categorized as positive or negative for El Niño phenomenon (the latter were classified as neutral and La Niña). Linear regression models were used for determining the associations. Results indicated that for the studied period, 2,187 confirmed cases of dengue fever were recorded, and the annual mean was 268 cases (± 371). The highest case toll was in year 2000 (up to 214 cases per month), and this had a climatic correlation with La Niña. Years negative for El Niño had the highest number of cases (1999, 2000, 2001, and 2004) which was 60.26% higher than the mean number of cases. This compared with the years where El Niño phenomenon occurred (1998, 2002, 2003) where there was a reduction in the case number compared with the mean values (−67.56%) (χ^2=21.76; p<0.01). Linear regression models found a statistically significant association between dengue fever and rainfall abnormalities in Caracas (r^2=0.01199; F=4.635; p=0.032), as well as with maximum temperatures recorded (r^2=0.1345; F=59.37; p<0.001) (Rifakis et al, 2005). Other studies in Venezuela have reported similar results (Barrera et al, 2002; Herrera-Martinez et al, 2009). Annual variations in dengue/dengue hemorrhagic fever in Honduras and Nicaragua appear to be related to climate-driven fluctuations in the vector densities (temperature, humidity, solar radiation and rainfall) (Patz et al, 2005). In some coastal areas of the Gulf of Mexico, an increase in sea surface temperature (SST), minimum temperature and precipitation was associated with an increase in dengue transmission cycles (Hurtado-Díaz et al, 2007). Other studies in Mexico have reported similar results (Peterson et al, 2005). In Barbados, Puerto Rico and Dominica, climate variability has been linked to dengue incidence (Depradine and Lovell, 2004; Schreiber, 2001; Rodriguez-Morales, 2005).

3.7 Evidences regarding Climate Change and its Potential Effect on Disease: Other Viral Diseases

Parasitic and other infectious diseases in Latin America have been linked to climate variability and climate change. This is the case of other viral diseases that are different to dengue, such as yellow fever, influenza, Hantaviruses and rabies, among others. A study conducted during 2002-2004 linked rabies occurrences in Venezuela to climate variability. Rabies in Venezuela has been important in the last years, affecting dogs, cats, other animals and humans and it is a reportable disease. In Zulia state, it is considered a major public health concern. Recently, a considerable increase in the incidence of rabies has been occurring, involving many epidemiological, ecoepidemiological and social factors. These factors were analyzed in 416 rabies cases recorded during the study period. The occurrences have been increasingly significantly, affecting mainly dogs (88.94%). Given this epidemiology it was associated ecoepidemiological and social factors with rabies incidence in the most affected state, Zulia. This area has varied environmental conditions. It is composed mostly of lowlands bordered in the west by a mountain system and, in the south, by the Andes. The mean temperature is 27.8°C, and the mean yearly rainfall is 750 mm. ĉlimatologically, year 2002 corresponded with El Niño (drought), middle 2003 evolved to a Neutral period and 2004 corresponded to La Niña (rainy). This change may have affected many diseases, including rabies. Ecological analysis showed that most cases occurred in lowland areas of the state and during the rainy season (p<0.05) (Rifakis et al, 2006). For Hantaviruses, outbreaks of Hantavirus pulmonary syndrome have been reported for Argentina, Bolivia, Chile, Paraguay, Panama and Brazil after prolonged droughts (Williams et al., 1997; Magrin et al, 2007). This may be due to the intense rainfall and flooding following the droughts, which increases food availability for peri-domestic (living both indoors and outdoors), rodents (Magrin et al, 2007). In Brazil and Venezuela, yellow fever outbreaks have been linked to climate variability (Vasconcelos et al, 2001; Rodriguez-Morales et al, 2004).

3.8 Evidences regarding Climate Change and its Potential Effect on Disease: Bacterial Infections

Bacterial infections have been associated to an increase linked to climate variability, climate change and global warming. *Staphylococcus*, *Streptococcus*, and enteric bacteria tend to colonize humans more readily in warmer climates. In addition, some authors have studied the changes in incidence of Gram-negative carriage from three skin sites in a climate controlled chamber at 35°C and 90% humidity for 64 h. Their findings showed that high temperatures and humidity increased the overall frequency of isolation of Gram-negative bacteria, although there were individual differences. If global warming continues, health care workers may one day encounter outbreaks of infectious diseases with these pathogens. As these organisms have a significant potential for inherent resistance to antimicrobials or for the development of antimicrobial resistance and the treatment of these patients will impose a huge challenge to medical sciences (Thong & Maibach, 2008). A study attempted to link gram-positive cocci (GPC) to climate variability in Venezuela. During the study period (1992-2001), 501 GPC infections were diagnosed and identified. The year with the highest incidence was 1999 (La Niña year), while the year with lowest incidence was 1992 (El Niño year). It was observed that during La Niña years (1998-2001) a more significant number of cases occurred compared with El Niño years (1992–1994, 1997) (15%, χ^2=25.96, p<0.01). During annual rainy seasons we found significantly more incidences (months of July and August) than in dry seasons (January and February) (75%) (F=29.85, p<0.01). However, this was affected by ENSO classification, because comparing La Niña and El Niño years, incidence was higher for the first during January to June, and for October and November; while for the second, incidence was higher for July to September (Rodriguez-Morales et al, 2006). Other bacteria, such as *Leptospira* has been linked to climate variability. Flooding produces outbreaks of leptospirosis in Brazil, particularly in densely populated areas without adequate drainage (Kupek et al, 2000). In 1998, increased rainfall and flooding after hurricane Mitch in Nicaragua, Honduras, and Guatemala caused a leptospirosis outbreak, and an increased number of cases of malaria, dengue fever, and cholera (Costello et al, 2009). In Peru, an autochthonous disease, Carrion's disease (*Bartonella bacilliformis*) has been linked to climate variability (Huarcaya et al, 2004). *Vibrio cholerae* is another bacterial pathogen in which its incidence has been linked to climate variability. As ocean temperatures rise with global warming and more intense El Niños, cholera outbreaks might increase as a result of more plankton blooms providing nutrients for *Vibrio cholerae*. Studies in Peru, Ecuador, Colombia, Mexico and Venezuela have shown evidence of these relationships (Patz et al, 2005; Farfan et al, 2006; Chavez et al, 2005; Franco et al, 1997; Lama et al, 2004).

3.9 Evidences regarding Climate Change and its Potential Effect on Disease: Zoonoses

For veterinary public health, climate change may be associated with seasonal occurrence of diseases in animals rather than with spatial propagation. This is the case for pathogens or parasitic diseases, such as fascioliasis, in areas with higher temperatures. When disease seasonality is extended as a consequence of the increased survival of the parasite outside the host or, conversely, shortened by increased summer dryness that decreases their numbers. For other pathogens, such as parasites that spend part of their life cycle as free stages outside the host, temperature and humidity may affect the duration of survival. Climate change could modify the rate of development of parasites, increasing in some cases the number of generations and extending the temporal and geographical distribution. New World screwworm is frequently found in South America, with infestations increasing in spring and summer and decreasing in autumn and winter (Rodriguez-Morales, 2006; Paris et al, 2008). West Nile Virus is a disease in which both long-distance bird migration and insect population dynamics (*Culex*) are driven by climate conditions. Vesicular stomatitis (VS) affects horses, cattle and pigs and is caused by various vesiculoviruses of the family Rhabdoviridae. Seasonal variation is observed in the occurrence of VS; it disappears at the end of the rainy season in tropical areas and at the time of the first frosts in temperate zones (Pinto et al, 2008).

3.10 Climate Change and Communicable Diseases: Public Health Perspectives

Given the substantial burden of disease associated with climate change in developing tropical countries, such as most of Latin America, it is of utmost relevance to incorporate climate changes into public health thinking, including health authorities and systems, as well as the whole public health education and faculties.

Although many studies may have some limitations, such as a lack of incorporation of other meteorological factors into the analysis (temperature, rainfall, sun radiation, transpiration or evotranspiration, relative humidity, vegetation indexes [Normalized Difference Vegetation Index, NDVI and Enhanced Vegetation Index, EVI] among others) (Cárdenas et al, 2006), it has been suggested that such findings are relevant from a public health perspective to better understand the ecoepidemiology of different communicable diseases (Rodriguez-Morales, 2005). However, further research is needed in this region and other endemic areas to develop monitoring systems that will assist in predicting the impact of climate changes in the incidence of tropical diseases in endemic areas with various biological and social conditions.

4. Climate change and Non-communicable diseases in Latin America

4.1 General Aspects: Environmental context and Climate change

Anyone pursuing the science of medicine must proceed accordingly. First he ought to consider what effects each season of the year can produce. Seasons are not all alike and differ widely within themselves and their changes. The next point is the hot winds and the cold, especially those that are universal, but also those that are peculiar to each particular region (as described by Hippocrates, regard airs, waters, places) (PAHO, 1988).

Since ancient times, men have been aware of the importance of climate changes in their health. What is important from these ancient evidences from our prime medical doctors to our westernized world, we must pay much attention to climate changes which certainly has increased during last 50 years due to the greenhouse effect.

The following excerpt from the World Health Organization (WHO), collected from the web, about Facts on Climate Change on Health, December 2009, seems to indicate that non-communicable diseases (including injuries and malnutrition) represent an important burden of the human health around the World and will be increased in the following years even more dramatically due to the climate change. Aspects of quality of life and living conditions will be directly affected due to limited food production and shortage, and air pollution. Also it is important to highlight that disasters associated to climate changes will provoke death and more disability. Definitely this will diminish social-economic development, especially in the developing countries.

"Climate and weather already exert strong influences on health: through deaths in heat waves, and in natural disasters such as floods, as well as influencing patterns of life-threatening vector-borne diseases such as malaria. Continuing climate change will affect, in profoundly adverse ways, some of the most fundamental determinants of health: food, air and water, according to WHO Director-General Dr. Margaret Chan. Areas with weak health infrastructure - mostly in developing countries - will be the least able to cope without assistance to prepare and respond. From the tropics to the arctic, climate and weather have powerful direct and indirect impacts on human life. Weather extremes - such as heavy rains, floods, and disasters like Hurricane Katrina that devastated New Orleans, USA in August 2005 - endanger health as well as destroy property and livelihoods. Approximately 600 000 deaths occurred worldwide as a result of weather-related natural disasters in the 1990s, some 95% of which took place in developing countries. Pollen and other aeroallergen levels are also higher in extreme heat. These can trigger asthma, which affects around 300 million people. Ongoing temperature increases are expected to increase this burden. Water scarcity encourages people to transport water long distances and store supplies in their homes. This can increase the risk of household water contamination, causing illnesses. Increasing temperatures on the planet and more variable rainfalls are expected to reduce crop yields in many tropical developing regions, where food security is already a problem. Steps to reduce greenhouse gas emissions or lessen the health impacts of climate change could have positive health effects. For example, promoting the safe use of public transportation and active movement - such as biking or walking as alternatives to using private vehicles - could reduce carbon dioxide emissions and improve public health. They can not only cut traffic injuries, but also air pollution and associated respiratory and cardiovascular diseases. Increased levels of physical activity can lower overall mortality rates" (WHO, 2009).

There is a positive message in the above paragraph: human prevention and information can help to reduce the climate change impact in order to improve public health, from individual and collective efforts.

In the Region of the Americas, human and health indicators have advanced over the past decades. Life expectancy at Birth has gone from 68.8 in 1980-1985 to 74.9 in 2005-2010; fertility rate (children/woman) from 3.1 to 2.6, infant mortality (per 1,000 live births) 37.8 to 16.5; urban population (%) from 69 to 79. An important aspect that has changed, showing a clear epidemiological transition, is that rates of mortality from communicable diseases (rate/100,000 inhabitants) dropped from 109 in 1980-1984 to 55.9 in 2000-2004. Meanwhile, mortality from diseases of the circulatory system (rate/100,000 inhabitants), a most important representative of non-communicable disease, dropped from 280 to 229.2, clearly a minor change in comparison to the former group of communicable diseases.

The Region of the Americas continues and will continue in the next years to experience three major demographic shifts: population growth, urbanization, and aging. Weather disasters are increasing in a significant proportion due to the climate change. These disaster pattern changes will be contributed to by the increase of the climate change and will be implicated with strong human activities–climate interaction with a heavy anthropogenic origin (PAHO, 2007).

Assessment of human health during the 21st century, without considering the environment and its implications, is forgetting the social and environmental determinants of health and quality of life of human beings. During the last few decades, the damage caused by human activity and demographic explosion has accelerated the degradation, although they are not the only reasons for this situation.

Deterioration of the environment, already made, recognition of global environmental changes, the hard lessons from human-caused disasters or mistakes such as incidents like Seveso, Bhopal, Minamata, the Chernobyl nuclear accident, the Exxon Valdez accident in Alaska, cholera epidemics in Latin America, globalization, migrations and tourism, travelling across the globe, obligate us to consider the relationship between human health and a changing environment.

Public policies are difficult to manage in order to warranty the basic need for human health as a human right which includes potable water for hygiene and cleanliness. Ethics are needed to guide decision making to gain human rights (Tavares, 2005).

Main contributors to climate change and its consequences in health include now: degradation of good quality water, deposits of toxins and chemical pollutants, the impossibility to treat these products of human activities, and the wide spectrum of synthetic chemical substances in circulation in the environment without control or even knowing their consequences to human beings in the long term.

In terms of the environment, a WHO report found that most of the major diseases were at least, partially caused by exposure to environmental risks and that environmental causes contributed to about one-fourth of disability-adjusted life years lost (DALYs) and one-fourth of associated deaths (Prüss-Üstün & Corvalán, 2006).

A study done by the Pan American Health Organization (PAHO) about inequality in the access, distribution and expenditure in potable water in Latin America and the Caribbean found interesting results. With more than 11 countries of the region in 2001, the study showed that the poorest are generally those that do not have water systems and have to pay more to get potable water. Also, urban populations have more access to intra-domiciliary water than rural communities. But more concerning is that associated with poor disposition of potable water, the climate changes in terms of drought accompanied by fires and air contamination, and heavy rainy seasons produce even more problems with tropical rain, flooding, mudslides, and contamination of water (OPS, 2001).

4.2 Non-communicable Diseases Globally and in Latin America

Last decades, non-communicable diseases (NCDs) are spreading around the world and imposing their predominance in developing countries. Of real concern is that this changing of morbidity and mortality will put more pressure and a heavier burden of infectious and non infectious diseases in a poor environment characterized by poor health systems. Non-communicable diseases will cause 7 out of every 10 deaths in developing countries. Many of these diseases can be prevented by attacking associated risk factors (Boutayeb & Boutayeb, 2005).

According the World Health Organization's statistics, chronic NCDs such Cardiovascular Diseases (CVDs), diabetes, cancers, obesity and respiratory diseases, account for about 60% of the 56.5 million deaths each year and almost half of the global burden of disease. In 1990, 47% of all mortality related to NCDs was in developing countries, as was 85% of the global burden of disease and 86% of the Disability Adjusted Life Years (DALYs) attributable to CVDs. An increasing burden will be born, mostly by these countries, in the next two decades. The socio-economic transition and the ageing trend of the population in developing countries will induce further demands and exacerbate the burden of NCDs in these countries. If the present trend is maintained, it is predicted that by 2020, NCDs will account for about 70 percent of the global burden of disease, causing seven out of every 10 deaths in developing countries, compared with less than half today (Boutayeb & Boutayeb, 2005).

In 1990, approximately 1.3 billion DALYs were lost as a result of new cases of disease and injury, with the major part in developing countries. In 2002, these countries supported 80% of the global Years Lived with Disability (YLDs) due to the double burden of communicable and non-communicable diseases. Consequently, their people are not only facing higher risk of premature life (lower life expectancy) but also living a longer part of their life in poor health. These remarks indicate that NCDs are exacerbating health inequities existing between developed and developing countries and making the gap more profound between rich and poor within low and middle-income countries (Boutayeb & Boutayeb, 2005).

Globally, non-communicable diseases are more important in terms of frequency, absolute and relative, representing the vast majority of deaths. According to a main prediction from WHO data globally, Project Global Burden Disease (GBD) from Murray and Lopez have provided an important contribution to understanding mortality and burden of disease projections into the future and selected indicators and DALY. Most recently, Colin D. Mathers and Dejan Loncar made further projections of Global Mortality and Burden of Disease from 2002 to 2030 taking into account HIV/AIDS which, according to these authors, were underestimated by Murray and all (Boutayeb & Boutayeb, 2005; WHO, 2009).

A traditional typology of disease is tripartite—communicable disease, non-communicable disease and injury. A first generation of diseases is linked to poverty—common infections, malnutrition and reproductive health hazards mostly affecting women and children. These mostly (but not entirely) communicable diseases are concentrated among the poor in developing countries. A second generation of primarily chronic and degenerative diseases—such as cardiovascular disease, cancer, stroke and diabetes—predominate among the middle-aged and elderly in all countries. Susceptibility to these non-communicable diseases is linked to lifestyle and health-related behaviour. Injury should be added to these two groups of diseases and is prevalent in both rich and poor countries.

Disease and injury causes of death are classified (simplified) in the GBD using a tree structure in which the first level comprises three broad cause groups: Group I (communicable, maternal, perinatal, and nutritional conditions), Group II (non-communicable diseases), and Group III (injuries). A large decline of all causes of Group I with the exception of HIV is projected between 2002 and 2030. Although age-specific death rates for most Group II conditions are projected to decline, ageing of the population will result in significantly increasing total deaths due to most Group II conditions over the next 30 years (Figure 3). Global cancer deaths are projected to increase from 7.1 million in 2002 to 11.5 million in 2030, and global cardiovascular deaths from 16.7 million in 2002 to 23.3 million in 2030. Overall, Group II conditions will account for almost 70% of all deaths in 2030 under the baseline scenario (Mathers & Loncar, 2006).

Another important group (external causes of death) project to increase (40%) due to injury between 2002 and 2030 mainly due to the increasing numbers of road traffic accident deaths, together with increases in population numbers more than offsetting small declines in age-specific death rates for other causes of injury. Road traffic accident deaths are projected to

increase from 1.2 million in 2002 to 2.1 million in 2030, primarily due to increased motor vehicle fatalities associated with economic growth in low- and middle-income countries. (6)
The rapid rise of non-communicable diseases (NCDs), mental disorders, injuries and violence represents one of the major health challenges to global development in the 21st century. However, the staff across WHO's Cluster for Non-communicable Diseases and Mental Health believes that affordable solutions exist today to prevent millions of premature deaths each year in developing countries, mainly through policy change, effective surveillance and monitoring, initiatives to reduce common risk factors, and the strengthening of health systems. A stronger commitment to tackle NCDs and malnutrition and forge new partnerships are critical to making progress (WHO, 2009).
Socio-economic deterioration meaning poverty, rapid urbanization and social fragmentation has contributed to greater inequalities and unhealthier environments. Urban areas are characterized by violence, poor housing conditions and lack of basic sanitation. Directly in terms of climate change and global warming, the 2001 report of the United Nations Intergovernmental Panel on Climate Change mentioned that two countries of Latin America are among the world's largest carbon-dioxide emitters. These countries are Brazil and Mexico (PAHO, 2007).
Among the most important emerging challenges confronting the Americas are non-communicable diseases and violence due to aging of the population and unhealthy lifestyles and risky behaviours. Overweight and obesity, diabetes, alcoholism, malignant neoplasm, diseases of the circulatory system, mental health problems, road traffic and violence injuries and death are consequences of this unhealthy and unsafe social environment and lifestyle (PAHO, 2007.

4.3 Chronic Non-communicable diseases in the Americas

Cardiovascular diseases, chronic obstructive respiratory diseases, cancer and diabetes are the chronic non-communicable diseases of greatest interest for public health in Latin America and the Caribbean. In both sub-regions, non-communicable diseases are responsible for two out of three deaths in the general population and nearly one-half of deaths among those under 70 years old. Of the 3,537,000 deaths registered in Latin America and the Caribbean in 2000, 67% were caused by these chronic diseases. Ischemic heart disease and cancer accounted for the majority of deaths in those 20-50 years old. Non-communicable diseases contributed 76% of DALYs to the overall disease burden. In addition, early mortality and complications, sequels and disability limit functionality and productivity. These represent huge medical and health expenditures with financial and social costs that undermine resources in both the health systems and social security. (13)
Cardiovascular diseases (which include ischemic heart disease, cerebrovascular disease, hypertensive disease, and heart failure) represent 31% of the mortality. Data from 2000-2004 shows that mortality from diseases of the circulatory system was higher in men (223.9 per 100,000 population) than in women (179.3 per 100,000), although there are important difference in magnitude between sub regions. Many of these deaths are consequences of improper diet, obesity, lack of physical activity, and smoking, and include ineffective hypertension control and disease management. Hypertensive diseases are a major risk factor for heart disease and cerebrovascular disease and an important cause of mortality. However, there are major differences between countries of mortality rates from above 30 (age and sex adjusted rate per 100,000 population) in Bahamas, Dominican Republic and

Trinidad and Tobago to mortality rates lower than 10 in Panama, Uruguay, El Salvador and Canada (PAHO, 2007).

The most common malignant neoplasm's are of bronchus and lung, stomach, cervix, the breast and prostate. Diabetes was the fourth cause of death in Latin America and the Caribbean in 2001 accounting for 5% of total deaths. In Mexico, diabetes was the leading cause of death in the total population in 2002, with 12.8% of deaths (PAHO, 2007).

Chronic respiratory diseases caused 3% of all deaths, mortality incidence (2000) range between 16 and 25 per 100,000 populations; and, in most countries it was higher in men. Asthma, chronic obstructive pulmonary disease, emphysema, and chronic bronchitis are the most common (PAHO, 2007).

External causes of deaths: Violence, intentional and unintentional. Homicide is the most common intentional violent cause of deaths. A complex interaction of individual, relational, social, cultural, and environmental factors makes the Regions of the Americas one of the most violent in the world. Traffic accidents are most common, unintentional violence cause of death (PAHO, 2007).

Other violent, unintentional causes of death are related to disasters. The Americas constitutes one of the World's regions most exposed to natural disasters, and this vulnerability increases the potential risk of destructive effects caused by events of any nature (PAHO, 2007).

Between 2001 and 2005 the impact of these destructive phenomena has left a toll of some 20,000 death, 28 million victims, and billions of dollars in property losses. It is estimated that every year an average of 130 natural disasters of varying degrees of magnitude occur in the Region. Around 79% of the population is at high risk of damages and death, mainly due to living in large urban areas not prepared to cope with disasters such as earthquake, rain falls, flooding, and mudslides. Most dwellings are in high risk locations, unplanned, poorly built , and do not follow appropriate construction standards (PAHO, 2007).

4.4 Integrative Communion: Climate Change and Non-communicable Diseases

The most recent report of the Intergovernmental Panel on Climate Change (WHO, 2009) confirmed that there is overwhelming evidence that humans are affecting the global climate, and highlighted a wide range of implications for human health. Climate variability and change cause death and disease through natural disasters, such as heat waves, floods and droughts. Many important diseases are highly sensitive to changing temperatures and precipitation. Climate change already contributes to the global burden of disease including malnutrition, and this contribution is expected to grow in the future. Continuing climate change will affect, in profoundly adverse ways, contamination and shifting some of the most fundamental determinants of health in lacking or adverse way: food, air and water. Also, individuals and social mental health will be affected due to the very fast dynamics of change, traumatic experiences which moves at a different scale the ways of life and the surrounding environment. Sanitary infrastructure deterioration and destruction will challenge governments and populations to provide the minimum of medical and health care services, and probably even access to sanitation and health care will be an issue for most in developing countries. Most advances done in this area within developing countries during the last few decades would be confronted earlier with a fast climate change and its consequences represented by natural disasters. Surveillance and information systems are essential for prevention and medical preparation to cope, appropriately, with consequences of disasters related to climate change

(Alfaro & Rivera, 2008). There are many public and private organizations that bring support and information about disasters and health, including the *Red Centroamericana de Información sobre Desastres y Salud* which provides scientific information, research, products and services related to its mission (Cespedes, 2007).

Many authors have reported regional assessments of health impacts due to climate change in the Americas and show that the main concerns are heat stress, malaria, dengue, cholera and other water-borne diseases. Other investigators have published and estimated relative risks (the ratio of risk of disease/outcome or death among the exposed to the risk among the unexposed) of different health outcomes in the year 2030 in Central America and South America. The highest relative risks are for coastal flooding deaths (drowning), followed by diarrhea, malaria and dengue (Kovats & Haines, 1995; Parry et al, 2007). Air pollution or contamination is an important issue related to urban areas and modes of transportation and automobiles are implicated as the main cause in cities like Buenos Aires and Argentina where some research has been done in comparison with other cities of Latin America (Fernando et al, 2001).

Climate change is likely to increase the risk of forest fires. In some countries, wildfires and intentional forest fires have been associated with an increased risk of out-patient visits to hospital for respiratory diseases and an increased risk of breathing problems (WHO, 2009; Mills, 2009). However, it is not easy to register, but a study in Florida during 1998 made this important remark after studying a fire event that occurred in several counties: rapid surveillance of non-reportable diseases and conditions is possible during a public health disaster. In urban areas exposed to the 'heat island' effect and located in the vicinity of topographical features which encourage stagnant air mass conditions and the ensuing air pollution, health problems would be exacerbated, particularly those resulting from surface ozone concentrations (PAHO, 2003; PAHO, 1988; PAHO, 2007).

Natural and technological disasters involved many casualties occurring in a short period of time and emergency actions and controls, pre and post events, (planning and preparation) are crucial for good management. The increasing number of disasters in the region has been noticed during the last three decades, events such as electric tropical storms and hurricanes, flooding, mudslides and droughts has affected more frequently. There are bases to consider the effects in terms of deaths, displacements and lost properties of disaster related to poor housing, unplanned settings, low income, lack of social security and labour, deforestation, primitive practice of agriculture, lack of planning territorial order, and poverty (Alfaro & Rivera, 2008).

During the last decades, due to intense rain precipitation in short periods of time, there have been described many events such as mudslides and landslides all over Latin America (described in Guatemala, Nicaragua, San Salvador, Brazil, Bolivia, Peru, Columbia, Argentina, Ecuador, Venezuela, Mexico) and the Caribbean. Especially vulnerable are unplanned urban settlements located on mountains and hilly ground, where soil texture is loose and not prepared for construction. Hundreds of people living in poor-quality buildings and houses have been affected by injuries, losing their properties and even death. These situations have become common since main infrastructure projects have been planned but not done. However, a lot of effort has been made to educate and inform people about pre, during and post natural disasters due to abundant rain occurring (Parry et al, 2007; PAHO, 2007). Also, these types of events and their frequency will affect tourism to certain areas that are very popular in Latin America such as Machu Picchu, Amazonas River and plains and will be even worse for the populations that live close to this natural monuments, predominantly indigenous and minority groups.

Problems related to skin control of temperature and water, cancer, photosensitivity and damage related to exposure to ultraviolet radiation or ozone have been described in other areas of the world. In Latin America, at least one reported example has been reported: Highly unusual stratospheric ozone loss and UV-B increases have occurred in the Punta Arenas (Chile) area over the past two decades, resulting in the non-photo adapted population being repeatedly exposed to an altered solar UV spectrum. This causes a greater risk of erythema and photocarcinogenesis. The rate of non-melanoma skin cancer, 81% of the total, has increased from 5.43 to 7.94 per 100,000 (46%) (PAHO, 2007).

Some investigators, environmentalists, nature protectors, biodiversity advocates, public health practitioners and many others are concerned that not much attention has been given to the roots of the problem of climate change, and ecosystems that should be protected are forgotten in the care of this natural barrier to many natural disasters related to tropical storms, hurricanes and flooding. A more integrative, environmental approach should be taken by national and regional governments. Plans to protect all natural barriers such as seas, rivers, water, forests, mangroves, biological diversity, and nature in general are supported by an increasing number of politicians, representative, communities, scientists, laypeople. It is the focus of attention nowadays for this new millennium and evidenced by world agreements such as those made in Kioto, Japan and Brazil.

5. Figures

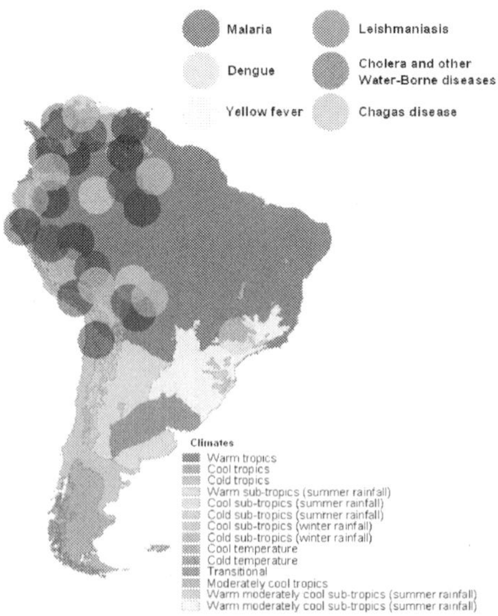

Fig. 1. Climates of South America and spots for the distribution of reports of endemic ties of communicable diseases that are prone to be affected by climate variability and climate change (adapted from World Book (2007). - South American climates. Available at: http://www.worldbook.com/wb/Students?content_spotlight/climates/south_american_climate).

6. References

Alfaro, W. & Rivera, L. (2008). *Cambio Climático en Mesoamérica: Temas para la creación de capacidades y la reducción de la vulnerabilidad*. The International Development Research Centre (IDRC) y Department for International Development (DFID-UK), London.

Araújo, C.A., Waniek, P.J., Jansen, A.M. (2009). An overview of Chagas disease and the role of triatomines on its distribution in Brazil. *Vector borne and zoonotic diseases*, Vol.9, No.3, (June 2009) 227-234, ISSN 1530-3667

Arria, M.; Rodríguez-Morales, A.J. & Franco-Paredes, C. (2005). Ecoepidemiología de las Enfermedades Tropicales en Países de la Cuenca Amazónica. *Revista Peruana de Medicina Experimental y Salud Publica*, Vol.22, No.3, (July 2005) 236-240, ISSN 1726-4634

Beck, L.R., Lobitz, B.M. & Wood, B.L. (2000). Remote sensing and human health: new sensors and new opportunities. *Emerging Infectious Diseases*, Vol.6, No.3, (2000) 217-227, ISSN 1080-6040

Benítez, J.A., Rodríguez-Morales, A.J., Sojo, M., Lobo, H., Villegas, C., Oviedo, L. & Brown, E. (2004). Descripción de un Brote Epidémico de Malaria de Altura en un área originalmente sin Malaria del Estado Trujillo, Venezuela. *Boletín de Malariología y Salud Ambiental*, Vol.44, No.2, (August 2004) 93-100, ISSN 1690-4648

Benítez, J.A., Sierra, C. & Rodríguez-Morales, A.J. (2005). Macroclimatic Variations and Ascaridiasis Incidence in Venezuela. *American Journal of Tropical Medicine & Hygiene*, Vol.73, No.(6 Suppl), (November 2005) 96, ISSN 0002-9637

Barrera, R., Delgado, N., Jimenez M. & Valero S. (2002). Eco-epidemiological factors associated with hyperendemic dengue hemorrhagic fever in Maracay city, Venezuela. *Dengue Bulletin*, Vol.26, No.1, (December 2002) 84–95, ISBN 9290222565

Botto, C., Escalona, E., Vivas-Martinez, S., Behm, V., Delgado, L. & Coronel, P. (2005). Geographical patterns of onchocerciasis in southern Venezuela: relationships between environment and infection prevalence. *Parassitologia*, Vol.47, No.1, (March 2005) 145–150, ISSN 0048-2951

Boutayeb, A. & Boutayeb, S. (2005). The burden of non communicable diseases in developing countries. *International Journal for Equity in Health*, Vol4., No., (2005), ISSN 1475-9276

Cabaniel, G., Rada, L., Blanco, J.J., Rodríguez-Morales, A.J. & Escalera, J.P. (2005). Impacto de Los Eventos de El Niño Southern Oscillation (ENSO) sobre la Leishmaniosis Cutánea en Sucre, Venezuela, a través del Uso de Información Satelital, 1994 - 2003. *Revista Peruana de Medicina Experimental y Salud Pública*, Vol.22, No.1, (January 2005) 32-38, ISSN 1726-4634

Cárdenas, R., Sandoval, C.M., Rodriguez-Morales, A.J. & Franco-Paredes, C. (2006). Impact of Climate Variability in the Occurrence of Leishmaniasis in Northeastern Colombia. *American Journal of Tropical Medicine & Hygiene*, Vol.75, No.2, (August 2006) 273-277, ISSN 0002-9637

Cárdenas, R., Sandoval, C.M., Rodriguez-Morales, A.J. & Vivas, P. (2008). Zoonoses and Climate Variability: the example of Leishmaniasis in Southern Departments of Colombia. *Annals of the New York Academy of Sciences*, Vol.1149, No.1, (January 2008) 326-330, ISSN 0077-8923

Céspedes, V.M. (2007). Los desastres, la información y el Centro Latinoamericano de Medicina de Desastres. *ACIMED*, Vol.16, No.2, (2007) 0-0, ISSN 1024-9435

Chaves, L. F. & Pascual, M. (2006). Climate cycles and forecasts of cutaneous leishmaniasis, a nonstationary vector-borne disease. *Plos Medicine,* Vol.3, No.8, (August 2006) e295, ISSN 1549-1277

Chavez, M.R.C., Sedas, V.P., Borunda, E.O. & Reynoso, F.L. (2005). Influence of water temperature and salinity on seasonal occurrences of *Vibrio cholerae* and enteric bacteria in oyster producing areas of Veracruz, Mexico. *Marine Pollution Bulletin,* Vol.50, No.12, (December 2005) 1641–1648, ISSN 0025-326X

Confalonieri, U. (2003). Variabilidade climática, vulnerabilidade social e saúde no Brasil. *Terra Livre*, Vol.1, No.20, (January 2003) 193-204, ISSN 0102-8030.

Costello, A., Abbas, M., Allen, A., Ball, S., Bell, S., Bellamy, R., Friel, S., Groce, N., Johnson, A., Kett, M., Lee, M., Levy, C., Maslin, M., McCoy, D., McGuire, B., Montgomery, H., Napier, D., Pagel, C., Patel, J., de Oliveira, J.A., Redclift, N., Rees, H., Rogger, D., Scott, J., Stephenson, J., Twigg, J., Wolff, J. & Patterson, C. (2009). Managing the health effects of climate change Lancet and University College London Institute for Global Health Commission. *Lancet*, Vol.373, No.9676, (May 2009) 1693-1733, ISSN 0140-6736

Depradine, C. & Lovell, E. (2004). Climatological variables and the incidence of Dengue fever in Barbados. *International Journal of Environmental Health Research,* Vol.14, No.6, (December 2004) 429–441, ISSN 0960-3123

Diaz, J.H. (2006). Global climate changes, natural disasters, and travel health risks. *Journal of Travel Medicine*, Vol.13, No.6, (November 2006), 361-72, ISSN 1195-1982

Ebi, K.L. & Paulson, J.A. (2007). Climate change and children. *Pediatrics Clinics of North America*, Vol.54, No.2, (April 2007) 213-226, ISSN 0031-3955

Farfan, R., Gomez, C., Escalera, J.P., Guerrero, L., Aragundy, J., Solano, E., Benitez, J.A., Rodriguez-Morales, A.J. & Franco-Paredes C. (2006). Climate Variability and Cholera in the Americas. *International Journal of Infectious Diseases,* Vol.10, No.Suppl 1, (June 2006) S12-S13, ISSN 1201-9712

Fernando, J., Brunstein J., Fernando, J. & Jankilevich, S.S. (2001). *Disyuntivas para el diseño de políticas de mitigación de la contaminación atmosférica global y local. El caso de la Ciudad de Buenos Aires. Documento de Trabajo N° 69,* Universidad de Belgrano, Buenos Aires.

Franke, C.R., Ziller, M., Staubach, C. & Latif, M. (2002). Impacts of the El Niño/Southern Oscillation on visceral leishmaniasis, Brazil. *Emerging Infectious Diseases*, Vol.8, No., (September 2002) 914-917, ISSN 1080-6040

Galvis-Ovallos, F., Espinosa, Y., Gutiérrez-Marín, R., Fernández, N., Rodriguez-Morales, A.J. & Sandoval, C. (2009). Climate variability and *Lutzomyia spinicrassa* abundance in an area of cutaneous leishmaniasis transmission in Norte de Santander, Colombia. *International Journal of Antimicrobial Agents,* Vol.34, No.Suppl 2, (July 2009) S4, ISSN 0924-8579

Gomez, C., Rodríguez-Morales, A.J. & Franco-Paredes, C. (2006). Impact of Climate Variability in the Occurrence of Leishmaniasis in Bolivia. *American Journal of Tropical Medicine & Hygiene,* Vol.75, No.(5 Suppl), (November 2006) 42, ISSN 0002-9637

Gubler, D.J., Suharyono, W., Lubis, I., Eram, S. & Gunarso, S. Epidemic dengue 3 in central Java, associated with low viremia in man. *American Journal of Tropical Medicine & Hygiene,* Vol.30, No.5, (September 1981) 1094-1099, ISSN 0002-9637

Halstead, S.B. (2006). Dengue in the Americas and Southeast Asia: do they differ? *Revista Panamericana de Salud Publica,* Vol.20, No.6, (December 2006) 407-415, ISSN 1020-4989

Herrera-Martinez, A. & Rodriguez-Morales, A.J. (2009). Potential influence of climate variability on dengue incidence in a western pediatric hospital of Venezuela, 2001–2008. *Tropical Medicine & International Health,* Vol.14, No.S2, (September 2009) 164-165, ISSN 1360-2276

Huarcaya, E., Chinga, E., Chávez, J.M., Chauca, J., Llanos, A., Maguiña, C., Pachas, P. & Gotuzzo, E. (2004). Influencia del fenómeno de El Niño en la epidemiología de la bartonelosis humana en los departamentos de Ancash y Cusco entre 1996 y 1999. *Revista Médica Herediana,* Vol.15, No., (2004) 4-10, ISSN 1018-130X

Hurtado-Diaz, M., Riojas-Rodriguez, H., Rothenberg, S., Gomez-Dantes, H. & Cifuentes, E. (2007). Impact of climate variability on the incidence of dengue in Mexico. *Tropical Medicine & International Health,* Vol.12, No.11, (October 2007) 1327-1337, ISSN 1360-2276

Kelly-Hope, L. & Thomson, M.C. (2008). Climate and Infectious Diseases (Chapter 3), In: *Seasonal Forecasts, Climatic Change and Human Health,* Thomson, M.C., Garcia-Herrera, R. & Beniston, M. (Ed), 31-70, Springer Science, ISBN 978-1-4020-6876-8, New York.

Kovats, S. & Haines, A. (1995). The potential health impacts of climate change: an overview. *Medicine and War,* Vol.11, No.4, (October 1995), 168-78, ISSN 0748-8009

Kupek, E., de Sousa Santos Faversani, M.C. & de Souza Philippi, J.M. (2000). The relationship between rainfall and human leptospirosis in Florianópolis, Brazil, 1991–1996. *Brazilian Journal of Infectious Diseases,* Vol.4, No.3, (June 2000) 131-134, ISSN 1413-8670

Lama, J.R., Seas, C.R., Leon-Barua, R., Gotuzzo, E. & Sack, R.B. (2004). Environmental temperature, cholera, and acute diarrhoea in adults in Lima, Peru. *Journal of Health, Population and Nutrition,* Vol.22, No.4, (December 2004) 399–403, ISSN 1606-0997

Lapola, D.M., Oyama, M.D., Nobre, C.A. & Sampaio, G. (2008). A new world natural vegetation map for global change studies. *Anais da Academia Brasileira de Ciências,* Vol.80, No.2, (June 2008) 397-408, ISSN 0001-3765

Liverman, D. (2009). *Suffering the Science. Climate change, people, and poverty,* Oxfam, Boston

Magrin, G., Gay García, C., Cruz Choque, D., Giménez, J.C., Moreno, A.R., Nagy, G.J., Nobre, C. & Villamizar, A. (2007). Latin America, In: *Climate Change 2007: Impacts, Adaptation and Vulnerability. Contribution of Working Group II to the Fourth Assessment Report of the Intergovernmental Panel on Climate Change,* Parry, M.L., Canziani, O.F., Palutikof, J.P., van der Linden, P.J. & Hanson, C.E., (Ed), 581-615, Cambridge University Press, ISBN 978 0521 88009-1, Cambridge, UK.

Martens, W.J., Slooff, R. & Jackson, E.K. (1997). Climate change, human health, and sustainable development. *Bulletin of the World Health Organization,* Vol.75, No.6, (1997) 583-588, ISSN 0042-9686

Mathers, C.D., Loncar, D. (2006) Projections of global mortality and burden of disease from 2002 to 2030. *PLoS Medicine,* Vol.3, No.11, (2006) e442

McMichael, A.J., Campbell-Lendrum, D.H., Corvalan, C.F., Ebi, K.L., Scheraga, J.D. & Woodwards, A. (2003). *Climate change and human health. Risk and responses*, World Health Organization, ISBN 92-4-156248-X, Geneva.

Mills, D.M. (2009). Climate change, extreme weather events, and us health impacts: what can we say? *Journal of Occupational & Environmental Medicine*, Vol.51, No.1, (January 2009) 26-32, ISSN 1076-2752

OPS. (2001). *Desigualdades en el acceso, uso y gasto con el agua potable en América Latina y el Caribe*, OPS, Washington, D.C.

Ortega García, J.A. (2007). El pediatra ante el cambio climático: desafíos y oportunidades. *Boletín de la Sociedad de Pediatría de Asturias, Cantabria, Castilla y León*, Vol.47, No.202, (January 2007) 331-343, ISSN 0214-2597

PAHO. (1988). *Hippocrates. Airs, waters, places. Pag. 18 Part I Historical development. The challenger of epidemiology. Issues and selected readings*, PAHO, Washington, D.C.

PAHO. (2003). *Protecting New Health Facilities from Natural Disasters: Guidelines for the Promotion of Disaster Mitigation*, PAHO, ISBN 92 75 124841, Washington, D.C.

PAHO. (2007). *Health in the Americas 2007. Volume I. Regional. Scientific and Thecnical Publication No. 622*, PAHO, Washington, D.C.

PAHO. (2008). Climate Change and Disaster Programs in the Health Sector. *Disasters: Preparedness and Mitigation in the Americas*, Vol.110, No.1, (October 2008) 1, 11, ISSN 1564-0701

Paris, L.A., Viscarret, M., Uban, C., Vargas, J. & Rodríguez-Morales, A.J. (2008). Pin-site myiasis: a rare complication of a treated open fracture of tibia. *Surgical Infections*, Vol.9, No.3, (June 2008) 403-406, ISSN 1096-2964

Parry, M.L., Canziani, O.F., Palutikof, J.P., van der Linden, P.J. & Hanson, C.E. (2007). *Climate Change 2007: Impacts, Adaptation and Vulnerability. Contribution of Working Group II to the Fourth Assessment Report of the Intergovernmental Panel on Climate Change*, Cambridge University Press, ISBN 9780521705974, Cambridge, United Kingdom and New York, NY, USA.

Patz, J.A., Campbell-Lendrum, D., Holloway T. & Foley, J.A. (2005). Impact of regional climate change on human health. *Nature*, Vol.438, No.7066, (November 2005) 310–317, ISSN 0028-0836

Peterson, A.T., Martinez-Campos, C., Nakazawa, Y. & Martinez-Meyer, E. (2005). Time-specific ecological niche modeling predicts spatial dynamics of vector insects and human dengue cases. *Transactions of the Royal Society of Tropical Medicine and Hygiene*, Vol.99, No.9, (September 2005) 647–655, ISSN 0035-9203

Pinto, J., Bonacic, C., Hamilton-West, C., Romero, J. & Lubroth J. (2008). Climate change and animal diseases in South America. *Revue scientifique et technique (International Office of Epizootics)*, Vol.27, No.2, (August 2008) 599-613, ISSN 0253-1933

Poveda, G.J., Rojas, W., Quiñones, M.L., Vélez, I.D., Mantilla, R.I., Ruiz, D., Zuluaga, J.S. & Rua, G.L. (2001). Coupling between annual and ENSO theme scales in the malaria climate association in Colombia. *Environmental Health Perspectives*, Vol.109, No., (May 2001) 489-493, ISSN 0091-6765

Prüss-Üstün, A. & Corvalán, C. (1988). *Preventing disease through healthy environments*, WHO, Geneva.

Ramal, C., Vásquez, J., Magallanes, J. & Carey, C. (2009). Variabilidad climática y transmisión de malaria en Loreto, Perú: 1995-2007. *Revista Peruana de Medicina Experimental y Salud Pública,* Vol.26, No.1, (January 2009) 9-14, ISSN 1726-4634

Rifakis, P., Gonçalves, N., Omaña, W., Manso, M., Espidel, A., Intingaro, A., Hernández, O. & Rodríguez-Morales, A.J. (2005). Asociación entre las Variaciones Climáticas y los Casos de Dengue en un Hospital de Caracas, Venezuela, 1998-2004. *Revista Peruana de Medicina Experimental y Salud Pública,* Vol.22, No.3, (July 2005) 183-190, ISSN 1726-4634

Rifakis, P.M., Benitez, J.A., Rodriguez-Morales, A.J., Dickson, S.M. & De-La-Paz-Pineda, J. (2006). Ecoepidemiological and social factors related to rabies incidence in Venezuela during 2002-2004. *International Journal of Biomedical Science,* Vol.2, No.1, (February 2006) 3-7, ISSN 1550-9702

Rodríguez-Morales, A.J., Barbella, R.A., Cabaniel, G., Gutiérrez, G. & Blanco J.J. (2004). Influence of Climatic Variations on Yellow Fever Outbreaks In Venezuela, 2002-2003, *Proceedings of 20th Clinical Virology Symposium and Annual Meeting Pan American Society for Clinical Virology,* pp. TM12, ISBN 0000-0000, Clearwater Beach, Florida, USA, april 2004, Pan American Society for Clinical Virology, Clearwater Beach, Florida, USA

Rodríguez-Morales, A.J. (2005). Ecoepidemiología y Epidemiología Satelital: Nuevas Herramientas en el Manejo de Problemas en Salud Pública. *Revista Peruana de Medicina Experimental y Salud Pública,* Vol.22, No.1, (January 2005) 54-63, ISSN 1726-4634

Rodríguez-Morales, A.J. (2006). Enfermedades Olvidadas: Miasis. *Revista Peruana de Medicina Experimental y Salud Pública,* Vol.23, No.2, (April 2006) 143-144, ISSN 1726-4634

Rodriguez-Morales, A.J., Rodríguez, C. & Meijomil P. (2006). Climate Variability Influence and Seasonal Patterns of Gram-positive Cocci Infections in Western Caracas, 1992–2001. *International Journal of Infectious Diseases,* Vol.10, No.Suppl 1, (June 2006) S13-S14, ISSN 1201-9712

Rodríguez-Morales, A.J. (2008). Impacto potencial para la salud pública latinoamericana del lanzamiento y puesta en órbita del satélite VENESAT-1. *Revista Peruana de Medicina Experimental y Salud Pública,* Vol.25, No.4, (October 2008) 444-445, ISSN 1726-4634

Rodríguez-Morales, A.J. (2009). Cambio climático y salud humana: enfermedades transmisibles y América Latina. *Revista Peruana de Medicina Experimental y Salud Pública,* Vol.26, No.2, (April 2009) 268-269, ISSN 1726-4634

Schreiber, K. V. (2001). An investigation of relationships between climate and dengue using a water budgeting technique. *International Journal of Biometeorology,* Vol.45, No.2, (July 2001) 81–89, ISSN 0020-7128

Sukri, N.C., Laras, K., Wandra, T., Didi, S., Larasati, R.P., Rachdyatmaka, J.R., Osok, S., Tjia, P., Saragih, J.M., Hartati, S., Listyaningsih, E., Porter, K.R., Beckett, C.G., Prawira, I.S., Punjabi, N., Suparmanto, S.A., Beecham, H.J., Bangs, M.J. & Corwin, A.L. (2003). Transmission of epidemic dengue hemorrhagic fever in easternmost Indonesia. *American Journal of Tropical Medicine & Hygiene,* Vol.68, No.5, (May 2003) 529-535, ISSN 0002-9637

Tavares S. (2005). *La bioética, el agua y el saneamiento,* Editorial Disinlimed, Caracas.

Thong, H.Y. & Maibach, H.I. (2008). Global warming and its dermatologic implications. *International Journal of Dermatology,* Vol.47, No.5, (May 2008) 522-524, ISSN 0011-9059.

Thomson, M.C., Garcia-Herrera, R. & Beniston, M. (2008). *Seasonal Forecasts, Climatic Change and Human Health,* Springer Science, ISBN 978-1-4020-6876-8, New York.

United Nations. (2006). *Global Survey of Early Warning Systems,* United Nations, ISBN 9789027725523, New York.

United Nations Development Programme. (2008). Fighting climate change: Human solidarity in a divided world. New York: Oxford University Press; 2006.

van der Meide, W.F., Jensema, A.J., Akrum, R.A.E., Sabajo, L.O.A., Lai, A., Fat, R.F.M., Lambregts, L., Schallig, H.D.F.H., van der Paardt, M. & Faber, W.R. (2006). Epidemiology of Cutaneous Leishmaniasis in Suriname: A Study Performed in 2006. *American Journal of Tropical Medicine & Hygiene,* Vol.79, No.2, (February 2006) 192-197, ISSN 0002-9637

Vasconcelos, P.F., Costa, Z.G., Travassos Da Rosa, E.S., Luna, E., Rodrigues, S.G., Barros, V.L., Dias, J.P., Monteiro, H.A., Oliva, O.F., Vasconcelos, H.B., Oliveira, R.C., Sousa, M.R., Barbosa Da Silva, J., Cruz, A.C., Martins, E. C. & Travassos Da Rosa, J.F. (2001). Epidemic of jungle yellow fever in Brazil, 2000: implications of climatic alterations in disease spread. *Journal of Medical Virology,* Vol.65, No.3, (November 2001) 598–604, ISSN 0146-6615

WHO. (2009). *Facts on climate change and health,* PAHO, Washington, D.C.

Williams, R.J., Bryan, R.T., Mills, J.N., Palma, R.E., Vera, I. & de Velásquez, F. (1997). An outbreak of hantavirus pulmonary syndrome in western Paraguay. *American Journal of Tropical Medicine & Hygiene,* Vol.57, No.3, (September 1997) 274-282, ISSN 0002-9637

Woolhouse, M.E. & Gowtage-Sequeria, S. (2005). Host range and emerging and reemerging pathogens. *Emerging Infectious Diseases*, Vol.11, No.12, (December 2005) 1842-1847, ISSN 1080-6040

Zeilhofer, P., Kummer, O.P., dos Santos, E.S., Ribeiro, A.L. & Missawa, N.A. (2008). Spatial modelling of *Lutzomyia (Nyssomyia) whitmani s.l.* (Antunes & Coutinho, 1939) (Diptera: Psychodidae: Phlebotominae) habitat suitability in the state of Mato Grosso, Brazil. *Memorias do Instituto Oswaldo Cruz,* Vol.103, No.7, (November 2008) 653-660, ISSN 0074-0276

7. Acknowledgements

We would like to thank Sharon Edwards and Diane Edrington (USA) for her review on the manuscript.